Introduction to Risk Parity and Budgeting

CHAPMAN & HALL/CRC
Financial Mathematics Series

Aims and scope:
The field of financial mathematics forms an ever-expanding slice of the financial sector. This series aims to capture new developments and summarize what is known over the whole spectrum of this field. It will include a broad range of textbooks, reference works and handbooks that are meant to appeal to both academics and practitioners. The inclusion of numerical code and concrete real-world examples is highly encouraged.

Published Titles

American-Style Derivatives; Valuation and Computation, *Jerome Detemple*

Analysis, Geometry, and Modeling in Finance: Advanced Methods in Option Pricing, *Pierre Henry-Labordère*

Computational Methods in Finance, *Ali Hirsa*

Credit Risk: Models, Derivatives, and Management, *Niklas Wagner*

Engineering BGM, *Alan Brace*

Financial Modelling with Jump Processes, *Rama Cont and Peter Tankov*

Interest Rate Modeling: Theory and Practice, *Lixin Wu*

Introduction to Credit Risk Modeling, Second Edition, *Christian Bluhm, Ludger Overbeck, and Christoph Wagner*

An Introduction to Exotic Option Pricing, *Peter Buchen*

Introduction to Risk Parity and Budgeting, *Thierry Roncalli*

Introduction to Stochastic Calculus Applied to Finance, Second Edition, *Damien Lamberton and Bernard Lapeyre*

Monte Carlo Methods and Models in Finance and Insurance, *Ralf Korn, Elke Korn, and Gerald Kroisandt*

Monte Carlo Simulation with Applications to Finance, *Hui Wang*

Nonlinear Option Pricing, *Julien Guyon and Pierre Henry-Labordère*

Numerical Methods for Finance, *John A. D. Appleby, David C. Edelman, and John J. H. Miller*

Option Valuation: A First Course in Financial Mathematics, *Hugo D. Junghenn*

Portfolio Optimization and Performance Analysis, *Jean-Luc Prigent*

Quantitative Fund Management, *M. A. H. Dempster, Georg Pflug, and Gautam Mitra*

Risk Analysis in Finance and Insurance, Second Edition, *Alexander Melnikov*

Robust Libor Modelling and Pricing of Derivative Products, *John Schoenmakers*

Stochastic Finance: A Numeraire Approach, *Jan Vecer*

Stochastic Financial Models, *Douglas Kennedy*

Stochastic Processes with Applications to Finance, Second Edition, *Masaaki Kijima*

Structured Credit Portfolio Analysis, Baskets & CDOs, *Christian Bluhm and Ludger Overbeck*

Understanding Risk: The Theory and Practice of Financial Risk Management, *David Murphy*

Unravelling the Credit Crunch, *David Murphy*

Proposals for the series should be submitted to one of the series editors above or directly to:
CRC Press, Taylor & Francis Group
3 Park Square, Milton Park
Abingdon, Oxfordshire OX14 4RN
UK

Chapman & Hall/CRC FINANCIAL MATHEMATICS SERIES

Introduction to Risk Parity and Budgeting

Thierry Roncalli

CRC Press
Taylor & Francis Group
Boca Raton London New York

CRC Press is an imprint of the
Taylor & Francis Group, an **informa** business
A CHAPMAN & HALL BOOK

CRC Press
Taylor & Francis Group
6000 Broken Sound Parkway NW, Suite 300
Boca Raton, FL 33487-2742

Printed on acid-free paper
Version Date: 20130827

International Standard Book Number-13: 978-1-4822-0715-6 (Hardback)

Library of Congress Cataloging-in-Publication Data

Roncalli, Thierry.
Introduction to risk parity and budgeting / Thierry Roncalli.
pages cm -- (Chapman & Hall/CRC financial mathematics series)
Includes bibliographical references and index.
ISBN 978-1-4822-0715-6 (hardcover : alk. paper) 1. Portfolio management. 2. Risk management. 3. Budget. I. Title.

HG4529.5.R656 2014
332.601--dc23 2013026141

Visit the Taylor & Francis Web site at
http://www.taylorandfrancis.com

and the CRC Press Web site at
http://www.crcpress.com

Introduction

The death of Markowitz optimization?

For a long time, investment theory and practice has been summarized as follows. The capital asset pricing model stated that the market portfolio is optimal. During the 1990s, the development of passive management confirmed the work done by William Sharpe. At that same time, the number of institutional investors grew at an impressive pace. Many of these investors used passive management for their equity and bond exposures. For asset allocation, they used the optimization model developed by Harry Markowitz, even though they knew that such an approach was very sensitive to input parameters, and in particular, to expected returns (Merton, 1980). One reason is that there was no other alternative model. Another reason is that the Markowitz model is easy to use and simple to explain. For expected returns, these investors generally considered long-term historical figures, stating that past history can serve as a reliable guide for the future. Management boards of pension funds were won over by this scientific approach to asset allocation.

The first serious warning shot came with the dot-com crisis. Some institutional investors, in particular defined benefit pension plans, lost substantial amounts of money because of their high exposure to equities (Ryan and Fabozzi, 2002). In November 2001, the pension plan of The Boots Company, a UK pharmacy retailer, decided to invest 100% in bonds (Sutcliffe, 2005). Nevertheless, the performance of the equity market between 2003 and 2007 restored confidence that standard financial models would continue to work and that the dot-com crisis was a non-recurring exception. However, the 2008 financial crisis highlighted the risk inherent in many strategic asset allocations. Moreover, for institutional investors, the crisis was unprecedentedly severe. In 2000, the internet crisis was limited to large capitalization stocks and certain sectors. Small capitalizations and value stocks were not affected, while the performance of hedge funds was flat. In 2008, the subprime crisis led to a violent drop in credit strategies and asset-backed securities. Equities posted negative returns of about −50%. The performance of hedge funds and alternative assets was poor. There was also a paradox. Many institutional investors diversified their portfolios by considering several asset classes and different regions. Unfortunately, this diversification was not enough to protect them. In the end, the 2008 financial crisis was more damaging than the dot-com crisis. This was particularly true for institutional investors in continental Europe, who were relatively well protected against the collapse of the internet bubble because of their low exposure to equities. This is why the 2008 financial crisis was a deep trauma for world-wide institutional investors.

Most institutional portfolios were calibrated through portfolio optimization. In this context, Markowitz's modern portfolio theory was strongly criticized by professionals, and several journal articles announced the death of the Markowitz model[1]. These extreme reactions can be explained by the fact that diversification is traditionally associated with Markowitz optimization, and it failed during the financial crisis. However, the problem was not entirely due to the allocation method. Indeed, much of the failure was caused by the input parameters. With expected returns calibrated to past figures, the model induced an overweight in equities. It also promoted assets that were supposed to have a low correlation to equities. Nonetheless, correlations between asset classes increased significantly during the crisis. In the end, the promised diversification did not occur.

Today, it is hard to find investors who defend Markowitz optimization. However, the criticisms concern not so much the model itself but the way it is used. In the 1990s, researchers began to develop regularization techniques to limit the impact of estimation errors in input parameters and many improvements have been made in recent years. In addition, we now have a better understanding of how this model works. Moreover, we also have a theoretical framework to measure the impact of constraints (Jagannathan and Ma, 2003). More recently, robust optimization based on the lasso approach has improved optimized portfolios (DeMiguel *et al.*, 2009). So the Markowitz model is certainly not dead. Investors must understand that it is a fabulous tool for combining risks and expected returns. The goal of Markowitz optimization is to find arbitrage factors and build a portfolio that will play on them. By construction, this approach is an aggressive model of active management. In this case, it is normal that the model should be sensitive to input parameters (Green and Hollifield, 1992). Changing the parameter values modifies the implied bets. Accordingly, if input parameters are wrong, then arbitrage factors and bets are also wrong, and the resulting portfolio is not satisfied. If investors want a more defensive model, they have to define less aggressive parameter values. This is the main message behind portfolio regularization. In consequence, reports of the death of the Markowitz model have been greatly exaggerated, because it will continue to be used intensively in active management strategies. Moreover, there are no other serious and powerful models to take into account return forecasts.

[1]See for example the article "*Is Markowitz Dead? Goldman Thinks So*" published in December 2012 by AsianInvestor.

The rise of risk parity portfolios

There are different ways to obtain less aggressive active portfolios. The first one is to use less aggressive parameters. For instance, if we assume that expected returns are the same for all of the assets, we obtain the minimum variance (or MV) portfolio. The second way is to use heuristic methods of asset allocation. The term 'heuristic' refers to experience-based techniques and trial-and-error methods to find an acceptable solution, which does not correspond to the optimal solution of an optimization problem. The equally weighted (or EW) portfolio is an example of such non-optimized 'rule of thumb' portfolio. By allocating the same weight to all the assets of the investment universe, we considerably reduce the sensitivity to input parameters. In fact, there are no active bets any longer. Although these two allocation methods have been known for a long time, they only became popular after the collapse of the internet bubble.

Risk parity is another example of heuristic methods. The underlying idea is to build a balanced portfolio in such a way that the risk contribution is the same for different assets. It is then an equally weighted portfolio in terms of risk, not in terms of weights. Like the minimum variance and equally weighted portfolios, it is impossible to date the risk parity portfolio. The term risk parity was coined by Qian (2005). However, the risk parity approach was certainly used before 2005 by some CTA and equity market neutral funds. For instance, it was the core approach of the All Weather fund managed by Bridgewater for many years (Dalio, 2004). At this point, we note that the risk parity portfolio is used, because it makes sense from a practical point of view. However, it was not until the theoretical work of Maillard *et al.* (2010), first published in 2008, that the analytical properties were explored. In particular, they showed that this portfolio exists, is unique and is located between the minimum variance and equally weighted portfolios.

Since 2008, we have observed an increasing popularity of the risk parity portfolio. For example, Journal of Investing and Investment and Pensions Europe (IPE) ran special issues on risk parity in 2012. In the same year, The Financial Times and Wall Street Journal published several articles on this topic[2]. In fact today, the term risk parity covers different allocation methods. For instance, some professionals use the term risk parity when the asset weight is inversely proportional to the asset return volatility. Others consider that the risk parity portfolio corresponds to the equally weighted risk contribution (or ERC) portfolio. Sometimes, risk parity is equivalent to a risk budgeting (or RB) portfolio. In this case, the risk budgets are not necessarily the same for all of the assets that compose the portfolio. Initially, risk parity

[2] "*New Allocation Funds Redefine Idea of Balance*" (February 2012), "*Same Returns, Less Risk*" (June 2012), "*Risk Parity Strategy Has Its Critics as Well as Fans*" (June 2012), "*Investors Rush for Risk Parity Shield*" (September 2012), etc.

only concerned a portfolio of bonds and equities. Today, risk parity is applied to all investment universes. Nowadays, risk parity is a *marketing term* used by the asset management industry to design a portfolio based on risk budgeting techniques.

More interesting than this marketing operation is the way risk budgeting portfolios are defined. Whereas the objective of Markowitz portfolios is to reach an expected return or to target ex-ante volatility, the goal of risk parity is to assign a risk budget to each asset. Like for the other heuristic approaches, the performance dimension is then absent and the risk management dimension is highlighted. In addition, this last point is certainly truer for the risk parity approach than for the other approaches. We also note that contrary to minimum variance portfolios, which have only seduced equity investors, risk parity portfolios concern not only different traditional asset classes (equities and bonds), but also alternative asset classes (commodities and hedge funds) and multi-asset classes (stock/bond asset mix policy and diversified funds). By placing risk management at the heart of these different management processes, risk parity represents a substantial break with respect to the previous period of Markowitz optimization. Over the last decades, the main objective of institutional investors was to generate performance well beyond the risk-free rate (sometimes approaching double-digit returns). After the 2008 crisis, investors largely revised their expected return targets. Their risk aversion level increased and they do not want to experience another period of such losses. In this context, risk management has become more important than performance management.

Nevertheless, like for many other hot topics, there is some exaggeration about risk parity. Although there are people who think that it represents a definitive solution to asset allocation problems, one should remain prudent. Risk parity remains a financial model of investment and its performance also depends on the investor's choice regarding parameters. Choosing the right investment universe or having the right risk budgets is as important as using the right allocation method. As a consequence, risk parity may be useful when defining a reliable allocation, but it cannot free investors of their duty of making their own choices.

About this book

The subject of this book is risk parity approaches. As noted above, risk parity is now a generic term used by the asset management industry to designate risk-based management processes. In this book, the term risk parity is used as a synonym of risk budgeting. When risk budgets are identical, we prefer to use the term ERC portfolio, which is more explicit and less overused by the investment industry. When we speak of a risk parity fund, it corresponds to an equally weighted risk contribution portfolio of equities and bonds.

This book comprises two parts. The first part is more theoretical. Its first chapter is dedicated to modern portfolio theory whereas the second chapter is a comprehensive guide to risk budgeting. The second part contains four chapters, each of which presents an application of risk parity to a specific asset class. The third chapter concerns risk-based equity indexation, also called smart indexing. In the fourth chapter, we show how risk budgeting techniques can be applied to the management of bond portfolios. The fifth chapter deals with alternative investments, such as commodities and hedge funds. Finally, the sixth chapter applies risk parity techniques to multi-asset classes. The book also contains two appendices. The first appendix provides the reader with technical materials on optimization problems, copula functions and dynamic asset allocation. The second appendix contains 30 tutorial exercises. The relevant solutions are not included in this book, but can be accessed at the following web page[3]:

`http://www.thierry-roncalli.com/riskparitybook.html`

This book began with an invitation by Professor Diethelm Würtz to present a tutorial on risk parity at the 6th R/Rmetrics Meielisalp Workshop & Summer School on Computational Finance and Financial Engineering. This seminar is organized every year at the end of June in Meielisalp, Lake Thune, Switzerland. The idea of tutorial sessions is to offer an overview on a specialized topic in statistics or finance. When preparing this tutorial, I realized that I had sufficient material to write a book on risk parity. First of all, I would like to thank Diethelm Würtz and the participants of the Meielisalp Summer School for their warm welcome and the different discussions we had about risk parity. I would also like to thank all of the people who have invited me to academic and professional conferences in order to speak about risk parity techniques and applications since 2008, in particular Yann Braouezec, Rama Cont, Nathalie Columelli, Felix Goltz, Marie Kratz, Jean-Luc Prigent, Fahd Rachidy and Peter Tankov. I would also like to thank Jérôme Glachant and my other colleagues of the Master of Science in Asset and Risk Management program at the Évry University where I teach the course on Risk Parity. I am also grateful to the CRC editorial staff, in particular Sunil Nair, for their support, encouragement and suggestions.

I would also like to thank my different co-authors on this subject, Benjamin Bruder, Pierre Hereil, Sébastien Maillard, Jérôme Teïletche and Guillaume Weisang, my colleagues at Lyxor Asset Management who work or have worked with me on risk parity strategies, in particular Cyrille Albert-Roulhac, Florence Barjou, Cédric Baron, Benjamin Bruder, Zélia Cazalet, Léo Culerier, Raphael Dieterlen, Nicolas Gaussel, Pierre Hereil, Julien Laplante, Guillaume

[3]This web page also provides readers and instructors other materials related to the book (errata, code, slides, etc.).

Lasserre, Sébastien Maillard, François Millet and Jean-Charles Richard. I am also grateful to Abdelkader Bousabaa, Jean-Charles Richard and Zhengwei Wu for their careful reading of the preliminary versions of this book. Special thanks to Zhengwei Wu who has been a helpful and efficient research assistant.

Last but not least, I express my deep gratitude to Théo, Eva, Sarah, Lucie and Nathalie for their support and encouragement during the writing of this book.

Paris, January 2013 Thierry Roncalli

Contents

List of Figures

List of Tables

List of Symbols and Notations

Symbol Description

Symbol	Description
$\cdot$	Scalar multiplication
$\circ$	Hadamard product: $(x \circ y)_i = x_i y_i$
$\otimes$	Kronecker product $A \otimes B$
$\lvert\mathcal{E}\rvert$	Cardinality of the set $\mathcal{E}$
$\mathbf{1}$	Vector of ones
$\mathbb{1}\{\mathcal{A}\}$	The indicator function is equal to 1 if $\mathcal{A}$ is true, 0 otherwise
$\mathbb{1}_{\mathcal{A}}\{x\}$	The characteristic function is equal to 1 if $x \in \mathcal{A}$, 0 otherwise
$\mathbf{0}$	Vector of zeros
$(A_{i,j})$	Matrix A with entry $A_{i,j}$ in row i and column j
A^{-1}	Inverse of the matrix A
$A^\top$	Transpose of the matrix A
A^+	Moore-Penrose pseudo-inverse of the matrix A
b	Vector of weights $(b_1, \ldots, b_n)$ for the benchmark b
$B_t(T)$	Price of the zero-coupon bond at time t for the maturity T
β_i	Beta of asset i with respect to portfolio x
$\beta_i(x)$	Another notation for the symbol β_i
$\beta(x \mid b)$	Beta of portfolio x when the benchmark is b
C (or ρ)	Correlation matrix
$\mathbf{C}$	Copula function
$C(t_m)$	Coupon paid at time t_m
$\text{cov}(X)$	Covariance of the random vector X
$C_n(\rho)$	Constant correlation matrix $(n \times n)$ with $\rho_{i,j} = \rho$
D	Covariance matrix of idiosyncratic risks
$\det(A)$	Determinant of the matrix A
$\mathcal{DR}(x)$	Diversification ratio of portfolio x
$\mathbf{e}_i$	The value of the vector is 1 for the row i and 0 elsewhere
$\mathbb{E}[X]$	Mathematical expectation of the random variable X
$\mathcal{E}(\lambda)$	Exponential probability distribution with parameter λ
$\text{ES}_\alpha(x)$	Expected shortfall of portfolio x at the confidence level α
$f(x)$	Probability density function (pdf)
$\mathbf{F}(x)$	Cumulative distribution function (cdf)
$\mathcal{F}$	Vector of risk factors $(\mathcal{F}_1, \ldots, \mathcal{F}_m)$
$\mathcal{F}_j$	Risk factor j
$F_t(T)$	Instantaneous forward rate at time t for the maturity T
$F_t(T, m)$	Forward interest rate at

	time t for the period $[T, T+m]$
$\mathcal{G}$	Gini coefficient
γ	Parameter $\gamma = \phi^{-1}$ of the Markowitz γ-problem
γ_1	Skewness
γ_2	Excess kurtosis
$\mathcal{H}$	Herfindahl index
i	Asset i
I_n	Identity matrix of dimension n
$\mathrm{IR}(x \mid b)$	Information ratio of portfolio x when the benchmark is b
$\ell(\theta)$	Log-likelihood function with θ the vector of parameters to estimate
ℓ_t	Log-likelihood function for the observation t
$L(x)$	Loss of portfolio x
$\mathcal{L}(x)$	Leverage measure of portfolio x
$\mathbb{L}(x)$	Lorenz function
λ	Parameter of exponential survival times
$\mathcal{MDD}$	Maximum drawdown
$\mathcal{MR}_i$	Marginal risk of asset i
μ	Vector of expected returns $(\mu_1, \ldots, \mu_n)$
μ_i	Expected return of asset i
$\hat{\mu}$	Empirical mean
$\hat{\mu}_{1\mathrm{Y}}$	Annualized return
$\mu(x)$	Expected return of portfolio x: $\mu(x) = x^\top \mu$
$\mu(x \mid b)$	Expected return of the tracking error of portfolio x when the benchmark is b
$\mathcal{N}(\mu, \sigma^2)$	Probability distribution of a Gaussian random variable with mean μ and standard deviation σ
$\mathcal{N}(\mu, \Sigma)$	Probability distribution of a Gaussian random vector with mean μ and covariance matrix Σ
Ω	Covariance matrix of risk factors
π	Vector of risk premia $(\pi_1, \ldots, \pi_n)$
$\tilde{\pi}$	Vector of implied risk premia $(\tilde{\pi}_1, \ldots, \tilde{\pi}_n)$
π_i	Risk premium of asset i: $\pi_i = \mu_i - r$
$\tilde{\pi}_i$	Implied risk premium of asset i
$\pi(y \mid x)$	Risk premium of portfolio y if the tangency portfolio is x: $\pi(y \mid x) = \beta(y \mid x)(\mu(x) - r)$
Π	P&L of the portfolio
ϕ	Risk aversion parameter of the quadratic utility function
$\phi(x)$	Probability density function of the standardized normal distribution
$\Phi(x)$	Cumulative distribution function of the standardized normal distribution
$\Phi^{-1}(\alpha)$	Inverse of the cdf of the standardized normal distribution
r	Return of the risk-free asset
$r^\star$	Yield to maturity
R	Vector of asset returns $(R_1, \ldots, R_n)$
R_i	Return of asset i
$R_{i,t}$	Return of asset i at time t
$R(x)$	Return of portfolio x: $R(x) = x^\top R$
$\mathcal{R}(x)$	Risk measure of portfolio x
$R_t(T)$	Zero-coupon rate at time t for the maturity T
$\mathcal{RC}_i$	Risk contribution of asset i
$\mathcal{RC}_i^\star$	Relative risk contribution of asset i
$\mathfrak{R}$	Recovery rate
ρ (or C)	Correlation matrix of asset returns

$\rho_{i,j}$	Correlation between asset returns i and j
$\rho(x,y)$	Correlation between portfolios x and y
$\mathfrak{s}$	Credit spread
$\mathbf{S}_t(x)$	Survival function at time t
Σ	Covariance matrix
$\hat{\Sigma}$	Empirical covariance matrix
σ_i	Volatility of asset i
σ_m	Volatility of the market portfolio
$\tilde{\sigma}_i$	Idiosyncratic volatility of asset i
$\hat{\sigma}$	Empirical volatility
$\hat{\sigma}_{1Y}$	Annualized volatility
$\sigma(x)$	Volatility of portfolio x: $\sigma(x) = \sqrt{x^\top \Sigma x}$
$\sigma(x \mid b)$	Standard deviation of the tracking error of portfolio x when the benchmark is b
$\sigma(x,y)$	Covariance between portfolios x and y
$\sigma(X)$	Standard deviation of the random variable X
SR_i	Sharpe ratio of asset i: $\mathrm{SR}_i = \mathrm{SR}(\mathbf{e}_i \mid r)$
$\mathrm{SR}(x \mid r)$	Sharpe ratio of portfolio x when the risk-free asset is r
$\mathbf{t}_v(x)$	Cumulative distribution function of the Student's t distribution with ν the number of degrees of freedom
$\mathbf{t}_v^{-1}(\alpha)$	Inverse of the cdf of the Student's t distribution with ν the number of degrees of freedom
$\mathbf{t}_{\rho,v}(x)$	Cumulative distribution function of the multivariate Student's t distribution with parameters ρ and ν
$\tau(x)$	Turnover of portfolio x
$\mathrm{tr}(A)$	Trace of the matrix A
$\mathrm{TR}(x \mid b)$	Treynor ratio of portfolio x when the benchmark is b
$\mathrm{VaR}_\alpha(x)$	Value-at-risk of portfolio x at the confidence level α
x	Vector of weights $(x_1, \ldots, x_n)$ for portfolio x
x_i	Weight of asset i in portfolio x
$x^\star$	Optimized portfolio

Portfolio Notation

ERC	Equally weighted risk contribution portfolio x_{erc}
EW	Equally weighted portfolio x_{ew}
MDP	Most diversified portfolio x_{mdp}
MSR	Max Sharpe ratio portfolio x_{msr}
MV	Minimum variance portfolio x_{mv}
MVO	Mean-variance optimized (or Markowitz) portfolio x_{mvo}
RB	Risk budgeting portfolio x_{rb}
RFP	Risk factor parity portfolio x_{rfp}
RP	Risk parity portfolio x_{rp}
WB	Weight budgeting portfolio x_{wb}

Part I

From Portfolio Optimization to Risk Parity

This part comprises two chapters. In the first chapter, we present the theoretical foundations of modern portfolio theory. We also show how this framework is implemented in practice and describe its limitations. The second chapter presents the risk budgeting approach. The main difference with the previous approach comes from the investor objective. Indeed, his objective is not to maximize a utility function or a risk-adjusted performance, but only to allocate the risk between assets. Consequently, the risk parity method does not need assumptions about expected returns and therefore constitutes a pure method of risk management.

Chapter 1

Modern Portfolio Theory

The concept of the market portfolio has a long history and dates back to the seminal work of Markowitz (1952). In his paper, Markowitz defined precisely what *portfolio selection* means: "*the investor does (or should) consider expected return a desirable thing and variance of return an undesirable thing*". Indeed, Markowitz showed that an efficient portfolio is the portfolio that maximizes the expected return for a given level of risk (corresponding to the variance of portfolio return). Markowitz concluded that there is not only one optimal portfolio, but a set of optimal portfolios which is called the efficient frontier.

By studying the liquidity preference, Tobin (1958) showed that the efficient frontier becomes a straight line in the presence of a risk-free asset. In this case, optimal portfolios correspond to a combination of the risk-free asset and one particular efficient portfolio named the tangency portfolio. Sharpe (1964) summarized the results of Markowitz and Tobin as follows: "*the process of investment choice can be broken down into two phases: first, the choice of a unique optimum combination of risky assets*[1]*; and second, a separate choice concerning the allocation of funds between such a combination and a single riskless asset*". This two-step procedure is today known as the *Separation Theorem* (Lintner, 1965).

One difficulty when computing the tangency portfolio is to precisely define the vector of expected returns of the risky assets and the corresponding covariance matrix of asset returns. In 1964, Sharpe developed the CAPM theory and highlighted the relationship between the risk premium of the asset (the difference between the expected return and the risk-free rate) and its beta (the systematic risk with respect to the tangency portfolio). By assuming that the market is at equilibrium, he showed that the prices of assets are such that the tangency portfolio is the market portfolio, which is composed of all risky assets in proportion to their market capitalization. This implies that we do not need assumptions about the expected returns, volatilities and correlations of assets to characterize the tangency portfolio. This major contribution of Sharpe led to the emergence of index funds and to the increasing development of passive management.

In the active management domain, fund managers use the Markowitz framework to optimize portfolios in order to take into account their views

[1]It is precisely the tangency portfolio.

and to play their bets. However, the implementation of portfolio theory is not simple. It requires the estimation of the covariance matrix and the forecasting of asset returns. One problem is that optimized portfolios are very sensitive to these inputs. Some stability issues make the practice of portfolio optimization less attractive than the theory (Michaud, 1989). In this case, regularization techniques may be employed to attenuate these problems. This approach is largely supported by Ledoit and Wolf (2003), who propose to combine different covariance matrix estimators to stabilize the solution. Today, the most promising approach consists in interpreting optimized portfolios as the solution of a linear regression problem and to use lasso or ridge penalization.

However, regularization is not sufficient to obtain satisfactory solutions, which is why practitioners introduce some constraints in the optimization problem. These constraints may be interpreted as a shrinkage method (Jagannathan and Ma, 2003). By imposing weight constraints, the portfolio manager implicitly changes the covariance matrix. This approach is then equivalent to having some views and is therefore related to the model of Black and Litterman (1992).

1.1 From optimized portfolios to the market portfolio

In this section, we review the seminal framework of Markowitz and the CAPM theory of Sharpe.

1.1.1 The efficient frontier

Sixty years ago, Markowitz introduced the concept of the efficient frontier. It was the first mathematical formulation of optimized portfolios. For him, "*the investor does (or should) consider expected return a desirable thing and variance of return an undesirable thing*". By translating these principles into a problem of mean-variance optimization, Markowitz (1952) showed that there is no one optimal portfolio, but a set of optimized portfolios.

We consider a universe of n assets. Let $x = (x_1, \ldots, x_n)$ be the vector of weights in the portfolio. We assume that the portfolio is fully invested meaning that $\sum_{i=1}^n x_i = \mathbf{1}^\top x = 1$. We denote $R = (R_1, \ldots, R_n)$ the vector of asset returns where R_i is the return of asset i. The return of the portfolio is then equal to $R(x) = \sum_{i=1}^n x_i R_i$. In a matrix form, we also obtain $R(x) = x^\top R$. Let $\mu = \mathbb{E}[R]$ and $\Sigma = \mathbb{E}\left[(R - \mu)(R - \mu)^\top\right]$ be the vector of expected returns and the covariance matrix of asset returns. The expected return of the portfolio is:

$$\mu(x) = \mathbb{E}[R(x)] = \mathbb{E}\left[x^\top R\right] = x^\top \mathbb{E}[R] = x^\top \mu$$

whereas its variance is equal to:

$$\begin{aligned}\sigma^2(x) &= \mathbb{E}\left[(R(x)-\mu(x))(R(x)-\mu(x))^\top\right] \\ &= \mathbb{E}\left[\left(x^\top R - x^\top \mu\right)\left(x^\top R - x^\top \mu\right)^\top\right] \\ &= \mathbb{E}\left[x^\top (R-\mu)(R-\mu)^\top x\right] \\ &= x^\top \mathbb{E}\left[(R-\mu)(R-\mu)^\top\right] x \\ &= x^\top \Sigma x\end{aligned}$$

We can then formulate the investor's financial problem as follows:

1. Maximizing the expected return of the portfolio under a volatility constraint (σ-problem):

$$\max \mu(x) \quad \text{u.c.} \quad \sigma(x) \leq \sigma^\star \tag{1.1}$$

2. Or minimizing the volatility of the portfolio under a return constraint (μ-problem):

$$\min \sigma(x) \quad \text{u.c.} \quad \mu(x) \geq \mu^\star \tag{1.2}$$

Example 1 *We consider four assets. Their expected returns are equal to* 5%, 6%, 8% *and* 6% *while their volatilities are equal to* 15%, 20%, 25% *and* 30%. *The correlation matrix of asset returns is given by the following matrix:*

$$C = \begin{pmatrix} 1.00 & & & \\ 0.10 & 1.00 & & \\ 0.40 & 0.70 & 1.00 & \\ 0.50 & 0.40 & 0.80 & 1.00 \end{pmatrix}$$

In Figure 1.1, we have simulated 1 000 portfolios and reported their expected return and their volatility (cross symbol). Let us consider the σ-problem with $\sigma^\star = 30\%$. Portfolio C could not be the solution even if it reached the volatility constraint, because it is dominated by portfolio B. However, this portfolio is not optimal, as we can find other portfolios with a higher expected return. Finally, the solution is portfolio A. In the same way, the optimal portfolio is D in the case of the μ-problem with $\mu^\star = 7\%$. The efficient frontier is then defined as the convex hull of the points $(\sigma(x), \mu(x))$ of all the possible portfolios. This convex hull may be computed numerically. In Figure 1.1, we have indicated the portfolios belonging to the convex hull by a solid circle symbol. In particular, the two optimal portfolios A and D are on the efficient frontier.

By considering all the portfolios belonging to the simplex set defined by $\left\{x \in [0,1]^n : \mathbf{1}^\top x = 1\right\}$, we can compute the expected return and volatility bounds of the portfolios: $\mu^- \leq \mu(x) \leq \mu^+$ and $\sigma^- \leq \sigma(x) \leq \sigma^+$. There is also a solution to the first problem if $\sigma^\star \geq \sigma^-$. The second problem has a solution if $\mu^\star \leq \mu^+$. If these two conditions are verified, the inequality constraints becomes $\sigma(x) = \min(\sigma^\star, \sigma^+)$ and $\mu(x) = \max(\mu^-, \mu^\star)$.

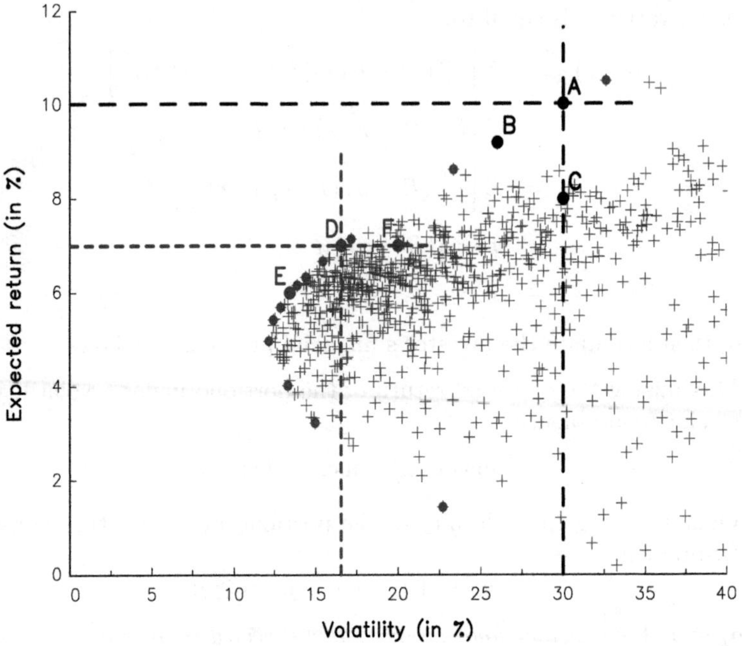

FIGURE 1.1: Optimized Markowitz portfolios

1.1.1.1 Introducing the quadratic utility function

The key idea of Markowitz (1956) was to transform the original non-linear optimization problem (1.1) into a quadratic optimization problem which is easier to solve numerically:

$$\begin{aligned} x^{\star}(\phi) &= \arg\max x^{\top}\mu - \frac{\phi}{2}x^{\top}\Sigma x \\ &\text{u.c.} \quad \mathbf{1}^{\top}x = 1 \end{aligned} \tag{1.3}$$

We can interpret ϕ as a risk-aversion parameter. If $\phi = 0$, the optimized portfolio is the one that maximizes the expected return and we have $\mu(x^{\star}(0)) = \mu^{+}$. If $\phi = \infty$, the optimization problem becomes:

$$\begin{aligned} x^{\star}(\infty) &= \arg\min \frac{1}{2}x^{\top}\Sigma x \\ &\text{u.c.} \quad \mathbf{1}^{\top}x = 1 \end{aligned}$$

The optimized portfolio is the one that minimizes the volatility and we have $\sigma(x^{\star}(\infty)) = \sigma^{-}$. It is called the minimum variance (or MV) portfolio.

Remark 1 *We note that the previous problem can also be written as follows:*

$$\begin{aligned} x^\star(\gamma) &= \arg\min \frac{1}{2} x^\top \Sigma x - \gamma x^\top \mu \\ &\quad u.c. \quad \mathbf{1}^\top x = 1 \end{aligned} \tag{1.4}$$

with $\gamma = \phi^{-1}$. From a numerical point of view, this formulation has the advantage to be a standard quadratic programming (QP) problem (see Appendix A.1.1 on page 301). In this case, the minimum variance portfolio corresponds to $\gamma = 0$. Depending on the objective, we will use either the ϕ-problem or the γ-problem to calculate optimized portfolios.

We consider Example 1. We have reported[2] in Table 1.1 the optimal portfolio for different values of ϕ. We verify that $\mu(x^\star(\phi))$ and $\sigma(x^\star(\phi))$ are two decreasing functions with respect to the parameter ϕ. It implies that the expected return $\mu(x^\star)$ is an increasing function of the volatility $\sigma(x^\star)$.

TABLE 1.1: Solving the ϕ-problem

ϕ	$+\infty$	5.00	2.00	1.00	0.50	0.20
$x_1^\star$	72.74	68.48	62.09	51.44	30.15	−33.75
$x_2^\star$	49.46	35.35	14.17	−21.13	−91.72	−303.49
$x_3^\star$	−20.45	12.61	62.21	144.88	310.22	806.22
$x_4^\star$	−1.75	−16.44	−38.48	−75.20	−148.65	−368.99
$\mu(x^\star)$	4.86	5.57	6.62	8.38	11.90	22.46
$\sigma(x^\star)$	12.00	12.57	15.23	22.27	39.39	94.57

The formulation (1.3) allows to give a new characterization of the efficient frontier. It is the parametric function $(\sigma(x^\star(\phi)), \mu(x^\star(\phi)))$ with $\phi \in \mathbb{R}_+$. If we consider the previous example, we obtain the efficient frontier in Figure 1.2. We note that optimized portfolios substantially improve the risk/return profile with respect to the four assets, which are represented by a cross symbol.

Solving the μ-problem or the σ-problem is equivalent to finding the optimal value of ϕ such that $\mu(x^\star(\phi)) = \mu^\star$ or $\sigma(x^\star(\phi)) = \sigma^\star$. We know that the functions $\mu(x^\star(\phi))$ and $\sigma(x^\star(\phi))$ are decreasing with respect to ϕ and are bounded. The optimal value of ϕ can then be easily computed using the Newton-Raphson algorithm. We have reported some numerical solutions in Tables 1.2 and 1.3. For example, if $\mu^\star$ is set to 7%, we obtain a portfolio with a volatility $\sigma(x^\star)$ equal to 16.54%. It corresponds to portfolio D in Figure 1.1. If we target a volatility equal to 30%, the expected return of the optimized portfolio is 10.02% and the solution is portfolio A in Figure 1.1.

[2] In this book, the values of weights, expected returns and volatilities are expressed in % except if another unit is specified.

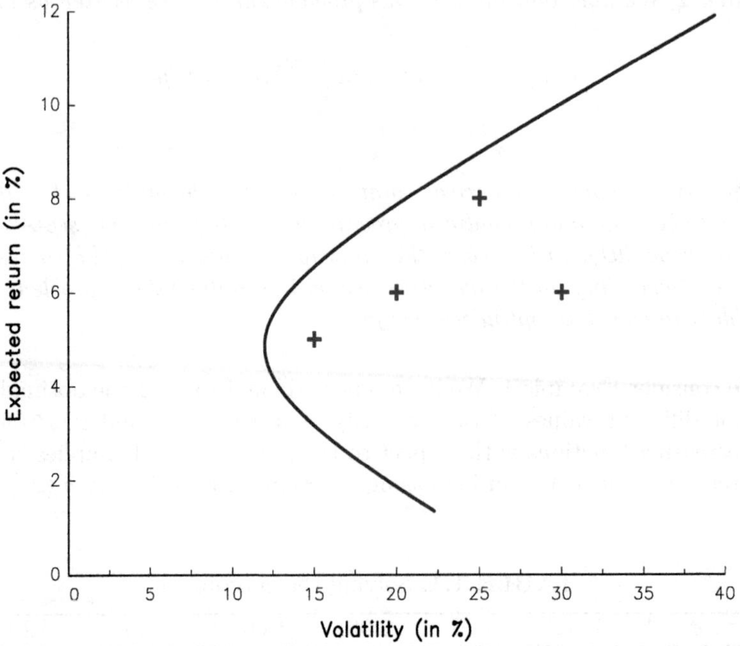

FIGURE 1.2: The efficient frontier of Markowitz

TABLE 1.2: Solving the unconstrained μ-problem

$\mu^\star$	5.00	6.00	7.00	8.00	9.00
$x_1^\star$	71.92	65.87	59.81	53.76	47.71
$x_2^\star$	46.73	26.67	6.62	−13.44	−33.50
$x_3^\star$	−14.04	32.93	79.91	126.88	173.86
$x_4^\star$	−4.60	−25.47	−46.34	−67.20	−88.07
$\sigma(x^\star)$	12.02	13.44	16.54	20.58	25.10
ϕ	25.79	3.10	1.65	1.12	0.85

TABLE 1.3: Solving the unconstrained σ-problem

$\sigma^\star$	15.00	20.00	25.00	30.00	35.00
$x_1^\star$	62.52	54.57	47.84	41.53	35.42
$x_2^\star$	15.58	−10.75	−33.07	−54.00	−74.25
$x_3^\star$	58.92	120.58	172.85	221.88	269.31
$x_4^\star$	−37.01	−64.41	−87.62	−109.40	−130.48
$\mu(x^\star)$	6.55	7.87	8.98	10.02	11.03
ϕ	2.08	1.17	0.86	0.68	0.57

1.1.1.2 Adding some constraints

The introduction of constraints consists in modifying the specification of the optimization problem (1.3):

$$\begin{aligned} x^{\star}(\phi) &= \arg\max x^{\top}\mu - \frac{\phi}{2}x^{\top}\Sigma x \\ & \text{u.c.} \quad \left\{ \begin{array}{l} \mathbf{1}^{\top}x = 1 \\ x \in \Omega \end{array} \right. \end{aligned} \tag{1.5}$$

where $x \in \Omega$ corresponds to the set of restrictions[3]. These restrictions may be linear or non-linear. In the latter case, the optimization problem cannot be solved by the standard quadratic programming algorithm, but by enhanced non-linear optimization algorithms. The imposition of constraints will impact the set of optimized portfolios by reducing opportunity arbitrages. It implies that the constrained efficient frontier is located at the right of the unconstrained efficient frontier in the mean-variance map.

The most frequent constraint is certainly the no short-selling restriction. In this case, $x_i \geq 0$ and $\Omega = [0,1]^n$. Let us define the leverage measure of the portfolio x as the sum of the absolute values of the weights:

$$\mathcal{L}(x) = \sum_{i=1}^{n} |x_i|$$

With the no short-selling restriction, the leverage measure is 100% because $\mathcal{L}(x) = \sum_{i=1}^{n} x_i = 1$ whereas it is larger than 100% without this constraint[4].

Let us introduce some constraints in Example 1. In Figure 1.3, we have reported two constrained efficient frontiers, the first one by imposing no short-selling and the second one by imposing that the weights are between 0% and 40%. We verify that we may substantially reduce opportunity arbitrages. Solutions of the σ-problem are given in Table 1.4. If we target a volatility equal of 15%, the expected return of the optimized portfolio is 6.55% for the unconstrained problem, 6.14% for the shortsale constrained problem and 6.11% if we impose an upper bound of 40%. So, by imposing no short positions, we have reduced the expected return by 41 bps. The impact of the upper bound is small. If the target volatility becomes 20%, the results become 7.87%, 7.15% and 6.74% and the impact is larger than in the previous case.

[3]The restriction $\mathbf{1}^{\top}x = 1$ is already a constraint influencing the optimized portfolio (DeMiguel *et al.*, 2009).

[4]Let $x_i^- = -\min(0, x_i)$ and $x_i^+ = \max(0, x_i)$ be respectively the negative and positive parts of the weight x_i. We have $x_i = x_i^+ - x_i^-$. It follows that $\mathcal{L}(x) = \sum_{i=1}^{n} \left| x_i^+ - x_i^- \right| = \sum_{i=1}^{n} x_i^+ + \sum_{i=1}^{n} x_i^-$ with $\sum_{i=1}^{n} x_i = \sum_{i=1}^{n} x_i^+ - \sum_{i=1}^{n} x_i^- = 1$. It implies that $\mathcal{L}(x) = 1 + 2\sum_{i=1}^{n} x_i^-$ meaning that the leverage measure is larger than 1 because $\sum_{i=1}^{n} x_i^- \geq 0$.

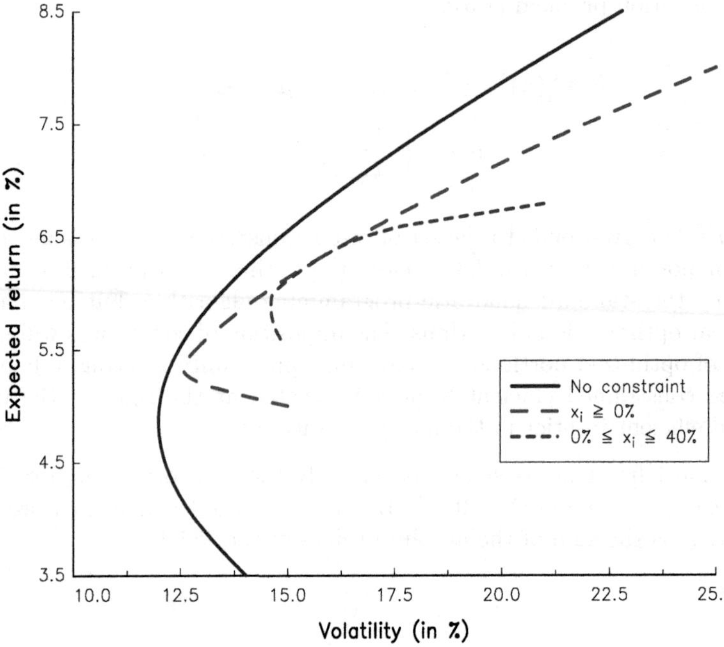

FIGURE 1.3: The efficient frontier with some weight constraints

TABLE 1.4: Solving the σ-problem with weight constraints

	$x_i \in \mathbb{R}$		$x_i \geq 0$		$0 \leq x_i \leq 40\%$	
$\sigma^\star$	15.00	20.00	15.00	20.00	15.00	20.00
$x_1^\star$	62.52	54.57	45.59	24.88	40.00	6.13
$x_2^\star$	15.58	−10.75	24.74	4.96	34.36	40.00
$x_3^\star$	58.92	120.58	29.67	70.15	25.64	40.00
$x_4^\star$	−37.01	−64.41	0.00	0.00	0.00	13.87
$\mu(x^\star)$	6.55	7.87	6.14	7.15	6.11	6.74
ϕ	2.08	1.17	1.61	0.91	1.97	0.28

1.1.1.3 Analytical solution

The Lagrange function of the optimization problem (1.3) is:

$$\mathcal{L}(x;\lambda_0) = x^\top \mu - \frac{\phi}{2} x^\top \Sigma x + \lambda_0 \left(\mathbf{1}^\top x - 1\right)$$

where λ_0 is the Lagrange coefficients associated with the constraint $\mathbf{1}^\top x = 1$. The solution $x^\star$ verifies the following first-order conditions:

$$\begin{cases} \partial_x \mathcal{L}(x;\lambda_0) = \mu - \phi \Sigma x + \lambda_0 \mathbf{1} = \mathbf{0} \\ \partial_{\lambda_0} \mathcal{L}(x;\lambda_0) = \mathbf{1}^\top x - 1 = 0 \end{cases}$$

We obtain $x = \phi^{-1}\Sigma^{-1}(\mu + \lambda_0 \mathbf{1})$. Because $\mathbf{1}^\top x - 1 = 0$, we have $\mathbf{1}^\top \phi^{-1}\Sigma^{-1}\mu + \lambda_0 \left(\mathbf{1}^\top \phi^{-1}\Sigma^{-1}\mathbf{1}\right) = 1$. It follows that:

$$\lambda_0 = \frac{1 - \mathbf{1}^\top \phi^{-1}\Sigma^{-1}\mu}{\mathbf{1}^\top \phi^{-1}\Sigma^{-1}\mathbf{1}}$$

The solution is then[5]:

$$x^\star(\phi) = \frac{\Sigma^{-1}\mathbf{1}}{\mathbf{1}^\top \Sigma^{-1}\mathbf{1}} + \frac{1}{\phi} \cdot \frac{\left(\mathbf{1}^\top \Sigma^{-1}\mathbf{1}\right)\Sigma^{-1}\mu - \left(\mathbf{1}^\top \Sigma^{-1}\mu\right)\Sigma^{-1}\mathbf{1}}{\mathbf{1}^\top \Sigma^{-1}\mathbf{1}} \tag{1.6}$$

We deduce also that the global minimum variance portfolio has the following expression:

$$x_{\mathrm{mv}} = x^\star(\infty) = \frac{\Sigma^{-1}\mathbf{1}}{\mathbf{1}^\top \Sigma^{-1}\mathbf{1}}$$

If we introduce other constraints, it is not possible to obtain a comprehensive analytical solution. Let us consider for example the no short-selling constraint. The Lagrange function becomes:

$$\mathcal{L}(x;\lambda_0,\lambda) = x^\top \mu - \frac{\phi}{2} x^\top \Sigma x + \lambda_0 \left(\mathbf{1}^\top x - 1\right) + \lambda^\top x$$

where $\lambda = (\lambda_1, \ldots, \lambda_n)$ is the vector of Lagrange coefficients associated with the constraints $x_i \geq 0$. The first-order condition is then $\mu - \phi\Sigma x + \lambda_0 \mathbf{1} + \lambda = \mathbf{0}$. It follows that $x = \phi^{-1}\Sigma^{-1}(\mu + \lambda_0 \mathbf{1} + \lambda)$. The Kuhn-Tucker conditions are $\min(\lambda_i, x_i) = 0$ for all $i = 1, \ldots, n$. It implies that if $x_i = 0$ then $\lambda_i > 0$ and if $x_i > 0$ then $\lambda_i = 0$. We find also a formula close to the previous one, but the universe is limited to assets which present positive weights. This formula is therefore endogenous.

[5]If we do not impose the constraint $\mathbf{1}^\top x = 1$, the solution becomes:

$$x^\star(\phi) = \frac{1}{\phi}\Sigma^{-1}\mu$$

1.1.2 The tangency portfolio

We recall that in the view of Markowitz, there is a set of optimized portfolios. However, Tobin showed in 1958 that one optimized portfolio dominates all the others if there is a risk-free asset.

Let us consider a combination of the risk-free asset and a portfolio x. We denote r the return of the risk-free asset. We have[6]:

$$R(y) = (1-\alpha) r + \alpha R(x)$$

where $y = \begin{pmatrix} \alpha x \\ 1-\alpha \end{pmatrix}$ is a vector of dimension $(n+1)$ and $\alpha \geq 0$ is the proportion of the wealth invested in the risky portfolio. It follows that:

$$\mu(y) = (1-\alpha) r + \alpha \mu(x) = r + \alpha(\mu(x) - r)$$

and:

$$\sigma^2(y) = \alpha^2 \sigma^2(x)$$

We deduce that:

$$\mu(y) = r + \frac{(\mu(x) - r)}{\sigma(x)} \sigma(y)$$

It is the equation of a linear function between the volatility and the expected return of the combined portfolio y. In Figure 1.4, we reported the previous (unconstrained) efficient frontier. The dashed line corresponds to the combination between the risk-free asset (r is equal to 1.5%) and the optimized portfolio A. Nevertheless this combination is suboptimal, because it is dominated by other combinations. We note that a straight line dominates all the other straight lines and the efficient frontier. This line is tangent to the efficient frontier and is called the capital market line. It implies that one optimized risky portfolio dominates all the other risky portfolios, namely the tangency portfolio.

Let $\mathrm{SR}(x \mid r)$ be the Sharpe ratio of portfolio x:

$$\mathrm{SR}(x \mid r) = \frac{\mu(x) - r}{\sigma(x)}$$

We note that we can write the previous equation as follows:

$$\frac{\mu(y) - r}{\sigma(y)} = \frac{\mu(x) - r}{\sigma(x)} \Leftrightarrow \mathrm{SR}(y \mid r) = \mathrm{SR}(x \mid r)$$

We deduce that the tangency portfolio is the one that maximizes the angle θ or equivalently $\tan\theta$ which is equal to the Sharpe ratio. The tangency portfolio is also the risky portfolio corresponding to the maximum Sharpe ratio. We

[6] We have $n+1$ assets in the universe where the first n assets correspond to the previous risky assets and the last asset is the risk-free asset.

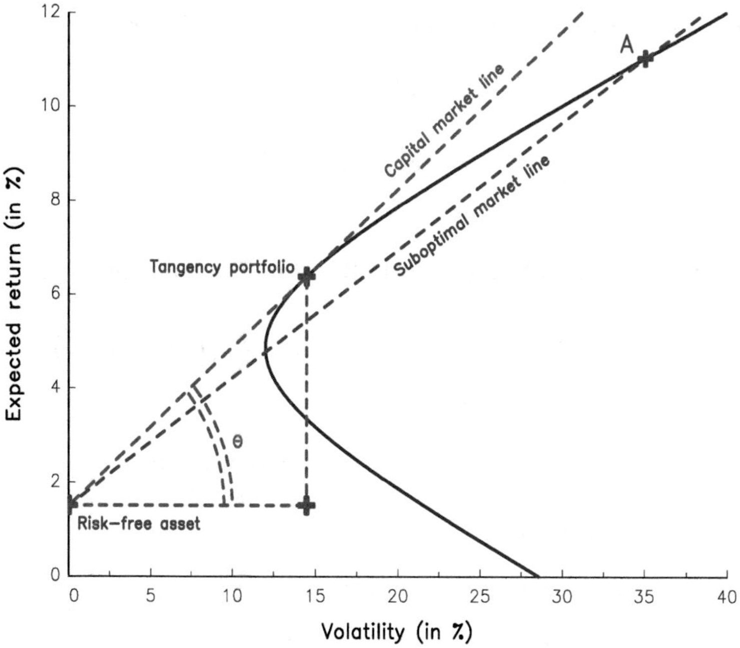

FIGURE 1.4: The capital market line

also note that any portfolio which belongs to the capital market line has the same Sharpe ratio[7].

Remark 2 *Let us consider a portfolio x of risky assets and a risk-free asset r. We denote by $\tilde{x}$ the augmented vector of dimension $n+1$ such that:*

$$\tilde{x} = \begin{pmatrix} x \\ x_r \end{pmatrix}$$

It follows that:

$$\tilde{\Sigma} = \begin{pmatrix} \Sigma & \mathbf{0} \\ \mathbf{0} & 0 \end{pmatrix} \quad \textit{and} \quad \tilde{\mu} = \begin{pmatrix} \mu \\ r \end{pmatrix}$$

If we include the risk-free asset, the Markowitz ϕ-problem becomes:

$$\begin{aligned} \tilde{x}^{\star}(\phi) &= \arg\max \tilde{x}^{\top}\tilde{\mu} - \frac{\phi}{2}\tilde{x}^{\top}\tilde{\Sigma}\tilde{x} \\ & \quad u.c. \quad \mathbf{1}^{\top}\tilde{x} = 1 \end{aligned}$$

[7]If $r = 1.5\%$, the composition of the tangency portfolio $x^{\star}$ is 63.63%, 19.27%, 50.28% and −33.17% and we have $\mu(x^{\star}) = 6.37\%$, $\sigma(x^{\star}) = 14.43\%$, $\mathrm{SR}(x^{\star} \mid r) = 0.34$ and $\theta(x^{\star}) =$ 18.64 degrees.

We note that the objective function can be written as follows:

$$\begin{aligned} f(\tilde{x}) &= \tilde{x}^\top \tilde{\mu} - \frac{\phi}{2} \tilde{x}^\top \tilde{\Sigma} \tilde{x} \\ &= x^\top \mu + x_r r - \frac{\phi}{2} x^\top \Sigma x \\ &= g(x, x_r) \end{aligned}$$

The constraint becomes $\mathbf{1}^\top x + x_r = 1$. *We deduce that the Lagrange function is:*

$$\mathcal{L}(x, x_r; \lambda_0) = x^\top \mu + x_r r - \frac{\phi}{2} x^\top \Sigma x - \lambda_0 \left(\mathbf{1}^\top x + x_r - 1\right)$$

The first-order conditions are:

$$\begin{cases} \partial_x \mathcal{L}(x, x_r; \lambda_0) = \mu - \phi \Sigma x - \lambda_0 \mathbf{1} = \mathbf{0} \\ \partial_{x_r} \mathcal{L}(x, x_r; \lambda_0) = r - \lambda_0 = 0 \\ \partial_{\lambda_0} \mathcal{L}(x, x_r; \lambda_0) = \mathbf{1}^\top x + x_r - 1 = 0 \end{cases}$$

The solution of the optimization problem is then:

$$\begin{cases} x^\star = \phi^{-1} \Sigma^{-1} (\mu - r\mathbf{1}) \\ \lambda_0^\star = r \\ x_r^\star = 1 - \phi^{-1} \mathbf{1}^\top \Sigma^{-1} (\mu - r\mathbf{1}) \end{cases}$$

Let $x_0^\star$ *be the following portfolio:*

$$x_0^\star = \frac{\Sigma^{-1} (\mu - r\mathbf{1})}{\mathbf{1}^\top \Sigma^{-1} (\mu - r\mathbf{1})}$$

We can then write the solution of the optimization problem in the following way:

$$\begin{cases} x^\star = \alpha x_0^\star \\ \lambda_0^\star = r \\ x_r^\star = 1 - \alpha \\ \alpha = \phi^{-1} \mathbf{1}^\top \Sigma^{-1} (\mu - r\mathbf{1}) \end{cases}$$

The first equation indicates that the relative proportions of risky assets in the optimized portfolio remain constant. If $\phi = \phi_0 = \mathbf{1}^\top \Sigma^{-1} (\mu - r\mathbf{1})$, *then* $x^\star = x_0^\star$ *and* $x_r^\star = 0$. *We deduce that* $x_0^\star$ *is the tangency portfolio. If* $\phi \neq \phi_0$, $x^\star$ *is proportional to* $x_0^\star$ *and the wealth invested in the risk-free asset is the complement* $(1 - \alpha)$ *to obtain a total exposure equal to 100%. We retrieve then the separation theorem:*

$$\tilde{x}^\star = \underbrace{\alpha \cdot \begin{pmatrix} x_0^\star \\ 0 \end{pmatrix}}_{\text{risky assets}} + \underbrace{(1 - \alpha) \cdot \begin{pmatrix} \mathbf{0} \\ 1 \end{pmatrix}}_{\text{risk-free asset}}$$

Let us consider our previous example. We include the risk-free asset in the universe of risky assets and we compute the efficient frontier using the QP problem. Results are reported in Figure 1.5. We verify that the efficient frontier with the risk-free asset is exactly the capital market line obtained in Figure 1.4. This analysis can be extended when we take into account some constraints, but with some limits. For example, the efficient frontier with positive weights is still a straight line. But, if $0 \leq x_i \leq 0.40$, it is only a straight line from the risk-free asset to the tangency portfolio. And if $0.20 \leq x_i \leq 0.70$, the efficient frontier corresponds to a straight line in a small region. The problem comes from the fact that building the capital market line geometrically by considering the tangency portfolio ignores completely the introduction of constraints. Let us consider the case $0 \leq x_i \leq 0.40$. The corresponding tangency portfolio is $(40.0\%, 34.7\%, 25.3\%, 0\%)$. When we geometrically build the capital market line, we assume that we can leverage this portfolio. However, it is possible we cannot because the weight of the first risky asset has reached the upper bound equal to 40%. In the same way, the solution is $(36.1\%, 23.9\%, 20.0\%, 20.0\%)$ if $0.20 \leq x_i \leq 0.70$. In this case, we can apply a leverage of 94% without reaching the constraint. However, we cannot deleverage this portfolio, because two assets of the tangency portfolio have reached their lower bounds. So, when we impose some constraints, the geometric construction of the capital market line is valid only on the restricted region that satisfies the constraints.

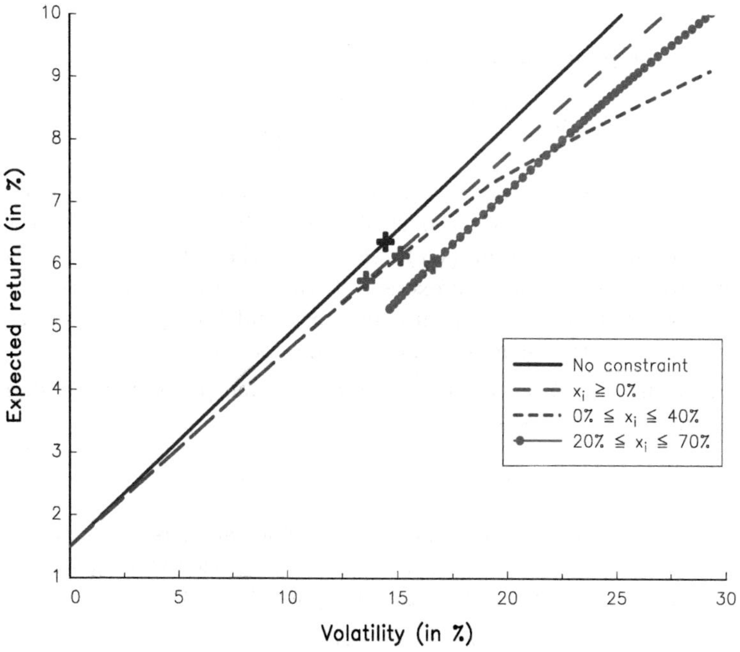

FIGURE 1.5: The efficient frontier with a risk-free asset

1.1.3 Market equilibrium and CAPM

In 1964, William Sharpe developed the capital asset pricing model (CAPM). Let $x^\star$ be the tangency portfolio. On the efficient frontier, we have:

$$\mu(y) = r + \frac{\sigma(y)}{\sigma(x^\star)}\left(\mu(x^\star) - r\right)$$

We consider a portfolio z with a proportion w invested in the asset i and a proportion $(1-w)$ invested in the tangency portfolio $x^\star$. We have[8] $\mu(z) = w\mu_i + (1-w)\,\mu(x^\star)$ and $\sigma^2(z) = w^2\sigma_i^2 + (1-w)^2\sigma^2(x^\star) + 2w(1-w)\,\rho(\mathbf{e}_i, x^\star)\,\sigma_i\sigma(x^\star)$. It follows that:

$$\frac{\partial\,\mu(z)}{\partial\,\sigma(z)} = \frac{\mu_i - \mu(x^\star)}{\left(w\sigma_i^2 + (w-1)\,\sigma^2(x^\star) + (1-2w)\,\rho(\mathbf{e}_i, x^\star)\,\sigma_i\sigma(x^\star)\right)\sigma^{-1}(z)}$$

When $w = 0$, the portfolio z is the tangency portfolio $x^\star$ and the previous derivative is equal to the Sharpe ratio $\mathrm{SR}(x^\star \mid r)$. We deduce that:

$$\frac{\left(\mu_i - \mu(x^\star)\right)\sigma(x^\star)}{\rho(\mathbf{e}_i, x^\star)\,\sigma_i\sigma(x^\star) - \sigma^2(x^\star)} = \frac{\mu(x^\star) - r}{\sigma(x^\star)}$$

which is equivalent to:

$$\pi_i = \mu_i - r = \beta_i\left(\mu(x^\star) - r\right) \tag{1.7}$$

with π_i the risk premium of the asset i and:

$$\beta_i = \frac{\rho(\mathbf{e}_i, x^\star)\,\sigma_i}{\sigma(x^\star)} = \frac{\mathrm{cov}\left(R_i, R(x^\star)\right)}{\mathrm{var}\left(R(x^\star)\right)} \tag{1.8}$$

The coefficient β_i is the ratio of the covariance between the return of asset i and the return of the tangency portfolio upon the variance of the tangency portfolio return. The equation (1.7) tells us that the risk premium of the asset i is equal to its beta times the excess return of the tangency portfolio. It is easy to show that this relationship remains valid for any portfolio and not only for the assets that compose the tangency portfolio.

Let $R_{i,t}$ and $R_t(x)$ be the returns of asset i and portfolio x at time t. We consider the linear regression:

$$R_{i,t} = \alpha_i + \beta_i R_t(x) + \varepsilon_{i,t}$$

with $\varepsilon_{i,t}$ a white noise process. The OLS coefficient $\hat{\beta}_i$ is an estimate of the beta β_i of the asset i. We can generalize this approach to estimate the beta of one portfolio y with respect to another portfolio x. We have:

$$R_t(y) = \alpha + \beta R_t(x) + \varepsilon_t$$

[8] $\mathbf{e}_i$ is the unit vector with 1 in the ith position and 0 elsewhere. It corresponds then to the portfolio fully invested in asset i.

Another way to compute the beta is to use the following relationship:

$$\beta(y \mid x) = \frac{\sigma(y, x)}{\sigma^2(x)} = \frac{y^\top \Sigma x}{x^\top \Sigma x}$$

We deduce that the expression of the beta of asset i is also:

$$\beta_i = \beta(\mathbf{e}_i \mid x) = \frac{\mathbf{e}_i^\top \Sigma x}{x^\top \Sigma x} = \frac{(\Sigma x)_i}{x^\top \Sigma x}$$

It follows that the beta of a portfolio is the weighted average of the beta of the assets that compose the portfolio:

$$\beta(y \mid x) = \frac{y^\top \Sigma x}{x^\top \Sigma x} = y^\top \frac{\Sigma x}{x^\top \Sigma x} = \sum_{i=1}^{n} y_i \beta_i$$

Let us consider Example 1 with $r = 1.5\%$. The tangency portfolio is $x^\star = (63.63\%, 19.27\%, 50.28\%, -33.17\%)$. We have $\mu(x^\star) = 6.37\%$. In Table 1.5, we have reported the beta of each asset and the beta of the equally weighted portfolio x_{ew}. We also compute the risk premium explained by the tangency portfolio $\pi(y \mid x^\star) = \beta(y \mid x^\star)(\mu(x^\star) - r)$ and verify the relationship (1.7). If we impose a lower bound $x_i \geq 0$, the tangency portfolio becomes $x^\star = (53.64\%, 32.42\%, 13.93\%, 0.00\%)$ and we have $\mu(x^\star) = 5.74\%$. Results are given in Table 1.6. We note that the relationship (1.7) does not hold for the fourth asset and for the portfolio x_{ew}. In these two cases, the beta is overestimated because the short position in the fourth asset is eliminated. This explains why we have $\mu_4 - r = \beta_4(\mu(x^\star) - r) + \pi_4^-$ where π_4^- represents a negative premium due to a lack of arbitrage on the fourth asset. This example shows that the relationship (1.7) is not valid if we impose some restrictions in portfolio optimization.

TABLE 1.5: Computation of the beta

Portfolio	$\mu(y)$	$\beta(y \mid x^\star)$	$\pi(y \mid x^\star)$
$\mathbf{e}_1$	3.50	0.72	3.50
$\mathbf{e}_2$	4.50	0.92	4.50
$\mathbf{e}_3$	6.50	1.33	6.50
$\mathbf{e}_4$	4.50	0.92	4.50
x_{ew}	4.75	0.98	4.75

The relationship (1.7) is very important and highlights the role of the beta coefficient. However, this result is not Sharpe's (1964) most important finding. In his article, Sharpe shows also that if the market is at the equilibrium, the prices of assets are such that the tangency portfolio is the market portfolio (or the market-cap portfolio). With this result, the characterization of the tangency portfolio does not depend upon the assumptions about expected returns, volatilities and correlations.

TABLE 1.6: Computation of the beta with a constrained tangency portfolio

Portfolio	$\mu(y)$	$\beta(y \mid x^{\star})$	$\pi(y \mid x^{\star})$
$\mathbf{e}_1$	3.50	0.83	3.50
$\mathbf{e}_2$	4.50	1.06	4.50
$\mathbf{e}_3$	6.50	1.53	6.50
$\mathbf{e}_4$	4.50	1.54	6.53
x_{ew}	4.75	1.24	5.26

In 1968, Jensen analyzed the performance of active management by using the following regression model:

$$R_{j,t} = \alpha_j + \beta_j R_t(x^{\star}) + \varepsilon_{j,t}$$

where $R_{j,t}$ is the return of the mutual fund j, $R_t(x^{\star})$ is the return of the market portfolio and $\varepsilon_{j,t}$ is an idiosyncratic risk. If the mutual fund outperforms the market portfolio, the assumption $\alpha_j > 0$ is not rejected. However, Jensen rejects this assumption for most mutual funds and concludes that active management does not outperform passive management on average[9]. The findings of Sharpe (1964) and Jensen (1969) explain the development of passive management, which starts with the launch of the first equity index fund by John McQuown in Wells Fargo Bank for the Samsonite luggage company (Bernstein, 1992).

Remark 3 *There is some confusion regarding the capital market and the security market lines. The capital market line corresponds to the efficient frontier in the presence of a risk-free asset, whereas the security market line derives from the CAPM theory. Let us formulate the equation (1.7) as follows:*

$$\frac{\pi_i}{\beta_i} = \mu(x^{\star}) - r$$

where $\pi_i = \mu_i - r$ is the risk premium of asset i. This means that the risk premium of the different assets adjusted for their systematic risk is constant. The relationship between β_i and π_i is then the security market line. It is the foundation of the Treynor ratio which corresponds to[10]*:*

$$\mathrm{TR}(x \mid x^{\star}) = \frac{\mu(x) - r}{\beta(x \mid x^{\star})}$$

where $\beta(x \mid x^{\star})$ is the beta of the portfolio x with respect to the market portfolio $x^{\star}$. The difference between the Sharpe ratio and the Treynor ratio comes from the specification of the risk measure. Sharpe uses the total risk $\sigma(x)$ whereas Treynor uses the systematic risk $\beta(x \mid x^{\star})$ (Dimson and Mussavian, 1999).

[9] It is interesting to note that the first paper developing the concept of alpha concludes that there is no alpha in active management, more precisely in equity mutual funds. Today, the concept of alpha is largely used to legitimate the hedge fund industry.

[10] By construction, the Treynor ratio of the market portfolio is equal to its excess return because we have $\beta(x^{\star} \mid x^{\star}) = 1$.

1.1.4 Portfolio optimization in the presence of a benchmark

Let us now consider a benchmark which is represented by a portfolio b. The tracking error between the active portfolio x and its benchmark b is the difference between the return of the portfolio and the return of the benchmark:

$$\begin{aligned} e &= R(x) - R(b) \\ &= \sum_{i=1}^{n} x_i R_i - \sum_{i=1}^{n} b_i R_i \\ &= x^\top R - b^\top R \\ &= (x-b)^\top R \end{aligned}$$

The expected excess return is:

$$\mu(x \mid b) = \mathbb{E}[e] = (x-b)^\top \mu$$

whereas the volatility of the tracking error is[11]:

$$\sigma(x \mid b) = \sigma(e) = \sqrt{(x-b)^\top \Sigma (x-b)}$$

The objective of the investor is to maximize the expected tracking error with a constraint on the tracking error volatility:

$$\begin{aligned} x^\star &= \arg\max (x-b)^\top \mu \\ & \text{u.c.} \quad \mathbf{1}^\top x = 1 \quad \text{and} \quad \sigma(e) \le \sigma^\star \end{aligned} \tag{1.9}$$

Like the Markowitz problem, we transform this σ-problem into a ϕ-problem:

$$x^\star(\phi) = \arg\max f(x \mid b)$$

with:

$$\begin{aligned} f(x \mid b) &= (x-b)^\top \mu - \frac{\phi}{2}(x-b)^\top \Sigma (x-b) \\ &= x^\top(\mu + \phi\Sigma b) - \frac{\phi}{2} x^\top \Sigma x - \left(\frac{\phi}{2} b^\top \Sigma b + b^\top \mu\right) \\ &= x^\top(\mu + \phi\Sigma b) - \frac{\phi}{2} x^\top \Sigma x + c \end{aligned}$$

where c is a constant which does not depend on the portfolio x. We recognize a quadratic program which could be solved easily using numerical algorithms. The efficient frontier is then the parametric curve $(\sigma(x^\star(\phi) \mid b), \mu(x^\star(\phi) \mid b))$ with $\phi \in \mathbb{R}_+$.

We consider Example 1 with the benchmark $b = (60\%, 40\%, 20\%, -20\%)$.

[11] In IOSCO and ESMA terminologies, $\mu(x \mid b)$ is called the tracking difference (TD) whereas $\sigma(x \mid b)$ is known as the tracking error (TE).

In Figure 1.6, we have represented the corresponding efficient frontier. We verify that it is a straight line when there is no restriction (Roll, 1992). If we impose that $x_i \geq -10\%$, the efficient frontier is moved to the right. For the third case, we assume that the weights are between a lower bound and an upper bound: $x_i^- \leq x_i \leq x_i^+$ with $x_i^+ = 50\%$. For the first three assets, the lower bound x_i^- is set to 0, whereas it is equal to -20% for the fourth asset.

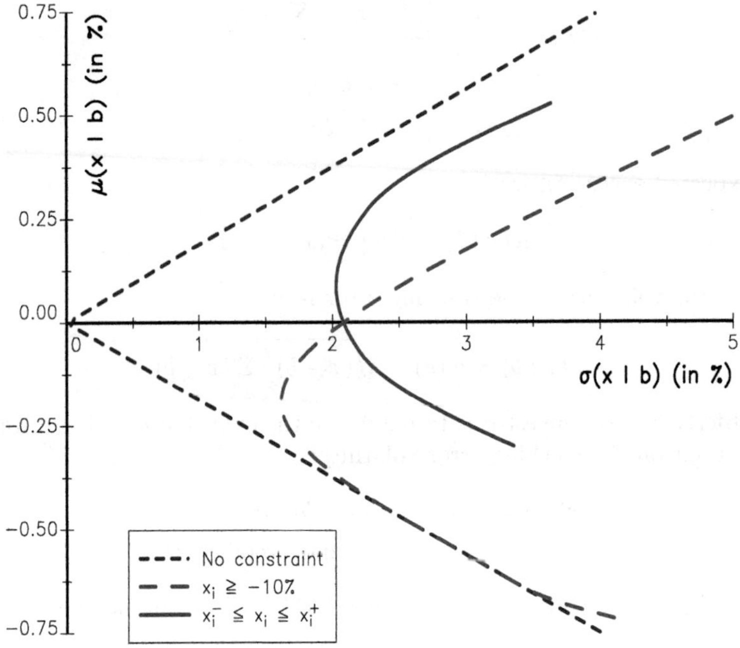

FIGURE 1.6: The efficient frontier with a benchmark

To compare the performance of different portfolios, a better measure than the Sharpe ratio is the information ratio which is defined as follows:

$$\mathrm{IR}(x \mid b) = \frac{\mu(x \mid b)}{\sigma(x \mid b)} = \frac{(x-b)^\top \mu}{\sqrt{(x-b)^\top \Sigma (x-b)}}$$

If we consider a combination of the benchmark b and the active portfolio x, the composition of the portfolio is:

$$y = (1-\alpha) b + \alpha x$$

with $\alpha \geq 0$ the proportion of wealth invested in the portfolio x. It follows that:

$$\mu(y \mid b) = (y-b)^\top \mu = \alpha \mu(x \mid b)$$

and:

$$\sigma^2(y \mid b) = (y-b)^\top \Sigma (y-b) = \alpha^2 \sigma^2(x \mid b)$$

We deduce that:

$$\mu(y \mid b) = \mathrm{IR}(x \mid b) \cdot \sigma(y \mid b)$$

It is the equation of a linear function between the tracking error volatility and the expected tracking error of the portfolio y. It implies that the efficient frontier is a straight line:

> "*If the manager is measured solely in terms of excess return performance, he or she should pick a point on the upper part of this efficient frontier. For instance, the manager may have a utility function that balances expected value added against tracking error volatility. Note that because the efficient set consists of a straight line, the maximal Sharpe ratio is not a usable criterion for portfolio allocation*" (Jorion, 2003, page 172).

If we add some other constraints to the portfolio optimization problem (1.9), the efficient frontier is no longer a straight line. In this case, one optimized portfolio dominates all the other portfolios. It is the portfolio which belongs to the efficient frontier and the straight line which is tangent to the efficient frontier. It is also the portfolio which maximizes the information ratio.

Let us look at the previous efficient frontier when we impose lower and upper bounds (third case). When we combine it with the benchmark, we obtain the straight line produced in Figure 1.7 and we obtain that the tangency portfolio is $(49.51\%, 29.99\%, 40.50\%, -20.00\%)$.

Remark 4 *It is interesting to note that the Markowitz-Sharpe approach is a special case of the above mentioned framework. Let us consider a portfolio x of risky assets and a risk-free asset r. We note $\tilde{x}$ and $\tilde{b}$ the two augmented vectors of dimension $n+1$ such that:*

$$\tilde{x} = \begin{pmatrix} x \\ 0 \end{pmatrix} \quad \text{and} \quad \tilde{b} = \begin{pmatrix} \mathbf{0} \\ 1 \end{pmatrix}$$

It follows that:

$$\tilde{\Sigma} = \begin{pmatrix} \Sigma & \mathbf{0} \\ \mathbf{0} & 0 \end{pmatrix} \quad \text{and} \quad \tilde{\mu} = \begin{pmatrix} \mu \\ r \end{pmatrix}$$

The objective function is then defined as follows:

$$\begin{aligned} f\left(\tilde{x} \mid \tilde{b}\right) &= \left(\tilde{x} - \tilde{b}\right)^\top \mu - \frac{\phi}{2}\left(\tilde{x} - \tilde{b}\right)^\top \Sigma \left(\tilde{x} - \tilde{b}\right) \\ &= \tilde{x}^\top \left(\tilde{\mu} + \phi\tilde{\Sigma}\tilde{b}\right) - \frac{\phi}{2}\tilde{x}^\top \tilde{\Sigma}\tilde{x} - \left(\frac{\phi}{2}\tilde{b}^\top \tilde{\Sigma}\tilde{b} + \tilde{b}^\top \tilde{\mu}\right) \\ &= x^\top \mu - \frac{\phi}{2} x^\top \Sigma x - r \\ &= x^\top (\mu - r\mathbf{1}) - \frac{\phi}{2} x^\top \Sigma x \end{aligned}$$

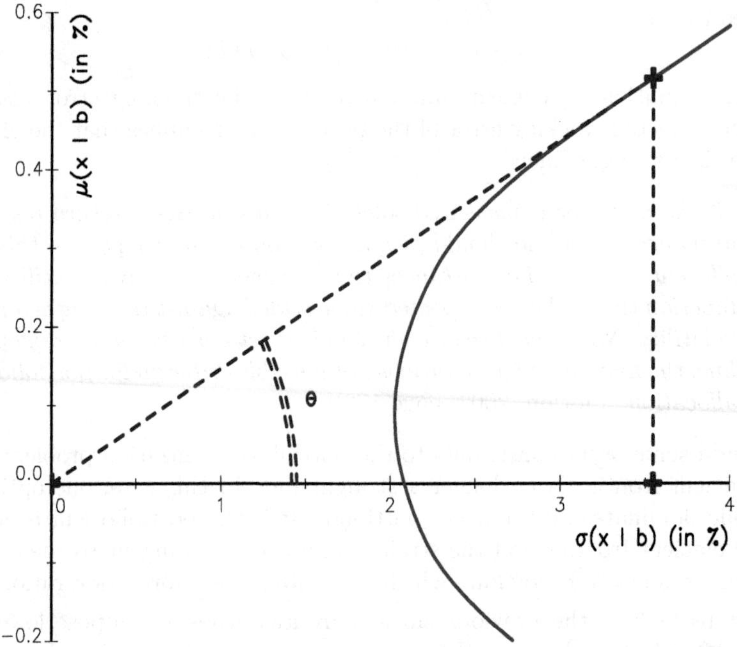

FIGURE 1.7: The tangency portfolio with respect to a benchmark

The solution of the QP problem $\tilde{x}^\star(\phi) = \arg\max f\left(\tilde{x} \mid \tilde{b}\right)$ *is related to the solution* $x^\star(\phi)$ *of the Markowitz* ϕ*-problem in the following way:*

$$\tilde{x}^\star(\phi) = \begin{pmatrix} x^\star(\phi) \\ 0 \end{pmatrix}$$

Because the volatility of the error tracking $\sigma\left(\tilde{x}^\star(\phi) \mid \tilde{b}\right)$ *is also the volatility of the portfolio* $\sigma(x^\star(\phi))$*, the only difference comes from the expected return side. In the Markowitz model, we consider the expected return of the portfolio whereas we consider the excess return in the benchmark approach. This explains why the Sharpe ratio is also an information ratio:*

$$\mathrm{SR}(x^\star(\phi) \mid r) = \mathrm{IR}\left(\tilde{x}^\star(\phi) \mid \tilde{b}\right)$$

1.1.5 The Black-Litterman model

The Black-Litterman model consists in incorporating portfolio manager's views in a strategic asset allocation. It is then a tactical asset allocation model. It could also be viewed as a synthesis of different approaches: the portfolio optimization of Harry Markowitz, the CAPM theory of William Sharpe, the introduction of constraints in asset allocation, etc. Starting with an initial allocation, the model computes the implied risk premia and then deduces the optimized portfolio which is coherent with the bets of the portfolio manager.

1.1.5.1 Computing the implied risk premia

Let us consider the optimization formulation (1.3) on page 6. If we omit the constraint $\mathbf{1}^\top x = 1$, the solution is:

$$x^\star = \frac{1}{\phi}\Sigma^{-1}\mu$$

In the Markowitz model, the unknown variable is the vector of weights x. We suppose now that an initial allocation x_0 is given. Black and Litterman (1992) assume that this allocation corresponds to an optimal solution. In this case, we have:

$$\tilde{\mu} = \phi\Sigma x_0 \tag{1.10}$$

We may interpret $\tilde{\mu}$ as the vector of expected returns which is coherent with the portfolio x_0. $\tilde{\pi} = \tilde{\mu} - r$ is the risk premium priced by the portfolio manager; it is also called the '*market risk premium*'.

Remark 5 *The concept of market risk premium comes from the fact that the authors use a general equilibrium approach. From the point of view of the portfolio manager, the current allocation x_0 is optimal. Indeed, if it is not optimal, the portfolio manager would certainly select another allocation. It implies that x_0 is his tangency portfolio. $\tilde{\pi}$ is then the risk premium which is compatible with the fact that x_0 is the manager's market portfolio and $\tilde{\mu}$ is the vector of expected returns which is implicitly priced by the portfolio manager.*

The computation of $\tilde{\mu}$ needs to specify Σ and ϕ. For the first parameter, we generally use the empirical covariance matrix $\hat{\Sigma}$. It is more difficult to calibrate the second parameter, which measures the portfolio manager's risk aversion. One idea consists in replacing ϕ by a parameter which is more tractable. Let us assume that the risk-free rate r is equal to 0. Using Equation (1.10), it follows that $x_0^\top \tilde{\mu} = \phi x_0^\top \Sigma x_0$. We deduce then:

$$\phi = \frac{x_0^\top \tilde{\mu}}{x_0^\top \Sigma x_0} = \frac{\mathrm{SR}\left(x_0 \mid r\right)}{\sqrt{x_0^\top \Sigma x_0}} \tag{1.11}$$

where $\mathrm{SR}\left(x_0 \mid r\right)$ is the portfolio's expected Sharpe ratio. We finally obtain[12]:

$$\tilde{\mu} = \mathrm{SR}\left(x_0 \mid r\right) \frac{\Sigma x_0}{\sqrt{x_0^\top \Sigma x_0}}$$

[12]In the general case when $r \neq 0$, this expression becomes:

$$\tilde{\mu} = r + \mathrm{SR}\left(x_0 \mid r\right) \frac{\Sigma x_0}{\sqrt{x_0^\top \Sigma x_0}} \tag{1.12}$$

Let us consider Example 1. We suppose that the initial allocation x_0 is $(40\%, 30\%, 20\%, 10\%)$. The ex-ante volatility of the portfolio is then $\sigma(x_0) = 15.35\%$. The objective of the portfolio manager is to target a Sharpe ratio equal to 0.25. If $r = 3\%$, we obtain $\phi = 1.63$ and the implied expected returns are $\tilde{\mu} = (5.47\%, 6.68\%, 8.70\%, 9.06\%)$.

1.1.5.2 The optimization problem

Black and Litterman (1992) state that μ cannot be known with certainty. In particular, they assume that μ is a Gaussian vector with expected returns $\tilde{\mu}$ and covariance matrix Γ:

$$\mu \sim \mathcal{N}(\tilde{\mu}, \Gamma)$$

The market risk premium $\tilde{\mu}$ is then the unconditional mathematical expectation of the asset returns R. To specify the portfolio manager's views, the authors assume that they are given by this relationship:

$$P\mu = Q + \varepsilon \tag{1.13}$$

where P is a $(k \times n)$ matrix, Q is a $(k \times 1)$ vector and $\varepsilon \sim \mathcal{N}(0, \Omega)$ is a Gaussian vector of dimension k.

Remark 6 *With the specification (1.13), we can express the views in an absolute or relative way. Let us consider the two-asset case. Suppose that the portfolio manager has an absolute view on the expected return of the first asset. If $\mu_1 = q + \varepsilon$, the matrix P is equal to the row vector $\begin{pmatrix} 1 & 0 \end{pmatrix}$ and the portfolio manager thinks that the expected return of the first asset is q. If $\Omega = 0$, the portfolio manager has a very high level of confidence. If $\Omega \neq 0$, his view is uncertain. Suppose now that he thinks that the outperformance of the first asset with respect to the second asset is q. We have $\mu_1 - \mu_2 = q + \varepsilon$. The matrix P becomes $\begin{pmatrix} 1 & -1 \end{pmatrix}$. If the portfolio manager has two views, the matrix P will have two rows. k is then the number of views. Ω is the covariance matrix of $P\mu - Q$, therefore it measures the uncertainty of the views.*

The Markowitz optimization problem becomes:

$$\begin{aligned} x^\star(\phi) &= \arg\max x^\top \bar{\mu} - \frac{\phi}{2} x^\top \Sigma x \\ &\text{u.c.} \quad \mathbf{1}^\top x = 1 \end{aligned} \tag{1.14}$$

where $\bar{\mu}$ is the vector of expected returns conditional to the views:

$$\bar{\mu} = \mathbb{E}[\mu \mid \text{views}] = \mathbb{E}[\mu \mid P\mu = Q + \varepsilon]$$

To compute $\bar{\mu}$, we consider the random vector $(\mu, \nu = P\mu - \varepsilon)$:

$$\begin{pmatrix} \mu \\ \nu \end{pmatrix} \sim \mathcal{N}\left(\begin{pmatrix} \tilde{\mu} \\ P\tilde{\mu} \end{pmatrix}, \begin{pmatrix} \Gamma & \Gamma P^\top \\ P\Gamma & P\Gamma P^\top + \Omega \end{pmatrix} \right)$$

We apply the conditional expectation formula[13]:

$$\begin{aligned}\bar{\mu} &= \mathbb{E}[\mu \mid \nu = Q] \\ &= \tilde{\mu} + \Gamma P^\top \left(P\Gamma P^\top + \Omega\right)^{-1}(Q - P\tilde{\mu}) \qquad (1.15)\end{aligned}$$

The conditional expectation $\bar{\mu}$ has two components:

1. The first component corresponds to the vector of implied expected returns $\tilde{\mu}$.

2. The second component is a correction term which takes into account the *disequilibrium* $(Q - P\tilde{\mu})$ between the manager views and the market views.

1.1.5.3 Numerical implementation of the model

From a practical point of view, the five steps to implement the Black-Litterman model are:

1. We estimate the empirical covariance matrix $\hat{\Sigma}$.

2. Given the current portfolio, we deduce the vector $\tilde{\mu}$ of implied expected returns using Equation (1.12).

3. We specify the views and the associated level of confidence; this is equivalent to defining the P, Q and Ω matrices.

4. Given a matrix Γ, we compute the conditional expectation $\bar{\mu}$ using Equation (1.15).

5. We finally perform the portfolio optimization (1.14) with $\bar{\mu}$ and ϕ given by Equation (1.11).

The difficulty lies in specifying the covariance matrix Γ. One solution is to define $\Gamma = \tau\Sigma$ and to calibrate τ in order to target a tracking error volatility (Meucci, 2005).

We consider again our previous example. We suppose that the portfolio manager has an absolute view on the first asset and a relative view on the second and third assets. We specify the matrices P, Q and Ω as follows:

$$P = \begin{pmatrix} 1 & 0 & 0 & 0 \\ 0 & 1 & -1 & 0 \end{pmatrix}, Q = \begin{pmatrix} q_1 \\ q_{2-3} \end{pmatrix} \text{ and } \Omega = \begin{pmatrix} \varpi_1^2 & 0 \\ 0 & \varpi_{2-3}^2 \end{pmatrix}$$

[13] Let us consider a Gaussian random vector (X, Y) of mean $\mu = (\mu_x, \mu_y)$ and covariance matrix $\Sigma = \begin{pmatrix} \Sigma_{xx} & \Sigma_{xy} \\ \Sigma_{yx} & \Sigma_{yy} \end{pmatrix}$. The conditional distribution of X given $Y = y$ is a multivariate normal distribution with mean $\mu_{x|y} = \mu_x + \Sigma_{xy}\Sigma_{yy}^{-1}(y - \mu_y)$ and covariance matrix $\Sigma_{xx|y} = \Sigma_{xx} - \Sigma_{xy}\Sigma_{yy}^{-1}\Sigma_{yx}$.

q_1 and q_{2-3} take the values 4% and −1%. This means that the portfolio manager believes that the expected return of the first asset is 4% whereas he believes that the third asset will outperform the second asset by 1% in average. We suppose also that $\varpi_1 = 10\%$ and $\varpi_{2-3} = 5\%$, meaning that the level of confidence is higher for the relative view than for the absolute view. We consider $\tau = 1$. The set of these values for the different parameters corresponds to the generic case #1. The other cases are just a modification of this generic case. In the case #2, we have $q_1 = 7\%$. The case #3 corresponds to $\varpi_1 = \varpi_{2-3} = 20\%$ whereas we use $\tau = 10\%$ and $\tau = 1\%$ for the cases #4 and #5. We also impose that the optimized weights are positive. Results are shown in Table 1.7. Portfolio #1 differs from the initial allocation x_0. The weight of the first asset decreases. This is coherent with the absolute view of the portfolio manger, who thinks that μ_1 is smaller than its market price (4% versus 5.47%). The weight difference between the second and third asset is 10% for the initial allocation whereas it becomes 46.1%. This is also coherent with the relative view. Indeed, the market believes implicitly that the third asset will outperform the second asset by 2.02%, whereas the portfolio manager thinks that this outperformance is only 1%. We note that if we increase the uncertainty of the views (portfolio #3) or if the covariance Γ of the expected return is smaller (portfolios #4 and #5), the differences between the optimized portfolio $x^\star$ and the current portfolio x_0 are reduced.

TABLE 1.7: Black-Litterman portfolios

	#0	#1	#2	#3	#4	#5
$x_1^\star$	40.00	33.41	51.16	36.41	38.25	39.77
$x_2^\star$	30.00	51.56	39.91	42.97	42.72	32.60
$x_3^\star$	20.00	5.46	0.00	10.85	9.14	17.65
$x_4^\star$	10.00	9.58	8.93	9.77	9.89	9.98
$\sigma(x^\star \mid x_0)$	0.00	3.65	3.67	2.19	2.18	0.45

To calibrate the parameter τ, we could target a tracking error volatility $\sigma^\star$. For example, if $\sigma^\star = 2\%$, the optimized portfolio is between portfolios #4 ($\sigma(x^\star \mid x_0) = 2.18\%$) and #5 ($\sigma(x^\star \mid x_0) = 0.45\%$) meaning that the optimal value of τ is between 10% and 1%. Using a bisection algorithm, we obtain $\tau = 5.2\%$ and the optimal portfolio is $x^\star = (36.80\%, 41.83\%, 11.58\%, 9.79\%)$ if we target a 2% tracking error volatility.

Remark 7 *There is a substantial body of literature on the Black-Litterman model. Readers who wish to deepen their knowledge on this topic may refer to the works of He and Litterman (1999), Satchell and Scowcroft (2000), Idzorek (2004), Meucci (2006) and Cheung (2010).*

1.2 Practice of portfolio optimization

In this section, we turn to the practical aspects of portfolio optimization. First, in order to implement the Markowitz approach, we have to estimate the covariance matrix and the vector of expected return. Once we have defined these input parameters, we can compute optimized portfolios. However, in most cases, the solution presents stability problems, which is why we have to regularize the input parameters or the objective function. The most common approach is based on shrinkage methods of the covariance matrix. However, even though these approaches are very interesting and improve the robustness of optimized portfolios, they are not enough. The problem comes from the fact that the most important quantity in portfolio optimization is the inverse of the covariance matrix, known as the information matrix, and that its regularization is particularly difficult. This is the reason why most portfolio managers prefers to use weight constraints. We will see that this approach is nevertheless similar to a shrinkage method. Moreover, by imposing weight constraints, the portfolio manager implicitly changes the covariance matrix, which is equivalent to having views on the assets' risk/return profile. In a sense, this approach is similar to the Black-Litterman approach.

1.2.1 Estimation of the covariance matrix

1.2.1.1 Empirical covariance matrix estimator

We assume that the vector of asset returns is a $n \times 1$ Gaussian vector with $R \sim \mathcal{N}(\mu, \Sigma)$. The log-likelihood function of the sample $\{R_1, \ldots, R_T\}$ is equal to:

$$\ell(\mu, \Sigma) = -\frac{nT}{2}\ln(2\pi) - \frac{T}{2}\ln|\Sigma| - \frac{1}{2}\sum_{t=1}^{T}(R_t - \mu)^\top \Sigma^{-1}(R_t - \mu)$$

The maximum likelihood (ML) estimator of μ satisfies $\partial_\mu \ell(\mu, \Sigma) = \mathbf{0}$. It is easy to show that $\hat{\mu} = \bar{R}$ where $\bar{R}$ is the vector of the means. By using the properties of the trace function, the concentrated log-likelihood function becomes:

$$\begin{aligned}
\ell(\Sigma) &= -\frac{nT}{2}\ln(2\pi) - \frac{T}{2}\ln|\Sigma| - \frac{1}{2}\sum_{t=1}^{T}\operatorname{tr}\left(\left(R_t - \bar{R}\right)^\top \Sigma^{-1}\left(R_t - \bar{R}\right)\right) \\
&= -\frac{nT}{2}\ln(2\pi) - \frac{T}{2}\ln|\Sigma| - \frac{1}{2}\sum_{t=1}^{T}\operatorname{tr}\left(\Sigma^{-1}\left(R_t - \bar{R}\right)\left(R_t - \bar{R}\right)^\top\right) \\
&= -\frac{nT}{2}\ln(2\pi) - \frac{T}{2}\ln|\Sigma| - \frac{1}{2}\operatorname{tr}\left(\Sigma^{-1}S\right)
\end{aligned}$$

where S is the $n \times n$ matrix defined in the following way:

$$S = \sum_{t=1}^{T} \left(R_t - \bar{R}\right)\left(R_t - \bar{R}\right)^{\top}$$

We deduce that the ML estimator of Σ satisfies this first-order condition:

$$\frac{\partial \,\ell(\Sigma)}{\partial \,\Sigma^{-1}} = \frac{T}{2}\Sigma - \frac{1}{2}S = 0$$

It follows that the ML estimator of Σ is the empirical covariance matrix:

$$\hat{\Sigma} = \frac{1}{T}S = \frac{1}{T}\sum_{t=1}^{T} \left(R_t - \bar{R}\right)\left(R_t - \bar{R}\right)^{\top}$$

This estimator has a very appealing property because it is equivalent to separately estimating the volatilities and the correlation matrix. This property is intensively used in finance to improve the robustness of this estimator.

Remark 8 *Covariance matrices and volatilities are expressed on a yearly basis, whereas $\hat{\Sigma}$ is expressed in the frequency of the asset returns (daily, weekly or monthly). To convert $\hat{\Sigma}$ into a yearly basis, we consider the square root approximation rule derived by the property of the geometric Brownian motion: volatilities are multiplied by a factor of $\sqrt{260}$ (resp. $\sqrt{52}$ and $\sqrt{12}$) when we use daily (resp. weekly and monthly) returns.*

Example 2 *Let us consider a universe with eight equity indices: S&P 500, Eurostoxx, FTSE 100, Topix, Bovespa, RTS, Nifty and HSI. The study period is January 2005–December 2011 and we use weekly returns.*

The estimated correlation matrix is:

$$\hat{C} = \left(\begin{array}{cccc|cccc}
1.00 & & & & & & & \\
0.88 & 1.00 & & & & & & \\
0.88 & 0.94 & 1.00 & & & & & \\
0.64 & 0.68 & 0.65 & 1.00 & & & & \\
\hline
0.77 & 0.76 & 0.78 & 0.61 & 1.00 & & & \\
0.56 & 0.61 & 0.61 & 0.50 & 0.64 & 1.00 & & \\
0.53 & 0.61 & 0.57 & 0.53 & 0.60 & 0.57 & 1.00 & \\
0.64 & 0.68 & 0.67 & 0.68 & 0.68 & 0.60 & 0.66 & 1.00
\end{array}\right) \qquad (1.16)$$

whereas the annual volatilities are respectively 20.5%, 23.4%, 20.7%, 21.1%, 28.4%, 40.6%, 26.4% and 25.0%. We note some differences between the indices of developed countries (DC) and the indices of emerging markets[14] (EM). In particular, developed equity indices present more intra-correlations and lower volatilities.

[14]These 4 emerging countries correspond to a financial thematic called BRIC (Brazil, Russia, India, China).

1.2.1.2 Hayashi-Yoshida estimator

The previous estimator assumes that the data are synchronous. This approach is only valid when we consider assets that are priced in the same market place or when assets are priced at the same closing time. However, if we consider a universe of assets in asynchronous markets, we have to correct the previous estimator as suggested by Hayashi and Yoshida (2005). Let $P(t) = (P_1(t), \ldots, P_n(t))$ be the price vector of n assets. We assume that $P(t)$ is a multi-dimensional geometric Brownian motion with:

$$\mathrm{d}P_i(t) = \mu_i P_i(t)\,\mathrm{d}t + \sigma_i P_i(t)\,\mathrm{d}W_i(t) \tag{1.17}$$

where $W_i(t)$ is a multivariate Wiener process with $\mathbb{E}[W_i(t)\,W_j(t)] = \rho_{i,j}t$. We suppose that the process (1.17) is observed at some discrete times $\tau_{i,k}$ with $k \in \mathbb{N}$. Let $\tau_i(t)$ be the last observation date before t. We have $\tau_i(t) = \tau_{i,k^\star}$ with $k^\star = \sup\{k : \tau_{i,k} \leq t\}$ meaning that the last price known at date t is $P_i(\tau_{i,k^\star})$ or equivalently $P_i(\tau_i(t))$. We are interested in computing the covariance matrix over the period $[0,T]$ with equally spaced times $0 = t_0 < t_1 < \cdots < t_M = T$. Let $S = (S_{i,j})$ be the $n \times n$ matrix with:

$$S_{i,j} = \sum_{m=1}^{M} \left(\tilde{P}_i(\tau_i(t_m)) - \tilde{P}_i(\tau_i(t_{m-1}))\right)\left(\tilde{P}_j(\tau_j(t_m)) - \tilde{P}_j(\tau_j(t_{m-1}))\right)$$

where $\tilde{P}_i(t) = \ln P_i(t)$ denotes the logarithm of the price and $\tilde{P}_i(\tau_i(t_m)) - \tilde{P}_i(\tau_i(t_{m-1}))$ is the logarithmic return of the asset i between t_m and t_{m-1}. By neglecting the mean, we have:

$$S_{i,j} = \sigma_i\sigma_j \sum_{m=1}^{M} \Delta W_i(\tau_i(t_m)) \cdot \Delta W_j(\tau_j(t_m))$$

where $\Delta W_i(\tau_i(t_m)) = W_i(\tau_i(t_m)) - W_i(\tau_i(t_{m-1}))$. The time period $[\min(\tau_i(t_{m-1}), \tau_j(t_{m-1})), \max(\tau_i(t_m), \tau_j(t_m))]$ may be decomposed into three subperiods:

1. $I_{i,j}^{-}(m) = [\min(\tau_i(t_{m-1}), \tau_j(t_{m-1})), \max(\tau_i(t_{m-1}), \tau_j(t_{m-1}))]$;
2. $I_{i,j}^{0}(m) = [\max(\tau_i(t_{m-1}), \tau_j(t_{m-1})), \min(\tau_i(t_m), \tau_j(t_m))]$;
3. $I_{i,j}^{+}(m) = [\min(\tau_i(t_m), \tau_j(t_m)), \max(\tau_i(t_m), \tau_j(t_m))]$;

For the subperiods $I_{i,j}^{-}(m)$ and $I_{i,j}^{+}(m)$, the product $\langle \mathrm{d}W_i(t), \mathrm{d}W_j(t)\rangle$ is equal to zero because one asset price is not observed. It follows that:

$$\begin{aligned}
\mathbb{E}[S_{i,j}] &= \sigma_i\sigma_j \sum_{m=1}^{M} \int_{\max(\tau_i(t_{m-1}),\tau_j(t_{m-1}))}^{\min(\tau_i(t_m),\tau_j(t_m))} \langle \mathrm{d}W_i(t), \mathrm{d}W_j(t)\rangle \\
&= \rho_{i,j}\sigma_i\sigma_j \sum_{m=1}^{M} I_{i,j}^{0}(m)
\end{aligned}$$

Let $\hat{\Sigma} = T^{-1}S$ be the covariance matrix estimator. We have:

$$\mathbb{E}\left[\hat{\Sigma}_{i,j}\right] = \frac{1}{T}\mathbb{E}\left[S_{i,j}\right] = \rho_{i,j}\sigma_i\sigma_j\frac{1}{T}\sum_{m=1}^{M} I_{i,j}^{0}(m)$$

If $i = j$, $\sum_{m=1}^{M} I_{i,j}^{0}(m) = T$ and $\mathbb{E}\left[\hat{\Sigma}_{i,i}\right] = \sigma_i^2$. If $i \neq j$, $\sum_{m=1}^{M} I_{i,j}^{0}(m) < T$ and $\mathbb{E}\left[\hat{\Sigma}_{i,i}\right] < \rho_{i,j}\sigma_i\sigma_j$. It implies that the estimator $\hat{\Sigma}$ is biased for the covariance terms, but not for the variance terms. Let $\hat{\rho}_{i,j}$ be the estimator of the correlation $\rho_{i,j}$. The correlations are always underestimated because the length of overlapping periods $\sum_{m=1}^{M} I_{i,j}^{0}(m)$ is smaller than the study period $[0, T]$:

$$\mathbb{E}\left[\hat{\rho}_{i,j}\right] = \frac{\mathbb{E}\left[\hat{\Sigma}_{i,j}\right]}{\sigma_i\sigma_j} = \rho_{i,j}\frac{\sum_{m=1}^{M} I_{i,j}^{0}(m)}{T} = \rho_{i,j}\overline{I_{i,j}^{0}}$$

In Figure 1.8, we have represented the trading hours of some equity indices. For each index, we indicate the opening and closing times. If we consider the Eurostoxx and S&P 500 indices, we note that $I_{i,j}^{-}(m) = 4^1/_2$ hours, $I_{i,j}^{0}(m) = 19^1/_2$ hours and $I_{i,j}^{+}(m) = 4^1/_2$ hours. In this case, the ratio $\overline{I_{i,j}^{0}}$ is equal to 19.5/24 or 81.25%. Using daily closing prices, we underestimate the correlation between Eurozone and US markets by 19% on average. The underestimation is more important for the Topix index. Indeed, the ratio $\overline{I_{i,j}^{0}}$ is equal to 60.42% with the Eurostoxx index and 41.67% with the S&P 500 index.

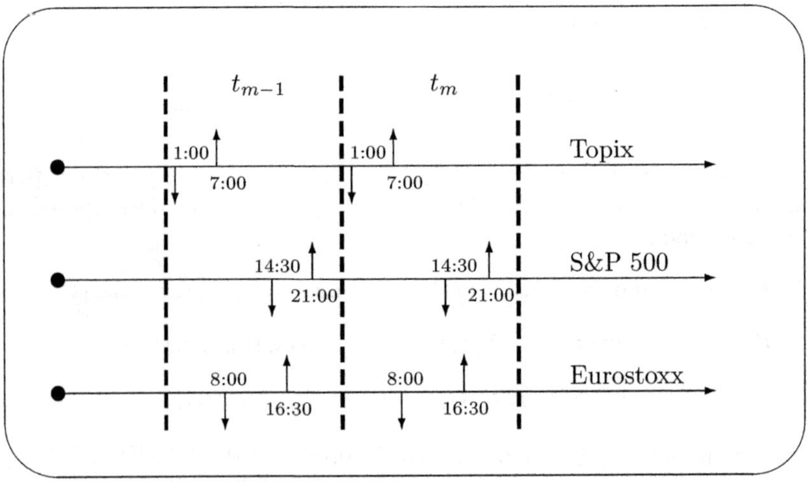

FIGURE 1.8: Trading hours of asynchronous markets (UTC time)

Let us consider a Monte Carlo experiment to illustrate the bias of the correlation estimator. In Figure 1.9, we have reported the density of the estimator $\hat{\rho}$ when the prices of the two assets follow a geometric Brownian motion without drift. The diffusion parameter is equal to 10% for the two assets whereas the cross-correlation ρ is set to 70%. The correlation estimator is based on 260 observations whereas its density is estimated using 2 000 simulations and a Gaussian kernel. The first case corresponds to synchronous daily returns. The second case corresponds to daily returns but with a lag of 6 hours between the two closing times. In the third case, we assume that the two prices are observed on a weekly basis but with a difference of one day[15]. We verify that $\hat{\rho}$ is biased in the last two cases. In our experiment, the average value of $\hat{\rho}$ is 69.9%, 46.4% and 55.8%. This means that the ratio $\rho/\hat{\rho}$ is respectively equal to 99.9%, 66.3% and 79.7% whereas the theoretical values taken by $\overline{I_{i,j}^{0}}$ are 100%, 66.7% and 80%.

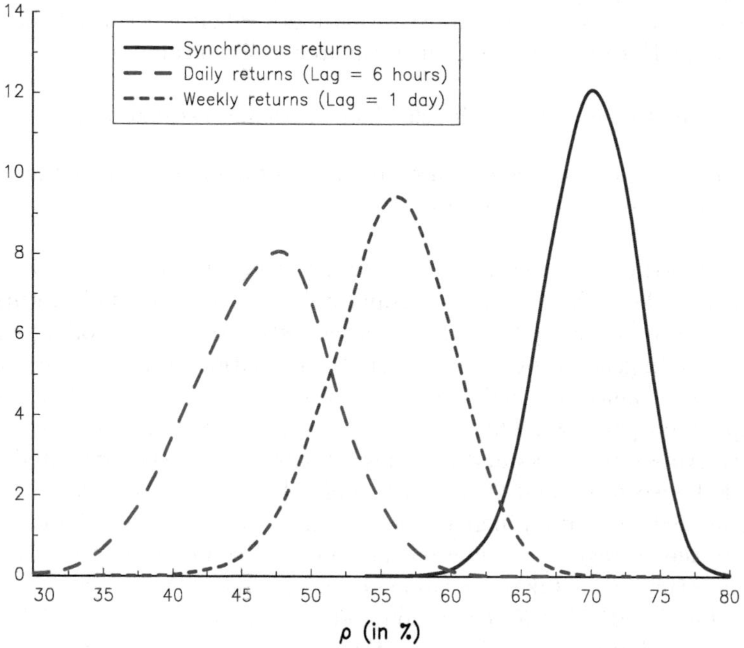

FIGURE 1.9: Density of the estimator $\hat{\rho}$ with asynchronous returns

The previous analysis led Hayashi and Yoshida (2005) to propose a new estimator of $S_{i,j}$:

$$S_{i,j} = \sum_{k_1=1}^{K} \sum_{k_2=1}^{K} \Delta\tilde{P}_i\left(\tau_{i,k_1}\right) \cdot \Delta\tilde{P}_j\left(\tau_{j,k_2}\right) \cdot \mathbb{1}\left\{I_i\left(k_1\right) \cap I_j\left(k_2\right) \neq \varnothing\right\}$$

where $\Delta\tilde{P}_i\left(\tau_{i,k}\right) = \tilde{P}_i\left(\tau_{i,k}\right) - \tilde{P}_i\left(\tau_{i,k-1}\right)$ and $I_i\left(k\right) = \left[\tau_{i,k-1}, \tau_{i,k}\right]$. This new estimator is called the cumulative covariance estimator because it is the prod-

[15]For example, we can assume that the price of the first asset is available on every Wednesday whereas the price of the second asset is only known on Thursday. We encounter such situations in the hedge fund industry.

uct of any pair of returns when observation intervals of assets i and j overlaps. If we consider our previous example with equity indices, the estimator of $\Sigma_{i,j}$ becomes:

$$\begin{aligned}\tilde{\Sigma}_{i,j} &= \frac{1}{T}\sum_{t=1}^{T}\left(R_{i,t}-\bar{R}_i\right)\left(R_{j,t}-\bar{R}_j\right)+ \\ & \quad \frac{1}{T}\sum_{t=1}^{T}\left(R_{i,t}-\bar{R}_i\right)\left(R_{j,t-1}-\bar{R}_j\right)\end{aligned}$$

where j is the equity index which has a closing time after the equity index i. In our case, j is necessarily the S&P 500 index whereas i can be the Topix index or the Eurostoxx index. This estimator has two components:

1. The first component is the classical covariance estimator $\hat{\Sigma}_{i,j}$.

2. The second component is a correction to take into account the lag between the two closing times.

Let us estimate the correlation between the S&P 500 index (SPX), the Eurostoxx index (SXXE) and the Topix index (TPX) with daily returns. In Figure 1.10, we have reported the empirical estimator with a rolling window of 260 days. In general, we observe that the correlation between equity indices of developed countries is high (see the results obtained with weekly returns in Example 2 on page 28). This is not the case when we consider daily returns between American stocks and Japanese stocks because the estimated correlation is between 0% and 20%. If we apply the Hayashi-Yoshida correction, the estimated correlation increases for the three pairs of equity indices. If we compute the average ratio between the empirical estimates and the Hayashi-Yoshida estimates, it takes the value 66.84% for the SPX/SXXE correlation, 19.53% for the SPX/TPX correlation and 41.02% for the SXXE/TPX correlation. These figures are smaller than the previous theoretical ratios, certainly because of the overnight effect.

Remark 9 *The estimator $\tilde{\Sigma}$ is not necessarily positive definite. In this case, we can use different numerical techniques based on the singular value decomposition to fix this problem (Roncalli, 2010).*

1.2.1.3 GARCH approach

In the previous paragraph, we assume that the variance of the asset return is constant. However, this hypothesis is generally not verified in finance. In 1982, Engle introduced a class of stochastic processes in order to take into account the heteroscedasticity of asset returns:

$$R_{i,t} = \mu_i + \varepsilon_t \quad \text{with} \quad \varepsilon_t = \sigma_t e_t \quad \text{and} \quad e_t \sim \mathcal{N}(0,1)$$

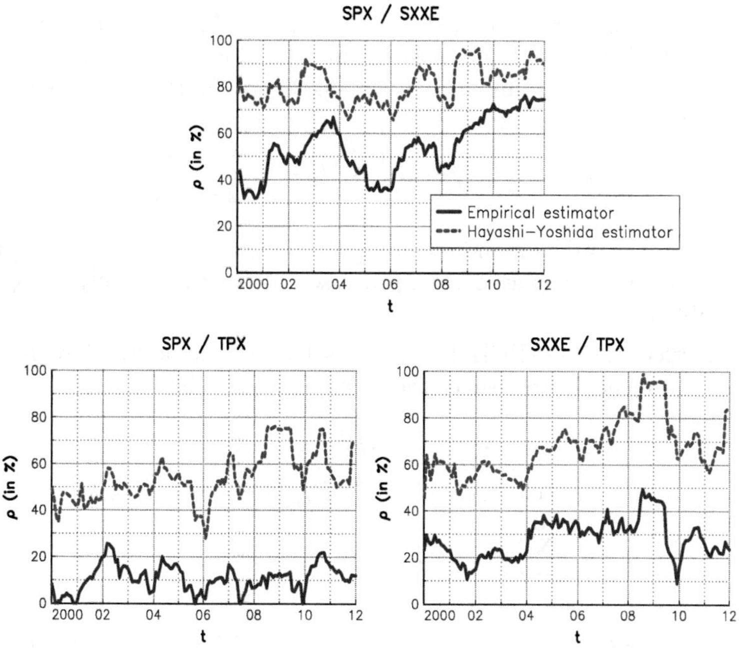

FIGURE 1.10: Hayashi-Yoshida estimator

where the time-varying variance σ_t^2 satisfies the following equation:

$$\sigma_t^2 = \kappa + \alpha_1 \varepsilon_{t-1}^2 + \alpha_2 \varepsilon_{t-2}^2 + \cdots + \alpha_q \varepsilon_{t-q}^2$$

with $\kappa > 0$ and $\alpha_j \geq 0$ for all $j > 0$. We note that the conditional variance of ε_t is not constant and depends on the past values of ε_t. A substantial impact on the asset return $R_{i,t}$ implies an increase of the conditional variance of ε_{t+1} at time $t+1$ and therefore an increase of the probability to observe another substantial impact on $R_{i,t+1}$. Therefore, this means that the volatility is persistent, which is a well-known stylized fact in finance (Chou, 1988).

This type of stochastic processes, known as ARCH models (Autoregressive Conditional Heteroscedasticity), has been extended by Bollerslev (1986) in the following way:

$$\begin{aligned} \sigma_t^2 &= \kappa + \delta_1 \sigma_{t-1}^2 + \delta_2 \sigma_{t-2}^2 + \cdots + \delta_p \sigma_{t-p}^2 + \\ &\quad \alpha_1 \varepsilon_{t-1}^2 + \alpha_2 \varepsilon_{t-2}^2 + \cdots + \alpha_q \varepsilon_{t-q}^2 \end{aligned}$$

In this case, the conditional variance depends also on its past values and we obtain a GARCH(p,q) model. If $\sum_{i=1}^{p} \delta_i + \sum_{i=1}^{q} \alpha_i = 1$, we may show that the process ε_t^2 has a unit root and the model is called an integrated GARCH (or IGARCH) process.

There are several methods to estimate a GARCH process. For example, we could use the generalized method of moments (GMM) or the two-stage least squares method. However, the most popular method remains the conditional maximum likelihood. In this case, the log-likelihood function has the following expression:

$$\ell = -\frac{T}{2}\ln(2\pi) - \frac{1}{2}\sum_{t=1}^{T}\ln\sigma_t^2 - \frac{1}{2}\sum_{t=1}^{T}\frac{\varepsilon_t^2}{\sigma_t^2}$$

The choice of the orders p and q can be determined by using the ACF/PACF method in a similar way as ARMA processes, although generally $p = q = 1$ is a reasonable choice.

In practice, the IGARCH(1,1) process has become a standard model in asset management. If we neglect the term μ_i, we obtain:

$$\begin{aligned}\sigma_t^2 &= (1-\alpha)\,\sigma_{t-1}^2 + \alpha R_{i,t-1}^2 \\ &= \alpha R_{i,t-1}^2 + (1-\alpha)\,\alpha R_{i,t-2}^2 + (1-\alpha)^2\,\sigma_{t-2}^2 \\ &= \sum_{j=1}^{\infty} w_{t-j} R_{i,t-j}^2\end{aligned}$$

where $w_{t-j} = \alpha(1-\alpha)^{j-1}$. We note that $W_m = \sum_{j=1}^{m} w_{t-j} = 1 - (1-\alpha)^m$ and $\lim_{m\to\infty} W_m = 1$. This estimator is then an exponentially weighted moving average (EWMA) with a factor λ equal to $1-\alpha$. In Figure 1.11, we have reported the cumulative weight function W_m for different values of λ. We verify that we give more importance to the present values than to the past values.

Remark 10 *In EMWA or GARCH models, we have to initialize the first value of the conditional volatility σ_0. Most of the time, we set it to the unconditional estimate $\hat{\sigma}$.*

In Figure 1.12, we have reported the volatility of the S&P 500 index estimated using the GARCH model (first panel). The ML estimates of the parameters are $\hat{\delta}_1 = 0.913$ and $\hat{\alpha}_1 = 0.080$. We verify that this estimated model is close to an IGARCH process. In the second and third panels, we compare the GARCH volatility with the EWMA volatility (with $\lambda = 0.90$) and a short volatility based on 20 trading days. We note that these two simple estimators give similar results to the GARCH estimator. This explains why some practitioners prefer to use a short volatility instead of a GARCH model, as the latter is more difficult to calibrate.

Remark 11 *The GARCH process could be extended in the multivariate case. Among the wide number of generalizations, the two main models are the VEC and BEKK models (Bauwens* et al.*, 2006). However, these models are not very useful in asset management. They are not robust and tractable because of the large number of parameters and the estimation step which is not cost*

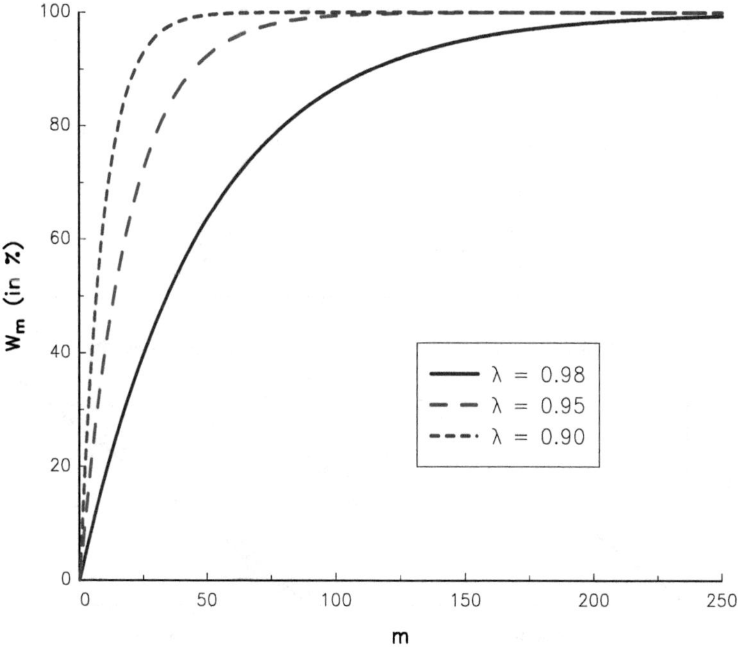

FIGURE 1.11: Cumulative weight W_m of the IGARCH model

efficient. This is why practitioners prefer to adopt a copula approach by first estimating the GARCH volatilities and then imposing a correlation structure (Jondeau and Rockinger, 2006).

1.2.1.4 Factor models

Imposing a correlation structure The ML estimator suggests that we could estimate the correlation matrix $C = (\rho_{i,j})$ associated to Σ from the centered returns $\tilde{R}_{i,t} = \hat{\sigma}_i^{-1}(R_{i,t} - \hat{\mu}_i)$. In this case, we have:

$$\begin{aligned} \hat{C} &= \arg\max \ell(C) \\ &\text{u.c.} \quad C \text{ is a correlation matrix} \end{aligned}$$

where $\ell(C)$ is the log-likelihood function defined as follows:

$$\ell(C) = -\frac{nT}{2}\ln(2\pi) - \frac{T}{2}\ln|C| - \frac{1}{2}\sum_{t=1}^{T}\tilde{R}_t^\top C_t^{-1}\tilde{R}_t$$

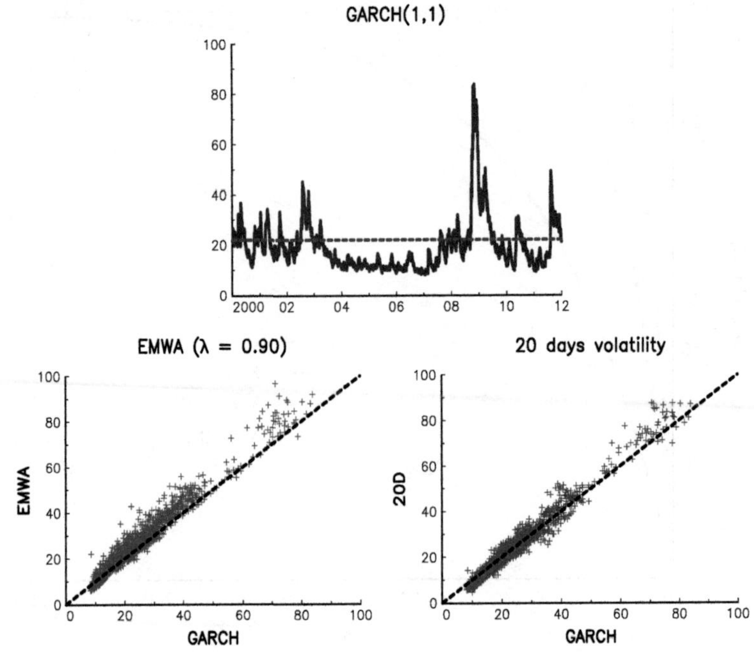

FIGURE 1.12: Estimation of the S&P 500 volatility

With this approach, we could take into account a correlation structure. For example, we may impose a uniform correlation matrix:

$$C = C_n(\rho) = \begin{pmatrix} 1 & & \rho \\ & \ddots & \\ \rho & & 1 \end{pmatrix}$$

In this case, we say that we have a one-factor risk model.

Remark 12 *To understand this correlation structure, we consider the following model:*

$$R_{i,t} = \sqrt{\rho}\mathcal{F}_t + \sqrt{1-\rho}\varepsilon_{i,t} \quad \text{for} \quad i = 1, \ldots, n \tag{1.18}$$

where $\mathcal{F}_t$ and $\varepsilon_{i,t}$ are two independent Gaussian random variables $\mathcal{N}(0,1)$. We interpret $\mathcal{F}_t$ as the common risk factor and $\varepsilon_{i,t}$ as the idiosyncratic risk factor. We deduce that:

$$\begin{aligned} \mathbb{E}[R_{i,t}R_{j,t}] &= \rho\mathbb{E}\left[\mathcal{F}_t^2\right] + (1-\rho)\mathbb{E}[\varepsilon_{i,t}\varepsilon_{j,t}] + \\ &\quad \sqrt{\rho(1-\rho)}\left(\mathbb{E}[\mathcal{F}_t\varepsilon_{i,t}] + \mathbb{E}[\mathcal{F}_t\varepsilon_{j,t}]\right) \\ &= \rho \end{aligned}$$

and $\mathbb{E}\left[R_{i,t}^2\right] = 1$. The correlation between the asset returns $R_{i,t}$ and $R_{j,t}$ is then equal to ρ.

The representation (1.18) is useful to obtain another expression of the log-likelihood function which is more easily to solve numerically. The conditional distribution of $R_{i,t}$ with respect to $\mathcal{F}_t$ is:

$$\begin{aligned}\Pr\left\{R_{i,t} \leq r \mid \mathcal{F}_t\right\} &= \Pr\left\{\sqrt{\rho}\mathcal{F}_t + \sqrt{1-\rho}\varepsilon_{i,t} \leq r \middle| \mathcal{F}_t\right\} \\ &= \Pr\left\{\varepsilon_{i,t} \leq \frac{r-\sqrt{\rho}\mathcal{F}_t}{\sqrt{1-\rho}} \middle| \mathcal{F}_t\right\} \\ &= \Phi\left(\frac{r-\sqrt{\rho}\mathcal{F}_t}{\sqrt{1-\rho}}\right)\end{aligned}$$

Because the random variables $(R_{1,t}, \ldots, R_{n,t})$ are independent conditionally to the factor $\mathcal{F}_t$, it follows that:

$$\Pr\left\{R_{1,t} \leq r_1, \ldots, R_{n,t} \leq r_n\right\} = \int_{-\infty}^{\infty} \prod_{i=1}^{n} \Phi\left(\frac{r_i - \sqrt{\rho}f}{\sqrt{1-\rho}}\right) \phi(f)\,\mathrm{d}f$$

We deduce that the log-likelihood function is:

$$\ell(\rho) = -\frac{nT}{2}\ln(1-\rho) + \sum_{t=1}^{T} \ln\left(\int_{-\infty}^{\infty} \prod_{i=1}^{n} \Phi\left(\tilde{R}_{i,t}\right) \phi\left(\tilde{R}_{i,t}\right) \phi(f)\,\mathrm{d}f\right)$$

with:

$$\tilde{R}_{i,t} = \frac{R_{i,t} - \sqrt{\rho}f}{\sqrt{1-\rho}}$$

In Figure 1.13, we have reported the density of the ML estimator $\hat{\rho}_{\mathrm{ML}}$ in the one-factor model for several values of n. The number of observations T is set to 500 whereas the true correlation is equal to 20%. We have also reported the density of the classical empirical estimator[16]. We verify that we reduce the variance of the estimator $\hat{\rho}_{\mathrm{ML}}$ by taking into account the correlation structure.

Remark 13 *The previous analysis can be extended in the following way:*

$$R_{i,t} = \sqrt{\rho}\mathcal{F}_t + \sqrt{\rho_{\varphi(i)} - \rho}\mathcal{F}_{\varphi(i),t} + \sqrt{1-\rho_{\varphi(i)}}\varepsilon_{i,t} \quad i = 1, \ldots, n$$

where $\varphi(i)$ indicates the specific factor of $R_{i,t}$. We have $j = \varphi(i) \in \{1, \ldots, m\}$. We also assume that the Gaussian random variables $\mathcal{F}_{j,t}$ are independent of $\mathcal{F}_t$ and $\varepsilon_{i,t}$. We have then a $(m+1)$-factor model. If $R_{i,t}$ and $R_{j,t}$ are sensitive to the same specific factor j, we verify that $\operatorname{cor}(R_{i,t}, R_{j,t}) = \rho_j$, otherwise we have $\operatorname{cor}(R_{i,t}, R_{j,t}) = \rho$. If we rank the returns $R_{i,t}$ with respect to the specific factors, the correlation matrix presents a block diagonal structure:

$$C = \begin{pmatrix} C_1 & C_0 & C_0 & C_0 \\ C_0 & C_2 & \ddots & C_0 \\ C_0 & \ddots & \ddots & C_0 \\ C_0 & C_0 & C_0 & C_m \end{pmatrix}$$

[16]This estimator is equivalent to the ML estimator in the one-factor model when n is equal to 2.

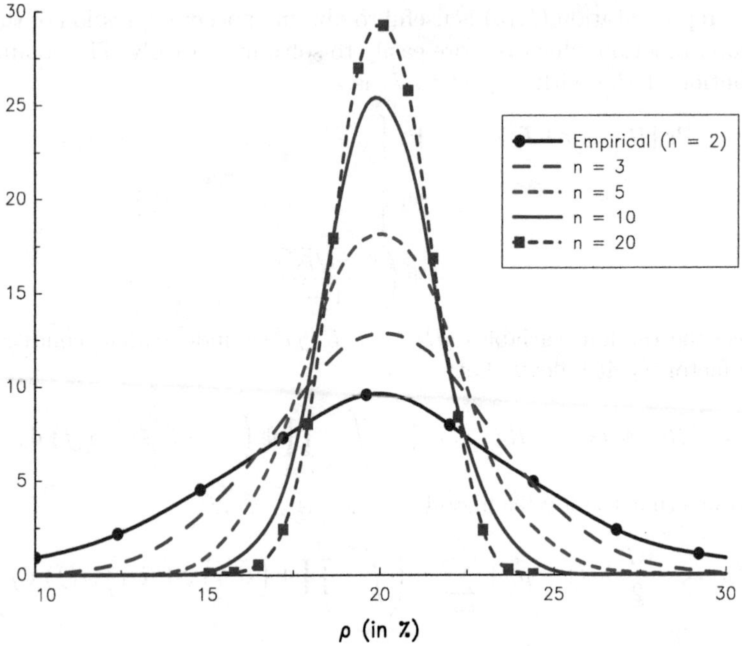

FIGURE 1.13: Density of the uniform correlation estimator

where C_j is a constant correlation matrix with parameter ρ_j and C_0 is a constant matrix with parameter ρ.

Linear factor model The linear factor model is defined by the following relationship:

$$R_{i,t} = A_i^\top \mathcal{F}_t + \varepsilon_{i,t}$$

where $\mathcal{F}_t$ is a Gaussian vector of m factors with $\mathbb{E}[\mathcal{F}_t] = \mu$ and $\operatorname{cov}(\mathcal{F}_t) = \Omega$. $A_i^\top \mathcal{F}_t$ and $\varepsilon_{i,t}$ represent respectively the factor and idiosyncratic components of the asset return. The matrix form of the previous equation is:

$$R_t = A\mathcal{F}_t + \varepsilon_t$$

We have $\mathbb{E}[\varepsilon_t] = \mathbf{0}$ and $\operatorname{cov}(\varepsilon_t) = D = \operatorname{diag}\left(\tilde{\sigma}_1^2, \ldots, \tilde{\sigma}_n^2\right)$. D is a diagonal matrix because the idiosyncratic risks are not correlated. Moreover, we assume also that $\mathbb{E}\left[\mathcal{F}_t \varepsilon_t^\top\right] = \mathbf{0}$. We deduce then that $\mathbb{E}[R_t] = A\mu$ and $\operatorname{cov}(R_t) = A\Omega A^\top + D$. If we assume that the asset returns are Gaussian, the log-likelihood of the observation t is:

$$\begin{aligned} \ell_t &= -\frac{n}{2}\ln(2\pi) - \frac{1}{2}\ln\left|A\Omega A^\top + D\right| - \\ & \frac{1}{2}(R_t - A\mu)^\top \left(A\Omega A^\top + D\right)^{-1}(R_t - A\mu) \end{aligned}$$

The estimation of the parameters is performed using the maximum likelihood principle. However, we face some identification problems, because we cannot estimate A, D and Ω simultaneously. This is why we generally impose that $\Omega = I_m$.

Let us consider our previous example with the eight equity indices. If we impose a uniform correlation, we obtain $\hat{\rho} = 66.24\%$, which is very close to the average correlation[17] of the matrix (1.16). However, we can think that a three-factor model with one common factor and two specific factors is more appropriate, where the first specific factor influence the DC indices and the second specific factor impacts the EM indices. In this case, the cross-correlation between DC indices (77%) is larger than between EM indices (59%) whereas the inter-correlation between DC and EM indices is 50%:

$$\hat{C} = \left(\begin{array}{cccc|cccc}
1.00 & & & & & & & \\
0.77 & 1.00 & & & & & & \\
0.77 & 0.77 & 1.00 & & & & & \\
0.77 & 0.77 & 0.77 & 1.00 & & & & \\
\hline
0.50 & 0.50 & 0.50 & 0.50 & 1.00 & & & \\
0.50 & 0.50 & 0.50 & 0.50 & 0.59 & 1.00 & & \\
0.50 & 0.50 & 0.50 & 0.50 & 0.59 & 0.59 & 1.00 & \\
0.50 & 0.50 & 0.50 & 0.50 & 0.59 & 0.59 & 0.59 & 1.00
\end{array}\right) \tag{1.19}$$

This factor model is different than a two-linear factor model, because it imposes that the sensitivity to the common and specific factors is the same for the different indices. In the case of a two-linear factor model, the sensitivities become:

$$A = \left(\begin{array}{cc}
0.9049 & -0.0671 \\
0.9638 & -0.0737 \\
0.9645 & -0.1131 \\
0.7105 & 0.2581 \\
0.8231 & 0.1594 \\
0.6603 & 0.2861 \\
0.6450 & 0.4027 \\
0.7432 & 0.4616
\end{array}\right) \tag{1.20}$$

whereas the estimated correlation matrix is:

$$\hat{C} = \left(\begin{array}{cccc|cccc}
1.00 & & & & & & & \\
0.88 & 1.00 & & & & & & \\
0.88 & 0.94 & 1.00 & & & & & \\
0.63 & 0.67 & 0.66 & 1.00 & & & & \\
\hline
0.73 & 0.78 & 0.78 & 0.63 & 1.00 & & & \\
0.58 & 0.62 & 0.60 & 0.54 & 0.59 & 1.00 & & \\
0.56 & 0.59 & 0.58 & 0.56 & 0.60 & 0.54 & 1.00 & \\
0.64 & 0.68 & 0.66 & 0.65 & 0.69 & 0.62 & 0.67 & 1.00
\end{array}\right) \tag{1.21}$$

[17] It is equal to 66.15%.

We note that the first factor is a market factor, because the weights are all positive. The second factor is a long-short portfolio (short in SPX, SXXE and UKX and long in the other indices[18]).

1.2.2 Designing expected returns

The estimation of expected returns is the most difficult step in portfolio construction. This explains why practitioners use a wide range of tools, models, recipes, data, etc. The purpose of this section is not to present an exhaustive survey of the different approaches[19], but to give the main concepts of the financial theory on this topic.

Before choosing an econometric model or a statistical procedure to forecast asset returns, we must specify the time horizon of the allocation. More precisely, we must distinguish market timing (MT), tactical asset allocation (TAA) and strategic asset allocation (SAA):

- Market timing refers to a very short-term investment horizon, typically from one day to one month. For example, a daily strategy consisting in playing the mean-reverting property of stock returns is a market timing strategy. Market timing is generally based on short-term market sentiment or behavioral model (Barberis and Thaler, 2003). Most of the time, it consists in finding empirical patterns or anomalies in the financial market.

- Tactical asset allocation is a short to medium-term investment horizon. These investment decisions are related to business cycles and/or medium-term market sentiment. Typically, the investor modifies the asset mix in the portfolio conditionally to the economic news flow or technical factors. TAA is related to the macro-finance models introduced by Lucas (1978) and popularized by the book of Cochrane (2001).

- Compared to TAA, strategic asset allocation is a long-term investment horizon (Brennan *et al.*, 1997). Over this horizon, the influence of financial crisis and business cycles is supposed to be less important. Long-run expectations adapt to structural factors such as population growth (or demographic change), government policies and productivity. Whereas TAA assumes that the risk premium of assets is time-varying, SAA is based on the stationary steady-state of the economy. In this case, risk premium is stationary[20].

In this section, we focus only on TAA and SAA, and not on market timing

[18]Curiously, the Topix index behaves differently than the three other DC indices.

[19]We invite the reader to refer to the book of Ilmanen (2011), which is a comprehensive guide to models of expected returns.

[20]We adopt here the point of view of long-term investors such as pension funds and sovereign wealth funds, even if this view is not completely shared by academic research.

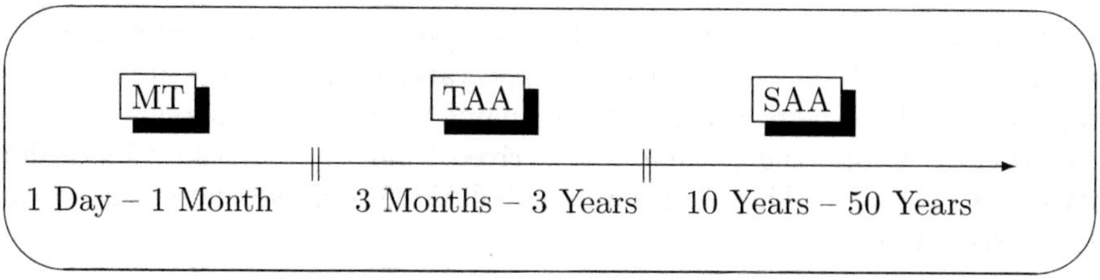

FIGURE 1.14: Time horizon of MT, TAA and SAA

which is very different from risk parity strategies. The estimation of expected returns can be performed in two steps. The first step consists in selecting the economic and financial variables that could help us forecast asset returns. The predictability power of such variables is generally small and depends on the time horizon. The selection procedure is a crucial step, because this approach is in contradiction with the efficient market hypothesis (Fama, 1970) and empirical evidence of predictability is never easy. The second step consists in considering an econometric model that could explain some stylized facts between the real economy and the financial market. Generally, we adopt the "fundamental fair value" approach. In the long run, the returns of financial assets converge towards the returns of physical assets. In the medium term, disequilibriums may be observed. However, restoring forces constrain the dynamics of asset prices, because of these long-run equilibrium relationships. In this case, the vector error correction model (VECM) is certainly the most appropriate econometric model for designing expected returns.

In the medium term, market risk premia may be driven by investors' intertemporal consumption preferences. In this case, they should vary with business cycles (Cochrane, 2001). In this context, many authors have proposed business cycle-related variables that could have a predictable component for financial asset returns. Among these different variables, Darolles *et al.* (2010) consider that the three significant factors are the dividend yield, the yield curve and the ratio of consumption over wealth. The dividend yield is the ratio between dividends and stock price. Since the seminal work of Campbell and Shiller (1988), dividend yield is certainly the most popular factor, because it is easily available on a daily basis, which is not the case for most leading economic activity variables (Stock and Watson, 1989). The second most popular factor corresponds to the term spread of the yield curve, i.e. the difference between the long-term and short-term interest rates. In economics, the term structure of interest rates depends on short rates expectations, implying that it contains information on the business cycle. This could justify the empirical evidence between the yield spread and the equity and bond future returns (Chen, 1991). The third factor was introduced by Lettau and Ludvigson (2001). They consider a general framework linking consumption (C), asset holdings (A), and labor income (Y) with expected returns. They find that the three economic variables should share a common trend over the long term, but may deviate substantially from one another in the short run. These deviations should summarize agents' expectations of future returns on the market portfolio. The factor, which they name CAY, is the difference between the observed C and the estimated C derived from the long-run relationship between C, A and Y. They find that a high CAY ratio is associated with rising equity risk premia.

In the long term, the usual method of long-term investors is to adopt the fair value approach. This means that long-run asset returns can be derived from the long-run path of the real economy. For example, Figure 1.15 corresponds to the framework adopted by Eychenne *et al.* (2011).

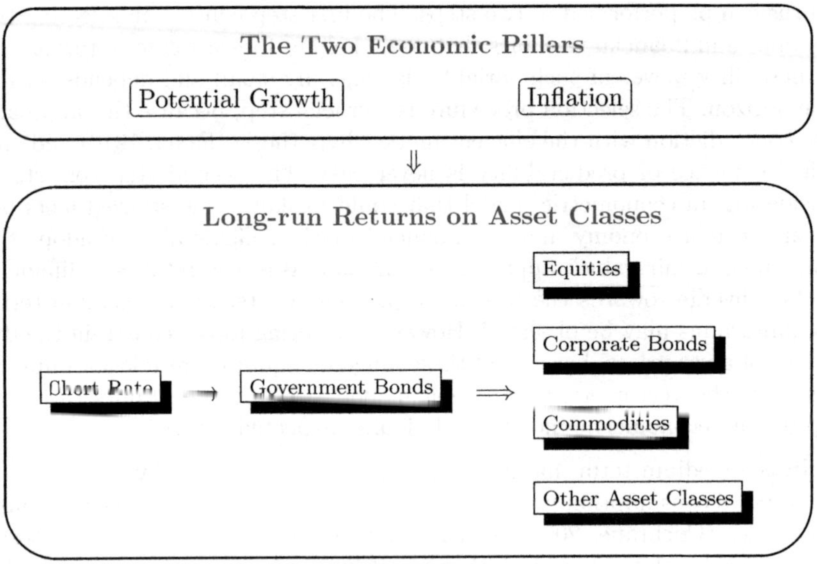

FIGURE 1.15: Fundamental approach of SAA

They explain their process as follows:

> "*First, we identify two representative economic fundamentals (output and inflation) and we build a long-run economic scenario for these two variables. Second, we derive long-run asset returns from this economic scenario. The long-run short rate is directly obtained by adding together output and inflation. To obtain the long-run government bond return, we add a bond risk premium to the long-run short rate. Finally, the expected returns of the other asset classes are derived from the specific risk premium associated with the nature of the asset class*" (Eychenne *et al.*, 2011).

More precisely, they derive long-run short rates r_∞ from the lower bound of the normative Golden rule:

$$r_\infty = g_\infty + \pi_\infty$$

where g_∞ is the long-run real potential output growth and π_∞ is the long-run inflation. The long-run value of the nominal bond yield R^b_∞ is equal to:

$$R^b_\infty = \mathcal{R}^{\mathrm{b}}_\infty + \pi_\infty$$

where $\mathcal{R}^{\mathrm{b}}_\infty$ is the long-run real bond yield. To estimate $\mathcal{R}^{\mathrm{b}}_\infty$, they consider the following regression model[21]:

$$\mathcal{R}^{\mathrm{b}}_t = \beta_0 + \beta_1 \mathfrak{r}_t + \beta_2 \sigma^\pi_t + \beta_3 \left(B/Y\right)_t + \varepsilon_t$$

where $\mathfrak{r}_t$ is the real short rate, σ^π_t is the inflation risk and $\left(B/Y\right)_t$ is the government balance on output ratio (proxy for debt risk). In the equity side, they assume that the long-run equity return is equal to:

$$R^{\mathrm{e}}_\infty = R^{\mathrm{b}}_\infty + \mathcal{R}^{\mathrm{e}}_\infty$$

where $\mathcal{R}^{\mathrm{e}}_\infty$ is the equity excess return. Then, they use the following regression model to compute the expected returns of equity asset classes:

$$\mathcal{R}^{\mathrm{e}}_{t+10} = \beta_0 + \beta_1 \,\mathrm{PE}_t + \beta_2 R^{\mathrm{b}}_t + \varepsilon_t$$

where PE_t is the price earning ratio and R^{b}_t is the 10-year bond yield.

When the economic and financial variables have been selected, we can use the VECM approach to design expected returns. Let y_t be a unit root stochastic process[22]. In unit root tests, we consider the model $y_t = \rho y_{t-1} + \varepsilon_t$ and we test the hypothesis $\mathcal{H}_0 : \rho = 1$ versus the alternative $\mathcal{H}_1 : \rho < 1$. If $\mathcal{H}_0$ is not rejected, we say that y_t is an integrated process of order one and we note $y_t \sim I(1)$. In this case, the process $y_t - y_{t-1}$ is stationary and we note $y_t - y_{t-1} \sim I(0)$. Let us now consider the multi-dimensional process $y_t = (y_{1,t}, \ldots, y_{m,t})$. We assume that $y_{j,t} \sim I(1)$ for $j = 1, \ldots, m$ and $z_t = \beta^\top y_t \sim I(0)$ with $\beta \neq \mathbf{0}$. In this case, we say that y_t is cointegrated. So, there exists at least one stationary linear combination of the time series. From an economic point of view, this implies that there is a long-run equilibrium between the different processes $y_{1,t}, \ldots, y_{m,t}$. In this case, Engle and Granger (1987) show that the cointegrated process y_t may be represented by an error correction model:

$$\Delta y_t = c_t + \gamma \beta^\top y_{t-1} + \zeta_1 \Delta y_{t-1} + \cdots + \zeta_{p-1} \Delta y_{t-p+1} + \eta_t \tag{1.22}$$

where c_t is a deterministic component. Δy_t is a vector autoregressive (VAR) process with an error correction term $\gamma \beta^\top y_{t-1}$ or γz_{t-1}. We note that $z_t \neq 0$ induces a correction on Δy_t and the magnitude of the correction is proportional to the magnitude of the long-run disequilibrium $|z_t|$.

For a long time, the VAR approach has been the standard approach for

[21]The time t is expressed in years.

[22]The concept of unit root is related to the random walk process. y_t is then a non-stationary process (Dickey and Fuller, 1981).

building expected returns of financial assets (Barberis, 2001; Campbell and Viveira, 2002; Campbell *et al.*, 2003). In this case, the forecasts are built using the equation (1.22) without the correction term $\gamma\beta^\top y_{t-1}$. However, the model of Bansal and Yaron (2004) highlights the importance of the long-run equilibrium[23]. The cointegration approach also offers the advantage of differentiating SAA and TAA. SAA is concerned with the long-run equilibrium $z_t = \beta^\top y_t$ whereas TAA corresponds to the short-run dynamics represented by the vector error correction model (1.22).

Remark 14 *In this paragraph, we have only considered the economic approach to design expected returns of asset classes. For individual securities, the traditional way to estimate expected returns is to use scoring methods with valuation and momentum factors (Lo and Patel, 2008). In the CTA industry, the forecasting techniques are generally based on trend filtering methods (Bruder et al., 2011). In the case of trading strategies, professionals intensively use the mean reversion pattern of assets (Poterba and Summers, 1988; Lehmann, 1990; Lo and MacKinlay, 1990) and build their anticipations with the machine learning approach (Hastie et al., 2001). However, the most popular approach for designing expected returns is the naive (or unconditional) approach, which consists in using historical figures as a guide for the future. By stating that past history will repeat itself similarly, the expected returns are based solely on historical returns, disregarding any structural economic changes that could arise. However, the economy and financial markets are more complex. This approach therefore appears unsatisfactory even though it is one of the most frequently used in asset management.*

1.2.3 Regularization of optimized portfolios

Once we have estimated the expected returns and the covariance matrix, we have the two input parameters of the Markowitz problem, and we could think that the job is complete. However, the story does not end here. In practice, portfolio optimization is a difficult task because optimized portfolios are not stable. This explains why many professionals do not use them:

> "*The indifference of many investment practitioners to mean-variance optimization technology, despite its theoretical appeal, is understandable in many cases. The major problem with mean-variance optimization is its tendency to maximize the effects of errors in the input assumptions. Unconstrained MV optimization can yield results that are inferior to those of simple equal-weighting schemes*" (Michaud, 1989).

Michaud explains that optimized portfolios are not optimal because they are not robust. By calling his article "Are Optimized Optimal?", Michaud makes

[23]It is confirmed by the empirical work of Bansal *et al.* (2009).

clearly the distinction between **mathematical optimization** and **financial optimality**. This stability issue may be illustrated easily because optimized portfolios are very sensitive to estimation errors. In order to stabilize the solution, we have to consider different regularization techniques like resampling techniques, denoising approaches or shrinkage methods.

1.2.3.1 Stability issues

Let us illustrate the stability problem of optimized portfolio with an example.

Example 3 *We consider a universe of three assets. The expected returns are respectively* $\mu_1 = \mu_2 = 8\%$ *and* $\mu_3 = 5\%$*. For the volatilities, we have* $\sigma_1 = 20\%$*,* $\sigma_2 = 21\%$ *and* $\sigma_3 = 10\%$*. Moreover, we assume that the cross-correlations are the same and we have* $\rho_{i,j} = \rho = 80\%$*.*

We suppose that the objective of the portfolio manager is to maximize the expected return of the portfolio for a 15% volatility target. The optimal portfolio $x^\star$ is $(38.3\%, 20.2\%, 41.5\%)$. In Table 1.8, we indicate how this solution differs when we slightly change the value of the input parameters. For example, if the correlation is equal to 90%, the weight of the second asset becomes 8.9% instead of 20.2% previously. In real life, we do not exactly know the true parameter values. For example, if the estimated correlation is 80%, the probability that the observed correlation takes this value is very small. This is the reason why we make the distinction between ex-ante estimation and ex-post estimation in finance. In our example, the realized correlation could be 90% and not 80%. In this case, the original portfolio $x^\star$ is far from the true optimal portfolio, i.e. it is sensitive to estimation errors.

TABLE 1.8: Sensitivity of the MVO portfolio to input parameters

ρ		70%	90%		90%	
σ_2				18%	18%	
μ_1						9%
x_1	38.3	38.3	44.6	13.7	−8.0	60.6
x_2	20.2	25.9	8.9	56.1	74.1	−5.4
x_3	41.5	35.8	46.5	30.2	34.0	44.8

1.2.3.2 Resampling techniques

A resampling technique is a set of Monte Carlo methods used to estimate the precision of sample statistics. Jorion (1992) was the first to apply these techniques to portfolio optimization[24]. The underlying idea is very simple. We consider a universe of n assets. Let $\hat{\mu}$ and $\hat{\Sigma}$ be the estimates of the

[24]See also the works of Broadie (1993) and Britten-Jones (1999).

vector of expected returns and the covariance matrix of asset returns. The efficient frontier computed with these statistics is an estimation of the true efficient frontier. In Figure 1.16, we have represented the efficient frontier corresponding to Example 4. If we resample the statistics $\hat{\mu}$ and $\hat{\Sigma}$ with 260 observations, we will not obtain the same optimized portfolios. In Figure 1.16, we have reported 100 simulations of resampled optimized portfolios estimated with the γ-problem[25] when γ takes the values 0, 0.025, 0.05, 0.075, 0.10, 0.15, 0.20, 0.30, 0.40 and 0.50. It is obvious that we have substantial uncertainty as regards the efficient frontier.

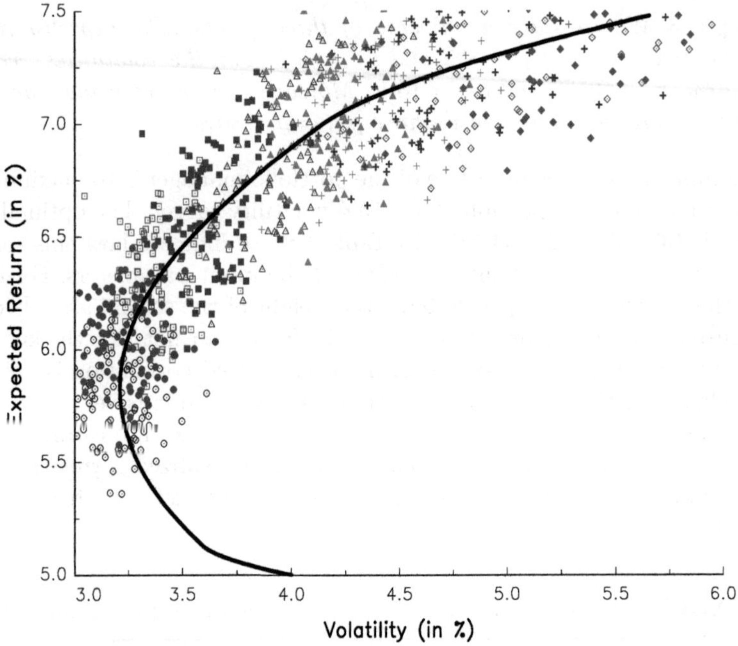

FIGURE 1.16: Uncertainty of the efficient frontier

Example 4 *We consider a universe of four assets. The estimated expected returns are* $\hat{\mu}_1 = 5\%$, $\hat{\mu}_2 = 9\%$, $\hat{\mu}_3 = 7\%$ *and* $\hat{\mu}_4 = 6\%$ *whereas the estimated volatilities are* $\hat{\sigma}_1 = 4\%$, $\hat{\sigma}_2 = 15\%$, $\hat{\sigma}_3 = 5\%$ *and* $\hat{\sigma}_4 = 10\%$. *For the estimated correlation matrix, we have:*

$$\hat{C} = \begin{pmatrix} 1.00 & & & \\ 0.10 & 1.00 & & \\ 0.40 & 0.20 & 1.00 & \\ -0.10 & -0.10 & -0.20 & 1.00 \end{pmatrix}$$

[25] Each symbol corresponds to a particular value of γ. For example, the circle symbols correspond to the minimum variance portfolio ($\gamma = 0$).

In 1998, Michaud proposed an algorithm to compute the resampled efficient frontier. The procedure consists in (1) sampling $\hat{\mu}$ and $\hat{\Sigma}$, (2) calculating optimized portfolios based on the sampled statistics, (3) averaging the weights of the resampled optimized portfolios and (4) estimating the resampled efficient frontier by using the average portfolios obtained in (3) and the statistics $\hat{\mu}$ and $\hat{\Sigma}$. The key point of this procedure is the second step. For example, we could compute the optimized portfolio with respect to a given value of ϕ (ϕ-problem) or to a given target of expected return (μ-problem) or volatility (σ-problem). Figure 1.17 illustrates this procedure in the case of Example 4 and the μ-problem when we impose that the weights are positive[26]. The estimated efficient frontier is the one obtained with the parameters $\hat{\mu}$ and $\hat{\Sigma}$ whereas the averaged efficient frontier corresponds to the average of simulated efficient frontier. They differ from the resampled efficient frontier in which we average portfolios[27]. To better understand the difference, we have reported an optimized portfolio of the resampled efficient frontier (solid circle). This portfolio is the average of the 300 portfolios designed by the cross symbol[28].

Resampling techniques have faced some criticisms (Scherer, 2002). The first point of criticism concerns the procedure itself. If $\hat{\mu}$ and $\hat{\Sigma}$ are not well estimated, the sampled efficient frontier will not correct these initial errors. The second point concerns the lack of theory. Resampling techniques is therefore more of an empirical method that seems to correct some biases, but they do not solve the robustness question concerning optimized portfolios.

1.2.3.3 Denoising the covariance matrix

The idea is to reduce the instability of the estimator $\hat{\Sigma}$ when the condition $T \gg n$ is not verified. We consider the eigendecomposition $\hat{\Sigma} = V \Lambda V^\top$ where $\Lambda = \operatorname{diag}(\lambda_1, \ldots, \lambda_n)$ is the diagonal matrix of eigenvalues with $\lambda_1 > \lambda_2 > \ldots > \lambda_n$ and V is an orthonormal matrix. In the principal component analysis, the endogenous factors are $\mathcal{F}_t = \Lambda^{-1/2} V^\top R_t$. By considering only the m first components, we can build an estimation of Σ with less noise. The principal

[26] We use the procedure described in U.S. patent #6,003,018 owned by Michaud Partners LLP. We consider 300 simulated samples. For each simulated efficient frontier, we compute the mean-variance portfolio with the smallest expected return (it is the minimum variance portfolio) and the mean-variance portfolio with the highest expected return. We build a grid of 60 portfolios which are equally spaced in terms of expected return between these two extreme portfolios. Each simulated optimized portfolio is then represented by an index j. The resampled optimized portfolio for the index j is then the average of the 300 simulated optimized portfolios which correspond to this index j.

[27] The difference is higher if we consider a larger number of assets, for example more than 100.

[28] It is the resampled optimized portfolio of index $j = 50$. Its composition is $\tilde{x}_1^\star = 3.76\%$, $\tilde{x}_2^\star = 15.54\%$, $\tilde{x}_3^\star = 63.86\%$ and $\tilde{x}_4^\star = 16.84\%$. It corresponds to an expected return $\mu(\tilde{x}^\star)$ equal to 7.07% and a volatility $\sigma(\tilde{x}^\star)$ equal to 4.35%. The corresponding estimated optimized portfolio is $x_1^\star = 0.00\%$, $x_2^\star = 12.17\%$, $x_3^\star = 70.20\%$ and $x_4^\star = 17.63\%$, which has the same expected return, but presents a small difference for the volatility ($\sigma(x^\star) = 4.27\%$). We note then that the resampled portfolio $\tilde{x}^\star$ is less concentrated than the optimized portfolio $x^\star$.

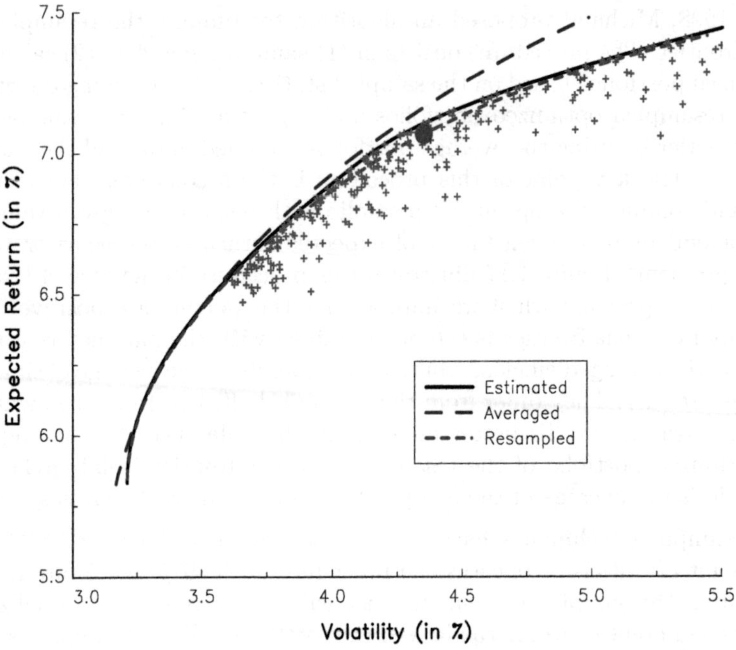

FIGURE 1.17: Resampled efficient frontier

difficulty lies in determining m. A first solution consists in keeping the factors which explain more than $1/n$ of the asset variance. In this case, we have:

$$m = \sup \{i : \lambda_i \geq (\lambda_1 + \ldots + \lambda_n) / n\}$$

Laloux *et al.* (1999) propose another solution by using results of the random matrix theory (RMT). They show that the eigenvalues of a random matrix M has a special probability distribution[29]. More precisely, the maximum value is asymptotically equal to the following expression:

$$\lambda_{\max} = \sigma^2 \left(1 + n/T + 2\sqrt{n/T}\right)$$

where σ^2 is the variance of M. If we apply this result to our problem, we find that:

$$m = \sup \{i : \lambda_i > \lambda_{\max}\}$$

[29]It is known as the Marcenko-Pastur distribution.

If we apply the previous method to our example of equity indices, we obtain:

$$\hat{C} = \left(\begin{array}{cccc|cccc} 1.00 & & & & & & & \\ 0.73 & 1.00 & & & & & & \\ 0.72 & 0.76 & 1.00 & & & & & \\ 0.61 & 0.64 & 0.64 & 1.00 & & & & \\ \hline 0.72 & 0.76 & 0.75 & 0.64 & 1.00 & & & \\ 0.71 & 0.75 & 0.74 & 0.63 & 0.74 & 1.00 & & \\ 0.63 & 0.66 & 0.65 & 0.56 & 0.66 & 0.65 & 1.00 & \\ 0.68 & 0.72 & 0.71 & 0.60 & 0.71 & 0.70 & 0.62 & 1.00 \end{array}\right) \quad (1.23)$$

1.2.3.4 Shrinkage methods

The general approach of shrinkage methods is to modify the empirical covariance matrix $\hat{\Sigma}$ by a new estimate $\tilde{\Sigma}$ in order to take into account some uncertainties. Generally, $\tilde{\Sigma}$ takes the form of a weighted sum of $\hat{\Sigma}$ and a correction term. This technique, which has been used in statistics for linear regressions for a long time, was introduced in asset management by Jorion (1986). He suggests using the Bayes-Stein estimator based on the one-factor model, which is defined by:

$$\tilde{\Sigma}_\alpha = \left(1 + \frac{1}{T+\alpha}\right)\hat{\Sigma} + \left(\frac{\alpha}{T(T+\alpha+1)}\right)\frac{\mathbf{1}\mathbf{1}^\top}{\mathbf{1}^\top\hat{\Sigma}^{-1}\mathbf{1}}$$

where $\alpha \geq 0$ is the shrinkage parameter. If $\alpha = 0$, we have $\tilde{\Sigma}_\alpha \simeq \hat{\Sigma}$. We note also that:

$$\lim_{\alpha\to\infty} \hat{\Sigma}_\alpha = \hat{\Sigma} + \frac{1}{T}\frac{\mathbf{1}\mathbf{1}^\top}{\mathbf{1}^\top\hat{\Sigma}^{-1}\mathbf{1}}$$

$\hat{\Sigma}_\alpha$ is the sum of two terms. The first one is the empirical covariance matrix whereas the second term is a correction due to the estimation error of the common mean.

Ledoit-Wolf approach The empirical covariance matrix $\hat{\Sigma}$ is an unbiased estimator of Σ, but its convergence is very slow in particular when n is large. We know also that the estimator $\hat{\Phi}$ based on factor models converges more quickly, but it is biased. Ledoit and Wolf (2003) propose then to combine the two estimators $\hat{\Sigma}$ and $\hat{\Phi}$ in order to obtain a more efficient estimator. Let $\tilde{\Sigma}_\alpha = \alpha\hat{\Phi} + (1-\alpha)\hat{\Sigma}$ be this new estimator. The statistical problem is to estimate the optimal value of α. Ledoit and Wolf consider the quadratic loss $L(\alpha)$ defined as follows:

$$L(\alpha) = \left\|\alpha\hat{\Phi} + (1-\alpha)\hat{\Sigma} - \Sigma\right\|^2$$

By solving the minimization problem $\alpha^\star = \arg\min \mathbb{E}[L(\alpha)]$, they show that the analytical expression of $\alpha^\star$ is:

$$\alpha^\star = \max\left(0, \min\left(\frac{1}{T}\frac{\pi-\varrho}{\gamma}, 1\right)\right)$$

where π is a function which only depends on $\hat{\Sigma}$, ϱ is a function which depends on the covariance between $\hat{\Phi}$ and $\hat{\Sigma}$ and γ is a function which depends on the difference between $\hat{\Phi}$ and $\hat{\Sigma}$. For example, if $\hat{\Phi}$ is the empirical covariance[30] with a constant correlation $\bar{\rho}$, we obtain:

$$\begin{aligned}
\pi &= \sum_{i=1}^{n}\sum_{j=1}^{n}\pi_{i,j} \\
\varrho &= \sum_{i=1}^{n}\pi_{i,i} + \sum_{i=1}^{n}\sum_{j\neq i}\frac{\bar{\rho}}{2}\left(\sqrt{\frac{\hat{\Sigma}_{j,j}}{\hat{\Sigma}_{i,i}}}\vartheta_{i,j} + \sqrt{\frac{\hat{\Sigma}_{i,i}}{\hat{\Sigma}_{j,j}}}\vartheta_{j,i}\right) \\
\gamma &= \sum_{i=1}^{n}\sum_{j=1}^{n}\left(\hat{\Phi}_{i,j} - \hat{\Sigma}_{i,j}\right)^2
\end{aligned}$$

with:

$$\begin{aligned}
\pi_{i,j} &= \frac{1}{T}\sum_{t=1}^{n}\left(\left(R_{i,t} - \bar{R}_i\right)\left(R_{j,t} - \bar{R}_j\right) - \hat{\Sigma}_{i,j}\right)^2 \\
\vartheta_{i,j} &= \frac{1}{T}\sum_{t=1}^{n}\left(\left(R_{i,t} - \bar{R}_i\right)^2 - \hat{\Sigma}_{i,j}\right)\left(\left(R_{i,t} - \bar{R}_i\right)\left(R_{j,t} - \bar{R}_j\right) - \hat{\Sigma}_{i,j}\right)
\end{aligned}$$

In the more general case of a single-factor model, Ledoit and Wolf (2003) show that the expressions of π and γ are the same whereas the function ϱ now becomes:

$$\varrho = \sum_{i=1}^{n}\sum_{j=1}^{n}\varrho_{i,j}$$

with $\varrho_{i,i} = \pi_{i,i}$ and:

$$\begin{aligned}
\varrho_{i,j} &= \frac{1}{T}\sum_{t=1}^{n}\varrho_{i,j,t} \\
\varrho_{i,j,t} &= \left(\hat{\Sigma}_{j,0}\hat{\Sigma}_{0,0}\left(R_{i,t} - \bar{R}_i\right) + \hat{\Sigma}_{i,0}\hat{\Sigma}_{0,0}\left(R_{j,t} - \bar{R}_j\right) - \hat{\Sigma}_{i,0}\hat{\Sigma}_{j,0}\left(\mathcal{F}_t - \bar{\mathcal{F}}\right)\right)\cdot \\
&\quad \left(R_{i,t} - \bar{R}_i\right)\left(R_{j,t} - \bar{R}_j\right)\left(\mathcal{F}_t - \bar{\mathcal{F}}\right)/\,\hat{\Sigma}_{0,0}^2 - \hat{\Phi}_{i,j}\hat{\Sigma}_{i,j}
\end{aligned}$$

In the expression $\varrho_{i,j,t}$, we use the augmented estimator $\hat{\Sigma}$ corresponding to the empirical covariance of $(\mathcal{F}_t, R_t)$ with the convention that the position of $\mathcal{F}_t$ is zero[31].

We continue to consider the example with the eight equity indices. If we

[30] It implies that $\hat{\Phi}_{i,i} = \hat{\Sigma}_{i,i}$ and $\hat{\Phi}_{i,j} = \bar{\rho}\sqrt{\hat{\Sigma}_{i,i}\hat{\Sigma}_{j,j}}$.

[31] $\hat{\Sigma}_{0,i}$ is then the empirical covariance between $\mathcal{F}_t$ and $R_{i,t}$ and $\hat{\Sigma}_{0,0}$ is the empirical variance of the factor $\mathcal{F}_t$.

apply a shrinkage method with a constant correlation matrix, we have:

$$\hat{C} = \left(\begin{array}{cccc|cccc} 1.00 & & & & & & & \\ 0.77 & 1.00 & & & & & & \\ 0.77 & 0.80 & 1.00 & & & & & \\ 0.65 & 0.67 & 0.65 & 1.00 & & & & \\ \hline 0.72 & 0.71 & 0.72 & 0.63 & 1.00 & & & \\ 0.61 & 0.64 & 0.63 & 0.58 & 0.65 & 1.00 & & \\ 0.60 & 0.64 & 0.62 & 0.60 & 0.63 & 0.62 & 1.00 & \\ 0.65 & 0.67 & 0.67 & 0.67 & 0.67 & 0.63 & 0.66 & 1.00 \end{array}\right) \quad (1.24)$$

The shrinkage parameter $\alpha^\star$ takes the value 51.2%. So, the shrinkage correlation matrix is an average of the empirical correlation matrix and the constant correlation matrix. We also note that it is closer to the RMT estimator (1.23) than to the empirical estimator (1.16).

Penalized regression techniques If we consider the quadratic utility function of the Markowitz model:

$$\mathcal{U}(x) = x^\top \mu - \frac{\phi}{2} x^\top \Sigma x$$

the solution of the utility maximization program is[32]:

$$x^\star(\gamma) = \left(X^\top X\right)^{-1} X^\top Y = \gamma \hat{\Sigma}^{-1} \hat{\mu}$$

where X is the matrix of asset returns[33] and $Y = \phi^{-1}\mathbf{1}$. We observe that $\hat{\Sigma} = T^{-1}\left(X^\top X\right)$ is the sample covariance matrix and $\hat{\mu} = T^{-1}\left(X^\top \mathbf{1}\right)$ is the sample mean. We can then interpret these optimized portfolios (also known as characteristic portfolios) as the solution of a linear regression problem (Scherer, 2007). This is the reason why we can use regularization methods of linear regression like the ridge approach (Hoerl and Kennard, 1970) or the lasso approach (Tibshirani, 1996) to improve the robustness of optimized portfolios.

If we consider the ridge regression, the optimization problem becomes:

$$\begin{aligned} x^\star(\gamma, \lambda) &= \arg\min \frac{1}{2} x^\top \Sigma x - \gamma x^\top \mu + \frac{\lambda}{2} x^\top A x \\ &= \arg\min \frac{1}{2} x^\top (\Sigma + \lambda A) x - \gamma x^\top \mu \end{aligned}$$

where λ is a positive scalar and A is a $n \times n$ matrix. Bruder *et al.* (2013) show that the solution is a linear combination of the MVO solution:

$$\begin{aligned} x^\star(\gamma, \lambda) &= \gamma (\Sigma + \lambda A)^{-1} \mu \\ &= \left(I_n + \lambda \Sigma^{-1} A\right)^{-1} x^\star(\gamma) \end{aligned}$$

---

[32]We recall that $\gamma = \phi^{-1}$ (see on page 7 for the definition of the γ-problem).

[33]We have $X_{t,i} = R_{i,t}$.

If $A = I_n$, the solution is equivalent to the MVO portfolio by replacing the covariance matrix by $\Sigma + \lambda I_n$. In this case, the penalized method reduces the correlations and the differences in terms of volatilities. This solution is then close to the shrinkage method of Ledoit and Wolf (2003). If $A = \operatorname{diag}\left(\sigma_1^2, \ldots, \sigma_n^2\right)$, we obtain:

$$\begin{aligned} x^\star(\gamma, \lambda) &= \gamma \left(\Sigma + \lambda A\right)^{-1} \mu \\ &= \frac{\gamma}{1+\lambda}\left(\frac{1}{1+\lambda}\Sigma + \left(1 - \frac{1}{1+\lambda}\right) A\right)^{-1} \mu \\ &= \eta \left(\eta I_n + (1-\eta) C^{-1}\right)^{-1} x^\star(\gamma) \end{aligned}$$

where $\eta = (1+\lambda)^{-1}$ and $C = \Sigma^{-1} A$. $x^\star(\gamma, \lambda)$ is then a combination of two MVO portfolios where the first one considers correlations while the other one ignores correlations.

We could extend the previous approach to the lasso regression[34]:

$$x^\star(\gamma, \lambda) = \arg\min \frac{1}{2} x^\top \Sigma x - \gamma x^\top \mu + \frac{\lambda}{2} A^\top |x| \tag{1.25}$$

where A is a $n \times 1$ vector. If $\gamma = 0$, we obtain a constrained minimum variance portfolio. In this case, the optimization program is close to the L_1 norm-constrained problem of DeMiguel *et al.* (2011):

$$\begin{aligned} x^\star(\delta) &= \arg\min \frac{1}{2} x^\top \Sigma x \\ &\text{u.c.} \quad \sum_{i=1}^{n} |x_i| \leq \delta \end{aligned}$$

These authors show that for the case $\delta = 1$, $x^\star(1)$ is the shortsale-constrained minimum variance portfolio. We retrieve the sparsity property of the lasso regression (Tibshirani, 1996). The optimization problem (1.25) is then particularly useful for asset selection (Bruder *et al.*, 2013). For example, if we assume that $\Sigma = I_n$ and $A = \mathbf{1}$, the solution is:

$$x_i^\star(\gamma, \lambda) = \operatorname{sgn}(\mu_i) \cdot \max\left(0, \gamma |\mu_i| - \lambda\right)$$

If λ increases, there are more and more assets that have a zero weight.

Remark 15 *The extension to dynamic asset allocation is straightforward. For the lasso approach, we have:*

$$x^\star(\gamma, \lambda) = \arg\min \frac{1}{2} x^\top \Sigma x - \gamma x^\top \mu + \frac{\lambda}{2} A^\top |x - x_0|$$

[34]See Appendix A.1.1 on page 303 to understand how to solve such problems using the QP algorithm.

where x_0 is the reference portfolio or the current portfolio. The solution could then be interpreted in terms of turnover and trading costs (Scherer, 2007). For the ridge approach, we have:

$$x^\star(\gamma, \lambda) = \arg\min \frac{1}{2} x^\top \Sigma x - \gamma x^\top \mu + \frac{\lambda}{2} (x - x_0)^\top A (x - x_0)$$

In this case, the parameter λ can be viewed as a measurement of the investor's level of risk aversion with respect to the MVO portfolio. We notice that the ridge solution is:

$$x^\star(\gamma, \lambda) = (\Sigma + \lambda A)^{-1} (\gamma \mu + \lambda A x_0)$$

The penalty modifies both the mean μ and the covariance matrix Σ. This is why this approach is similar to the double shrinkage method described in Candelon et al. (2012).

TABLE 1.9: Solutions of penalized mean-variance optimization

	MVO		Ridge		Lasso	
	(NC)	(C)	(S)	(D)	(S)	(D)
$x_1^\star$	112.29	62.09	38.88	51.62	24.41	25.00
$x_2^\star$	48.30	14.17	28.06	36.85	11.36	25.00
$x_3^\star$	48.10	62.21	27.34	29.34	27.78	25.00
$x_4^\star$	−39.69	−38.48	−1.57	−0.47	0.00	20.42

We consider Example 1 with $\gamma = 0.5$ and $\lambda = 3\%$. Results are reported in Table 1.9. The non-constrained (NC) MVO portfolio corresponds to the formula $x^\star = \gamma \Sigma^{-1} \mu$ whereas we impose that the sum of the weights is equal to 1 for the constrained (C) MVO portfolio. We note that introducing the constraint $\mathbf{1}^\top x = 1$ is already a shrinkage method (DeMiguel *et al.*, 2011). For the ridge and lasso approaches, we consider two cases: the static (S) case when the penalization concerns only the portfolio x and the dynamic (D) case when the penalization depends also on the current allocation, which corresponds to the equally weighted portfolio. In Figure 1.18, we report the dynamics of the weights for different values of λ. For the static case, we impose the constraint $\mathbf{1}^\top x = 1$ meaning that the solution does not converge to zero weights when λ increases.

1.2.4 Introducing constraints

The previous regularization techniques, despite being interesting and improving the robustness of optimized portfolios, are not sufficient. The problem comes from the fact that the most important quantity in portfolio optimization is not the covariance matrix, but the inverse of the covariance matrix, i.e. the

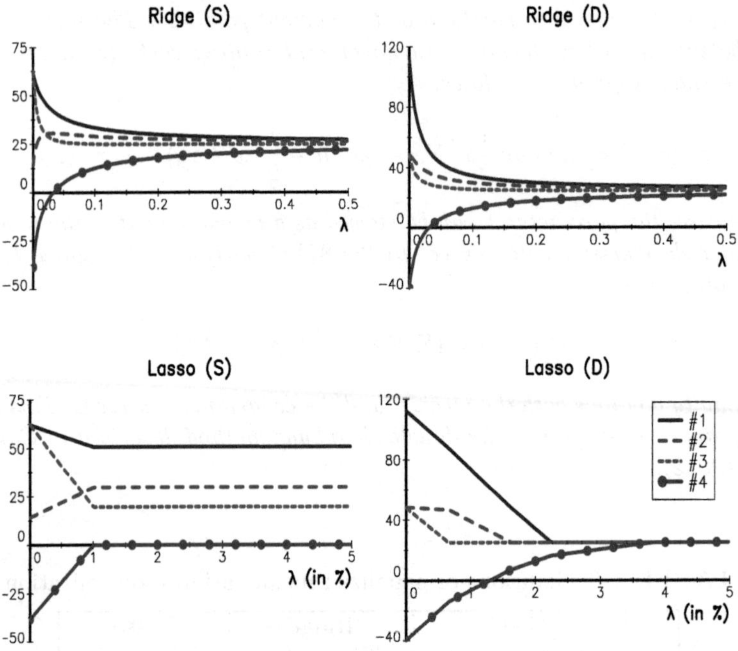

FIGURE 1.18: Weights of penalized MVO portfolios (in %)

information matrix in statistics. This information matrix presents some financial interpretation difficulties. This is the reason why most portfolio managers prefer to impose weight constraints.

1.2.4.1 Why regularization techniques are not sufficient

If we consider the analytical solution of the unconstrained optimization problem (Equation (1.6) on page 11), we note that optimal solutions are of the following form: $x^\star \propto f\left(\Sigma^{-1}\right)$. The important quantity is then the information matrix $\mathcal{I} = \Sigma^{-1}$ and not the covariance matrix itself (Stevens, 1998).

However, these two matrices are strongly related in terms of eigendecomposition. For the covariance matrix Σ, we have $\Sigma = V \Lambda V^\top$ where $V^{-1} = V^\top$ and $\Lambda = (\lambda_1, \ldots, \lambda_n)$ with $\lambda_1 \geq \ldots \geq \lambda_n$ the ordered eigenvalues, whereas the decomposition for the information matrix is $\mathcal{I} = U \Delta U^\top$ with $\delta_1 \geq \ldots \geq \delta_n$. Nevertheless, we have:

$$\begin{aligned}
\Sigma^{-1} &= \left(V \Lambda V^\top\right)^{-1} \\
&= \left(V^\top\right)^{-1} \Lambda^{-1} V^{-1} \\
&= V \Lambda^{-1} V^\top
\end{aligned}$$

It follows that $U = V$ and $\delta_i = 1/\lambda_{n-i+1}$. We note that the eigenvectors of the information matrix are the same as those of the covariance matrix, but the eigenvalues of the information matrix are the inverse of the eigenvalues of the covariance matrix. This means that the risk factors are the same, but they are in the reverse order.

TABLE 1.10: Principal component analysis of the covariance matrix Σ

Asset / Factor	1	2	3
1	65.35%	−72.29%	−22.43%
2	69.38%	69.06%	−20.43%
3	30.26%	−2.21%	95.29%
Eigenvalue	8.31%	0.84%	0.26%
% cumulated	88.29%	97.20%	100.00%

Let us consider Example 3 on page 45. It has been used to illustrate the lack of stability of optimized portfolios. The principal component analysis of the covariance matrix Σ is reported in Table 1.10. The first factor is a market factor because the asset weights are all positive. This factor represents 88.29% of the variance of the assets. The second factor is a long-short portfolio, short in asset one and long in asset two. This explains about 9% of the variance of the assets. The last factor is another long-short portfolio, short in assets one and two and long in asset three. Now, if we consider the eigendecomposition of the information matrix, we obtain the results given in Table 1.11. We verify that the factors are the same but in the reverse order. The smaller the eigenvalue of the covariance matrix, the higher the eigenvalue of the information matrix.

TABLE 1.11: Principal component analysis of the information matrix $\mathcal{I}$

Asset / Factor	1	2	3
1	−22.43%	−72.29%	65.35%
2	−20.43%	69.06%	69.38%
3	95.29%	−2.21%	30.26%
Eigenvalue	379.97	119.18	12.04
% cumulated	74.33%	97.65%	100.00%

Let us consider a large universe of assets, for example the 100 stocks of the FTSE 100 index. In Figure 1.19, we have reported the eigenvalues (in %) of the asset covariance matrix at the end of June 2012 using the last 260 daily returns. Because the universe is composed of 100 stocks, we have 100 PCA factors. One difficulty is to interpret these factors from a financial point of view. The first factor is recognized to be the market risk factor. Furthermore, we have some sector risk factors, in the sense that they are generally representative of long-short portfolios, long and short of stocks belonging to the same sector, or long in stocks of one sector and short in stocks of another sector. The last factors are called arbitrage factors, because they correspond to portfolios with a limited number of stocks. These last factors are the most important factors when performing portfolio optimization. One problem is that we do not know if these last factors are really true arbitrage factors or noise factors.

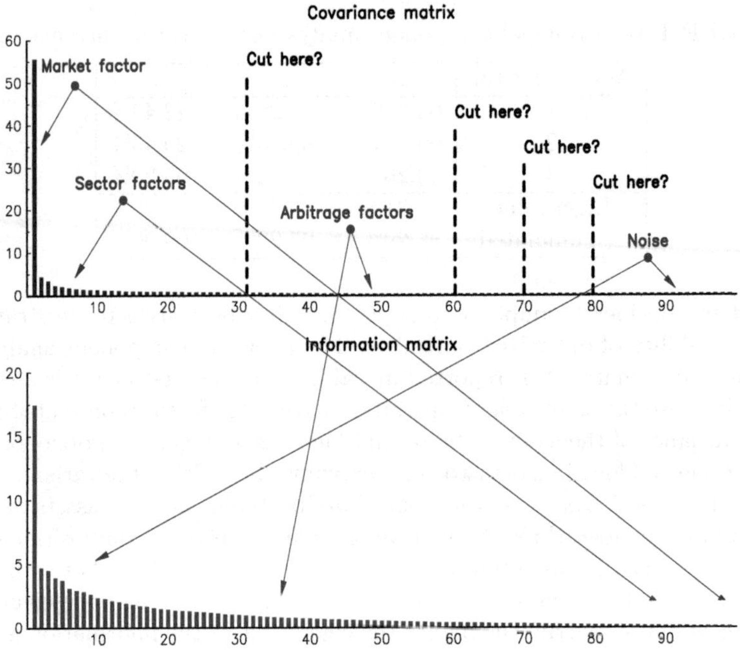

FIGURE 1.19: PCA applied to the stocks of the FTSE index (June 2012)

Example 5 *We consider a universe of 6 assets. The volatility is equal respectively to* 20%, 21%, 17%, 24%, 20% *and* 16%. *For the correlation matrix, we have:*

$$\rho = \begin{pmatrix} 1.00 & & & & & \\ 0.40 & 1.00 & & & & \\ 0.40 & 0.40 & 1.00 & & & \\ 0.50 & 0.50 & 0.50 & 1.00 & & \\ 0.50 & 0.50 & 0.50 & 0.60 & 1.00 & \\ 0.50 & 0.50 & 0.50 & 0.60 & 0.60 & 1.00 \end{pmatrix}$$

To illustrate this problem, we consider the previous example. We compute the minimum variance (MV) portfolio with a shortsale constraint. Then, we delete one PCA factor and compute again the optimized portfolio. The results are given in Table 1.12. We note that if we delete the first factor ($\lambda_1 = 0$), the solution changes less than if we delete the last factor[35]($\lambda_6 = 0$). The

[35]If we do not impose the constraint $x_i \geq 0$, the solution with $\lambda_6 = 0$ has nothing to do with the original MV portfolio.

factors with the smallest eigenvalues are then the most important factors when performing portfolio optimization[36].

TABLE 1.12: Effect of deleting a PCA factor

$x^\star$	MV	$\lambda_1 = 0$	$\lambda_2 = 0$	$\lambda_3 = 0$	$\lambda_4 = 0$	$\lambda_5 = 0$	$\lambda_6 = 0$
$x_1^\star$	15.29	15.77	20.79	27.98	0.00	13.40	0.00
$x_2^\star$	10.98	16.92	1.46	12.31	0.00	8.86	0.00
$x_3^\star$	34.40	12.68	35.76	28.24	52.73	53.38	2.58
$x_4^\star$	0.00	22.88	0.00	0.00	0.00	0.00	0.00
$x_5^\star$	1.01	17.99	2.42	0.00	15.93	0.00	0.00
$x_6^\star$	38.32	13.76	39.57	31.48	31.34	24.36	97.42

1.2.4.2 How to specify the constraints

We have seen that regularization is not enough to stabilize the solution. Moreover, we face a paradox. For a portfolio manager, the first PCA factors are the most important factors because he could interpret them whereas he is not confident with the last factors. For portfolio optimization, the contrary solution applies. This explains why most portfolio managers prefer to impose some constraints. In this case, the γ-problem becomes:

$$\begin{aligned} x^\star(\gamma) &= \arg\min \frac{1}{2} x^\top \Sigma x - \gamma x^\top \mu \\ &\text{u.c. } \left\{ \begin{array}{l} \mathbf{1}^\top x = 1 \\ x \in \Omega \end{array} \right. \end{aligned} \tag{1.26}$$

The constraints are specified through the space Ω. Optimized portfolios were previously obtained as a solution of a quadratic programming problem. Because QP algorithms are very powerful, we could easily obtain the solution even with a large universe of assets. However, by imposing the constraints $x \in \Omega$, the optimization problem cannot remain a QP problem. In this case, numerical solutions may be difficult to compute, because we have to use more complex algorithms. It is then important for the portfolio manager to keep in mind these numerical difficulties and to specify an equivalent QP program when possible.

No short-selling is a constraint frequently imposed in traditional asset management. In this case, we have $\Omega = [0,1]^n$. This constraint is easy to handle within the QP framework presented in Appendix A.1.1 by considering a lower

[36] To understand this result, we assume that there is no constraint. In this case, the optimized portfolio is $x^\star = \phi^{-1}\Sigma^{-1}\mu$. Using the eigendecomposition $\Sigma = V \Lambda V^\top$, we deduce that $x^\star = \phi^{-1} V \Lambda^{-1} V^\top \mu$. Let $y^\star = V^\top x^\star$ be the V-rotation of the solution. We then have $y^\star = \phi^{-1}\Lambda^{-1}\nu$ with $\nu = V\mu$. This means that we weight the asset by the expected return of each PCA portfolio divided by its eigenvalue. A small eigenvalue then has a large impact on the solution.

bound $x_- = \mathbf{0}$. However, the portfolio generally uses more complicated constraints. Let us consider a multi-asset universe of eight asset classes represented by the following indices[37]: four equity indices (S&P 500, Eurostoxx, Topix, MSCI EM), two bond indices (EGBI, US BIG) and two alternatives indices (GSCI, EPRA). Suppose the portfolio manager wants the following exposures: at least 50% bonds, less than 10% commodities and that emerging market equities may not represent more than one third of the total exposure on equities. The constraints are then expressed as follows[38]:

$$\begin{cases} x_5 + x_6 \geq 50\% \\ x_7 \leq 10\% \\ x_4 \leq \frac{1}{3}\left(x_1 + x_2 + x_3 + x_4\right) \end{cases}$$

We could also impose these constraints by specifying the generic constraint $Cx \geq D$ of the QP problem. We have:

$$\begin{pmatrix} 0 & 0 & 0 & 0 & 1 & 1 & 0 & 0 \\ 0 & 0 & 0 & 0 & 0 & 0 & -1 & 0 \\ 1/3 & 1/3 & 1/3 & -2/3 & 0 & 0 & 0 & 0 \end{pmatrix} \begin{pmatrix} x_1 \\ x_2 \\ x_3 \\ x_4 \\ x_5 \\ x_6 \\ x_7 \\ x_8 \end{pmatrix} \geq \begin{pmatrix} 0.50 \\ -0.10 \\ 0.00 \end{pmatrix}$$

Let us consider some constraints that are more difficult to handle. Suppose that we want to limit the turnover of the long-only optimized portfolio with respect to a current portfolio x^0. In this case, we have $\Omega = \left\{x \in [0,1]^n : \sum_{i=1}^n \left|x_i - x_i^0\right| \leq \tau^+\right\}$ where τ^+ is the maximum turnover. To transform this non-linear optimization problem into a QP problem, Scherer (2007) proposes to introduce some additional variables x_i^- and x_i^+ such that:

$$x_i = x_i^0 + x_i^+ - x_i^-$$

with $x_i^- \geq 0$ and $x_i^+ \geq 0$. x_i^+ indicates then a positive weight change with respect to the initial weight x_i^0 whereas x_i^- indicates a negative weight change. The expression of the turnover becomes:

$$\sum_{i=1}^n \left|x_i - x_i^0\right| = \sum_{i=1}^n \left|x_i^+ - x_i^-\right| = \sum_{i=1}^n x_i^+ + \sum_{i=1}^n x_i^-$$

[37]We have the following correspondence: the S&P 500 index concerns U.S. stocks; the Eurostoxx index represents stocks of the Eurozone; the Topix index is composed of Japanese stocks; the MSCI EM index is an exposure on emerging market equities; the EGBI index is a sovereign bond index restricted to the Eurozone; the US BIG index represents the performance of the investment grade bonds in the U.S.; the GSCI index is a commodity index; the EPRA index is a benchmark for listed real estate.

[38]The weights are indexed according to the list of the assets.

because one of the variables x_i^+ or x_i^- is necessarily equal to zero. We could also write the γ-problem of Markowitz as follows:

$$\begin{aligned} x^\star &= \arg\min \frac{1}{2} x^\top \Sigma x - \gamma x^\top \mu \\ & \text{u.c.} \quad \left\{ \begin{array}{l} \sum_{i=1}^n x_i = 1 \\ x_i = x_i^0 + x_i^+ - x_i^- \\ \sum_{i=1}^n x_i^+ + \sum_{i=1}^n x_i^- \leq \tau^+ \\ 0 \leq x_i \leq 1 \\ 0 \leq x_i^- \leq 1 \\ 0 \leq x_i^+ \leq 1 \end{array} \right. \end{aligned}$$

We note that we obtain an augmented QP problem of dimension $3n$ instead of n previously. Let $X = \left(x_1, \ldots, x_n, x_1^-, \ldots, x_n^-, x_1^+, \ldots, x_n^+\right)$. We have:

$$\begin{aligned} X^\star &= \arg\min \frac{1}{2} X^\top Q X - X^\top R \\ & \text{u.c.} \quad \left\{ \begin{array}{l} AX = B \\ CX \geq D \\ \mathbf{0} \leq X \leq \mathbf{1} \end{array} \right. \end{aligned}$$

with:

$$\begin{aligned} Q &= \begin{pmatrix} \Sigma & \mathbf{0} & \mathbf{0} \\ \mathbf{0} & \mathbf{0} & \mathbf{0} \\ \mathbf{0} & \mathbf{0} & \mathbf{0} \end{pmatrix}, \quad R = \begin{pmatrix} \mu \\ \mathbf{0} \\ \mathbf{0} \end{pmatrix}, \quad A = \begin{pmatrix} \mathbf{1}^\top & \mathbf{0} & \mathbf{0} \\ I_n & I_n & -I_n \end{pmatrix} \\ B &= \begin{pmatrix} 1 \\ x^0 \end{pmatrix}, \quad C = \begin{pmatrix} \mathbf{0} & -\mathbf{1}^\top & -\mathbf{1}^\top \end{pmatrix} \quad \text{and} \quad D = -\tau^+ \end{aligned}$$

Let us consider Example 1 on page 5. We impose that the weights are positive. On page 10, we found that the optimal long-only portfolio $x^\star$ for a 15% volatility target is $(45.59\%, 24.74\%, 29.67\%, 0.00\%)$. Let us suppose that the current portfolio x^0 is $(30\%, 45\%, 15\%, 10\%)$. If we move directly from portfolio x^0 to portfolio $x^\star$, the turnover is equal to 60.53%. If we want to limit the turnover, we could use the previous framework. In this case, we obtain the results in Table 1.13. If we impose a turnover of 5%, there is no solution to the optimization problem because it is not possible to target a 15% volatility. If the turnover is set to 10%, we rebalance the portfolio by selling 5% of the fourth asset and by buying 5% of the first asset. We verify that the solution converges to the optimal long-only portfolio when we allow more turnover.

Remark 16 *The previous technique which consists in introducing some additional variables in order to transform the optimization problem into a QP problem could also be used to take into account transaction costs (Scherer,*

TABLE 1.13: Limiting the turnover of MVO portfolios

τ^+	5.00	10.00	25.00	50.00	75.00	x^0
$x_1^\star$		35.00	36.40	42.34	45.59	30.00
$x_2^\star$		45.00	42.50	30.00	24.74	45.00
$x_3^\star$		15.00	21.10	27.66	29.67	15.00
$x_4^\star$		5.00	0.00	0.00	0.00	10.00
$\mu(x^\star)$		5.95	6.06	6.13	6.14	6.00
$\sigma(x^\star)$		15.00	15.00	15.00	15.00	15.69

2007). Let c_i^- and c_i^+ be the bid and ask transactions costs. The γ-problem of Markowitz now becomes:

$$\begin{aligned} x^\star &= \arg\min \frac{1}{2} x^\top \Sigma x - \gamma \left(\sum x_i \mu_i - \sum x_i^- c_i^- - \sum x_i^+ c_i^+ \right) \\ & u.c. \quad \left\{ \begin{array}{l} \sum x_i + \sum x_i^- c_i^- + \sum x_i^+ c_i^+ = 1 \\ x_i = x_i^0 + x_i^+ - x_i^- \\ 0 \leq x_i \leq 1 \\ 0 \leq x_i^- \leq 1 \\ 0 \leq x_i^+ \leq 1 \end{array} \right. \end{aligned}$$

As previously, we obtain an augmented QP problem with[39]*:*

$$R = \left(\begin{array}{c} \mu \\ -c^- \\ -c^+ \end{array} \right), \quad A = \left(\begin{array}{ccc} \mathbf{1}^\top & \left(c^-\right)^\top & \left(c^+\right)^\top \\ I_n & I_n & -I_n \end{array} \right) \quad \textit{and} \quad B = \left(\begin{array}{c} 1 \\ x^0 \end{array} \right)$$

In some cases, it is not possible to obtain a QP problem. For example, index sampling is a technique widely used by trackers. Contrary to the full replication method, which consists in holding all the stocks of the index in exact proportion to their weights in the index, the goal of index sampling is to invest in a portfolio with a smaller number of stocks than the index. Compared to the full replication technique, index sampling produces smaller transaction costs, but larger tracking errors. The choice of the replication method depends then on the tradeoff between transaction costs and tracking errors. From a mathematical point of view, index sampling could be written as a portfolio optimization problem with a benchmark:

$$\begin{aligned} x^\star &= \arg\min \frac{1}{2} (x - b)^\top \Sigma (x - b) \\ & \text{u.c.} \quad \left\{ \begin{array}{l} \mathbf{1}^\top x = 1 \\ x \geq \mathbf{0} \\ \sum_{i=1}^n \mathbb{1}\{x_i > 0\} \leq n_x \end{array} \right. \end{aligned} \qquad (1.27)$$

where b is the vector of index weights. The idea is to minimize the volatility

[39]The Q matrix does not change.

of the tracking error such that the number of stocks n_x in the portfolio is smaller than the number of stocks n_b in the benchmark ($n_x < n_b$). For example, one would like to replicate the S&P 500 index with only 50 stocks. The optimization problem (1.27) could be solved using a mixed integer non-linear optimization algorithm. However, programming such an algorithm is a difficult task and the convergence of the algorithm is hazardous. This is why most professionals prefer to solve the problem (1.27) by using a heuristic algorithm:

1. The algorithm is initialized with $\mathcal{N}_{(0)} = \emptyset$ and $x^\star_{(0)} = b$.

2. At the iteration k, we define a set $\mathcal{I}_{(k)}$ of stocks having the smallest positive weights in the portfolio $x^\star_{(k-1)}$. We then update the set $\mathcal{N}_{(k)}$ with $\mathcal{N}_{(k)} = \mathcal{N}_{(k-1)} \cup \mathcal{I}_{(k)}$. We define the upper bounds $x^+_{(k)}$ in the following way:

$$x^+_{(k),i} = \begin{cases} 0 & \text{if} \quad i \in \mathcal{N}_{(k)} \\ 1 & \text{if} \quad i \notin \mathcal{N}_{(k)} \end{cases}$$

3. We solve the QP problem by taking into account the upper bounds $x^+_{(k)}$:

$$\begin{aligned} x^\star_{(k)} &= \arg\min \frac{1}{2} \left(x_{(k)} - b\right)^\top \Sigma \left(x_{(k)} - b\right) \\ &\text{u.c.} \quad \begin{cases} \mathbf{1}^\top x_{(k)} = 1 \\ \mathbf{0} \leq x_{(k)} \leq x^+_{(k)} \end{cases} \end{aligned}$$

4. We iterate steps 2 and 3 until the convergence criterion:

$$\sum_{i=1}^{n} \mathbb{1}\left\{x^*_{(k),i} > 0\right\} \leq n_x$$

The heuristic algorithm is close to the '*backward elimination*' process of the stepwise regression, but the elimination process of the heuristic algorithm uses the criterion of the smallest weight. A more natural criterion would be the tracking error variance. We then obtain a second algorithm, which is more complex. Like the stepwise regression, we could also define a '*forward selection*' process. These two new algorithms are summarized in the following way:

- The backward elimination algorithm starts with all the stocks, computes the optimized portfolio, deletes the stock which presents the highest tracking error variance, and repeats this process until the number of stocks in the optimized portfolio reaches the target value n_x.

- The forward selection algorithm starts with no stocks in the portfolio, adds the stock which presents the smallest tracking error variance, and repeats this process until the number of stocks in the optimized portfolio reaches the target value n_x.

However, these algorithms increase the computation time. Indeed, the number of solved QP problems is respectively equal to $n_b - n_x$ for the heuristic algorithm, $(n_b - n_x)(n_b + n_x + 1)/2$ for the backward elimination algorithm and $n_x(2n_b - n_x + 1)/2$ for the forward selection algorithm. These last two algorithms induce a larger number of QP problems than the heuristic algorithm as illustrated below:

		Number of solved QP problems		
n_b	n_x	Heuristic	Backward	Forward
50	10	40	1 220	455
	40	10	455	1 220
500	50	450	123 975	23 775
	450	50	23 775	123 975
1 500	100	1 400	1 120 700	145 050
	1 000	500	625 250	1 000 500

This is why professionals prefer to use the heuristic algorithm[40]. In order to illustrate the differences between the three algorithms, we apply them[41] to the Eurostoxx 50 at the end of June 2012. Results are reported in Tables 1.14, 1.15 and 1.16. The first column corresponds to the iteration k whereas the second column indicates the stock which is selected or deleted. The weight of the stock in the index and the tracking error volatility $\sigma(x_{(k)} \mid b)$ of the optimized portfolio after the iteration k are given in the third and fourth columns. For the heuristic algorithm, we note that Nokia is the first stock to be deleted[42]. Then, the process deletes successively Carrefour, Repsol, etc. If we would like to replicate the Eurostoxx 50 with 5 stocks, the optimized portfolio is then composed by BBVA, Sanofi, Allianz, Total and Siemens and the corresponding tracking error volatility is 5.02%. If we consider the backward elimination algorithm, results are very different. It begins by eliminating Iberdrola, then France Telecom, Carrefour, etc. In this case, the optimized portfolio for $n_x = 5$ is composed by LVMH, Allianz, Sanofi, BBVA and Siemens. The tracking error volatility is close to the previous one[43]. In the case of the forward selection algorithm, the process begins by selecting Siemens, then Banco Santander, Bayer, Eni, Allianz, etc. We note that the rankings done by the three algorithms are not coherent. For example, Banco Santander and BBVA are the 5th and 18th stocks in terms of weights in the Eurostoxx 50 index at the end of June 2012. If we consider the heuristic (resp. the backward elimination) algorithm, they are ranked in the 17th and 5th (resp. 31st and 2nd) positions. For the forward selection algorithm, the order is reversed and they are ranked in the 2nd and 20th positions. The reason is that these two stocks are highly

[40]Even though the backward elimination and forward selection algorithms nowadays take only a few seconds with 500 stocks.

[41]We use the one-year empirical covariance matrix to estimate Σ.

[42]Because it has the smallest weight in the Eurostoxx 50.

[43]It is equal to 4.99%.

TABLE 1.14: Sampling the SX5E index with the heuristic algorithm

k	Stock	b_i	$\sigma\left(x_{(k)} \mid b\right)$
1	Nokia	0.45	0.18
2	Carrefour	0.60	0.23
3	Repsol	0.71	0.28
4	Unibail-Rodamco	0.99	0.30
5	Muenchener Rueckver	1.34	0.32
6	RWE	1.18	0.36
7	Koninklijke Philips	1.07	0.41
8	Generali	1.06	0.45
9	CRH	0.82	0.51
10	Volkswagen	1.34	0.55
42	LVMH	2.39	3.67
43	Telefonica	3.08	3.81
44	Bayer	3.51	4.33
45	Vinci	1.46	5.02
46	BBVA	2.13	6.53
47	Sanofi	5.38	7.26
48	Allianz	2.67	10.76
49	Total	5.89	12.83
50	Siemens	4.36	30.33

TABLE 1.15: Sampling the SX5E index with the backward elimination algorithm

k	Stock	b_i	$\sigma\left(x_{(k)} \mid b\right)$
1	Iberdrola	1.05	0.11
2	France Telecom	1.48	0.18
3	Carrefour	0.60	0.22
4	Muenchener Rueckver	1.34	0.26
5	Repsol	0.71	0.30
6	BMW	1.37	0.34
7	Generali	1.06	0.37
8	RWE	1.18	0.41
9	Koninklijke Philips	1.07	0.44
10	Air Liquide	2.10	0.48
42	GDF Suez	1.92	3.49
43	Bayer	3.51	3.88
44	BNP Paribas	2.26	4.42
45	Total	5.89	4.99
46	LVMH	2.39	5.74
47	Allianz	2.67	7.15
48	Sanofi	5.38	8.90
49	BBVA	2.13	12.83
50	Siemens	4.36	30.33

correlated. So, if one of these stocks is first selected, the other will be selected later.

TABLE 1.16: Sampling the SX5E index with the forward selection algorithm

k	Stock	b_i	$\sigma\left(x_{(k)} \mid b\right)$
1	Siemens	4.36	12.83
2	Banco Santander	3.65	8.86
3	Bayer	3.51	6.92
4	Eni	3.32	5.98
5	Allianz	2.67	5.11
6	LVMH	2.39	4.55
7	France Telecom	1.48	3.93
8	Carrefour	0.60	3.62
9	BMW	1.37	3.35
41	Société Générale	1.07	0.50
42	CRH	0.82	0.45
43	Air Liquide	2.10	0.41
44	RWE	1.18	0.37
45	Nokia	0.45	0.33
46	Unibail-Rodamco	0.99	0.28
47	Repsol	0.71	0.24
48	Essilor	1.17	0.18
49	Muenchener Rueckver	1.04	0.11
50	Iberdrola	1.05	0.00

In Figure 1.20, we report the results of the sampling method for the Eurostoxx 50 index and the S&P 500 index using the heuristic algorithm. The first and second panel corresponds to the relationship between the number of stocks n_x and the volatility of the tracking error $\sigma(x \mid b)$. The third panel compares the two indices by scaling the number of selected stocks n_x by the number of stocks n_b in the benchmark. Thus, the tracking error volatility is equal to 3.2% and 52 bps if the ratio is fixed at 20%. From a relative point of view, it is then more efficient to replicate the S&P 500 index than the Eurostoxx 50 index with the sampling approach.

Remark 17 *The previous heuristic algorithm could be extended to non long-only portfolios. A 130/30 strategy consists in a long position of 130% combined by a short position of 30%. The net position of the portfolio remains 100%, but the strategy benefits from more flexibility than the traditional long-only strategy (Lo and Patel, 2008). In the latter, the negative bets are more constrained than the positive bets. For example, if one stock has a weight of 1% in the benchmark, the weight difference between the portfolio and the benchmark is between a range of −1% and +99%. This means that if the portfolio manager has a negative view on this stock, he could only underweight it by 1%. This*

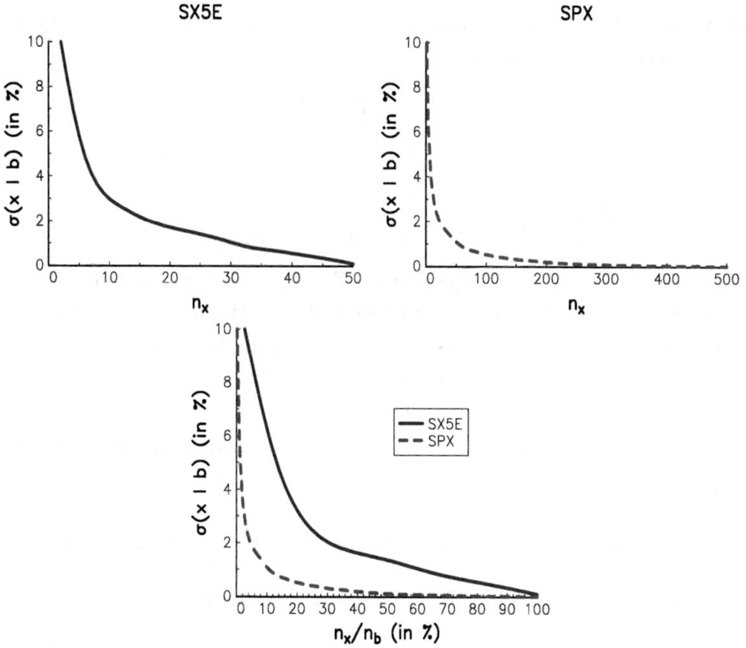

FIGURE 1.20: Sampling the SX5E and SPX indices

implies that the portfolio manager is highly constrained for his negative bets, especially for stocks with small weights. With a 130/30 strategy, the portfolio manager has more flexibility to underweight the stocks which present negative views. Such portfolio construction could be computed using a heuristic algorithm similar to the sampling algorithm (Roncalli, 2010).

1.2.4.3 Shrinkage interpretation of the constrained solution

Let us specify the Markowitz problem in the following way:

$$\min \frac{1}{2} x^\top \Sigma x$$
$$\text{u.c.} \left\{ \begin{array}{l} \mathbf{1}^\top x = 1 \\ \mu^\top x \geq \mu^\star \\ x \in \mathcal{C} \end{array} \right.$$

where $\mathcal{C}$ is the set of weight constraints. We consider two optimized portfolios:

- The first one is the unconstrained portfolio $x^\star$ or $x^\star(\mu, \Sigma)$ with $\mathcal{C} = \mathbb{R}^n$.
- The second one is the constrained portfolio $\tilde{x}$ when the weight of asset i is between a lower bound x_i^- and an upper bound x_i^+:

$$\mathcal{C}\left(x^-, x^+\right) = \left\{ x \in \mathbb{R}^n : x_i^- \leq x_i \leq x_i^+ \right\}$$

Jagannathan and Ma (2003) show that the constrained portfolio is the solution of the unconstrained problem:

$$\tilde{x} = x^{\star}\left(\tilde{\mu}, \tilde{\Sigma}\right)$$

with:

$$\left\{ \begin{array}{l} \tilde{\mu} = \mu \\ \tilde{\Sigma} = \Sigma + \left(\lambda^{+} - \lambda^{-}\right)\mathbf{1}^{\top} + \mathbf{1}\left(\lambda^{+} - \lambda^{-}\right)^{\top} \end{array} \right. \tag{1.28}$$

where λ^{-} and λ^{+} are the Lagrange coefficients vectors associated with the lower and upper bounds. Introducing weight constraints is then equivalent to using another covariance matrix $\tilde{\Sigma}$. It could also be viewed as a shrinkage method or as a Black-Litterman approach when the portfolio manager has some views on the covariance matrix.

Let us prove this result. Without weight constraints, the expression of the Lagrangian is:

$$\mathcal{L}\left(x;\lambda_0,\lambda_1\right) = \frac{1}{2}x^{\top}\Sigma x - \lambda_0\left(\mathbf{1}^{\top}x - 1\right) - \lambda_1\left(\mu^{\top}x - \mu^{\star}\right)$$

with $\lambda_0 \geq 0$ and $\lambda_1 \geq 0$. The first-order conditions are:

$$\left\{ \begin{array}{l} \Sigma x - \lambda_0\mathbf{1} - \lambda_1\mu = \mathbf{0} \\ \mathbf{1}^{\top}x - 1 = 0 \\ \mu^{\top}x - \mu^{\star} = 0 \end{array} \right.$$

We deduce that the solution $x^{\star}$ depends on the vector of expected return μ and the covariance matrix Σ and we note $x^{\star} = x^{\star}\left(\mu,\Sigma\right)$. If we impose now the weight constraints $\mathcal{C}\left(x^{-},x^{+}\right)$, we have:

$$\begin{array}{rcl} \mathcal{L}\left(x;\lambda_0,\lambda_1,\lambda^{-},\lambda^{+}\right) & = & \frac{1}{2}x^{\top}\Sigma x - \lambda_0\left(\mathbf{1}^{\top}x - 1\right) - \lambda_1\left(\mu^{\top}x - \mu^{\star}\right) - \\ & & \lambda^{-\top}\left(x - x^{-}\right) - \lambda^{+\top}\left(x^{+} - x\right) \end{array}$$

with $\lambda_0 \geq 0$, $\lambda_1 \geq 0$, $\lambda_i^{-} \geq 0$ and $\lambda_i^{+} \geq 0$. In this case, the Kuhn-Tucker conditions are:

$$\left\{ \begin{array}{l} \Sigma x - \lambda_0\mathbf{1} - \lambda_1\mu - \lambda^{-} + \lambda^{+} = \mathbf{0} \\ \mathbf{1}^{\top}x - 1 = 0 \\ \mu^{\top}x - \mu^{\star} = 0 \\ \min\left(\lambda_i^{-}, x_i - x_i^{-}\right) = 0 \\ \min\left(\lambda_i^{+}, x_i^{+} - x_i\right) = 0 \end{array} \right.$$

It is not possible to obtain an analytic solution but we may numerically solve the optimization problem using a quadratic programming algorithm. Given a constrained portfolio $\tilde{x}$, it is possible to find a covariance matrix $\tilde{\Sigma}$ such that $\tilde{x}$ is the solution of unconstrained mean-variance portfolio. Let $\mathcal{E} = \left\{\tilde{\Sigma} > 0 : \tilde{x} = x^{\star}\left(\mu,\tilde{\Sigma}\right)\right\}$ denote the corresponding set. We have:

$$\mathcal{E} = \left\{\tilde{\Sigma} > 0 : \tilde{\Sigma}\tilde{x} - \lambda_0\mathbf{1} - \lambda_1\mu = \mathbf{0}\right\}$$

Of course, the set $\mathcal{E}$ contains several solutions. From a financial point of view, we are interested in covariance matrices $\tilde{\Sigma}$ that are close to Σ. Jagannathan and Ma (2003) note that the matrix $\tilde{\Sigma}$ defined by:

$$\tilde{\Sigma} = \Sigma + \left(\lambda^{+} - \lambda^{-}\right) \mathbf{1}^{\top} + \mathbf{1} \left(\lambda^{+} - \lambda^{-}\right)^{\top}$$

is a solution of $\mathcal{E}$. Indeed, we have:

$$\begin{aligned}
\tilde{\Sigma}\tilde{x} &= \Sigma\tilde{x} + \left(\lambda^{+} - \lambda^{-}\right) \mathbf{1}^{\top}\tilde{x} + \mathbf{1} \left(\lambda^{+} - \lambda^{-}\right)^{\top} \tilde{x} \\
&= \Sigma\tilde{x} + \left(\lambda^{+} - \lambda^{-}\right) + \mathbf{1} \left(\lambda^{+} - \lambda^{-}\right)^{\top} \tilde{x} \\
&= \lambda_0 \mathbf{1} + \lambda_1 \mu + \mathbf{1} \left(\lambda_0 \mathbf{1} + \lambda_1 \mu - \Sigma\tilde{x}\right)^{\top} \tilde{x} \\
&= \lambda_0 \mathbf{1} + \lambda_1 \mu + \mathbf{1} \left(\lambda_0 + \lambda_1 \mu^{\star} - \tilde{x}^{\top} \Sigma \tilde{x}\right) \\
&= \left(2\lambda_0 - \tilde{x}^{\top} \Sigma \tilde{x} + \lambda_1 \mu^{\star}\right) \mathbf{1} + \lambda_1 \mu
\end{aligned}$$

It proves that $\tilde{x}$ is the solution of the unconstrained optimization problem[44]. Moreover, $\tilde{\Sigma}$ is generally a positive definite matrix[45].

The implied covariance matrix defined by the equation (1.28) is very interesting because it is easy to compute and has a natural interpretation. Indeed, we have:

$$\tilde{\Sigma}_{i,j} = \Sigma_{i,j} + \Delta_{i,j}$$

where the elements of the perturbation matrix are:

$\Delta_{i,j}$	x_i^{-}	$\left]x_i^{-}, x_i^{+}\right[$	x_i^{+}
x_j^{-}	$-\left(\lambda_i^{-} + \lambda_j^{-}\right)$	$-\lambda_j^{-}$	$\lambda_i^{+} - \lambda_j^{-}$
$\left]x_j^{-}, x_j^{+}\right[$	$-\lambda_i^{-}$	0	λ_i^{+}
x_j^{+}	$\lambda_j^{+} - \lambda_i^{-}$	λ_j^{+}	$\lambda_i^{+} + \lambda_j^{+}$

The perturbation $\Delta_{i,j}$ may be negative, nul or positive. It is nul when the optimized weights do not reach the constraints $\tilde{x}_i \neq \left(x_i^{-}, x_i^{+}\right)$ and $\tilde{x}_j \neq \left(x_j^{-}, x_j^{+}\right)$. It is positive (resp. negative) when one asset reaches its upper (resp. lower) bound whereas the second asset does not reach its lower (resp. upper) bound. Introducing weight constraints is also equivalent to apply a shrinkage method to the covariance matrix (Ledoit and Wolf, 2003). Lower bounds have a negative impact on the volatility whereas upper bounds have a positive impact on the volatility:

$$\tilde{\sigma}_i = \sqrt{\sigma_i^2 + \Delta_{i,i}}$$

[44]The Lagrange coefficients $\lambda_0^{\star}$ and $\lambda_1^{\star}$ for the unconstrained problem are respectively $2\tilde{\lambda}_0 - \tilde{x}^{\top}\Sigma\tilde{x} + \tilde{\lambda}_1\mu^{\star}$ and $\tilde{\lambda}_1$ where $\tilde{\lambda}_0$ and $\tilde{\lambda}_1$ are the Lagrange coefficient for the constrained problem.

[45]Jagannathan and Ma (2003) prove this result only in the case of the minimum variance portfolio when the lower bound x^{-} is exactly equal to $\mathbf{0}$. In other cases, it is extremely difficult to show that $\tilde{\Sigma}$ is a positive definite matrix. Nevertheless, this assumption is verified in practice, except for some special situations (see Exercise B.1.12 on page 349).

The impact on the correlation coefficient is more complex. In the general case, we have:

$$\tilde{\rho}_{i,j} = \frac{\rho_{i,j}\sigma_i\sigma_j + \Delta_{i,j}}{\sqrt{\left(\sigma_i^2 + \Delta_{i,i}\right)\left(\sigma_j^2 + \Delta_{j,j}\right)}}$$

The correlation may increase or decrease depending on the magnitude of the Lagrange coefficients with respect to the parameters $\rho_{i,j}$, σ_i and σ_j.

We consider Example 1. We recall that the global minimum variance portfolio is:

$$x^\star = \begin{pmatrix} 72.742\% \\ 49.464\% \\ -20.454\% \\ -1.753\% \end{pmatrix}$$

In this portfolio, we have two long positions on the first and second assets and two short positions on the third and fourth assets. Suppose that the portfolio manager is not satisfied by this optimized portfolio and decides to impose some constraints. For example, he could decide that the portfolio must contain at least 10% of all the assets. Results are given in Table 1.17. In the constrained optimized portfolio, the weights of the third and fourth assets are set to 10%. To obtain this solution, one must decrease (implicitly) the volatility of these two assets. Indeed, the implied volatility $\tilde{\sigma}_3$ is equal to 19.67% whereas its true volatility σ_3 is equal to 25%. Concerning the correlations, we note also that they are lower than the original ones. We continue our illustration and we suppose that the portfolio manager is not yet satisfied by this portfolio. In particular, he would like to limit the exposure on the first asset and he imposes that the weights must be smaller than 40%. We obtain then the portfolio given in Table 1.18. It is then equivalent to increase the volatility and the cross-correlations of the first asset in order to reduce its weight in the portfolio. Finally, the solution is completely defined by the weight constraints imposed by the portfolio manager.

TABLE 1.17: Minimum variance portfolio when $x_i \geq 10\%$

$\tilde{x}_i$	λ_i^-	λ_i^+	$\tilde{\sigma}_i$	$\tilde{\rho}_{i,j}$			
56.195	0.000	0.000	15.00	100.00			
23.805	0.000	0.000	20.00	10.00	100.00		
10.000	1.190	0.000	19.67	10.50	58.71	100.00	
10.000	1.625	0.000	23.98	17.38	16.16	67.52	100.00

TABLE 1.18: Minimum variance portfolio when $10\% \leq x_i \leq 40\%$

$\tilde{x}_i$	λ_i^-	λ_i^+	$\tilde{\sigma}_i$	$\tilde{\rho}_{i,j}$			
40.000	0.000	0.915	20.20	100.00			
40.000	0.000	0.000	20.00	30.08	100.00		
10.000	0.915	0.000	21.02	35.32	61.48	100.00	
10.000	1.050	0.000	26.27	39.86	25.70	73.06	100.00

If we target an expected return $\mu^\star$ equal to 6%, the optimized portfolio is[46]:

$$x^\star = \begin{pmatrix} 65.866\% \\ 26.670\% \\ 32.933\% \\ -25.470\% \end{pmatrix}$$

By imposing the constraints $10\% \leq x_i \leq 40\%$, we obtain the portfolio specified in Table 1.19. We note that this constrained portfolio $\tilde{x}$ is far from the unconstrained portfolio $x^\star$. Implicitly, the constraints are equivalent to decrease the parameters σ_4, $\rho_{1,4}$ and $\rho_{2,4}$ while the other parameters are very close to the original values.

TABLE 1.19: Mean-variance portfolio when $10\% \leq x_i \leq 40\%$ and $\mu^\star = 6\%$

$\tilde{x}_i$	λ_i^-	λ_i^+	$\tilde{\sigma}_i$	$\tilde{\rho}_{i,j}$			
40.000	0.000	0.125	15.81	100.00			
30.000	0.000	0.000	20.00	13.44	100.00		
20.000	0.000	0.000	25.00	41.11	70.00	100.00	
10.000	1.460	0.000	24.66	23.47	19.06	73.65	100.00

Remark 18 *The previous framework could be used in the case of the tangency portfolio. We have:*

$$\begin{aligned} x^\star &= \arg\max \frac{x^\top (\mu - r\mathbf{1})}{\sqrt{x^\top \Sigma x}} \\ \text{u.c.} & \left\{ \begin{array}{l} \mathbf{1}^\top x = 1 \\ x \in \mathcal{C} \end{array} \right. \end{aligned}$$

We have seen that this portfolio is also the solution of a ϕ-problem. We can then show that the tangency portfolio $\tilde{x}$ for the constrained problem with $\mathcal{C} = \mathcal{C}(x^-, x^+)$ is the solution of the unconstrained problem:

$$\tilde{x} = x^\star \left(\tilde{\mu}, \tilde{\Sigma}, \tilde{\phi}\right)$$

[46]See results in Table 1.2 on page 8.

where $\tilde{\phi}$ is the optimal value of ϕ for the constrained optimization program. The implied parameters $\tilde{\mu}$ and $\tilde{\Sigma}$ are the same as previously (Equation (1.28) on page 66). Let us consider the special case where all the assets have the same Sharpe ratio (Martellini, 2008), that is when expected excess return is proportional to volatility. This tangency portfolio is known as the MSR portfolio. Using the previous example, the solution is:

$$x^{\star} = \begin{pmatrix} 51.197\% \\ 50.784\% \\ -21.800\% \\ 19.818\% \end{pmatrix}$$

If we impose that $10\% \leq x_i \leq 40\%$, we obtain the results in Table 1.20. These results are interesting because they illustrate how imposing weight constraints may modify the underlying assumptions of portfolio theory. In the case of the MSR portfolio, the central assumption is that all the assets have the same Sharpe ratio. However, this assumption is only true in the unconstrained problem. If we impose some weight constraints, it is obvious that this assumption does not hold. The question is how far is the optimized portfolio from the key assumption. If we consider the optimized portfolio given in Table 1.20 and if we assume that the Sharpe ratio is 0.50 for all the assets, the implied Sharpe ratio does not change for the second and fourth assets, but is respectively equal to 0.44 and 0.53 for the first and third assets.

TABLE 1.20: MSR portfolio when $10\% \leq x_i \leq 40\%$

$\tilde{x}_i$	λ_i^-	λ_i^+	$\tilde{\sigma}_i$	$\tilde{\rho}_{i,j}$			
40.000	0.000	0.342	17.13	100.00			
39.377	0.000	0.000	20.00	18.75	100.00		
10.000	0.390	0.000	23.39	36.25	66.49	100.00	
10.623	0.000	0.000	30.00	50.44	40.00	79.96	100.00

In conclusion of this paragraph, we note that by using weight constraints, the portfolio manager may change (implicitly):

1. the value and/or the ordering of the volatilities;
2. the value, the sign and/or the ordering of the correlations; and
3. the underlying assumption of the theory itself.

One major issue concerns the myopic behavior of the portfolio manager, because he is generally not aware of and in agreement with these changes.

Chapter 2

Risk Budgeting Approach

By using shrinkage methods or by imposing weight constraints, we introduce discretionary decisions on the portfolio solution. In this case, the performance of an allocation method may be strongly different depending on the choice of the constraints. In the end, we don't know if one allocation method is better than another, because of the portfolio method itself or because of the constraints. In this chapter, we present an allocation method requiring less discretionary inputs, namely the risk budgeting approach or the '*risk parity*' approach. The term risk parity is the technical term[1] used in the asset management industry.

The risk parity is one of the three budgeting methods in asset allocation. The other two are the weight budgeting and the performance budgeting approaches. Figure 2.1 illustrates the differences between these allocation methods. In the weight budgeting (WB) portfolio, we directly define the weights. For example, we consider two assets and a 30/70 policy rule, meaning that the weight of asset one is 30% and the weight of asset two is 70%. In the risk budgeting (RB) approach, we choose the risk budgets of the assets. With the 30/70 policy rule, and if we want the risk measure of the portfolio to be equal to 20%, this implies that the risk budget of asset one is 6% and the risk budget of asset two is 14%. The performance budgeting (PB) approach consists in calibrating the weights of the portfolio in order to achieve some performance contributions. In our example, if we target a portfolio return of 10%, we would like the performance contribution for the assets to be 3% and 7% respectively. In this chapter, we only focus our analysis on the risk budgeting method, but we will see that it is related to the other two approaches.

This chapter is organized as follows. In the first section, we present the Euler decomposition of a convex risk measure and apply this allocation principle to volatility, value-at-risk and expected shortfall. We study the properties of risk budgeting portfolios in the second section. The third section is dedicated to the ERC portfolio. Subsequently, we compare the risk budgeting approach with the weight budgeting approach. Finally, we extend the notion of risk parity portfolios when we analyze the risk measure with respect to risk factors.

[1] or the '*marketing*' term.

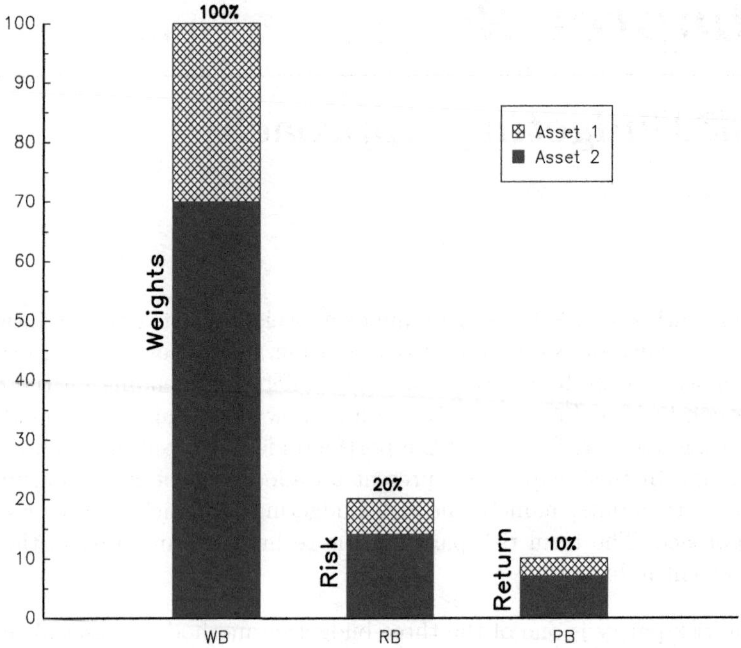

FIGURE 2.1: Three budgeting methods with a 30/70 policy rule

2.1 Risk allocation principle

Measuring the risk of a portfolio is very different from managing it. In particular, a risk measure is a single number that is not very helpful for understanding the diversification of the portfolio. To go further, we must define precisely the notion of risk contribution in order to propose risk allocation principles. We note that most of the materials presented here are not specific to asset management and have been generally developed for capital allocation.

2.1.1 Properties of a risk measure

Let $\mathcal{R}(x)$ be the risk measure of portfolio x. In this section, we define the different properties that should satisfy risk measure $\mathcal{R}(x)$ in order to be acceptable in terms of a risk allocation principle.

2.1.1.1 Coherency and convexity of risk measures

Following Artzner *et al.* (1999), $\mathcal{R}$ is said to be '*coherent*' if it satisfies the following properties:

1. Subadditivity

$$\mathcal{R}(x_1 + x_2) \leq \mathcal{R}(x_1) + \mathcal{R}(x_2)$$

The risk of two portfolios should be less than adding the risk of the two separate portfolios.

2. Homogeneity

$$\mathcal{R}(\lambda x) = \lambda \mathcal{R}(x) \quad \text{if } \lambda \geq 0$$

Leveraging or deleveraging of the portfolio increases or decreases the risk measure in the same magnitude.

3. Monotonicity

$$\text{if } x_1 \prec x_2, \text{ then } \mathcal{R}(x_1) \geq \mathcal{R}(x_2)$$

If portfolio x_2 has a better return than portfolio x_1 under all scenarios, risk measure $\mathcal{R}(x_1)$ should be higher than risk measure $\mathcal{R}(x_2)$.

4. Translation invariance

$$\text{if } m \in \mathbb{R}, \text{ then } \mathcal{R}(x + m) = \mathcal{R}(x) - m$$

Adding a cash position of amount m to the portfolio reduces the risk by m.

The definition of coherent risk measures led to a considerable interest in the quantitative management of risk. Thus, Föllmer and Schied (2002) propose to replace the homogeneity and subadditivity conditions by a weaker condition called the convexity property:

$$\mathcal{R}(\lambda x_1 + (1 - \lambda) x_2) \leq \lambda \mathcal{R}(x_1) + (1 - \lambda) \mathcal{R}(x_2)$$

This condition means that diversification should not increase the risk.

By definition, the loss of the portfolio is $L(x) = -R(x)$ where $R(x)$ is the return of the portfolio. We consider then different risk measures:

- Volatility of the loss

$$\mathcal{R}(x) = \sigma(L(x)) = \sigma(x)$$

The volatility of the loss is the portfolio's volatility.

- Standard deviation-based risk measure

$$\mathcal{R}(x) = \mathrm{SD}_c(x) = \mathbb{E}[L(x)] + c \cdot \sigma(L(x)) = -\mu(x) + c \cdot \sigma(x)$$

To obtain this measure, we scale the volatility by factor $c > 0$ and subtract the expected return of the portfolio.

- Value-at-risk

$$\mathcal{R}(x) = \text{VaR}_\alpha(x) = \inf\{\ell : \Pr\{L(x) \leq \ell\} \geq \alpha\}$$

The value-at-risk is the α-quantile of the loss distribution $\mathbf{F}$ and we note it $\mathbf{F}^{-1}(\alpha)$.

- Expected shortfall

$$\mathcal{R}(x) = \text{ES}_\alpha(x) = \frac{1}{1-\alpha}\int_\alpha^1 \text{VaR}_u(x)\, du$$

The expected shortfall is the average of the VaRs at level α and higher (Acerbi and Tasche, 2002). We note that it is also equal to the expected loss given that the loss is beyond the value-at-risk:

$$\text{ES}_\alpha(x) = \mathbb{E}[L(x) \mid L(x) \geq \text{VaR}_\alpha(x)]$$

We can show that the standard deviation-based risk measure and the expected shortfall satisfy the previous coherency and convexity conditions. For the value-at-risk, the subadditivity property does not hold in general. This is a problem because the portfolio's risk may have be meaningful in this case. More curiously, the volatility is not a coherent risk measure because it does not verify the translation invariance axiom. However, such axioms were designed based on an economic (and regulatory) capital allocation perspective for the banking system. The translation invariance axiom is not well adapted for portfolio management. This is the reason why we consider that volatility is a coherent and convex risk measure.

Let us assume that the asset returns are normally distributed: $R \sim \mathcal{N}(\mu, \Sigma)$. We have $\mu(x) = x^\top \mu$ and $\sigma(x) = \sqrt{x^\top \Sigma x}$. It follows that the standard deviation-based risk measure is:

$$\text{SD}_c(x) = -x^\top \mu + c \cdot \sqrt{x^\top \Sigma x} \tag{2.1}$$

For the value-at-risk, we have $\Pr\{L(x) \leq \text{VaR}_\alpha(x)\} = \alpha$. Because $L(x) = -R(x)$ and $\Pr\{R(x) \geq -\text{VaR}_\alpha(x)\} = \alpha$, we deduce that:

$$\Pr\left\{\frac{R(x) - x^\top \mu}{\sqrt{x^\top \Sigma x}} \leq \frac{-\text{VaR}_\alpha(x) - x^\top \mu}{\sqrt{x^\top \Sigma x}}\right\} = 1 - \alpha$$

It follows that:

$$\frac{-\text{VaR}_\alpha(x) - x^\top \mu}{\sqrt{x^\top \Sigma x}} = \Phi^{-1}(1-\alpha)$$

We finally obtain:

$$\text{VaR}_\alpha(x) = -x^\top \mu + \Phi^{-1}(\alpha)\sqrt{x^\top \Sigma x} \tag{2.2}$$

This is a special case of the standard deviation-based risk measure with $c = \Phi^{-1}(\alpha)$. It implies that the value-at-risk is a coherent and convex risk measure if the asset returns are normally distributed. The expression of the expected shortfall is:

$$\mathrm{ES}_\alpha(x) = \frac{1}{1-\alpha}\int_{-\mu(x)+\sigma(x)\Phi^{-1}(\alpha)}^{\infty} \frac{u}{\sigma(x)\sqrt{2\pi}} \exp\left(-\frac{1}{2}\left(\frac{u+\mu(x)}{\sigma(x)}\right)^2\right) \mathrm{d}u$$

With the variable change $t = \sigma(x)^{-1}(u + \mu(x))$, we obtain:

$$\begin{aligned}
\mathrm{ES}_\alpha(x) &= \frac{1}{1-\alpha}\int_{\Phi^{-1}(\alpha)}^{\infty} (-\mu(x) + \sigma(x)t) \frac{1}{\sqrt{2\pi}} \exp\left(-\frac{1}{2}t^2\right) \mathrm{d}t \\
&= -\frac{\mu(x)}{1-\alpha}\left[\Phi(t)\right]_{\Phi^{-1}(\alpha)}^{\infty} + \frac{\sigma(x)}{(1-\alpha)\sqrt{2\pi}}\int_{\Phi^{-1}(\alpha)}^{\infty} t\exp\left(-\frac{1}{2}t^2\right) \mathrm{d}t \\
&= -\mu(x) + \frac{\sigma(x)}{(1-\alpha)\sqrt{2\pi}}\left[-\exp\left(-\frac{1}{2}t^2\right)\right]_{\Phi^{-1}(\alpha)}^{\infty} \\
&= -\mu(x) + \frac{\sigma(x)}{(1-\alpha)\sqrt{2\pi}} \exp\left(-\frac{1}{2}\left[\Phi^{-1}(\alpha)\right]^2\right)
\end{aligned}$$

The expected shortfall of portfolio x is then:

$$\mathrm{ES}_\alpha(x) = -x^\top \mu + \frac{\sqrt{x^\top \Sigma x}}{(1-\alpha)}\phi\left(\Phi^{-1}(\alpha)\right) \tag{2.3}$$

Like the value-at-risk, it is a standard deviation-based risk measure with $c = \phi\left(\Phi^{-1}(\alpha)\right)/(1-\alpha)$.

In a Gaussian world, the different risk measures are then based on the volatility risk measure. Moreover, if we neglect the term of expected return, they are equivalent. Generally, the portfolio manager builds the allocation in order to have a positive return, meaning that $x^\top \mu \geq 0$. If the portfolio manager has very optimistic forecasts, component $x^\top \mu$ may substantially reduce the risk measure. This explains why omitting the mean component is standard practice in the asset management industry.

Example 6 *We consider three stocks A, B and C. Their current prices are respectively* 244, 135 *and* 315 *dollars. We assume that their expected returns are equal to* 50 *bps,* 30 *bps and* 20 *bps on a daily basis, whereas their daily volatilities are* 2%, 3% *and* 1%. *The correlation of asset returns is given by the following matrix:*

$$\rho = \begin{pmatrix} 1.00 & & \\ 0.50 & 1.00 & \\ 0.25 & 0.60 & 1.00 \end{pmatrix}$$

We consider Portfolio #1 composed of two stocks A, one stock B and one stock C. The value of the portfolio is then equal to 938 dollars. We deduce that the weights x are $(52.03\%, 14.39\%, 33.58\%)$. Computation gives $\mu(x) = 37.0$ bps and $\sigma(x) = 1.476\%$. In Table 2.1, we report the values taken by the value-at-risk and the expected shortfall for different confidence levels α. Because these risk measures are computed using Formulas (2.2) and (2.3) with the portfolio weights, the values are expressed as a percentage. For example, the value-at-risk of the portfolio for $\alpha = 99\%$ is equal to[2]:

$$\mathrm{VaR}_{99\%}(x) = -0.370\% + 2.326 \times 1.476\% = 3.06\%$$

For the expected shortfall, we obtain[3]:

$$\mathrm{ES}_{99\%}(x) = -0.370\% + 2.667 \times 1.476\% = 3.56\%$$

TABLE 2.1: Computation of risk measures $\mathrm{VaR}_{\alpha}(x)$ and $\mathrm{ES}_{\alpha}(x)$

Portfolio	$\mathcal{R}(x)$		α 90%	95%	99%	99.5%
#1	VaR	(in %)	1.52	2.06	3.06	3.43
		(in \$)	14.27	19.30	28.74	32.20
	ES	(in %)	2.22	2.67	3.56	3.90
		(in \$)	20.83	25.09	33.44	36.58
#2	VaR	(in %)	5.68	7.45	10.76	11.98
		(in \$)	14.94	19.59	28.31	31.50
	ES	(in %)	7.98	9.48	12.41	13.52
		(in \$)	21.00	24.94	32.64	35.54

We can also express the risk measures in dollars. In this case, it suffices to replace the weights x by the nominal exposures in the previous formulas. In our example, these nominal exposures are (\$488, \$135, \$315). For $\alpha = 99\%$, the value-at-risk (resp. the expected shortfall) is equal to 28.74 dollars (resp. 33.44 dollars). Another way to compute the risk measure in dollars is to note that the loss expressed in nominal terms is:

$$\begin{aligned} L(x) &= P_{t+1}(x) - P_t(x) \\ &= P_t(x) \cdot (1 + R_{t+1}(x)) - P_t(x) \\ &= P_t(x) \cdot R_{t+1}(x) \end{aligned}$$

where $P_t(x)$ is the value of the portfolio at time t. By using the homogeneity property, we deduce that the risk measure in dollars is equal to the product of the risk measure in % by the current value of the portfolio. Thus we verify

[2] We have $\Phi^{-1}(99\%) = 2.326$.
[3] Because $\phi(\Phi^{-1}(99\%)) = \phi(2.326) = 2.667\%$.

that $3.06\% \times 938 = \$28.74$. We now consider Portfolio #2 composed of two stocks A, one stock C and a short position in stock B. The results are also given in Table 2.1. We note that the second portfolio is more risky than the first portfolio in relative terms. However, they present a similar risk in dollars. The problem is that measuring the risk in percent is not appropriate for long-short portfolios. Suppose for example that the long position matches perfectly the short position in dollars. In this case, the value of the portfolio is zero and the relative risk measure is infinite. This is why it is better to reserve the computation of risk measures in percent to long-only portfolios.

2.1.1.2 Euler allocation principle

Measuring risk is only the first step of portfolio risk management. It must be completed by a second step consisting in decomposing the risk portfolio into a sum of risk contributions by assets. This second step is also called *risk allocation* (Litterman, 1996). The concept of risk contribution is key in identifying concentrations and understanding the risk profile of the portfolio, and there are different methods for defining them. As illustrated by Denault (2001), some methods are more pertinent than others and the Euler principle is certainly the most used and accepted one.

Let Π be the P&L of the portfolio. We decompose it as the sum of the n asset P&L:

$$\Pi = \sum_{i=1}^{n} \Pi_i$$

We note $\mathcal{R}(\Pi)$ the risk measure[4] associated with the P&L. Let us consider the risk-adjusted performance measure (RAPM) defined by:

$$\text{RAPM}(\Pi) = \frac{\mathbb{E}[\Pi]}{\mathcal{R}(\Pi)}$$

Tasche (2008) considers the portfolio-related RAPM of the ith asset defined by:

$$\text{RAPM}(\Pi_i \mid \Pi) = \frac{\mathbb{E}[\Pi_i]}{\mathcal{R}(\Pi_i \mid \Pi)}$$

Based on the notion of RAPM, Tasche (2008) states two properties of risk contributions that are desirable from an economic point of view:

1. Risk contributions $\mathcal{R}(\Pi_i \mid \Pi)$ to portfolio-wide risk $\mathcal{R}(\Pi)$ satisfy the full allocation property if:

$$\sum_{i=1}^{n} \mathcal{R}(\Pi_i \mid \Pi) = \mathcal{R}(\Pi) \tag{2.4}$$

[4]In the case where the risk measure is defined by the loss of the portfolio, we use the relationship $\Pi = -L$.

2. Risk contributions $\mathcal{R}(\Pi_i \mid \Pi)$ are RAPM compatible if there are some $\varepsilon_i > 0$ such that[5]:

$$\text{RAPM}(\Pi_i \mid \Pi) > \text{RAPM}(\Pi) \Rightarrow \text{RAPM}(\Pi + h\Pi_i) > \text{RAPM}(\Pi) \tag{2.5}$$

for all $0 < h < \varepsilon_i$.

Tasche (2008) shows therefore that if there are risk contributions that are RAPM compatible in the sense of the two previous properties (2.4) and (2.5), then $\mathcal{R}(\Pi_i \mid \Pi)$ is uniquely determined as:

$$\mathcal{R}(\Pi_i \mid \Pi) = \frac{\mathrm{d}}{\mathrm{d}h}\mathcal{R}(\Pi + h\Pi_i)\bigg|_{h=0} \tag{2.6}$$

and the risk measure is homogeneous of degree 1. In the case of a subadditive risk measure, one can also show that:

$$\mathcal{R}(\Pi_i \mid \Pi) \leq \mathcal{R}(\Pi_i) \tag{2.7}$$

This means that the risk contribution of asset i is always smaller than its stand-alone risk measure. The difference is related to the risk diversification.

Let us return to risk measure $\mathcal{R}(x)$ defined in terms of weights. The previous framework implies that the risk contribution of asset i is uniquely defined as:

$$\mathcal{RC}_i = x_i \frac{\partial\, \mathcal{R}(x)}{\partial\, x_i} \tag{2.8}$$

and the risk measure satisfies the Euler decomposition:

$$\mathcal{R}(x) = \sum_{i=1}^{n} x_i \frac{\partial\, \mathcal{R}(x)}{\partial\, x_i} = \sum_{i=1}^{n} \mathcal{RC}_i \tag{2.9}$$

This relationship is also called the Euler allocation principle. It is the core of risk parity portfolios and is currently used intensively by practitioners.

Remark 19 *We can always define the risk contributions of a risk measure by using Equation (2.8). However, this does not mean that it satisfies the Euler decomposition.*

Remark 20 *Kalkbrener (2005) develops an axiomatic approach to risk contributions. In particular, he shows that the Euler allocation principle is the only risk allocation method compatible with diversification principle (2.7) if the risk measure is subadditive.*

---

[5] This property means that assets with a better risk-adjusted performance than the portfolio continue to have a better RAPM if their allocation increases in a small proportion.

2.1.2 Risk contribution of portfolio assets

2.1.2.1 Computing the risk contributions

In the case of Gaussian asset returns, risk measures such as value-at-risk or expected shortfall are related to the volatility. Let us first consider the two-asset case. We have:

$$\sigma(x) = \sqrt{x_1^2\sigma_1^2 + 2x_1x_2\rho\sigma_1\sigma_2 + x_2^2\sigma_2^2}$$

It follows that the marginal risk of the first asset is:

$$\begin{aligned}\frac{\partial\,\sigma(x)}{\partial\,x_1} &= \frac{2x_1\sigma_1^2 + 2x_2\rho\sigma_1\sigma_2}{2\sqrt{x_1^2\sigma_1^2 + 2x_1x_2\rho\sigma_1\sigma_2 + x_2^2\sigma_2^2}} \\ &= \frac{x_1\sigma_1^2 + x_2\rho\sigma_1\sigma_2}{\sqrt{x_1^2\sigma_1^2 + 2x_1x_2\rho\sigma_1\sigma_2 + x_2^2\sigma_2^2}} \\ &= \frac{\operatorname{cov}(R_1, R(x))}{\operatorname{var}(R(x))}\end{aligned}$$

We then deduce that the risk contribution of the first asset is:

$$\begin{aligned}\mathcal{RC}_1 &= x_1\frac{\partial\,\sigma(x)}{\partial\,x_1} \\ &= \frac{x_1^2\sigma_1^2 + x_1x_2\rho\sigma_1\sigma_2}{\sqrt{x_1^2\sigma_1^2 + 2x_1x_2\rho\sigma_1\sigma_2 + x_2^2\sigma_2^2}}\end{aligned}$$

We verify that the sum of the two risk contributions is equal to the portfolio's volatility:

$$\begin{aligned}\mathcal{RC}_1 + \mathcal{RC}_2 &= \frac{x_1^2\sigma_1^2 + x_1x_2\rho\sigma_1\sigma_2}{\sqrt{x_1^2\sigma_1^2 + 2x_1x_2\rho\sigma_1\sigma_2 + x_2^2\sigma_2^2}} + \\ &\quad \frac{x_2x_1\rho\sigma_1\sigma_2 + x_2^2\sigma_2^2}{\sqrt{x_1^2\sigma_1^2 + 2x_1x_2\rho\sigma_1\sigma_2 + x_2^2\sigma_2^2}} \\ &= \frac{x_1^2\sigma_1^2 + 2x_1x_2\rho\sigma_1\sigma_2 + x_2^2\sigma_2^2}{\sqrt{x_1^2\sigma_1^2 + 2x_1x_2\rho\sigma_1\sigma_2 + x_2^2\sigma_2^2}} \\ &= \sqrt{x_1^2\sigma_1^2 + 2x_1x_2\rho\sigma_1\sigma_2 + x_2^2\sigma_2^2} \\ &= \sigma(x)\end{aligned}$$

The previous formulas can be extended to the case $n > 2$. Because $\sigma(x) = \sqrt{x^\top \Sigma x}$, it follows that the vector of marginal volatilities is:

$$\begin{aligned}\frac{\partial\,\sigma(x)}{\partial\,x} &= \frac{1}{2}\left(x^\top \Sigma x\right)^{-1}(2\Sigma x) \\ &= \frac{\Sigma x}{\sqrt{x^\top \Sigma x}}\end{aligned}$$

The risk contribution of the ith asset is then:

$$\mathcal{RC}_i = x_i \cdot \frac{(\Sigma x)_i}{\sqrt{x^\top \Sigma x}}$$

Like the two-asset case, we verify the full allocation property:

$$\begin{aligned}\sum_{i=1}^{n} \mathcal{RC}_i &= \sum_{i=1}^{n} x_i \cdot \frac{(\Sigma x)_i}{\sqrt{x^\top \Sigma x}} \\ &= x^\top \frac{\Sigma x}{\sqrt{x^\top \Sigma x}} \\ &= \sqrt{x^\top \Sigma x} \\ &= \sigma(x)\end{aligned}$$

We can then deduce the risk contribution of the three risk measures $\mathrm{SD}_c(x)$, $\mathrm{VaR}_\alpha(x)$ and $\mathrm{ES}_\alpha(x)$. For the standard deviation-based risk measure $\mathrm{SD}_c(x)$, we have:

$$\mathcal{RC}_i = x_i \cdot \left(-\mu_i + c\frac{(\Sigma x)_i}{\sqrt{x^\top \Sigma x}}\right) \tag{2.10}$$

In the case of the value-at-risk, the risk contribution becomes:

$$\mathcal{RC}_i = x_i \cdot \left(-\mu_i + \Phi^{-1}(\alpha)\frac{(\Sigma x)_i}{\sqrt{x^\top \Sigma x}}\right) \tag{2.11}$$

whereas in the case of the expected shortfall, it is:

$$\mathcal{RC}_i = x_i \cdot \left(-\mu_i + \frac{(\Sigma x)_i}{(1-\alpha)\sqrt{x^\top \Sigma x}}\phi\left(\Phi^{-1}(\alpha)\right)\right) \tag{2.12}$$

Example 7 *We consider three assets. We assume that their expected returns are equal to zero whereas their volatilities are equal to 30%, 20% and 15%. The correlation of asset returns is given by the following matrix:*

$$\rho = \begin{pmatrix} 1.00 & & \\ 0.80 & 1.00 & \\ 0.50 & 0.30 & 1.00 \end{pmatrix}$$

Let us consider the portfolio $(50\%, 20\%, 30\%)$ in the case of Example 7. Using the relationship $\Sigma_{i,j} = \rho_{i,j}\sigma_i\sigma_j$, we deduce that the covariance matrix is[6]:

$$\Sigma = \begin{pmatrix} 9.00 & 4.80 & 2.25 \\ 4.80 & 4.00 & 0.90 \\ 2.25 & 0.90 & 2.25 \end{pmatrix} \times 10^{-2}$$

[6]The covariance term between assets 1 and 2 is equal to $\Sigma_{1,2} = 80\% \times 30\% \times 20\% = 4.80\%$.

It follows that the variance of the portfolio is:

$$\begin{aligned}\sigma^2(x) &= 0.50^2 \times 0.09 + 0.20^2 \times 0.04 + 0.30^2 \times 0.0225 + \\ &\quad 2 \times 0.50 \times 0.20 \times 0.0480 + 2 \times 0.50 \times 0.30 \times 0.0225 + \\ &\quad 2 \times 0.20 \times 0.30 \times 0.0090 \\ &= 4.3555\%\end{aligned}$$

The volatility is then $\sigma(x) = \sqrt{4.3555\%} = 20.8698\%$. The computation of the marginal volatilities gives:

$$\frac{\Sigma x}{\sqrt{x^\top \Sigma x}} = \frac{1}{20.8698\%}\begin{pmatrix} 6.1350\% \\ 3.4700\% \\ 1.9800\% \end{pmatrix} = \begin{pmatrix} 29.3965\% \\ 16.6269\% \\ 9.4874\% \end{pmatrix}$$

Finally, we obtain the risk contributions by multiplying the weights by the marginal volatilities:

$$x \circ \frac{\Sigma x}{\sqrt{x^\top \Sigma x}} = \begin{pmatrix} 50\% \\ 20\% \\ 30\% \end{pmatrix} \circ \begin{pmatrix} 29.3965\% \\ 16.6269\% \\ 9.4874\% \end{pmatrix} = \begin{pmatrix} 14.6982\% \\ 3.3254\% \\ 2.8462\% \end{pmatrix}$$

We verify that the sum of risk contributions is equal to the volatility:

$$\sum_{i=1}^{3} \mathcal{RC}_i = 14.6982\% + 3.3254\% + 2.8462\% = 20.8698\%$$

TABLE 2.2: Risk decomposition of the volatility

Asset	x_i	$\mathcal{MR}_i$	$\mathcal{RC}_i$	$\mathcal{RC}_i^\star$
1	50.00	29.40	14.70	70.43
2	20.00	16.63	3.33	15.93
3	30.00	9.49	2.85	13.64
$\mathcal{R}(x)$			20.87	

These different results[7] are summarized in Table 2.2. For the value-at-risk or the expected shortfall, the marginal risks and the risk contributions are scaled by the corresponding value of c. We obtain results given in Table 2.3 and Table 2.4. We note that the first asset represents 50% of the weight, but about 70% of the portfolio's risk. Similarly, the third asset has a weight of 30%, but only a relative risk contribution of 13.64%. Contrary to the first asset, it generates less risk than its weight.

[7] $\mathcal{MR}_i$ and $\mathcal{RC}_i$ correspond to the marginal risk and the risk contribution of asset i whereas $\mathcal{RC}_i^\star$ is the risk contribution expressed in percent of the risk measure:

$$\mathcal{RC}_i^\star = \frac{\mathcal{RC}_i}{\mathcal{R}(x)}$$

TABLE 2.3: Risk decomposition of the value-at-risk

Asset	x_i	$\mathcal{MR}_i$	$\mathcal{RC}_i$	$\mathcal{RC}_i^\star$
1	50.00	68.39	34.19	70.43
2	20.00	38.68	7.74	15.93
3	30.00	22.07	6.62	13.64
$\mathcal{R}(x)$			48.55	

TABLE 2.4: Risk decomposition of the expected shortfall

Asset	x_i	$\mathcal{MR}_i$	$\mathcal{RC}_i$	$\mathcal{RC}_i^\star$
1	50.00	78.35	39.17	70.43
2	20.00	44.31	8.86	15.93
3	30.00	25.29	7.59	13.64
$\mathcal{R}(x)$			55.62	

Remark 21 *Even if the risk measure is coherent and convex, it does not necessarily satisfy the Euler allocation principle. The most famous example is the variance of the portfolio. We have* $\operatorname{var}(x) = x^\top \Sigma x$ *and* $\partial_x \operatorname{var}(x) = 2\Sigma x$. *It follows that* $\sum_{i=1}^n x_i \cdot \partial_x \operatorname{var}(x) = \sum_{i=1}^n x_i \cdot (2\Sigma x)_i = 2x^\top \Sigma x = 2\operatorname{var}(x) > \operatorname{var}(x)$. *In the case of the portfolio variance, the sum of the risk contributions is then always larger than the risk measure itself.*

2.1.2.2 Interpretation of risk contributions

There are two ways of interpreting the risk contributions. The first one is based on marginal analysis and was proposed by Litterman (1996) and Garman (1997). This approach is related to the sensitivity analysis of the risk measure. Indeed, the marginal risk of asset i is defined by:

$$\frac{\partial\,\mathcal{R}(x)}{\partial\,x_i} = \lim_{h \to 0} \frac{\mathcal{R}(x + h\mathbf{e}_i) - \mathcal{R}(x)}{h}$$

If h is small, we have:

$$\mathcal{R}(x + h\mathbf{e}_i) \simeq \mathcal{R}(x) + h\frac{\partial\,\mathcal{R}(x)}{\partial\,x_i} \tag{2.13}$$

This means that if we increase the weight of asset i by a small amount h, the risk measure is increased by the product of h and the marginal risk. The second way to interpret the risk contributions considers performance analysis. If we define the utility function by $\mathcal{U}(x) = \mathbb{E}[R(x)] - \frac{1}{2}\phi\mathcal{R}(x)$, the optimized portfolio satisfies:

$$\mu_i = \frac{\phi}{2}\frac{\partial\,\mathcal{R}(x)}{\partial\,x_i}$$

This means that performance contributions are equal to risk contributions:

$$\frac{x_i \cdot \mu_i}{\sum_{i=1}^n x_i \cdot \mu_i} = \frac{x_i \cdot \partial_{x_i} \mathcal{R}(x)}{\sum_{i=1}^n x_i \cdot \partial_{x_i} \mathcal{R}(x)} \tag{2.14}$$

There is then a duality between return and risk. The performance analysis has its roots in the Black-Litterman model (or the CAPM). However, Formula (2.14) involves an equilibrium argument. In the case of the value-at-risk, we can obtain the duality without this argument. Qian (2006) notes that risk contributions are related to loss contributions[8]:

$$\mathcal{RC}_i = \mathbb{E}\left[L_i \mid L = \mathrm{VaR}_\alpha(L)\right]$$

The risk contribution of the ith asset is then the loss contribution of the ith asset when the portfolio loss reaches the value-at-risk.

Let us go back to Example 7. The marginal volatility for the first asset is:

$$\frac{\partial\,\sigma(x)}{\partial\,x_1} = \lim_{h\to 0} \frac{\sigma(x_1+h, x_2, x_3) - \sigma(x_1, x_2, x_3)}{(x_1+h) - x_1}$$

If $h = 1\%$, we have:

$$\frac{\partial\,\sigma(x)}{\partial\,x_1} \simeq \frac{21.1639\% - 20.8698\%}{1\%} = 29.4050\%$$

If $h = 0.1\%$, we have:

$$\frac{\partial\,\sigma(x)}{\partial\,x_1} \simeq \frac{20.8992\% - 20.8698\%}{0.1\%} = 29.3974\%$$

If $h = 0.01\%$, we have:

$$\frac{\partial\,\sigma(x)}{\partial\,x_1} \simeq \frac{20.8728\% - 20.8698\%}{0.01\%} = 29.3966\%$$

We verify that the finite difference converges to the true value which is 29.3965%. In Table 2.5, we have measured the impact of h on $\mathcal{R}(x + h\mathbf{e}_i)$. For example, the volatility is equal to 21.1639% if h is set to 1% for the first asset. In Table 2.6, we have reported the approximated value of $\mathcal{R}(x + h\mathbf{e}_i)$ given by Equation (2.13). For example, the approximated value of the volatility is equal to 21.1638% if h is set to 1% for the first asset. It is therefore very close to the true value.

Remark 22 *Marginal analysis is key to understanding micro-economic analysis, in particular general equilibrium between production and consumption (Varian, 1992). In this perspective, the allocation of scarce resources is generally based on the use of marginal quantities such as marginal productivity, marginal cost, marginal utility, marginal consumption, etc.*

[8]This result will be explained in the next section.

TABLE 2.5: Sensitivity analysis of the volatility with respect to the factor h

Asset	1 bp	10 bp	1%	10%	50%	$-x_i$
1	20.8728	20.8992	21.1639	23.8170	35.6938	6.8593
2	20.8715	20.8865	21.0364	22.5599	29.7077	17.6847
3	20.8708	20.8793	20.9650	21.8495	26.2640	18.3576

TABLE 2.6: Marginal analysis of the volatility with respect to the factor h

Asset	1 bp	10 bp	1%	10%	50%	$-x_i$
1	20.8728	20.8992	21.1638	23.8095	35.5681	6.1716
2	20.8715	20.8865	21.0361	22.5325	29.1833	17.5445
3	20.8708	20.8793	20.9647	21.8186	25.6135	18.0236

2.1.3 Application to non-normal risk measures

2.1.3.1 Non-normal value-at-risk and expected shortfall

In the previous section, we provided formulas for when asset returns are Gaussian. However, the previous expressions can be extended in the general case. For the value-at-risk, Gourieroux *et al.* (2000) show that the risk contribution is equal to[9]:

$$\begin{aligned}\mathcal{RC}_i &= \mathcal{R}(\Pi_i \mid \Pi) \\ &= -\mathbb{E}\left[\Pi_i \mid \Pi = -\operatorname{VaR}_\alpha(\Pi)\right] \\ &= \mathbb{E}\left[L_i \mid L(x) = \operatorname{VaR}_\alpha(L)\right] \qquad (2.15)\end{aligned}$$

Formula (2.15) is more general than Expression (2.11) obtained in the Gaussian case. Indeed, we can retrieve the latter if we assume that the returns are Gaussian. We recall that the portfolio return is $R(x) = \sum_{i=1}^n x_i R_i = x^\top R$. The portfolio loss is defined by $L(x) = -R(x)$. We deduce that:

$$\begin{aligned}\mathcal{RC}_i &= \mathbb{E}\left[-x_i R_i \mid -R(x) = \operatorname{VaR}_\alpha(x)\right] \\ &= -x_i \cdot \mathbb{E}\left[R_i \mid R(x) = -\operatorname{VaR}_\alpha(x)\right]\end{aligned}$$

Because $R(x)$ is a linear combination of R, the random vector $(R, R(x))$ is Gaussian and we have:

$$\begin{pmatrix} R \\ R(x) \end{pmatrix} \sim \mathcal{N}\left(\begin{pmatrix} \mu \\ x^\top \mu \end{pmatrix}, \begin{pmatrix} \Sigma & \Sigma x \\ x^\top \Sigma & x^\top \Sigma x \end{pmatrix}\right)$$

[9]See also Hallerbach (2003).

We know that $\mathrm{VaR}_\alpha(x) = -x^\top \mu + \Phi^{-1}(\alpha)\sqrt{x^\top \Sigma x}$. It follows that[10]:

$$\begin{aligned}
\mathbb{E}\left[R \mid R(x) = -\mathrm{VaR}_\alpha(x)\right] &= \mathbb{E}\left[R \mid R(x) = x^\top \mu - \Phi^{-1}(\alpha)\sqrt{x^\top \Sigma x}\right] \\
&= \mu + \Sigma x \left(x^\top \Sigma x\right)^{-1} \cdot \\
&\quad \left(x^\top \mu - \Phi^{-1}(\alpha)\sqrt{x^\top \Sigma x} - x^\top \mu\right) \\
&= \mu - \Phi^{-1}(\alpha)\Sigma x \frac{\sqrt{x^\top \Sigma x}}{\left(x^\top \Sigma x\right)^{-1}} \\
&= \mu - \Phi^{-1}(\alpha)\frac{\Sigma x}{\sqrt{x^\top \Sigma x}}
\end{aligned}$$

We obtain finally the same expression as Equation (2.11):

$$\begin{aligned}
\mathcal{RC}_i &= -x_i \cdot \left(\mu - \Phi^{-1}(\alpha)\frac{\Sigma x}{\sqrt{x^\top \Sigma x}}\right)_i \\
&= x_i \cdot \left(-\mu_i + \Phi^{-1}(\alpha)\frac{(\Sigma x)_i}{\sqrt{x^\top \Sigma x}}\right)
\end{aligned}$$

In the same way, Tasche (2002) shows that the general expression of the risk contributions for the expected shortfall is:

$$\begin{aligned}
\mathcal{RC}_i &= \mathcal{R}(\Pi_i \mid \Pi) \\
&= -\mathbb{E}\left[\Pi_i \mid \Pi \leq -\mathrm{VaR}_\alpha(\Pi)\right] \\
&= \mathbb{E}\left[L_i \mid L(x) \geq \mathrm{VaR}_\alpha(L)\right] \qquad (2.16)
\end{aligned}$$

Using Bayes' theorem, it follows that:

$$\mathcal{RC}_i = \frac{\mathbb{E}\left[L_i \cdot \mathbb{1}\left\{L \geq \mathrm{VaR}_\alpha(L)\right\}\right]}{1-\alpha}$$

If we apply the previous formula to the Gaussian case, we obtain:

$$\mathcal{RC}_i = -\frac{x_i}{1-\alpha}\mathbb{E}\left[R_i \cdot \mathbb{1}\left\{R(x) \leq -\mathrm{VaR}_\alpha(L)\right\}\right]$$

We know that the random vector $(R_i, R(x))$ has a bivariate normal distribution with:

$$\begin{pmatrix} R_i \\ R(x) \end{pmatrix} \sim \mathcal{N}\left(\begin{pmatrix} \mu_i \\ x^\top \mu \end{pmatrix}, \begin{pmatrix} \Sigma_{i,i} & (\Sigma x)_i \\ (\Sigma x)_i & x^\top \Sigma x \end{pmatrix}\right)$$

Let $I = \mathbb{E}\left[R_i \cdot \mathbb{1}\left\{R(x) \leq -\mathrm{VaR}_\alpha(L)\right\}\right]$. We note f the density function of the

[10]We use the formula of conditional expectation given in Footnote 13 on page 25.

random vector $(R_i, R(x))$ and $\rho = \Sigma_{i,i}^{-1/2} \left(x^\top \Sigma x\right)^{-1/2} (\Sigma x)_i$ the correlation between R_i and $R(x)$. It follows that:

$$\begin{aligned} I &= \int_{-\infty}^{+\infty} \int_{-\infty}^{+\infty} r \cdot \mathbb{1}\left\{ s \leq -\operatorname{VaR}_\alpha (L) \right\} f(r,s) \,\mathrm{d}r \,\mathrm{d}s \\ &= \int_{-\infty}^{+\infty} \int_{-\infty}^{-\operatorname{VaR}_\alpha (L)} r f(r,s) \,\mathrm{d}r \,\mathrm{d}s \end{aligned}$$

Let $t = (r - \mu_i) / \sqrt{\Sigma_{i,i}}$ and $u = \left(s - x^\top \mu\right) / \sqrt{x^\top \Sigma x}$. We deduce that[11]:

$$I = \int_{-\infty}^{+\infty} \int_{-\infty}^{\Phi^{-1}(1-\alpha)} \frac{\mu_i + \sqrt{\Sigma_{i,i}} t}{2\pi \sqrt{1-\rho^2}} \exp\left(-\frac{t^2 + u^2 - 2\rho t u}{2(1-\rho^2)} \right) \mathrm{d}t \,\mathrm{d}u$$

By considering the change of variables $(t, u) = \varphi(t, v)$ such that $u = \rho t + \sqrt{1-\rho^2} v$, we obtain[12]:

$$\begin{aligned} I &= \int_{-\infty}^{+\infty} \int_{-\infty}^{g(t)} \frac{\mu_i + \sqrt{\Sigma_{i,i}} t}{2\pi} \exp\left(-\frac{t^2 + v^2}{2} \right) \mathrm{d}t \,\mathrm{d}v \\ &= \mu_i \int_{-\infty}^{+\infty} \int_{-\infty}^{g(t)} \frac{1}{2\pi} \exp\left(-\frac{t^2 + v^2}{2} \right) \mathrm{d}t \,\mathrm{d}v + \\ &\quad \sqrt{\Sigma_{i,i}} \int_{-\infty}^{+\infty} \int_{-\infty}^{g(t)} \frac{t}{2\pi} \exp\left(-\frac{t^2 + v^2}{2} \right) \mathrm{d}t \,\mathrm{d}v + \\ &= \mu_i I_1 + \sqrt{\Sigma_{i,i}} I_2 \end{aligned}$$

where the bound $g(t)$ is defined as follows:

$$g(t) = \frac{\Phi^{-1}(1-\alpha) - \rho t}{\sqrt{1-\rho^2}}$$

For the first integral, we have[13]:

$$\begin{aligned} I_1 &= \int_{-\infty}^{+\infty} \frac{1}{\sqrt{2\pi}} \exp\left(-\frac{t^2}{2} \right) \left(\int_{-\infty}^{g(t)} \frac{1}{\sqrt{2\pi}} \exp\left(-\frac{v^2}{2} \right) \mathrm{d}v \right) \mathrm{d}t \\ &= \int_{-\infty}^{+\infty} \Phi\left(\frac{\Phi^{-1}(1-\alpha) - \rho t}{\sqrt{1-\rho^2}} \right) \phi(t) \,\mathrm{d}t \\ &= 1 - \alpha \end{aligned}$$

---

[11] Because we have $\Phi^{-1}(1-\alpha) = -\Phi^{-1}(\alpha)$.

[12] We use the fact that $\mathrm{d}t \,\mathrm{d}v = \sqrt{1-\rho^2} \,\mathrm{d}t \,\mathrm{d}u$ because the determinant of the Jacobian matrix containing the partial derivatives $D\varphi$ is $\sqrt{1-\rho^2}$.

[13] We use the fact that:

$$\mathbb{E}\left[\Phi\left(\frac{\Phi^{-1}(p) - \rho T}{\sqrt{1-\rho^2}} \right) \right] = p$$

where $T \sim \mathcal{N}(0,1)$. This result is well-known in the Basle II framework of credit risk (Roncalli, 2009, page 182).

The computation of the second integral I_2 is slightly more tedious. Integration by parts with the derivative function $t\phi(t)$ gives:

$$\begin{aligned} I_2 &= \int_{-\infty}^{+\infty} \Phi\left(\frac{\Phi^{-1}(1-\alpha)-\rho t}{\sqrt{1-\rho^2}}\right) t\phi(t)\, \mathrm{d}t \\ &= -\frac{\rho}{\sqrt{1-\rho^2}} \int_{-\infty}^{+\infty} \phi\left(\frac{\Phi^{-1}(1-\alpha)-\rho t}{\sqrt{1-\rho^2}}\right) \phi(t)\, \mathrm{d}t \\ &= -\frac{\rho}{\sqrt{1-\rho^2}} \phi\left(\Phi^{-1}(1-\alpha)\right) \int_{-\infty}^{+\infty} \phi\left(\frac{t-\rho\Phi^{-1}(1-\alpha)}{\sqrt{1-\rho^2}}\right) \mathrm{d}t \\ &= -\rho\phi\left(\Phi^{-1}(1-\alpha)\right) \end{aligned}$$

We can then deduce the value of I:

$$\begin{aligned} I &= \mu_i(1-\alpha) - \rho\sqrt{\Sigma_{i,i}}\phi\left(\Phi^{-1}(1-\alpha)\right) \\ &= \mu_i(1-\alpha) - \frac{(\Sigma x)_i}{\sqrt{x^\top \Sigma x}}\phi\left(\Phi^{-1}(\alpha)\right) \end{aligned}$$

We finally obtain that:

$$\mathcal{RC}_i = x_i \cdot \left(-\mu_i + \frac{(\Sigma x)_i}{(1-\alpha)\sqrt{x^\top \Sigma x}}\phi\left(\Phi^{-1}(\alpha)\right)\right)$$

This is the same expression as found previously.

When returns are not Gaussian, we use Equations (2.15) and (2.16) to compute the risk contributions. Because an analytical expression is generally difficult to find, we consider Monte Carlo simulations. Let m and $L_i^{(j)}$ be the number of simulations and the jth simulated loss of the ith asset. The loss of the portfolio for the simulation j is then equal to $L^{(j)} = \sum_{i=1}^n L_i^{(j)}$. We consider the order statistics[14]:

$$\min\left(L^{(j)}\right) = L^{(1:m)} \leq L^{(2:m)} \leq \cdots \leq L^{(m:m)} = \max\left(L^{(j)}\right)$$

The value-at-risk is then equal to the m_α order statistic:

$$\mathrm{VaR}_\alpha(L) = L^{(m_\alpha:m)}$$

where[15] $m_\alpha = m\alpha + 1$. Using Equation (2.15), we deduce that an estimate of the risk contribution is then:

$$\mathcal{RC}_i = L_i^{(m_\alpha:m)}$$

[14]To simplify the expressions, we assume that the inequalities are strict.

[15]We use the financial convention that the $\mathrm{VaR}_{99\%}(L)$ corresponds to the smallest value of 100 sampled P&Ls, that is the largest value of 100 sampled losses. This definition differs from the traditional statistical convention of empirical quantiles.

where $L_i^{(j:m)}$ is the individual loss of the ith asset for the m_α order statistic of the portfolio loss. For the expected shortfall, we have:

$$\begin{aligned}\mathrm{ES}_\alpha(L) &= \frac{1}{m(1-\alpha)}\sum_{j=m_\alpha}^{m} L^{(j:m)} \\ &= \frac{1}{m(1-\alpha)}\sum_{j=1}^{m} L^{(j)} \cdot \mathbb{1}\left\{L^{(j)} \geq \mathrm{VaR}_\alpha(L)\right\}\end{aligned}$$

It corresponds to the average of the losses larger or equal than the value-at-risk. The risk contribution is then computed according to Equation (2.16):

$$\mathcal{RC}_i = \frac{1}{m(1-\alpha)}\sum_{j=m_\alpha}^{m} L_i^{(j:m)}$$

Remark 23 *We note that the risk contribution estimator uses one observation for the value-at-risk and $m(1-\alpha)$ observations for the expected shortfall. This implies that it is less efficient in the case of the value-at-risk. This is the reason why we can use a regularization method (Scaillet, 2004). The idea is to estimate the value-at-risk by weighting the order statistics:*

$$\mathrm{VaR}_\alpha(L) = \sum_{j=1}^{m} \varpi_i^m(\alpha)\, L^{(j:m)}$$

where $\varpi_j^m(\alpha)$ is a weight function dependent on the confidence level α. The expression of the risk contribution then becomes:

$$\mathcal{RC}_i = \sum_{j=1}^{m} \varpi_j^m(\alpha)\, L_i^{(j:m)}$$

Of course, this naive method can be improved by using more sophisticated approaches such as importance sampling (Glasserman, 2005).

Example 8 *We consider a portfolio with two assets. Here are the losses of each asset for 20 observations:*

j	1	2	3	4	5	6	7	8	9	10
$L_1^{(j)}$	14	3	−4	5	6	8	12	25	23	−9
$L_2^{(j)}$	10	−3	8	7	2	17	14	22	−8	−2
j	11	12	13	14	15	16	17	18	19	20
$L_1^{(j)}$	−50	−17	18	−9	−6	−2	0	17	19	1
$L_2^{(j)}$	−10	12	−12	−19	25	−10	4	12	36	−5

We illustrate the previous estimators using the above example. We compute the portfolio loss as the sum of the losses of the two assets:

j	1	2	3	4	5	6	7	8	9	10
$L^{(j)}$	24	0	4	12	8	25	26	47	15	−11
j	11	12	13	14	15	16	17	18	19	20
$L^{(j)}$	−60	−5	6	−28	19	−12	4	29	55	−4

We deduce the order statistics:

j	1	2	3	4	5	6	7	8	9	10
$L^{(j:m)}$	−60	−28	−12	−11	−5	−4	0	4	4	6
j	11	12	13	14	15	16	17	18	19	20
$L^{(j:m)}$	8	12	15	19	24	25	26	29	47	55

For example, the first order statistic corresponds to the minimum value of the different losses. It is equal to −60 and is obtained for the 11th observation. If α is equal to 80%, the value-at-risk corresponds to the 16th order statistic and is equal to 26. We note that this value-at-risk occurs for the 7th observation. We deduce that the risk contributions are $\mathcal{RC}_1 = 12$ and $\mathcal{RC}_2 = 14$. The computation of the expected shortfall is:

$$\mathrm{ES}_\alpha(L) = \frac{26 + 29 + 47 + 55}{4} = 39.25$$

To compute the risk contributions, it is convenient to rank the individual losses according to the order statistics. We have:

j	1	2	3	4	5	6	7	8	9	10
$L_1^{(j:m)}$	−50	−9	−2	−9	−17	1	3	−4	0	18
$L_2^{(j:m)}$	−10	−19	−10	−2	12	−5	−3	8	4	−12
j	11	12	13	14	15	16	17	18	19	20
$L_1^{(j:m)}$	6	5	23	−6	14	8	12	17	25	19
$L_2^{(j:m)}$	2	7	−8	25	10	17	14	12	22	36

We deduce that:

$$\mathcal{RC}_1 = \frac{12 + 17 + 25 + 19}{4} = 18.25$$

and:

$$\mathcal{RC}_2 = \frac{14 + 12 + 22 + 36}{4} = 21.00$$

We verify that the sum of these two risk contributions is equal to the expected shortfall. Let us now assume that the value-at-risk is estimated using a regularization method. We consider the following weight function:

$$\varpi_j^m(\alpha) = \frac{1}{2h+1} \cdot \mathbb{1}\{|j - m_\alpha| \leq h\}$$

It corresponds to a uniform kernel on the range $[m_\alpha - h, m_\alpha + h]$. In our case, we obtain for $h = 1$:

$$\varpi_j^m(\alpha) = \begin{cases} 1/3 & \text{if } j = 16, 17, 18 \\ 0 & \text{elsewhere} \end{cases}$$

Computing the 80% value-at-risk is then equivalent to average the 75%, 80% and 85% quantiles. We obtain $\mathrm{VaR}_\alpha(L) = 26.67$, $\mathcal{RC}_1 = 12.33$ and $\mathcal{RC}_2 = 14.33$.

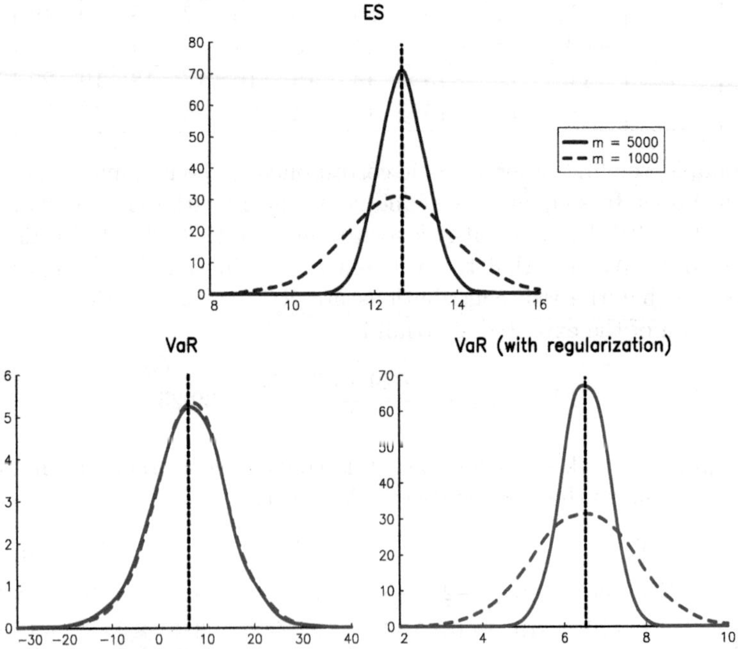

FIGURE 2.2: Density of the risk contribution estimator $\mathcal{RC}_1$

To illustrate the importance of the regularization method, we consider an equally weighted portfolio with two assets. We assume that the asset returns $R_i(t)$ are distributed according to the Student's t distribution:

$$\frac{R_i(t) - \mu_i}{\sigma_i} \sim t_{\nu_i}$$

The dependence function between the asset returns is given by the Clayton copula:

$$\mathbf{C}(u_1, u_2) = \left(u_1^{-\theta} + u_2^{-\theta} - 1\right)^{-1/\theta}$$

For the numerical illustration, we consider the following values: $\mu_1 = 10\%$, $\sigma_1 = 20\%$, $\nu_1 = 6$, $\mu_2 = 10\%$, $\sigma_2 = 25\%$, $\nu_2 = 4$ and $\theta = 2$. The confidence level α is set to 90%. All the results are expressed in percent with respect to the initial value of the portfolio. In Figure 2.2, we have reported the density[16] of the risk contribution estimator for the first asset. We verify that the variance of the estimator is larger for the value-at-risk than for the expected shortfall. Not surprisingly, the number of simulations m has little influence on the variance in the case of the value-at-risk. If we employ the regularization technique with the previous uniform kernel[17], we improve the efficiency of the estimator. For example, we obtain the following results for $m = 5\,000$:

	Vol	VaR	ES	VaR-R
$\mathcal{R}(x)$	24.51	18.32	35.99	18.49
	(0.57)	(0.69)	(1.18)	(0.66)
$\mathcal{RC}_1(x)$	8.95	6.26	12.67	6.51
	(0.23)	(7.65)	(0.55)	(0.54)
$\mathcal{RC}_2(x)$	15.56	12.05	23.32	11.98
	(0.55)	(7.65)	(0.88)	(0.63)

For each estimator, we provide the expected value whereas the figure in parenthesis indicates the standard deviation of the estimator. For the value-at-risk, the risk contribution estimator for the first asset has an expected value and a standard deviation equal to 6.26% and 7.65%. This implies that it is not statistically significant. If we consider the estimator with the regularization technique (VaR-R), these figures become 6.51% and 0.54%. The standard deviation has been reduced by a factor larger than 10.

Remark 24 *In the previous example, $\mathcal{RC}_1(x)$ (measured in percent of the risk measure) is equal respectively to 36.53%, 35.22% and 35.20% for the volatility, the value-at-risk and the expected shortfall. We verify that these different risk contributions are different, because the asset returns are not Gaussian. The non-Gaussian property of asset returns could also be illustrated by noticing that the scaling factor κ defined by:*

$$\kappa = \frac{c\left(\mathrm{ES}_\alpha(L)\right)}{c\left(\mathrm{VaR}_\alpha(L)\right)} = \frac{\mathrm{ES}_\alpha(L) + \mu(x)}{\mathrm{VaR}_\alpha(L) + \mu(x)}$$

is different from the Gaussian case:

$$\kappa = \frac{\phi\left(\Phi^{-1}(\alpha)\right)}{(1-\alpha)\,\Phi^{-1}(\alpha)}$$

In our example, κ takes the value 1.62 whereas it is 1.37 in the Gaussian case. Moreover, we can show that this difference increases with α because of the impact of the characteristics of the fat-tailed asset returns.

[16] It is estimated using the Gaussian kernel with 2 000 replications.

[17] The parameter h is equal to $2.5\% \times m$.

2.1.3.2 Historical value-at-risk

The historical value-at-risk consists in computing the quantile of the empirical distribution of losses using a set of historical scenarios, typically the last 260 trading days. It is also called the non-parametric approach, because it does not involve estimating parameters like in the Gaussian VaR. Let $\hat{\mathbf{F}}$ be the empirical distribution of loss portfolio $L(x)$. The historical VaR is then:

$$\mathrm{VaR}_\alpha(L) = \hat{\mathbf{F}}^{-1}(\alpha)$$

To estimate this quantity, we can proceed as in the case of Monte Carlo simulations. In this case, m, $L_i^{(j)}$ and $L^{(j)}$ denote respectively the number of historical scenarios, the loss of the ith asset for the jth scenario and the portfolio loss for the jth scenario. We have seen that the value-at-risk is then equal to the m_α order statistic: $\mathrm{VaR}_\alpha(L) = L^{(m_\alpha:m)}$. If m_α is not an integer, we consider an interpolation between the two order statistics enclosing the confidence level α. We have:

$$\mathrm{VaR}_\alpha(L) = L^{(m_\alpha:m)} + (m\alpha + 1 - m_\alpha)\left(L^{(m_\alpha+1:m)} - L^{(m_\alpha:m)}\right)$$

where $m_\alpha = \lfloor m\alpha + 1 \rfloor$ is the integer part of $m\alpha + 1$. For example, if $\alpha = 99\%$ and $m = 250$, we obtain:

$$\begin{aligned}
\mathrm{VaR}_{99\%}(L) &= L^{(248:250)} + \frac{1}{2}\left(L^{(249:250)} - L^{(248:250)}\right) \\
&= -\left(\Pi^{(3:250)} + \frac{1}{2}\left(\Pi^{(2:250)} - \Pi^{(3:250)}\right)\right) \\
&= -\left(\frac{1}{2}\Pi^{(2:250)} + \frac{1}{2}\Pi^{(3:250)}\right)
\end{aligned}$$

Computing the 99% value-at-risk with 250 historical scenarios is then equal to the opposite of the average of the second and third smallest P&Ls.

The risk contribution estimation faces the same problem as in the Monte Carlo case. If asset returns are assumed to be elliptically distributed, Carroll *et al.* (2001) shows that[18]:

$$\mathcal{RC}_i = \mathbb{E}[L_i] + \frac{\mathrm{cov}(L, L_i)}{\sigma^2(L)}\left(\mathrm{VaR}_\alpha(L) - \mathbb{E}[L]\right) \tag{2.17}$$

[18] We verify that the sum of the risk contributions is equal to the value-at-risk:

$$\begin{aligned}
\sum_{i=1}^n \mathcal{RC}_i &= \sum_{i=1}^n \mathbb{E}[L_i] + \left(\mathrm{VaR}_\alpha(L) - \mathbb{E}[L]\right)\sum_{i=1}^n \frac{\mathrm{cov}(L, L_i)}{\sigma^2(L)} \\
&= \mathbb{E}[L] + \left(\mathrm{VaR}_\alpha(L) - \mathbb{E}[L]\right) \\
&= \mathrm{VaR}_\alpha(L)
\end{aligned}$$

Estimating the risk contributions with the historical value-at-risk is then straightforward. It suffices to apply Formula (2.17) by replacing the statistical moments by their sample statistics:

$$\mathcal{RC}_i = \bar{L}_i + \frac{\sum_{j=1}^{m} \left(L^{(j)} - \bar{L}\right)\left(L_i^{(j)} - \bar{L}_i\right)}{\sum_{j=1}^{m} \left(L^{(j)} - \bar{L}\right)^2} \left(\mathrm{VaR}_\alpha(L) - \bar{L}\right)$$

where $\bar{L}_i = m^{-1} \sum_{j=1}^{m} L_i^{(j)}$ and $\bar{L} = m^{-1} \sum_{j=1}^{m} L^{(j)}$.

Result (2.17) can be viewed as the estimation of the conditional expectation $\mathbb{E}\left[L_i \mid L = \mathrm{VaR}_\alpha(L)\right]$ in a linear regression framework:

$$L_i = \beta L + \varepsilon_i$$

Because the least squared estimator is $\hat{\beta} = \mathrm{cov}(L, L_i) / \sigma^2(L)$, we deduce that:

$$\begin{aligned}
\mathbb{E}\left[L_i \mid L = \mathrm{VaR}_\alpha(L)\right] &= \hat{\beta}\,\mathrm{VaR}_\alpha(L) + \mathbb{E}\left[\varepsilon_i\right] \\
&= \hat{\beta}\,\mathrm{VaR}_\alpha(L) + \left(\mathbb{E}\left[L_i\right] - \hat{\beta}\mathbb{E}\left[L\right]\right) \\
&= \mathbb{E}\left[L_i\right] + \hat{\beta}\left(\mathrm{VaR}_\alpha(L) - \mathbb{E}\left[L\right]\right)
\end{aligned}$$

Epperlein and Smillie (2006) extend Formula (2.17) in the case of non-elliptical distributions. If we consider the generalized conditional expectation $\mathbb{E}\left[L_i \mid L = x\right] = f(x)$ where the function f is unknown, the estimate is given by the kernel regression[19]:

$$\hat{f}(x) = \frac{\sum_{j=1}^{m} \mathcal{K}\left(L^{(j)} - x\right) L_i^{(j)}}{\sum_{j=1}^{m} \mathcal{K}\left(L^{(j)} - x\right)}$$

where $\mathcal{K}(u)$ is a kernel function. We deduce that:

$$\mathcal{RC}_i = \hat{f}\left(\mathrm{VaR}_\alpha(L)\right)$$

Epperlein and Smillie (2006) note however that this risk decomposition does not satisfy the Euler principle. This is why they propose the following correction:

$$\begin{aligned}
\mathcal{RC}_i &= \frac{\mathrm{VaR}_\alpha(L)}{\sum_{i=1}^{n} \mathcal{RC}_i} \hat{f}\left(\mathrm{VaR}_\alpha(L)\right) \\
&= \mathrm{VaR}_\alpha(L) \frac{\sum_{j=1}^{m} \mathcal{K}\left(L^{(j)} - \mathrm{VaR}_\alpha(L)\right) L_i^{(j)}}{\sum_{i=1}^{n} \sum_{j=1}^{m} \mathcal{K}\left(L^{(j)} - \mathrm{VaR}_\alpha(L)\right) L_i^{(j)}} \\
&= \mathrm{VaR}_\alpha(L) \frac{\sum_{j=1}^{m} \mathcal{K}\left(L^{(j)} - \mathrm{VaR}_\alpha(L)\right) L_i^{(j)}}{\sum_{j=1}^{m} \mathcal{K}\left(L^{(j)} - \mathrm{VaR}_\alpha(L)\right) L^{(j)}} \qquad (2.18)
\end{aligned}$$

[19] $\hat{f}(x)$ is called the Nadaraya-Watson estimator.

Professionals prefer to use historical value-at-risk instead of Gaussian value-at-risk because it depends (implicitly) on all the moments of the loss distribution. Another approach consists in correcting the Gaussian value-at-risk by taking into account the third and fourth moments. Let $\mu_r = \mathbb{E}\left[(L - \mu(L))^r\right]$ be the centered r-order moment of the portfolio loss[20]. The skewness $\gamma_1 = \mu_3 / \mu_2^{3/2}$ is the measure of the asymmetry of the loss distribution[21]. To characterize whether the loss is peaked or flat relative to the normal distribution, we consider the excess kurtosis[22] $\gamma_2 = \mu_4 / \mu_2^2 - 3$. Using the Cornish-Fisher expansion of the distribution, Zangari (1996) proposes to estimate the value-at-risk in the following way:

$$\mathrm{VaR}_\alpha(L) = -x^\top \mu + z \cdot \sqrt{x^\top \Sigma x}$$

where:

$$z = z_\alpha + \frac{1}{6}\left(z_\alpha^2 - 1\right)\gamma_1 + \frac{1}{24}\left(z_\alpha^3 - 3z_\alpha\right)\gamma_2 - \frac{1}{36}\left(2z_\alpha^3 - 5z_\alpha\right)\gamma_1^2$$

with $z_\alpha = \Phi^{-1}(\alpha)$. This is the same formula as the one used for the Gaussian value-at-risk but with another scaling parameter[23]. We deduce that the expression of the risk contribution is:

$$\mathcal{RC}_i = x_i \cdot \left(-\mu_i + z \cdot \frac{(\Sigma x)_i}{\sqrt{x^\top \Sigma x}} + \partial_{x_i} z \cdot \sqrt{x^\top \Sigma x}\right)$$

In a first time, we can assume that $\partial_{x_i} z = 0$ because local perturbations of the portfolio have little influence on the parameter z. We can also follow Boudt *et al.* (2008) who compute the exact value of $\partial_{x_i} z$. Jondeau and Rockinger (2006) show that the third and fourth centered moments of the portfolio loss are:

$$\begin{aligned}\mu_3 &= -x^\top M_3 (x \otimes x) \\ \mu_4 &= x^\top M_4 (x \otimes x \otimes x)\end{aligned}$$

where M_3 and M_4 are the co-skewness and co-kurtosis matrices of asset returns:

$$\begin{aligned}M_3 &= \mathbb{E}\left[(R-\mu)(R-\mu)^\top \otimes (R-\mu)^\top\right] \\ M_4 &= \mathbb{E}\left[(R-\mu)(R-\mu)^\top \otimes (R-\mu)^\top \otimes (R-\mu)^\top\right]\end{aligned}$$

We obtain:

$$\partial_{x_i} z = \frac{1}{6}\left(z_\alpha^2 - 1\right)\partial_{x_i}\gamma_1 + \frac{1}{24}\left(z_\alpha^3 - 3z_\alpha\right)\partial_{x_i}\gamma_2 - \frac{1}{18}\left(2z_\alpha^3 - 5z_\alpha\right)\gamma_1 \partial_{x_i}\gamma_1$$

[20] The mathematical expectation $\mu(L)$ is equal to $-x^\top \mu$.

[21] If $\gamma_1 < 0$ (resp. $\gamma_1 > 0$), the loss distribution is left-skewed (resp. right-skewed) because the left (resp. right) tail is longer. For the Gaussian distribution, γ_1 is equal to zero.

[22] If $\gamma_2 > 0$, the loss distribution presents heavy tails.

[23] If $\gamma_1 = \gamma_2 = 0$, we retrieve the Gaussian value-at-risk with $z = \Phi^{-1}(\alpha)$.

with:

$$\partial_{x_i}\gamma_1 = 3\mu_3\frac{(\Sigma x)_i}{(x^\top\Sigma x)^{5/2}} - 3\frac{(M_3(x\otimes x))_i}{(x^\top\Sigma x)^{3/2}}$$

$$\partial_{x_i}\gamma_2 = 4\frac{(M_4(x\otimes x\otimes x))_i}{(x^\top\Sigma x)^2} - 4\mu_4\frac{(\Sigma x)_i}{(x^\top\Sigma x)^3}$$

Example 9 *We consider the skew normal distribution* $Y \sim \mathcal{SN}(\xi,\omega^2,\beta)$. *This distribution generalizes the normal distribution to take into account non-zero skewness. Its probability density function is:*

$$f(y) = \frac{2}{\omega}\phi\left(\frac{y-\xi}{\omega}\right)\Phi\left(\beta\frac{y-\xi}{\omega}\right)$$

Azzalini (1985, 1986) shows that:

$$\begin{aligned}\mathbb{E}[Y] &= \xi+\omega\mu \\ \sigma^2(Y) &= \omega^2(1-\mu^2) \\ \gamma_1(Y) &= \frac{4-\pi}{2}\frac{\mu^3}{(1-\mu^2)^{3/2}} \\ \gamma_2(Y) &= 2(\pi-3)\frac{\mu^4}{(1-\mu^2)^2}\end{aligned}$$

with $\mu = \delta\sqrt{2/\pi}$ *and:*

$$\delta = \frac{\beta}{\sqrt{1+\beta^2}}$$

We assume that the P&L Π is distributed according to a skew normal distribution $\mathcal{SN}(\xi,\omega^2,\beta)$. We consider the following three sets of parameters:

Set	ξ	ω	β
#1	1%	5%	0
#2	−5%	10%	2
#3	15%	20%	−5

In Figure 2.3, we have represented the probability density function of Π for these different sets of parameters. We deduce that the corresponding moments are:

Set	$\mathbb{E}[\Pi]$	$\sigma^2(\Pi)$	$\gamma_1(\Pi)$	$\gamma_2(\Pi)$
#1	0.01000	0.00250	0.00000	0.00000
#2	0.02136	0.00491	0.45383	0.30505
#3	−0.00648	0.01551	−0.85097	0.70535

We can then compute the Cornish-Fisher VaR[24] and compare it with the

[24]Because $L = -\Pi$, we have $\mathbb{E}[L] = -\mathbb{E}[\Pi]$ and $\gamma_1(L) = -\gamma_1(\Pi)$ whereas the even moments are the same.

normal VaR and the exact value by inverting the cumulative distribution function of the $\mathcal{SN}$ distribution. The results are reported in Table 2.7. We note that the normal VaR can be far from the exact VaR value, whereas the Cornish-Fisher correction improves the result.

TABLE 2.7: Value-at-risk (in %) when the P&L is skew normal distributed

α		80%	85%	90%	95%	99%
Normal	#1	3.21	4.18	5.41	7.22	10.63
	#2	3.76	5.12	6.84	9.39	14.16
	#3	11.13	13.56	16.61	21.14	29.62
Cornish-Fisher	#1	3.21	4.18	5.41	7.22	10.63
	#2	3.80	4.94	6.34	8.34	11.95
	#3	10.63	13.79	17.90	24.20	36.52
Skew normal	#1	3.21	4.18	5.41	7.22	10.63
	#2	3.86	5.03	6.43	8.41	11.78
	#3	10.67	13.70	17.66	23.80	36.08

Remark 25 *The skew normal distribution is very appealing in portfolio management. It can be generalized to the multivariate skew normal distribution $\mathcal{MSN}$ (Azzalini and Dalla Valle, 1996), which is closed under affine transformation. If we assume that the asset returns follow a $\mathcal{MSN}$ distrbution, the portfolio return is then distributed according to a $\mathcal{SN}$ distribution. Moreover, Genton et al. (2001) provide various formulas for computing the moments matrices M_2, M_3 and M_4. We then have all the materials needed to easily compute the value-at-risk and the risk contributions.*

Example 10 *We consider a portfolio with two assets: the MSCI World index and the Citigroup World Government Bond index (WGBI). The study period begins at July 2009 and ends at June 2012. The investment horizon is one week.*

To compute the VaR, we first compute the weekly returns. Then we estimate the first four moments matrices. We obtain:

$$M_1 = \begin{pmatrix} 22.5124 \\ 9.9389 \end{pmatrix} \times 10^{-4}$$

$$M_2 = \begin{pmatrix} 7.0343 & 0.2931 \\ 0.2931 & 0.7641 \end{pmatrix} \times 10^{-4}$$

$$M_3 = \begin{pmatrix} -6.5540 & 0.8548 & 0.8548 & -0.2178 \\ 0.8548 & -0.2178 & -0.2178 & -0.0179 \end{pmatrix} \times 10^{-6}$$

$$M_4 = \begin{pmatrix} 2.099 & 0.022 & 0.022 & 0.049 & 0.022 & 0.049 & 0.049 & 0.006 \\ 0.022 & 0.049 & 0.049 & 0.006 & 0.049 & 0.006 & 0.006 & 0.017 \end{pmatrix} \times 10^{-6}$$

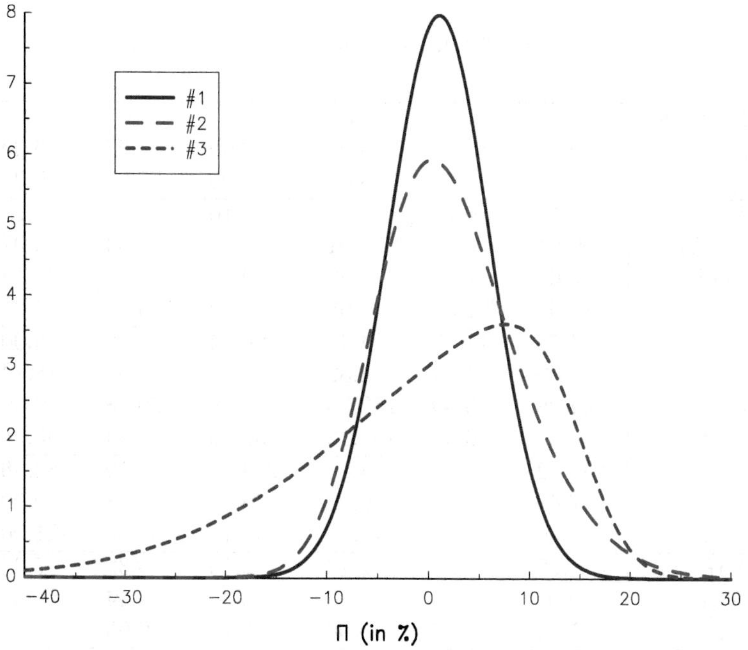

FIGURE 2.3: Density of the P&L with a skew normal distribution

For a given portfolio x, we can also compute the mean, variance, skewness and kurtosis of the loss. We deduce finally the different derivatives and the risk contributions. Table 2.8 presents those computations in the case of several portfolios. For example, if $x_1 = 50\%$ and $x_2 = 50\%$, z takes the value 2.5895 and the value-at-risk is equal to 3.59%. The main contribution comes from the first asset, which represents 97% of the risk. We also note that the normal value-at-risk $\mathrm{VaR}^{\star}_{\alpha}(L)$ underestimates the risk when the proportion of equities is large. If $x_1 = 75\%$ and $x_2 = 25\%$, the Cornish Fisher value-at-risk is equal to 5.39% whereas the normal value-at-risk is equal to 4.53%.

2.2 Analysis of risk budgeting portfolios

In this section, we define precisely what is a risk budgeting (or RB) portfolio. Then, we study the main properties of such portfolios, in particular their existence and uniqueness. We show also under which conditions these portfolios are optimal and illustrate why they are more stable than optimized portfolios.

TABLE 2.8: Statistics (in %) to compute the Cornish-Fisher risk contributions

Portfolio	#1	#2	#3	#4	#5	#6
x_1	0.00	10.00	25.00	50.00	75.00	100.00
x_2	100.00	90.00	75.00	50.00	25.00	0.00
$\mathbb{E}[L]$	−0.10	−0.11	−0.13	−0.16	−0.19	−0.23
$\sigma^2(L)$	0.01	0.01	0.01	0.02	0.04	0.07
$\gamma_1(L)$	2.67	7.73	8.42	19.20	29.18	35.13
$\gamma_2(L)$	−13.05	−21.50	−16.08	58.12	103.97	124.21
$\partial_{x_1}\gamma_1$	94.76	11.12	14.46	24.44	7.68	0.00
$\partial_{x_2}\gamma_1$	0.00	−1.24	−4.82	−24.44	−23.05	−18.14
$\partial_{x_1}\gamma_2$	−44.29	−90.59	159.39	131.49	29.58	0.00
$\partial_{x_2}\gamma_2$	0.00	10.07	−53.13	−131.49	−88.74	−53.28
z	231.52	233.07	234.80	258.95	275.19	282.86
$\partial_{x_1}z$	57.41	−13.65	46.98	45.18	10.88	0.00
$\partial_{x_2}z$	0.00	1.52	−15.66	−45.18	−32.63	−21.00
$\mathrm{VaR}_\alpha(L)$	1.92	1.90	2.19	3.59	5.39	7.28
$\mathcal{RC}_1$	0.00	0.23	1.23	3.49	5.44	7.28
$\mathcal{RC}_2$	1.92	1.67	0.96	0.10	−0.05	0.00
$\mathrm{VaR}^\star_\alpha(L)$	1.93	1.89	2.17	3.21	4.53	5.94

2.2.1 Definition of a risk budgeting portfolio

We consider a set of given risk budgets $\{B_1, \ldots, B_n\}$. Here, B_i is an amount of risk measured in dollars. We note $\mathcal{RC}_i(x_1, \ldots, x_n)$ the risk contribution of asset i with respect to portfolio $x = (x_1, \ldots, x_n)$. The risk budgeting portfolio is then defined by the following constraints:

$$\left\{\begin{array}{l} \mathcal{RC}_1(x_1, \ldots, x_n) = B_1 \\ \vdots \\ \mathcal{RC}_i(x_1, \ldots, x_n) = B_i \\ \vdots \\ \mathcal{RC}_n(x_1, \ldots, x_n) = B_n \end{array}\right. \tag{2.19}$$

It is therefore the portfolio x such that the risk contributions match the risk budgets. We note that there are two main differences between a risk budgeting portfolio and an optimized portfolio:

1. A risk budgeting portfolio is not based on the maximization of a utility function.
2. A risk budgeting portfolio does not depend (explicitly) on the expected performance of the portfolio.

Contrary to the Markowitz approach, we only consider the risk dimension. The main idea is that the performance dimension is too complicated to encompass, because the forecasting step is generally not robust. This is why we focus on the patterns of the portfolio risk.

Remark 26 *This does not mean that we do not need any assumption about the asset returns to build the risk budgeting portfolio. Indeed, some risk measures, such as the normal value-at-risk, depend on the vector μ of asset returns. Nevertheless μ is set to zero in most cases in order to have a conservative risk measure.*

Example 11 *We consider three asset classes. Their expected returns are equal to* 10%, 5% *and* 9% *whereas their volatilities are equal to* 30%, 20% *and* 15%. *The correlation of asset returns is given by the following matrix:*

$$\rho = \begin{pmatrix} 1.00 & & \\ 0.80 & 1.00 & \\ 0.50 & 0.30 & 1.00 \end{pmatrix}$$

We consider the example above. The risk measure of the investor is the normal expected shortfall and his risk limit is an annual loss of \$500 000. Additionally to this global risk limit, the investor has other risk limits fixed by asset classes: \$250 000 for the first asset class, \$100 000 for the second asset class and \$150 000 for the third asset class. We face here a risk budgeting problem with $B_1 = \$250\,000$, $B_2 = \$100\,000$ and $B_3 = \$150\,000$. If we consider the confidence levels 95% and 99%, we obtain the results given in Tables 2.9 and 2.10. For each portfolio, we have indicated the nominal exposure x_i, the weight w_i, the marginal risk $\mathcal{MR}_i$, the nominal risk contribution $\mathcal{RC}_i$ and the normalized risk contribution $\mathcal{RC}_i^\star$. If α is set to 95%, the total exposure is equal to 1 894 142 and the composition of the portfolio is $(28.21\%, 19.68\%, 52.11\%)$. We note that these figures change if α is equal to 99%. The reason for this is that the expected shortfall is not homogeneous with respect to α because of the expected loss term.

TABLE 2.9: Risk budgeting portfolio when the risk measure is the expected shortfall ($\alpha = 95\%$)

Asset	x_i	w_i	$\mathcal{MR}_i$	$\mathcal{RC}_i$	$\mathcal{RC}_i^\star$
1	534 430	28.21%	46.78%	250 000	50.00%
2	372 705	19.68%	26.83%	100 000	20.00%
3	987 007	52.11%	15.20%	150 000	30.00%
sum	1 894 142			500 000	

2.2.1.1 The right specification of the RB portfolio

The previous example suggests that using nominal exposures is equivalent to considering the weights of the portfolio. Let x be the nominal exposure of

TABLE 2.10: Risk budgeting portfolio when the risk measure is the expected shortfall ($\alpha = 99\%$)

Asset	x_i	w_i	$\mathcal{MR}_i$	$\mathcal{RC}_i$	$\mathcal{RC}_i^\star$
1	391 926	29.00%	63.79%	250 000	50.00%
2	273 737	20.26%	36.53%	100 000	20.00%
3	685 779	50.74%	21.87%	150 000	30.00%
sum	1 351 441			500 000	

the portfolio. We note c a scalar and we consider a new portfolio y defined by $y = cx$. The homogeneity property of coherent risk measures implies that $\mathcal{R}(y) = c\mathcal{R}(x)$. We deduce that:

$$\begin{aligned}\sum_{i=1}^n y_i \frac{\partial\,\mathcal{R}(y)}{\partial\,y_i} &= c\sum_{i=1}^n x_i \frac{\partial\,\mathcal{R}(x)}{\partial\,x_i} \\ &= \sum_{i=1}^n c x_i \frac{\partial\,\mathcal{R}(x)}{\partial\,x_i} \\ &= \sum_{i=1}^n y_i \frac{\partial\,\mathcal{R}(x)}{\partial\,x_i}\end{aligned}$$

This proves that the marginal risks with respect to portfolio y are the same as those with respect to portfolio x:

$$\frac{\partial\,\mathcal{R}(y)}{\partial\,y_i} = \frac{\partial\,\mathcal{R}(x)}{\partial\,x_i}$$

It is also easy to show that the solution of the system (2.19) with rescaled risk budgets $b_i = cB_i$ is $y = cx$. Bruder and Roncalli (2012) propose to simplify the problem by using weights instead of nominal exposures[25] and defining the problem with the relative risk budgets b_i expressed in %. These authors note also that risk budgeting techniques are used to build diversified portfolios. Specifying that some assets have a negative risk contribution implies that the risk is highly concentrated in the other assets of the portfolio. Generally, we prefer to obtain a long-only portfolio meaning that all the weights are positive. This is why we define a proper RB portfolio by the following non-linear system:

$$\left\{\begin{array}{l}\mathcal{RC}_i(x) = b_i\mathcal{R}(x) \\ b_i \geq 0 \\ x_i \geq 0 \\ \sum_{i=1}^n b_i = 1 \\ \sum_{i=1}^n x_i = 1\end{array}\right. \tag{2.20}$$

[25] In this case, we have $c = 1/\sum_{i=1}^n x_i$.

Another difficulty may appear when one specifies that some risk budgets are equal to zero. Let us consider the volatility as the risk measure. We note Σ the covariance matrix of asset returns. We have $\Sigma_{i,j} = \rho_{i,j}\sigma_i\sigma_j$ where $\sigma_i > 0$ is the volatility of asset i and $\rho_{i,j}$ is the cross-correlation[26] between assets i and j. It follows that:

$$\frac{\partial\,\sigma(x)}{\partial\,x_i} = \frac{x_i\sigma_i^2 + \sigma_i\sum_{j\neq i} x_j\rho_{i,j}\sigma_j}{\sigma(x)}$$

Suppose that the risk budget b_k is equal to zero. This means that:

$$x_k\left(x_k\sigma_k^2 + \sigma_k\sum_{j\neq k} x_j\rho_{k,j}\sigma_j\right) = 0$$

We obtain two solutions. The first one is $x'_k = 0$ whereas the second one verifies:

$$x''_k = -\frac{\sum_{j\neq k} x_j\rho_{k,j}\sigma_j}{\sigma_k}$$

If $\rho_{k,j} \geq 0$ for all j, we have $\sum_{j\neq k} x_j\rho_{k,j}\sigma_j \geq 0$ because $x_j \geq 0$ and $\sigma_j > 0$. This implies that $x''_k \leq 0$ meaning that $x'_k = 0$ is the unique positive solution. Finally the only way to have $x''_k > 0$ is to have some negative correlations $\rho_{k,j}$. In this case, this implies that:

$$\sum_{j\neq k} x_j\rho_{k,j}\sigma_j < 0$$

If we consider a universe of three assets, this constraint is verified for $k = 3$ and a covariance matrix such that $\rho_{1,3} < 0$ and $\rho_{2,3} < 0$. For example, if $\sigma_1 = 20\%$, $\sigma_2 = 10\%$, $\sigma_3 = 5\%$, $\rho_{1,2} = 50\%$, $\rho_{1,3} = -25\%$ and $\rho_{2,3} = -25\%$, the two solutions are $(33.33\%, 66.67\%, 0\%)$ and $(20\%, 20\%, 40\%)$ if the risk budgets are $(50\%, 50\%, 0\%)$.

In practice, this second solution may not satisfy the investor. When he sets one risk budget to zero, he expects that he will not have the corresponding asset in his portfolio. This is why it is important to impose the strict constraint $b_i > 0$. To summarize, the RB portfolio is the solution to the following non-linear problem:

$$x^\star = \left\{x \in [0,1]^n : \textstyle\sum_{i=1}^n x_i = 1, x_i \cdot \partial_{x_i}\mathcal{R}(x) = b_i\mathcal{R}(x)\right\} \tag{2.21}$$

where $b \in \left]0,1\right]^n$ and $\sum_{i=1}^n b_i = 1$.

Remark 27 *If we would like to impose that some risk budgets are equal to zero, we can first reduce the universe of assets by excluding the assets corresponding to these zero risk contributions and then solve Problem (2.21).*

---

[26]We have of course $\rho_{i,i} = 1$.

2.2.1.2 Solving the non-linear system of risk budgeting contraints

With the exception of some trivial cases, it is not possible to find an analytical solution to Problem (2.21). However, we can always find a numerical solution. The first approach is to consider Broyden or SQRF algorithms to solve the non-linear system (2.21), but they do not always converge. This is why it is better to transform the non-linear system into an optimization problem:

$$\begin{aligned} x^{\star} &= \arg\min f(x;b) \\ &\text{u.c.} \quad \mathbf{1}^{\top}x = 1 \quad \text{and} \quad \mathbf{0} \leq x \leq \mathbf{1} \end{aligned} \tag{2.22}$$

For example, we may specify the function $f(x;b)$ as follows:

$$f(x;b) = \sum_{i=1}^{n} \left(x_i \cdot \partial_{x_i} \mathcal{R}(x) - b_i \mathcal{R}(x) \right)^2$$

If $x^{\star}$ is the solution of (2.22) and if $f(x^{\star};b) = 0$, it is obvious that $x^{\star}$ is also the solution of (2.21). We note that there are other appropriate functions $f(x;b)$ to define the RB portfolio, for example:

$$f(x;b) = \sum_{i=1}^{n} \sum_{j=1}^{n} \left(\frac{x_i \cdot \partial_{x_i} \mathcal{R}(x)}{b_i} - \frac{x_j \cdot \partial_{x_j} \mathcal{R}(x)}{b_j} \right)^2$$

Remark 28 *To solve Problem (2.22), we consider the sequential quadratic programming (or SQP) algorithm*[27] *presented in Appendix A.1.3. Computing the analytical Jacobian and Hessian of the function $f(x;b)$ helps reduce computational times, especially when the number of assets is large.*

2.2.2 Some properties of the RB portfolio

In this section, we derive some particular solutions in the case of the volatility risk measure. Then we show that the RB portfolio always exists and is unique. We also discuss the optimality of such a portfolio and illustrate what the solution becomes when some risk budgets are set to zero.

2.2.2.1 Particular solutions with the volatility risk measure

The two-asset case ($n = 2$) We begin by analyzing the RB portfolio in the bivariate case. Let ρ be the correlation, $x = (w, 1-w)$ be the vector of weights and $(b, 1-b)$ be the vector of risk budgets. The risk contributions are:

$$\begin{pmatrix} \mathcal{RC}_1 \\ \mathcal{RC}_2 \end{pmatrix} = \frac{1}{\sigma(x)} \begin{pmatrix} w^2\sigma_1^2 + w(1-w)\rho\sigma_1\sigma_2 \\ (1-w)^2\sigma_2^2 + w(1-w)\rho\sigma_1\sigma_2 \end{pmatrix}$$

[27] From a numerical point of view, it is preferable to solve the system without the constraint $\mathbf{1}^{\top}x = 1$ and then to rescale the solution.

The unique solution satisfying $0 \leq w \leq 1$ is:

$$w^{\star} = \frac{(b - 1/2)\,\rho\sigma_1\sigma_2 - b\sigma_2^2 + \sigma_1\sigma_2\sqrt{(b - 1/2)^2\,\rho^2 + b\,(1-b)}}{(1-b)\,\sigma_1^2 - b\sigma_2^2 + 2\,(b - 1/2)\,\rho\sigma_1\sigma_2} \tag{2.23}$$

We note that the solution is a complex function of the volatilities σ_1 and σ_2, the correlation ρ and the risk budget b.

In order to have some intuitions about the behavior of the solution, we consider some special cases. For example, if $\rho = 0$, we obtain:

$$w^{\star} = \frac{\sigma_2\sqrt{b}}{\sigma_1\sqrt{1-b} + \sigma_2\sqrt{b}}$$

The weight of asset i is then proportional to the square root of its risk budget and inversely proportional to its volatility. If $\rho = 1$, the solution reduces to:

$$w^{\star} = \frac{\sigma_2 b}{\sigma_1\,(1-b) + \sigma_2 b}$$

In this case, the weight of asset i is directly proportional to its risk budget. If $\rho = -1$, we note also that the solution does not depend on the risk budgets:

$$w^{\star} = \frac{\sigma_2}{\sigma_1 + \sigma_2}$$

because the portfolio's volatility is equal to zero. We may also show that the weight of asset i is an increasing function of its risk budget and a decreasing function of its volatility. In Table 2.11, we report some values taken by $w^{\star}$ (in %) with respect to the parameters ρ and b, whereas the behavior of $w^{\star}$ is illustrated in Figure 2.4.

TABLE 2.11: Weights $w^{\star}$ in the RB portfolio with respect to some values of b and ρ

	$\sigma_2 = \sigma_1$				$\sigma_2 = 3 \times \sigma_1$			
ρ/b	20%	50%	70%	90%	20%	50%	70%	90%
−50%	41.9	50.0	55.2	61.6	68.4	75.0	78.7	82.8
0%	33.3	50.0	60.4	75.0	60.0	75.0	82.1	90.0
25%	29.3	50.0	63.0	80.6	55.5	75.0	83.6	92.6
50%	25.7	50.0	65.5	84.9	51.0	75.0	85.1	94.4
75%	22.6	50.0	67.8	87.9	46.7	75.0	86.3	95.6
90%	21.0	50.0	69.1	89.2	44.4	75.0	87.1	96.1

The general case ($n > 2$) In the case $n > 2$, the number of parameters increases quickly, with n individual volatilities and $n(n-1)/2$ bivariate correlations. Because the two-asset case leads to a complex solution, finding an

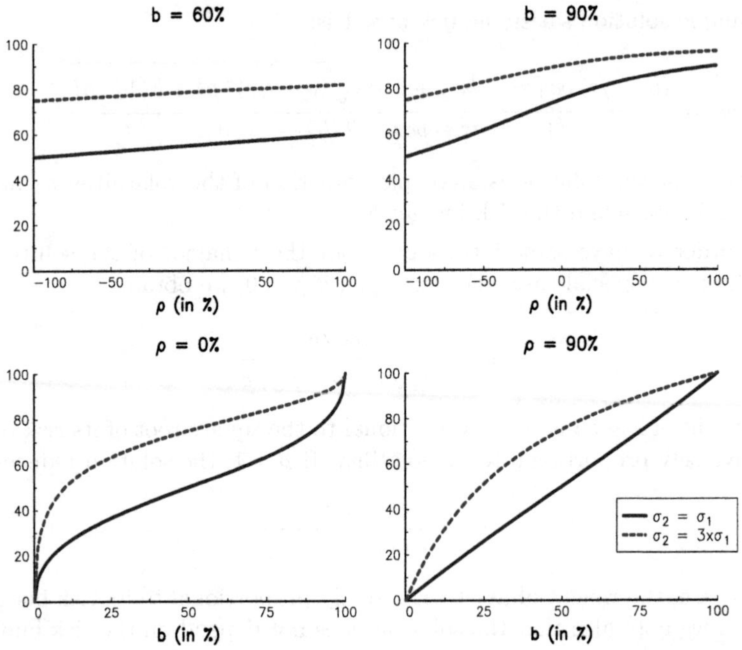

FIGURE 2.4: Evolution of the weight $w^{\star}$ in the RB portfolio with respect to b and ρ

explicit solution for the general case is certainly impossible. However, we can find some results that help us to understand the behavior of the RB portfolio.

We recall that the expression of the risk contribution is:

$$\begin{aligned}\mathcal{RC}_i &= \frac{x_i \cdot (\Sigma x)_i}{\sigma(x)} \\ &= \frac{x_i^2\sigma_i^2 + \sum_{j\neq i} x_i x_j \rho_{i,j}\sigma_i\sigma_j}{\sigma(x)}\end{aligned}$$

If we assume that the correlation matrix of assets is constant ($\rho_{i,j} = \rho$), we obtain:

$$\begin{aligned}\mathcal{RC}_i &= \frac{x_i^2\sigma_i^2 + \rho\sum_{j\neq i} x_i x_j \sigma_i\sigma_j}{\sigma(x)} \\ &= \frac{x_i^2\sigma_i^2 + \rho\sum_{j=1}^{n} x_i x_j \sigma_i\sigma_j - \rho x_i^2\sigma_i^2}{\sigma(x)} \\ &= \frac{x_i\sigma_i\left((1-\rho)\,x_i\sigma_i + \rho\sum_{j=1}^{n} x_j\sigma_j\right)}{\sigma(x)} \qquad (2.24)\end{aligned}$$

The RB portfolio verifies the following non-linear system:

$$x_i\sigma_i\left((1-\rho)\,x_i\sigma_i+\rho\left(\sum_{j=1}^{n}x_j\sigma_j\right)\right)=b_i\sigma^2\left(x\right)$$

or equivalently:

$$(1-\rho)\,x_i^2\sigma_i^2+\frac{x_i\sigma_i}{x_j\sigma_j}\left(b_j\sigma^2\left(x\right)-(1-\rho)\,x_j^2\sigma_j^2\right)=b_i\sigma^2\left(x\right) \tag{2.25}$$

It is not possible to solve this problem in the general case, but we can find an analytical expression in three cases: $\rho=0$, $\rho=1$ and $\rho=-1/\left(n-1\right)$ which defines the lower bound of the constant correlation matrix.

If we have no correlation, i.e. if $\rho=0$, the risk contribution of asset i becomes $\mathcal{RC}_i=x_i^2\sigma_i^2/\sigma\left(x\right)$. The RB portfolio being defined by $\mathcal{RC}_i=b_i\sigma\left(x\right)$ for all i, some simple algebra shows that this is here equivalent to $\sqrt{b_j}x_i\sigma_i=\sqrt{b_i}x_j\sigma_j$. Coupled with the (normalizing) budget constraint $\sum_{i=1}^{n}x_i=1$, we deduce that:

$$x_i=\frac{\sqrt{b_i}\sigma_i^{-1}}{\sum_{j=1}^{n}\sqrt{b_j}\sigma_j^{-1}} \tag{2.26}$$

The weight allocated to component i is thus proportional to the square root of its risk budget and the inverse of its volatility. The higher the volatility of a component, the lower its weight in the RB portfolio. In the case of perfect correlation, i.e. $\rho=1$, the risk contribution of asset i becomes $\mathcal{RC}_i=x_i\sigma_i\left(\sum_{j=1}^{n}x_j\sigma_j\right)/\sigma\left(x\right)$. It follows that $b_jx_i\sigma_i=b_ix_j\sigma_j$. We then deduce that:

$$x_i=\frac{b_i\sigma_i^{-1}}{\sum_{j=1}^{n}b_j\sigma_j^{-1}} \tag{2.27}$$

The opposite of perfect correlation corresponds to the lower bound of the constant correlation matrix, which is reached for $\rho=-1/\left(n-1\right)$. If we suppose that $\sigma\left(x\right)=0$, Equation (2.25) becomes:

$$x_i\sigma_i=x_j\sigma_j=\varpi$$

Let us compute the variance of the portfolio:

$$\begin{aligned}\sigma^2\left(x\right)&=\rho\sum_{i=1}^{n}x_i\sigma_i\left(\sum_{j=1}^{n}x_j\sigma_j\right)+(1-\rho)\sum_{i=1}^{n}x_i^2\sigma_i^2\\&=\rho n^2\varpi^2+(1-\rho)\,n\varpi^2\\&=\left(-\frac{n^2}{n-1}+\left(1+\frac{1}{n-1}\right)n\right)\varpi^2\\&=0\end{aligned}$$

We verify that the portfolio's volatility is equal to zero meaning that the solution is given by $x_i\sigma_i = x_j\sigma_j$. When the constant correlation reaches its lower bound, the solution is then:

$$x_i = \frac{\sigma_i^{-1}}{\sum_{j=1}^{n} \sigma_j^{-1}} \tag{2.28}$$

Unfortunately, there is no analytical solution in the general case when the constant correlation differs from 0, 1 and $-1/(n-1)$, but we can find an implicit form. We recall that the RB portfolio verifies the following equation in the case of the constant correlation matrix:

$$x_i\sigma_i \left((1-\rho)\, x_i\sigma_i + \rho \left(\sum_{j=1}^{n} x_j\sigma_j \right) \right) = b_i\sigma^2(x)$$

If we note $X_i = x_i\sigma_i$ and $B_i = b_i\sigma^2(x)$, the previous equation becomes:

$$(1-\rho)\, X_i^2 + \rho X_i \left(\sum_{j=1}^{n} X_j \right) = B_i$$

This implies that the general form of the solution is:

$$x_i = \frac{f_i(\rho, b)\, \sigma_i^{-1}}{\sum_{j=1}^{n} f_j(\rho, b)\, \sigma_j^{-1}}$$

The function $f_i(\rho, b)$ depends on the constant correlation ρ and the vector b of risk budgets. In particular, it verifies $f_i\left(-(n-1)^{-1}, b\right) = 1$, $f_i(0, b) = \sqrt{b_i}$, $f_i(1, b) = b_i$ and $f_i\left(\rho, n^{-1}\mathbf{1}\right) = 1$.

To illustrate the case of the constant correlation matrix, we have reported some simulations in Figure 2.5. We consider a universe of n assets with identical volatilities. We note b_1 the risk budget of the first asset. The risk budget of the other assets is uniform and equal to $(1-b_1)/(n-1)$. We note the effect of the constant correlation ρ on the weight x_1, in particular when the number n of assets is large. When the number of assets is small (less than 10), the correlation ρ has an impact only when it is small (less than 10%). In other cases, Formula (2.27) is a good approximation of the solution[28].

In other cases, it is not possible to find explicit solutions of the RB portfolio. Nonetheless we can find a financial interpretation of the RB portfolio.

[28] Let $x_i(\rho)$ be the weight of asset i in the RB portfolio when the constant correlation is equal to ρ. A better approximation is given by the following rule:

$$x_i(\rho) \simeq \left(1-\sqrt{\rho}\right) x_i(0) + \sqrt{\rho}\, x_i(1)$$

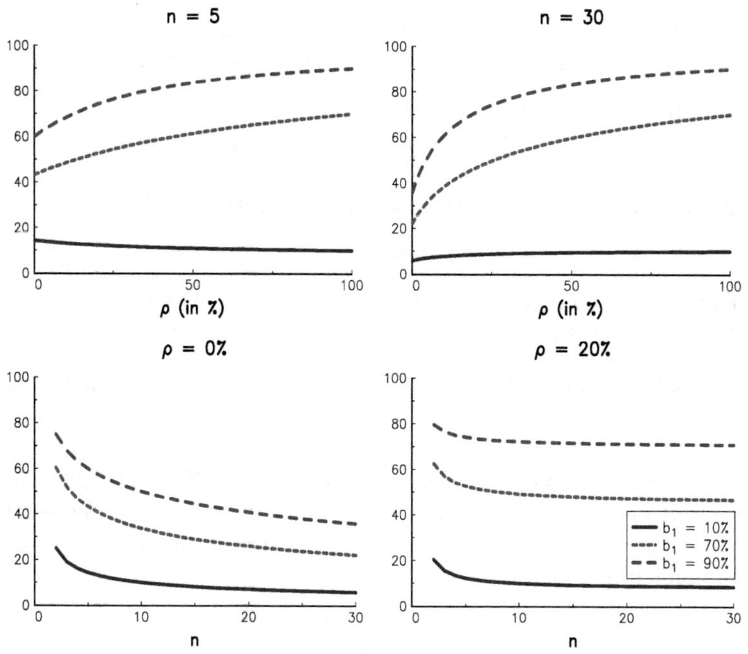

FIGURE 2.5: Simulation of the weight x_1 when the correlation is constant

We recall that the covariance between the returns of assets and portfolio x is equal to Σx. The beta β_i of asset i with respect to portfolio x is then defined as the ratio between the covariance term $(\Sigma x)_i$ and the variance $x^\top \Sigma x$ of the portfolio:

$$\beta_i = \frac{\operatorname{cov}(R_i, R(x))}{\sigma^2(x)} = \frac{(\Sigma x)_i}{\sigma^2(x)}$$

β_i indicates the sensitivity of asset i to the systematic risk, which is represented here by portfolio x. This means that the risk contribution $\mathcal{RC}_i$ is equal to:

$$\mathcal{RC}_i = x_i \beta_i \sigma(x)$$

Finding the RB portfolio such that $\mathcal{RC}_i / b_i = \mathcal{RC}_j / b_j$ implies then:

$$b_j x_i \beta_i = b_i x_j \beta_j$$

We finally deduce that:

$$x_i = \frac{b_i \beta_i^{-1}}{\sum_{j=1}^{n} b_j \beta_j^{-1}} \tag{2.29}$$

The weight allocated to component i is thus inversely proportional to its beta. However, contrary to the solutions (2.26), (2.27) and (2.28), this one is endogenous since β_i is the beta of asset i with respect to the RB portfolio[29].

Remark 29 *Other properties will be derived in Section 2.3 concerning the special case when the risk budgets are the same (ERC portfolio).*

2.2.2.2 Existence and uniqueness of the RB portfolio

In this section, we consider two situations:

- The first one corresponds to the RB portfolio as defined by the non-linear system (2.21). In this case, the risk budgets are strictly positive.
- The second situation covers the case when some risk budgets are equal to zero.

The case with strictly positive risk budgets To study this question, the optimization problem (2.22) is not very appropriate. Following Maillard *et al.* (2010), Bruder and Roncalli (2012) suggest considering a new optimization problem defined as follows:

$$\begin{aligned} y^\star &= \arg\min \mathcal{R}(y) \\ &\text{u.c.} \quad \left\{ \begin{array}{l} \sum_{i=1}^n b_i \ln y_i \geq c \\ y \geq \mathbf{0} \end{array} \right. \end{aligned} \tag{2.30}$$

with c an arbitrary constant. The associated Lagrange function is:

$$\mathcal{L}(y;\lambda,\lambda_c) = \mathcal{R}(y) - \lambda^\top y - \lambda_c \left(\sum_{i=1}^n b_i \ln y_i - c \right)$$

with $\lambda \in \mathbb{R}^n$ and $\lambda_c \in \mathbb{R}$. The solution $y^\star$ verifies the following first-order condition:

$$\frac{\partial \mathcal{L}(y;\lambda,\lambda_c)}{\partial y_i} = \frac{\partial \mathcal{R}(y)}{\partial y_i} - \lambda_i - \lambda_c \frac{b_i}{y_i} = 0$$

while the Kuhn-Tucker conditions are:

$$\left\{ \begin{array}{l} \min(\lambda_i, y_i) = 0 \\ \min\left(\lambda_c, \sum_{i=1}^n b_i \ln y_i - c\right) = 0 \end{array} \right.$$

Because $\ln y_i$ is not defined for $y_i = 0$, it follows that $y_i > 0$ and $\lambda_i = 0$. We note that the constraint $\sum_{i=1}^n b_i \ln y_i = c$ is necessarily reached (because the solution cannot be $y^\star = \mathbf{0}$), then $\lambda_c > 0$ and we have:

$$y_i \frac{\partial \mathcal{R}(y)}{\partial y_i} = \lambda_c b_i$$

[29] With Equation (2.29), we have a new interpretation of the function $f_i(\rho, b)$:

$$f_i(\rho, b) = \frac{b_i}{\beta_i} \sigma_i$$

It is the product of the risk budget and the volatility scaled by the beta.

We verify that the risk contributions are proportional to the risk budgets:

$$\mathcal{RC}_i = \lambda_c b_i$$

Moreover, we note that we face a well-known optimization problem (minimizing a convex function subject to lower convex bounds). We deduce that the optimization program (2.30) has a solution and is unique.

Portfolio $y^\star$ is a RB portfolio, but it does not correspond to the normalized RB portfolio defined by the non-linear system (2.21) because there is no reason that $\sum_{i=1}^n y_i^\star = 1$. However, we can easily deduce the normalized RB portfolio by scaling the solution $y^\star$:

$$x_i^\star = \frac{y_i^\star}{\sum_{j=1}^n y_j^\star}$$

In this case, portfolio $x^\star$ is the unique solution of the non-linear system (2.22).

Note that the solution $x^\star$ may be found directly from the optimization problem (2.30) by using a constant $c^\star = c - \ln\left(\sum_{i=1}^n y_i^\star\right)$ where c is the constant used to find $y^\star$. It is therefore tempting to define the RB portfolio $x^\star$ as the solution to the modified problem:

$$\begin{aligned} x^\star(c) &= \arg\min \mathcal{R}(x) \\ &\text{u.c.} \quad \begin{cases} \sum_{i=1}^n b_i \ln x_i \geq c \\ \mathbf{1}^\top x = 1 \\ x \geq \mathbf{0} \end{cases} \end{aligned} \tag{2.31}$$

The Lagrange function becomes:

$$\mathcal{L}(x;\lambda,\lambda_c) = \mathcal{R}(x) - \lambda_0\left(\mathbf{1}^\top x - 1\right) - \lambda^\top x - \lambda_c\left(\sum_{i=1}^n b_i \ln x_i - c\right)$$

with $\lambda_0 \in \mathbb{R}$. At the optimum, we have:

$$\mathcal{RC}_i = x_i \frac{\partial\,\mathcal{R}(x)}{\partial\,x_i} = \lambda_0 x_i + \lambda_c b_i$$

However, there is no reason that $\lambda_0 = 0$ meaning that the solution to the optimization problem (2.31) is not necessarily a RB portfolio. Indeed, there is only one value c such that $\lambda_0 = 0$. It corresponds to the previous value $c^\star$ and in this case defines the RB portfolio.

Remark 30 *We note that the convexity property of the risk measure is essential to the existence and uniqueness of the RB portfolio. If $\mathcal{R}(x)$ is not convex, the preceding analysis becomes invalid.*

Effect on the solution of setting risk budgets to zero Following Bruder and Roncalli (2012), the analysis of the optimization problem (2.30) also clarifies the disturbing point discussed in Section 2.2.1.1 on page 101 about the possibility of several solutions when some assets have a zero risk budget. The previous analysis is valid because $b_i > 0$. If one or several risk budgets are set to zero, the solution is modified as follows. Let $\mathcal{N}$ be the set of assets such that $b_i = 0$. In this case, the Lagrange function becomes:

$$\mathcal{L}(y; \lambda, \lambda_c) = \mathcal{R}(y) - \lambda^\top y - \lambda_c \left(\sum_{i \notin \mathcal{N}} b_i \ln y_i - c \right)$$

The solution $y^\star$ verifies the following first-order condition:

$$\frac{\partial \mathcal{L}(y; \lambda, \lambda_c)}{\partial y_i} = \begin{cases} \partial_{y_i} \mathcal{R}(y) - \lambda_i - \lambda_c b_i y_i^{-1} = 0 & \text{if} \quad i \notin \mathcal{N} \\ \partial_{y_i} \mathcal{R}(y) - \lambda_i = 0 & \text{if} \quad i \in \mathcal{N} \end{cases}$$

If $i \notin \mathcal{N}$, the previous analysis is valid and we verify that risk contributions are proportional to the risk budgets:

$$y_i \frac{\partial \mathcal{R}(y)}{\partial y_i} = \lambda_c b_i$$

If $i \in \mathcal{N}$, we must distinguish two cases. If $y_i = 0$, it implies that $\lambda_i > 0$ and $\partial_{y_i} \mathcal{R}(y) > 0$. In the other case, if $y_i > 0$, it implies that $\lambda_i = 0$ and $\partial_{y_i} \mathcal{R}(y) = 0$. The solution $y_i = 0$ or $y_i > 0$ if $i \in \mathcal{N}$ will then depend on the structure of the covariance matrix Σ.

We conclude that the solution $y^\star$ of the optimization problem (2.30) exists and is unique even if some risk budgets are set to zero. As previously, we deduce the normalized RB portfolio $x^\star$ by scaling $y^\star$. This solution, noted $\mathcal{S}_1$, satisfies the following relationships:

$$\begin{cases} \mathcal{RC}_i = x_i \cdot \partial_{x_i} \mathcal{R}(x) = b_i & \text{if} \quad i \notin \mathcal{N} \\ \begin{cases} x_i = 0 \text{ and } \partial_{x_i} \mathcal{R}(x) > 0 \quad (i) \\ \text{or} \\ x_i > 0 \text{ and } \partial_{x_i} \mathcal{R}(x) = 0 \quad (ii) \end{cases} & \text{if} \quad i \in \mathcal{N} \end{cases} \tag{2.32}$$

The conditions (i) and (ii) are mutually exclusive for one asset $i \in \mathcal{N}$, but not necessarily for all the assets $i \in \mathcal{N}$.

However, the previous analysis implies also that there may be several solutions to the non-linear system (2.20) when $b_i = 0$ for $i \in \mathcal{N}$. Let $\mathcal{N} = \mathcal{N}_1 \bigsqcup \mathcal{N}_2$ where $\mathcal{N}_1$ is the set of assets verifying the condition (i) and $\mathcal{N}_2$ is the set of assets verifying the condition (ii) in Equation (2.32). The number of solutions[30] is equal to 2^m where $m = |\mathcal{N}_2|$ is the cardinality of $\mathcal{N}_2$. Indeed, it is

[30]To be more exact, 2^m is the maximum number of solutions. However, the case when fixing some weights of $\mathcal{N}_2$ to zero implies that other assets of $\mathcal{N}_2$ have a weight equal to zero has a small probability to occur.

the number of k-combinations for all $k = 0, 1, \ldots, m$. Suppose that $m = 0$. In this case, the solution $\mathcal{S}_1$ verifies $x_i = 0$ if $b_i = 0$. It corresponds to the solution expected by the investor. If $m \geq 1$, there are several solutions, in particular the solution $\mathcal{S}_1$ given by the optimization problem (2.30) and another solution $\mathcal{S}_2$ with $x_i = 0$ for all assets such that $b_i = 0$. We then obtain a paradox because even if $\mathcal{S}_2$ is the solution expected by the investor, the only acceptable solution is $\mathcal{S}_1$. The argument is very simple. Suppose we impose $b_i = \varepsilon_i$ with $\varepsilon_i > 0$ a small number for $i \in \mathcal{N}$. In this case, we obtain a unique solution. If $\varepsilon_i \to 0$, this solution will converge to $\mathcal{S}_1$, not to $\mathcal{S}_2$ or all the others solutions $\mathcal{S}_j$ for $j = 2, \ldots, 2^m$.

Example 12 *We consider a universe of three assets with $\sigma_1 = 20\%$, $\sigma_2 = 10\%$ and $\sigma_3 = 5\%$. The correlation of asset returns is given by the following matrix:*

$$\rho = \begin{pmatrix} 1.00 & & \\ 0.50 & 1.00 & \\ \rho_{1,3} & \rho_{2,3} & 1.00 \end{pmatrix}$$

We would like to build a RB portfolio such that the risk budgets with respect to the volatility risk measure are $(50\%, 50\%, 0\%)$.

This example has already been used in Section 2.2.1.1 on page 101 to illustrate the case of multiple solutions. The results are reported in Table 2.12. If $\rho_{1,3} = \rho_{2,3} = -25\%$, we obtain two solutions $\mathcal{S}_1$ and $\mathcal{S}_2$. For the first solution, we have $x_1 = 20\%$, $x_2 = 40\%$ and $x_3 = 40\%$, whereas the weights for the second solution are respectively $x_1 = 33.33\%$, $x_2 = 66.67\%$ and $x_3 = 0\%$. We verify that the marginal risk $\partial_{x_3} \sigma(x)$ is negative for the second solution indicating that it cannot be a solution to the optimization problem (2.30). This is why the volatility of this second solution (11.55%) is larger than the volatility of the first solution (6.63%). We also note that if we slightly alter the risk budgets ($b_1 = 49.5\%$, $b_2 = 49.5\%$ and $b_3 = 1\%$), the solution $\mathcal{S}_1'$ is closer to $\mathcal{S}_1$ than to $\mathcal{S}_2$. It confirms that the first solution $\mathcal{S}_1$ is more acceptable. If all the correlations are positive ($\rho_{1,3} = \rho_{2,3} = 25\%$), we do not face this problem. The unique solution is $x_1 = 33.33\%$, $x_2 = 66.67\%$ and $x_3 = 0\%$ and we verify that the marginal risk $\partial_{x_3} \sigma(x)$ is positive. In Figure 2.6, we have represented the evolution of the volatility of the portfolio $(50\%, 50\%, x_3)$ with respect to the weight x_3. If $\rho_{1,3} = \rho_{2,3} = -25\%$, the volatility decreases around $x_3 = 0$ meaning that the minimum is after $x_3 = 0$. If $\rho_{1,3} = \rho_{2,3} = 25\%$, the volatility increases around $x_3 = 0$ meaning that the minimum is before $x_3 = 0$. This figure illustrates that the solution verifies $x_3 > 0$ in the first case and $x_3 = 0$ in the second case.

Let us add a fourth asset to the previous universe with $\sigma_4 = 10\%$, $\rho_{1,3} = \rho_{2,3} = \rho_{1,4} = \rho_{2,4} = -25\%$ and $\rho_{3,4} = 50\%$. Table 2.13 gives the results when the risk budgets are $b_1 = 50\%$, $b_2 = 50\%$, $b_3 = 0\%$ and $b_4 = 0\%$. If we analyze the solution $\mathcal{S}_1$, we note that the number m of assets such that both the marginal risk and the risk budgets are zero is equal to 2. This is why we

TABLE 2.12: RB solutions when the risk budget b_3 is equal to 0

$\rho_{1,3} = \rho_{2,3}$	Solution		1	2	3	$\sigma(x)$
−25%	$\mathcal{S}_1$	x_i	20.00%	40.00%	40.00%	
		$\mathcal{MR}_i$	16.58%	8.29%	0.00%	6.63%
		$\mathcal{RC}_i$	50.00%	50.00%	0.00%	
	$\mathcal{S}_2$	x_i	33.33%	66.67%	0.00%	
		$\mathcal{MR}_i$	17.32%	8.66%	−1.44%	11.55%
		$\mathcal{RC}_i$	50.00%	50.00%	0.00%	
	$\mathcal{S}'_1$	x_i	19.23%	38.46%	42.31%	
		$\mathcal{MR}_i$	16.42%	8.21%	0.15%	6.38%
		$\mathcal{RC}_i$	49.50%	49.50%	1.00%	
25%	$\mathcal{S}_1$	x_i	33.33%	66.67%	0.00%	
		$\mathcal{MR}_i$	17.32%	8.66%	1.44%	11.55%
		$\mathcal{RC}_i$	50.00%	50.00%	0.00%	

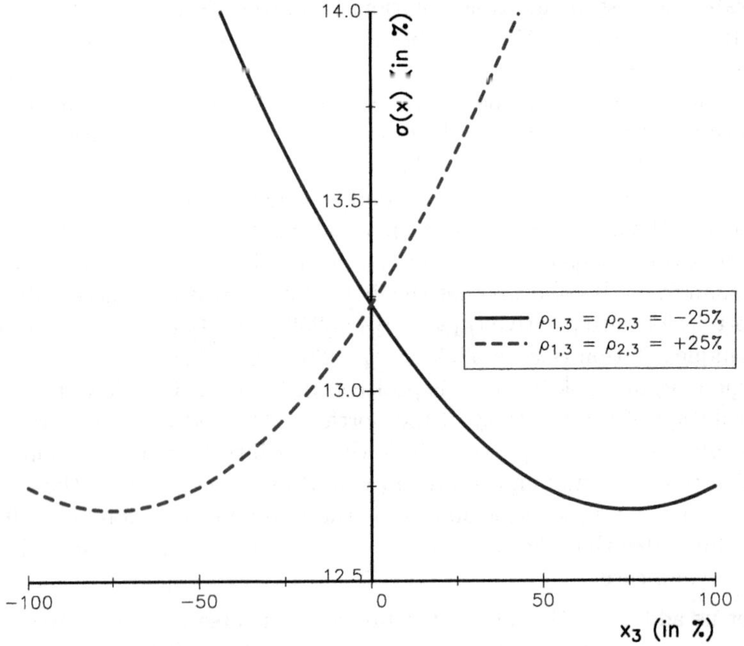

FIGURE 2.6: Evolution of the portfolio's volatility with respect to x_3

obtain $2^2 = 4$ solutions. Another way to determine the number of solutions is to compute the solution $\mathcal{S}_2$ desired by the investor and set m equal to the number of assets such that the marginal risk is strictly negative when the risk budget is equal to zero.

TABLE 2.13: RB solutions when the risk budgets b_3 and b_4 are equal to 0

Solution		1	2	3	4	$\sigma(x)$
	x_i	20.00%	40.00%	26.67%	13.33%	
$\mathcal{S}_1$	$\mathcal{MR}_i$	16.33%	8.16%	0.00%	0.00%	6.53%
	$\mathcal{RC}_i$	50.00%	50.00%	0.00%	0.00%	
	x_i	33.33%	66.67%	0.00%	0.00%	
$\mathcal{S}_2$	$\mathcal{MR}_i$	17.32%	8.66%	−1.44%	−2.89%	11.55%
	$\mathcal{RC}_i$	50.00%	50.00%	0.00%	0.00%	
	x_i	20.00%	40.00%	40.00%	0.00%	
$\mathcal{S}_3$	$\mathcal{MR}_i$	16.58%	8.29%	0.00%	−1.51%	6.63%
	$\mathcal{RC}_i$	50.00%	50.00%	0.00%	0.00%	
	x_i	25.00%	50.00%	0.00%	25.00%	
$\mathcal{S}_4$	$\mathcal{MR}_i$	16.58%	8.29%	−0.75%	0.00%	8.29%
	$\mathcal{RC}_i$	50.00%	50.00%	0.00%	0.00%	

RB portfolio with negative risk budgets We have stated previously that it is not natural to set risk budgets to be negative. The following analysis is therefore more theoretical, but it can help us understand why negative risk budgets cause issues to define the RB portfolio.

Let $\mathcal{P}$ and $\mathcal{N}$ be the set of assets such that $b_i > 0$ and $b_i < 0$. In this case, the Kuhn-Tucker condition of the optimization portfolio (2.30) with $y_i > 0$ becomes:

$$\min\left(\lambda_c, \sum_{i\in\mathcal{P}} b_i \ln y_i - \sum_{i\in\mathcal{N}} b_i \ln y_i - c\right) = 0$$

The constraint is not convex anymore and we do not necessarily have $\lambda_c > 0$. This is why the existence of the solution is not guaranteed. If we impose $y_i \leq 0$ if $b_i < 0$, we can modify the constraint:

$$\sum_{i\in\mathcal{P}} b_i \ln y_i + \sum_{i\in\mathcal{N}} b_i \ln(-y_i) \geq c$$

We face the same problem as previously because the constraint is not convex.

2.2.3 Optimality of the risk budgeting portfolio

The risk budgeting approach is a heuristic asset allocation method. In finance, we like to derive optimal portfolios from the maximization problem of a utility function. For example, Maillard *et al.* (2010) show that the ERC portfolio corresponds to the tangency portfolio when the correlations are the same and when the assets have the same Sharpe ratio. For the RB portfolio, it is more difficult to find such properties. However, if we consider the dual of the optimization problem (2.30), we obtain:

$$
\begin{aligned}
x^\star &= \arg\max \sum_{i=1}^{n} b_i \ln x_i \\
&\text{u.c.} \quad \begin{cases} \mathcal{R}(x) \geq c \\ \mathbf{1}^\top x = 1 \\ x \geq \mathbf{0} \end{cases}
\end{aligned}
$$

The objective function is then equal to $\ln\left(\prod_{i=1}^{n} x_i^{b_i}\right)$. As noted by Bruder and Roncalli (2012), "*interpreting this problem as a utility maximization problem is attractive, but the objective function does not present the right properties to be a utility function*".

Another route for exploring the optimality of the RB portfolio is to consider the utility function[31]:

$$
\mathcal{U}(x) = \mu(x) - \frac{\phi}{2}\mathcal{R}^2(x)
$$

This is a generalization of the Makowitz utility function[32]. Following Black and Litterman (1992), portfolio x is optimal if the vector of expected returns satisfies this relationship[33]:

$$
\tilde{\mu} = \frac{\partial\, \mu(x)}{\partial\, x} = \phi\mathcal{R}(x)\frac{\partial\, \mathcal{R}(x)}{\partial\, x}
$$

If the RB portfolio is optimal, we deduce that the performance contribution $\mathcal{PC}_i$ of asset i is proportional to its risk budget:

$$
\begin{aligned}
\mathcal{PC}_i &= x_i\tilde{\mu}_i \\
&= \phi\mathcal{R}(x)\cdot\mathcal{RC}_i \\
&\propto b_i
\end{aligned}
$$

The specification of the risk budgets allows us to decide not only which amount of risk to invest in an asset, but also which amount of expected performance to attribute to the asset. If the risk measure is the volatility, the previous equation becomes:

$$
\begin{aligned}
\mathcal{PC}_i &= \phi\mathcal{R}(x)\cdot\mathcal{RC}_i \\
&= \phi x_i\cdot(\Sigma x)_i
\end{aligned}
$$

[31]We assume implicitly that the risk-free rate is equal to 0. The generalization to non-zero risk-free rate is straightforward.

[32]Note that if we specify the utility function as $\mathcal{U}(x) = \mu(x) - \frac{\phi}{2}\mathcal{R}(x)$, we obtain the same result.

[33]We use the notation $\tilde{\mu}$ to specify that it is the market price of expected returns with respect to the current portfolio.

This is the relationship (1.10) exhibited in the Black-Litterman model. The absolute performance contribution will depend on the specification of ϕ. Using Equation (1.11) on page 23, another expression of the performance contribution is:

$$\mathcal{PC}_i = \mathrm{SR}\left(x \mid r\right) \frac{x_i \cdot \left(\Sigma x\right)_i}{\sqrt{x^\top \Sigma x}}$$

where $\mathrm{SR}\left(x \mid r\right)$ is the expected Sharpe ratio of the RB portfolio.

Remark 31 *From an ex-ante point of view, performance budgeting and risk budgeting are equivalent. This result is not new, as it was already the core idea behind the Black-Litterman model.*

Example 13 *We consider a universe of four assets. The volatilities are respectively* 10%, 20%, 30% *and* 40%. *The correlation of asset returns is given by the following matrix:*

$$\rho = \begin{pmatrix} 1.00 & & & \\ 0.80 & 1.00 & & \\ 0.20 & 0.20 & 1.00 & \\ 0.20 & 0.20 & 0.50 & 1.00 \end{pmatrix}$$

The results are reported in Tables 2.14 and 2.15 for two sets of risk budgets and the volatility risk measure. We assume that the investor targets a Sharpe ratio equal to 0.50. In the case $b = (20\%, 25\%, 40\%, 15\%)$, the RB portfolio is $x = (40.91\%, 25.12\%, 25.26\%, 8.71\%)$ and its volatility is 14.53%. Using these figures, we deduce that $\phi = 0.58$. We can also compute the implied return $\tilde{\mu}$, which is equal to $(3.55\%, 7.23\%, 11.50\%, 12.52\%)$. It follows that the (implied) expected return of the portfolio is:

$$\mu\left(x\right) = \sum_{i=1}^{4} x_i \tilde{\mu}_i = 7.27\%$$

We can decompose this expected return by performance contributions. We obtain $\mathcal{PC}_1 = 1.45\%$, $\mathcal{PC}_2 = 1.82\%$, $\mathcal{PC}_3 = 2.91\%$ and $\mathcal{PC}_4 = 1.09\%$. We verify that performance contributions are equal to risk contributions in relative value. In the case $b = (10\%, 10\%, 10\%, 70\%)$, we note that the fourth asset presents an expected return equal to 18.37%. We can easily understand this high expected return: If the investor would like to budget 70% of the risk in the fourth asset, which is the riskier asset of the universe, he believes that it has necessarily a high expected return.

Remark 32 *In this section, we assume that the portfolio is optimal from the point of view of the investor. We can consider it to be a strong hypothesis. However, when the investor puts his money into a portfolio, he thinks necessarily that it is the best investment for him*[34]. *So, the investment is optimal for him, and the implied returns $\tilde{\mu}$ can be viewed as the performance of each returns expected by the investor.*

---

[34]If he didn't think that, he would not invest in this portfolio.

TABLE 2.14: Implied risk premia when $b = (20\%, 25\%, 40\%, 15\%)$

Asset	x_i	$\mathcal{MR}_i$	$\tilde{\mu}_i$	$\mathcal{PC}_i$	$\mathcal{PC}_i^\star$
1	40.91	7.10	3.55	1.45	20.00
2	25.12	14.46	7.23	1.82	25.00
3	25.26	23.01	11.50	2.91	40.00
4	8.71	25.04	12.52	1.09	15.00
Expected return				7.27	

TABLE 2.15: Implied risk premia when $b = (10\%, 10\%, 10\%, 70\%)$

Asset	x_i	$\mathcal{MR}_i$	$\tilde{\mu}_i$	$\mathcal{PC}_i$	$\mathcal{PC}_i^\star$
1	35.88	5.27	2.63	0.94	10.00
2	17.94	10.53	5.27	0.94	10.00
3	10.18	18.56	9.28	0.94	10.00
4	35.99	36.75	18.37	6.61	70.00
Expected return				9.45	

2.2.4 Stability of the risk budgeting approach

We have seen that optimized portfolios are very sensitive to input parameters. In practice, they are not very stable. This is why the authors have proposed so many regularization approaches, which have been presented in Section 1.2.3 in the first chapter. Our experience shows that we do not face this problem with RB portfolios.

Example 14 *We consider a universe of three assets. The expected returns are respectively* $\mu_1 = \mu_2 = 8\%$ *and* $\mu_3 = 5\%$*. For the volatilities, we have* $\sigma_1 = 20\%$, $\sigma_2 = 21\%$, $\sigma_3 = 10\%$*. Moreover, we assume that the cross-correlations are the same and we have* $\rho_{i,j} = \rho = 80\%$*.*

This example has already been studied on page 45 to illustrate the lack of robustness of optimized portfolios. If the objective is to target a volatility equal to 15%, the optimized (or MVO) portfolio is equal to $x_1 = 38.3\%$, $x_2 = 20.2\%$ and $x_3 = 41.5\%$. In this case, the risk budgets are respectively $b_1 = 49.0\%$, $b_2 = 25.8\%$ and $b_3 = 25.2\%$. Of course, the RB portfolio corresponding to these risk budgets is exactly the optimized portfolio. We would like to evaluate how these two portfolios are impacted by slightly changes in the input parameters. For that, the objective is respectively:

- to maximize the expected return with a volatility constraint of 15% for the MVO portfolio[35]; and

[35]We also impose the long-only constraint in order to compare the MVO portfolio to the RB portfolio.

- to match the risk budgets (49.0%, 25.8%, 25.2%) for the RB portfolio.

We obtain the results given in Tables 2.16 and 2.17. For example, if the uniform correlation is equal to 90%, the MVO and RB portfolios become (44.6%, 8.9%, 46.5%) and (38.9%, 20.0%, 41.1%). We have a substantial decrease in the weight of the second asset for the MVO portfolio. We observe similar behavior if we change the parameter σ_2, etc.

TABLE 2.16: Sensitivity of the MVO portfolio to input parameters

ρ		70%	90%		90%	
σ_2				18%	18%	
μ_1						9%
x_1	38.3	38.3	44.6	13.7	0.0	56.4
x_2	20.2	25.9	8.9	56.1	65.8	0.0
x_3	41.5	35.8	46.5	30.2	34.2	43.6

TABLE 2.17: Sensitivity of the RB portfolio to input parameters

ρ		70%	90%		90%	
σ_2				18%	18%	
μ_1						9%
x_1	38.3	37.7	38.9	37.1	37.7	38.3
x_2	20.2	20.4	20.0	22.8	22.6	20.2
x_3	41.5	41.9	41.1	40.1	39.7	41.5

The MVO portfolio is thus very sensitive to the input parameters whereas the RB portfolio is more robust. For MVO portfolios, the risk approach is marginal and the quantity of interest to study is the marginal volatility. For RB portfolios, the risk approach becomes more global by mixing the weight and the marginal volatility. It is a first explanation of the better behavior of the RB portfolio.

A second explanation concerns the nature of the optimization problem. We have seen that the RB portfolio is a minimum risk portfolio subject to a constraint of weight diversification. So, the RB portfolio optimization problem contains implicitly a regularization constraint. Let us consider the case of the volatility risk measure. On page 109, we have seen that the solution to the following optimization problem[36]:

$$
\begin{aligned}
x^{\star} &= \arg\min \frac{1}{2} x^{\top} \Sigma x \\
\text{u.c.} & \quad \sum_{i=1}^{n} b_i \ln x_i \geq c^{\star}
\end{aligned}
$$

[36]We omit the linear constraint $\mathbf{1}^{\top} x = 1$ because we consider the optimal value $c^{\star}$.

is the RB portfolio $x^\star$. In this case, we can use the results of Jagannathan and Ma (2003) to build a shrinkage covariance matrix $\tilde{\Sigma} = \Sigma + \Delta$ such that $x^\star$ is the minimum variance portfolio[37]. Let $\lambda_{c^\star}$ be the Lagrange multiplier associated with the diversification constraint at the optimum. One solution is:

$$\tilde{\Sigma}^{(1)} = \Sigma + \alpha_1 \mathbf{1}\mathbf{1}^\top - \left(\Lambda \mathbf{1}^\top + \mathbf{1}\Lambda^\top\right)$$

where:

$$\Lambda = \lambda_{c^\star} \begin{pmatrix} b_1 / x_1^\star \\ \vdots \\ b_n / x_n^\star \end{pmatrix}$$

with $\alpha_1 > n\lambda_{c^\star}$. However, this solution is not unique. For example, another candidate for $\tilde{\Sigma}$ is:

$$\tilde{\Sigma}^{(2)} = \Sigma + \alpha_2 D - \left(\Lambda \mathbf{1}^\top + \mathbf{1}\Lambda^\top\right)$$

with $D = \operatorname{diag}(\Lambda_1, \ldots, \Lambda_n)$ and $\alpha_2 > n$. Using these two previous solutions, we can always find other solutions $\tilde{\Sigma}^{(3)} = \Sigma + \alpha_1 \mathbf{1}\mathbf{1}^\top + \alpha_2 D - \left(\Lambda \mathbf{1}^\top + \mathbf{1}\Lambda^\top\right)$ or $\tilde{\Sigma}^{(4)} = \alpha_1 \tilde{\Sigma}^{(1)} + \alpha_2 \tilde{\Sigma}^{(2)}$ such that the Frobenious norm $\left\|\tilde{\Sigma} - \Sigma\right\|$ reaches its minimum.

Example 15 *We consider an investment universe of three assets. The volatilities are equal to* 20%, 15% *and* 25% *and the correlation matrix is:*

$$\rho = \begin{pmatrix} 1.00 & & \\ 0.70 & 1.00 & \\ 0.40 & 0.50 & 1.00 \end{pmatrix}$$

With $b = (50\%, 30\%, 20\%)$, the composition of the RB portfolio is $x^\star = (44.35\%, 36.82\%, 18.83\%)$. Because we have $\lambda_{c^\star} = 2.593\%$, we obtain the shrinkage covariance matrices given in Tables 2.18 and 2.19.

TABLE 2.18: Shrinkage covariance matrix $\tilde{\Sigma}^{(1)}$ associated to the RB portfolio

Asset	$\tilde{\sigma}_i$	$\tilde{\rho}_{i,j}$		
1	19.13%	100.00%		
2	18.92%	82.54%	100.00%	
3	22.93%	57.69%	68.08%	100.00

[37] It implies that $\tilde{\Sigma}x^\star$ is a constant vector $\lambda_0 \mathbf{1}$.

TABLE 2.19: Shrinkage covariance matrix $\tilde{\Sigma}^{(3)}$ associated to the RB portfolio

Asset	$\tilde{\sigma}_i$	$\tilde{\rho}_{i,j}$		
1	18.26%	100.00%		
2	17.93%	67.67%	100.00%	
3	24.40%	33.25%	49.39%	100.00

2.3 Special case: the ERC portfolio

The ERC portfolio is a RB portfolio such that the risk budgets are the same for the different assets:

$$b_i = \frac{1}{n}$$

It can be viewed as a neutral portfolio when the manager has no views on the risk budgets. As a result, it has the same properties of the RB portfolio presented before, but it also has some specific properties which have been studied by Maillard *et al.* (2010) in the case of the volatility risk measure.

2.3.1 The two-asset case ($n = 2$)

We note by σ_i the volatility of the ith asset, ρ the correlation and $x = (w, 1 - w)$ the portfolio weights. The marginal risk contributions are:

$$\frac{\partial\,\sigma(x)}{\partial\,x} = \frac{\Sigma x}{\sqrt{x^\top \Sigma x}} = \frac{1}{\sigma(x)} \begin{pmatrix} w\sigma_1^2 + (1-w)\rho\sigma_1\sigma_2 \\ (1-w)\sigma_2^2 + w\rho\sigma_1\sigma_2 \end{pmatrix}$$

The ERC portfolio satisfies then:

$$w \cdot \frac{\left(w\sigma_1^2 + (1-w)\rho\sigma_1\sigma_2\right)}{\sigma(x)} = (1-w) \cdot \frac{\left((1-w)\sigma_2^2 + w\rho\sigma_1\sigma_2\right)}{\sigma(x)}$$

We deduce that:

$$w^2\sigma_1^2 = (1-w)^2\sigma_2^2$$

Because $0 \leq w \leq 1$, the unique solution is:

$$\begin{aligned} w^\star &= \frac{1}{\sigma_1} \Big/ \left(\frac{1}{\sigma_1} + \frac{1}{\sigma_2}\right) \\ &= \frac{\sigma_2}{\sigma_1 + \sigma_2} \end{aligned} \tag{2.33}$$

The weights of the ERC portfolio are inversely proportional to the volatilities. Curiously, they do not depend on the correlation ρ.

Remark 33 *Another way to find these optimal weights is to consider the solution of the RB portfolio given in Equation (2.23) with $b = 1/2$. We obtain:*

$$\begin{aligned} w^{\star} &= \frac{-1/2\sigma_2^2 + 1/2\sigma_1\sigma_2}{1/2\sigma_1^2 - 1/2\sigma_2^2} \\ &= \frac{\sigma_2 (\sigma_1 - \sigma_2)}{(\sigma_1 - \sigma_2)(\sigma_1 + \sigma_2)} \\ &= \frac{\sigma_2}{\sigma_1 + \sigma_2} \end{aligned}$$

Let us compare this portfolio with two other heuristic approaches, which are the equally weighted (or EW) portfolio and the minimum variance (or MV) portfolio. We use the parameterization $\sigma_1 = \sigma$ and $\sigma_2 = k\sigma$ with $k \geq 0$. The ERC portfolio volatility is:

$$\sigma(x_{\mathrm{erc}}) = \sqrt{2(1+\rho)} \frac{k}{1+k} \sigma$$

In comparison, the volatility of an equally weighted portfolio is:

$$\sigma(x_{\mathrm{ew}}) = \frac{1}{2} \sigma \sqrt{1 + k^2 + 2\rho k}$$

Using these expressions, we can check that $\sigma(x_{\mathrm{ew}}) \geq \sigma(x_{\mathrm{erc}})$. The equality is obtained for $k = 1$, that is when $x_{\mathrm{erc}} = x_{\mathrm{ew}}$. We recall that, in the two-asset case, the minimum variance portfolio is:

$$x_{\mathrm{mv}} = \frac{1}{\sigma_1^2 + \sigma_2^2 - 2\rho\sigma_1\sigma_2} \begin{pmatrix} \sigma_2^2 - \rho\sigma_1\sigma_2 \\ \sigma_1^2 - \rho\sigma_1\sigma_2 \end{pmatrix}$$

Therefore, its volatility is:

$$\sigma(x_{\mathrm{mv}}) = k\sigma \sqrt{\frac{1-\rho^2}{1+k^2-2\rho k}}$$

Maillard *et al.* (2010) deduce finally that the volatility of the ERC portfolio is located between the volatility of the minimum variance portfolio and the equally weighted portfolio:

$$\sigma_{\mathrm{mv}} \leq \sigma_{\mathrm{erc}} \leq \sigma_{\mathrm{ew}}$$

This property plays an important role in risk parity (see Section 2.4.1).

Example 16 *We consider two assets with $\sigma_1 = 20\%$ and $\sigma_2 = 30\%$. The correlation between the asset returns is set to 50%.*

In Table 2.20, we have reported the composition of the EW, ERC and MV portfolios. We have also indicated the risk contributions of each portfolio. We note that $\sigma_{\mathrm{ew}} = 21.79\%$, $\sigma_{\mathrm{erc}} = 20.78\%$ and $\sigma_{\mathrm{mv}} = 19.64\%$, meaning that the inequality $\sigma_{\mathrm{mv}} \leq \sigma_{\mathrm{erc}} \leq \sigma_{\mathrm{ew}}$ holds.

TABLE 2.20: Risk contributions of EW, ERC and MV portfolios

Portfolio	Asset	x_i	$\mathcal{MR}_i$	$\mathcal{RC}_i$	$\mathcal{RC}_i^\star$
EW	1	50.00	16.06	8.03	36.84
	2	50.00	27.53	13.76	63.16
ERC	1	60.00	17.32	10.39	50.00
	2	40.00	25.98	10.39	50.00
MV	1	85.71	19.64	16.83	85.71
	2	14.29	19.64	2.81	14.29

2.3.2 The general case $(n > 2)$

As in the case of the RB portfolio, it is not possible to find an analytical solution when $n > 2$. However, we have some closed-form solutions in some specific cases, for instance when the correlation or the volatility is the same for the different assets of the investment universe.

If we assume a constant correlation matrix with $\rho_{i,j} = \rho$ for all i, j, the expression of the risk contribution is given by Equation (2.24). It follows that the weights of the ERC portfolio satisfy the following relationship:

$$x_i\sigma_i \left((1-\rho)\, x_i\sigma_i + \rho \sum_{k=1}^{n} x_k\sigma_k \right) = x_j\sigma_j \left((1-\rho)\, x_j\sigma_j + \rho \sum_{k=1}^{n} x_k\sigma_k \right)$$

It follows that $x_i\sigma_i = x_j\sigma_j$. Because $\sum_{i=1}^{n} x_i = 1$, we deduce that:

$$x_i = \frac{\sigma_i^{-1}}{\sum_{j=1}^{n} \sigma_j^{-1}} \tag{2.34}$$

Like in the two-asset case, the weight allocated to asset i is inversely proportional to its volatility and does not depend on the value of the correlation. Moreover, we can show that this solution is connected to the minimum variance portfolio. Let $\Sigma = \sigma\sigma^\top \circ C_n(\rho)$ be the covariance matrix with $C_n(\rho)$ the constant correlation matrix. We have $\Sigma^{-1} = \Gamma \circ C_n^{-1}(\rho)$ with $\Gamma_{i,j} = \sigma_i^{-1}\sigma_j^{-1}$ and:

$$C_n^{-1}(\rho) = \frac{\rho \mathbf{1}\mathbf{1}^\top - ((n-1)\rho + 1) I_n}{(n-1)\rho^2 - (n-2)\rho - 1}$$

We saw that the minimum variance portfolio is $x = \left(\Sigma^{-1}\mathbf{1}\right) / \left(\mathbf{1}^\top\Sigma^{-1}\mathbf{1}\right)$. We deduce that the expression of the MV weights are:

$$x_i = \frac{-((n-1)\rho + 1)\,\sigma_i^{-2} + \rho \sum_{j=1}^{n} (\sigma_i\sigma_j)^{-1}}{\sum_{k=1}^{n} \left(-((n-1)\rho + 1)\,\sigma_k^{-2} + \rho \sum_{j=1}^{n} (\sigma_k\sigma_j)^{-1} \right)}$$

We know that the lower bound of $C_n(\rho)$ is achieved for $\rho = -(n-1)^{-1}$. In this case, the solution becomes:

$$x_i = \frac{\sum_{j=1}^{n} (\sigma_i \sigma_j)^{-1}}{\sum_{k=1}^{n} \sum_{j=1}^{n} (\sigma_k \sigma_j)^{-1}} = \frac{\sigma_i^{-1}}{\sum_{k=1}^{n} \sigma_k^{-1}}$$

The ERC portfolio is then equal to the MV portfolio when the correlation is at its lowest possible value.

If we assume now that all volatilities are equal, i.e. $\sigma_i = \sigma$ for all i, the risk contribution becomes:

$$\mathcal{RC}_i = \frac{\left(\sum_{k=1}^{n} x_i x_k \rho_{i,k}\right) \sigma^2}{\sigma(x)}$$

The ERC portfolio verifies then:

$$x_i \left(\sum_{k=1}^{n} x_k \rho_{i,k} \right) = x_j \left(\sum_{k=1}^{n} x_k \rho_{j,k} \right)$$

We deduce that:

$$x_i = \frac{\left(\sum_{k=1}^{n} x_k \rho_{i,k}\right)^{-1}}{\sum_{j=1}^{n} \left(\sum_{k=1}^{n} x_k \rho_{j,k}\right)^{-1}} \tag{2.35}$$

The weight of asset i is inversely proportional to the weighted average of correlations of asset i. Contrary to the previous case, this solution is endogenous since x_i is a function of itself directly. The same issue of endogeneity naturally arises in the general case where both the volatilities and the correlations differ. If we replace b_i by $1/n$ in Equation (2.29), we deduce that:

$$x_i = \frac{\beta_i^{-1}}{\sum_{j=1}^{n} \beta_j^{-1}}$$

Because $\sum_{i=1}^{n} x_i \beta_i = 1$, we finally obtain:

$$x_i = \frac{1}{n \beta_i} \tag{2.36}$$

Example 17 *We consider an investment universe of four assets with $\sigma_1 = 15\%$, $\sigma_2 = 20\%$, $\sigma_3 = 30\%$ and $\sigma_4 = 10\%$. The correlation of asset returns is given by the following matrix:*

$$\rho = \begin{pmatrix} 1.00 & & & \\ 0.50 & 1.00 & & \\ 0.00 & 0.20 & 1.00 & \\ -0.10 & 0.40 & 0.70 & 1.00 \end{pmatrix}$$

The composition of the ERC is reported in Table 2.21. We verify that the weight of asset i is inversely proportional to its beta. Knowing the volatility, we can also deduce the risk contribution of each asset because we have $\mathcal{RC}_i = \sigma(x)/n$.

TABLE 2.21: Composition of the ERC portfolio

Asset	x_i	$\mathcal{MR}_i$	β_i	$\mathcal{RC}_i$	$\mathcal{RC}_i^\star$
1	31.34	8.52	0.80	2.67	25.00
2	17.49	15.27	1.43	2.67	25.00
3	13.05	20.46	1.92	2.67	25.00
4	38.12	7.00	0.66	2.67	25.00
Volatility				10.68	

2.3.3 Optimality of the ERC portfolio

Maillard *et al.* (2010) show that the ERC portfolio corresponds to the tangency portfolio in the case where the Sharpe ratio is the same for all assets and the correlation matrix is uniform.

Let us consider the tangency portfolio defined by:

$$x^\star = \arg\max \frac{\mu(x) - r}{\sigma(x)}$$

with $\mu(x) = x^\top \mu$ and $\sigma(x) = \sqrt{x^\top \Sigma x}$. The first-order condition of the optimization program is:

$$\frac{\partial_{x_i} \mu(x) - r}{\partial_{x_i} \sigma(x)} = \frac{\mu(x) - r}{\sigma(x)}$$

As noted by Scherer (2007), this condition is sufficient. It is therefore equivalent to say that the Sharpe ratio is maximized when the ratio of the marginal excess return to the marginal risk is the same for all assets of the portfolio and equals the Sharpe ratio of the portfolio. At the optimum, portfolio $x^\star$ verifies the following relationship:

$$\mu - r\mathbf{1} = \left(\frac{\mu(x^\star) - r}{\sigma(x^\star)^2} \right) \Sigma x^\star$$

If we assume a constant correlation matrix, the ERC portfolio is defined by $x_i = c\sigma_i^{-1}$ with $c = \left(\sum_{j=1}^n \sigma_j^{-1} \right)^{-1}$. Using Equation (2.24), it follows that:

$$\frac{\partial\, \sigma(x)}{\partial\, x_i} = c\frac{\sigma_i \left((1-\rho) + \rho n \right)}{\sigma(x)}$$

and:

$$\sigma^2(x) = nc^2 \left((1-\rho) + \rho n \right)$$

The ERC portfolio is the tangency portfolio if and only if we verify:

$$\mu_i - r = \frac{c \sum_{j=1}^n (\mu_i - r)\, \sigma_i^{-1}}{nc} \sigma_i$$

We can write this condition as follows:

$$\mu_i = r + \mathrm{SR} \cdot \sigma_i$$

with:

$$\mathrm{SR} = \frac{c \sum_{j=1}^{n} (\mu_i - r)\, \sigma_i^{-1}}{nc}$$

This implies that the Sharpe ratio is the same for all the assets.

Example 18 *We consider an investment universe of five assets. The volatilities are respectively equal to* 5%, 7%, 9%, 10% *and* 15%.

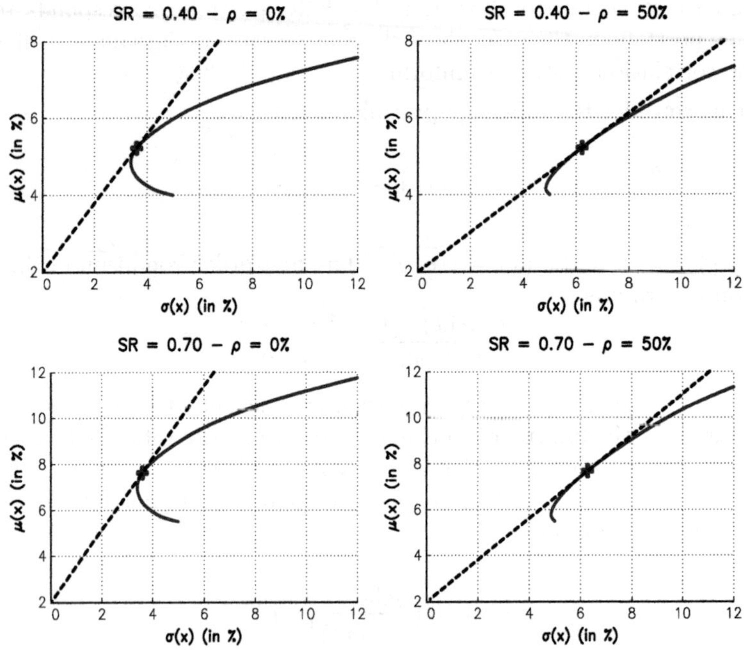

FIGURE 2.7: Location of the ERC portfolio in the mean-variance diagram when the Sharpe ratios are the same and the asset correlations are uniform

The results are reported in Figure 2.7 in the case of identical Sharpe ratios with a uniform correlation ρ and a risk-free rate equal to 2%. We note that the ERC portfolio is exactly equal to the tangency portfolio. In Figure 2.8, we assume a more general correlation matrix:

$$\rho = \begin{pmatrix} 1.00 & & & & \\ 0.50 & 1.00 & & & \\ 0.25 & 0.25 & 1.00 & & \\ 0.00 & 0.00 & 0.00 & 1.00 & \\ -0.25 & -0.25 & -0.25 & 0.00 & 1.00 \end{pmatrix}$$

In this case, the ERC portfolio (solid circle) is not equal to the tangency portfolio (star), but they are not very far from each other.

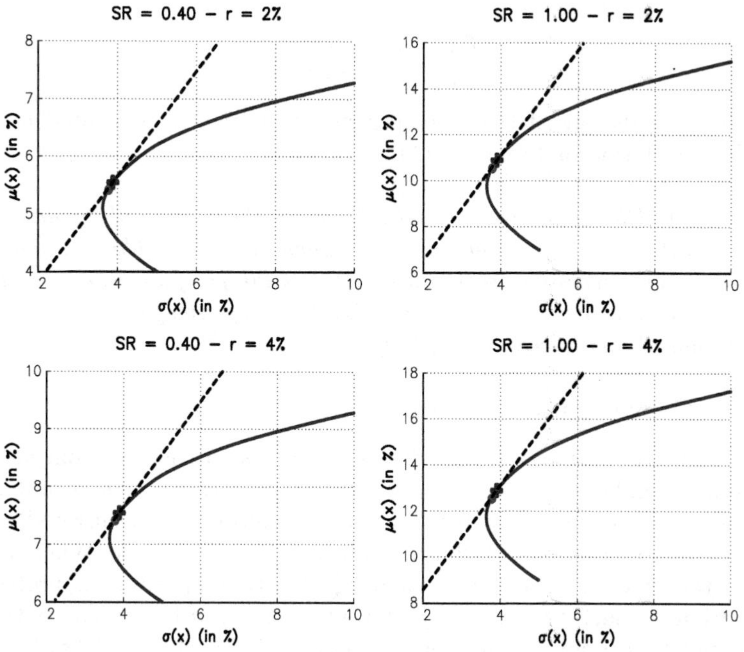

FIGURE 2.8: Location of the ERC portfolio in the mean-variance diagram when the Sharpe ratios are identical and the asset correlations are not uniform

2.3.4 Back to the notion of diversification

The goal of diversification is to optimize or reduce the risk of a portfolio. This is why diversification is generally associated with the efficient frontier although there is no precise definition. In this section, we study some related concepts that can help us better understand the notion of diversification.

2.3.4.1 Diversification index

Tasche (2008) defines the diversification index as follows:

$$\begin{aligned}\mathcal{D}(x) &= \frac{\mathcal{R}\left(\sum_{i=1}^{n} L_i\right)}{\sum_{i=1}^{n} \mathcal{R}(L_i)} \\ &= \frac{\mathcal{R}(x)}{\sum_{i=1}^{n} x_i \mathcal{R}(\mathbf{e}_i)}\end{aligned}$$

The diversification index is the ratio between the risk measure of portfolio x and the weighted risk measure of the assets. If $\mathcal{R}$ is a coherent risk measure, we have $\mathcal{D}(x) \leq 1$. If $\mathcal{D}(x) = 1$, it implies that the losses are comonotonic[38]. For instance, if $\mathcal{R}$ is the volatility risk measure, we obtain:

$$\mathcal{D}(x) = \frac{\sqrt{x^\top \Sigma x}}{\sum_{i=1}^n x_i \sigma_i} \tag{2.37}$$

It takes the value one if the asset returns are perfectly correlated meaning that the correlation matrix is $C_n(1)$.

Remark 34 *The diversification index (2.37) is the inverse of the diversification ratio introduced by Choueifaty and Coignard (2008). These authors also define the most diversified portfolio (MDP) as the portfolio which minimizes the diversification index (see Section 3.2.3.3 on page 168 for a comprehensive presentation of this allocation method).*

2.3.4.2 Concentration indices

Another way to measure the diversification is to consider concentration (or diversity) indices. Let $\pi \in \mathbb{R}_+^n$ such that $\mathbf{1}^\top \pi = 1$. π is then a probability distribution. The probability distribution π^+ is perfectly concentrated if there exists one observation i_0 such that $\pi_{i_0}^+ = 1$ and $\pi_i^+ = 0$ if $i \neq i_0$. When n tends to $+\infty$, the limit distribution is noted π_∞^+. On the opposite, the probability distribution π^- such that $\pi_i^- = 1/n$ for all $i = 1, \ldots, n$ has no concentration. A concentration index is a mapping function $\mathcal{C}(\pi)$ such that $\mathcal{C}(\pi)$ increases with concentration and verifies $\mathcal{C}(\pi^-) \leq \mathcal{C}(\pi) \leq \mathcal{C}(\pi^+)$. Eventually, this index can be normalized such that $\mathcal{C}(\pi^-) = 0$ and $\mathcal{C}(\pi_\infty^+) = 1$. Sometimes, one prefers to use the concept of diversity, which is the opposite of the notion of concentration.

For instance, if π represents the weights of the portfolio, $\mathcal{C}(\pi)$ measures then the weight concentration. By construction, $\mathcal{C}(\pi)$ reaches the minimum value if the portfolio is equally weighted. To measure the risk concentration of the portfolio, we define π as the distribution of the risk contributions. In this case, the portfolio corresponding to the lower bound $\mathcal{C}(\pi^-) = 0$ is the ERC portfolio.

The most popular methods to measure the concentration are the Herfindahl index and the Gini index. Another interesting statistic is the Shannon entropy which measures the diversity. Their definition is given below.

Herfindahl index The Herfindahl index associated with π is defined as:

$$\mathcal{H}(\pi) = \sum_{i=1}^n \pi_i^2$$

[38] The random variables $X_1, \ldots, X_n$ are said to be comonotonic if there exists a random variable X such that $X_i = f_i(X)$ for all i with $f_i(x)$ a non-decreasing function.

This index takes the value 1 for a probability distribution π^+ and $1/n$ for a distribution with uniform probabilities. To scale the statistics onto $[0, 1]$, we consider the normalized index $\mathcal{H}^\star(\pi)$ defined as follows:

$$\mathcal{H}^\star(\pi) = \frac{n\mathcal{H}(\pi) - 1}{n - 1}$$

Gini index The Gini index is based on the famous Lorenz curve of inequality. Let X and Y be two random variables. The Lorenz curve $y = \mathbb{L}(x)$ is defined by the following parameterization:

$$\begin{cases} x = \Pr\{X \leq x\} \\ y = \Pr\{Y \leq y \mid X \leq x\} \end{cases}$$

The Lorenz curve was first proposed to measure the inequality of wealth. In this case, it gives the proportion of the wealth y owned by the poorest x of the population. In asset management, the Lorenz curve is frequently used to measure the weight concentration (Hereil and Roncalli, 2011). An example is provided in Figure 2.9. We note that 70% of the assets represents only 34.3% of the total weight of the portfolio. The Lorenz curve admits two limit cases. If the portfolio is perfectly concentrated, the distribution of the weights corresponds to π_∞^+. On the opposite, the least concentrated portfolio is the equally weighted portfolio and the Lorenz curve is the bisecting line $y = x$. The Gini index is then defined as:

$$\mathcal{G}(\pi) = \frac{A}{A + B}$$

with A the area between $\mathbb{L}_{\pi^-}(x)$ and $\mathbb{L}(x)$, and B the area between $\mathbb{L}(x)$ and $\mathbb{L}_{\pi_\infty^+}(x)$. By construction, we have $\mathcal{G}(\pi^-) = 0$, $\mathcal{G}(\pi_\infty^+) = 1$ and:

$$\mathcal{G}(\pi) = 1 - 2\int_0^1 \mathbb{L}(x)\,\mathrm{d}x$$

In the case when π is a discrete probability distribution, we obtain:

$$\mathcal{G}(\pi) = \frac{2\sum_{i=1}^n i\pi_{i:n}}{n\sum_{i=1}^n \pi_{i:n}} - \frac{n+1}{n}$$

with $\{\pi_{1:n}, \ldots, \pi_{n:n}\}$ the ordered statistics of $\{\pi_1, \ldots, \pi_n\}$.

Remark 35 *The Gini coefficient may be viewed as a standardized statistic of dispersion. This is why it is sometimes used as a substitute for the standard deviation. For instance, Shalit and Yitzhaki (1984) propose to replace the mean-variance framework by the mean-Gini framework advocating that the latter is more robust as regards the non-normality of asset returns.*

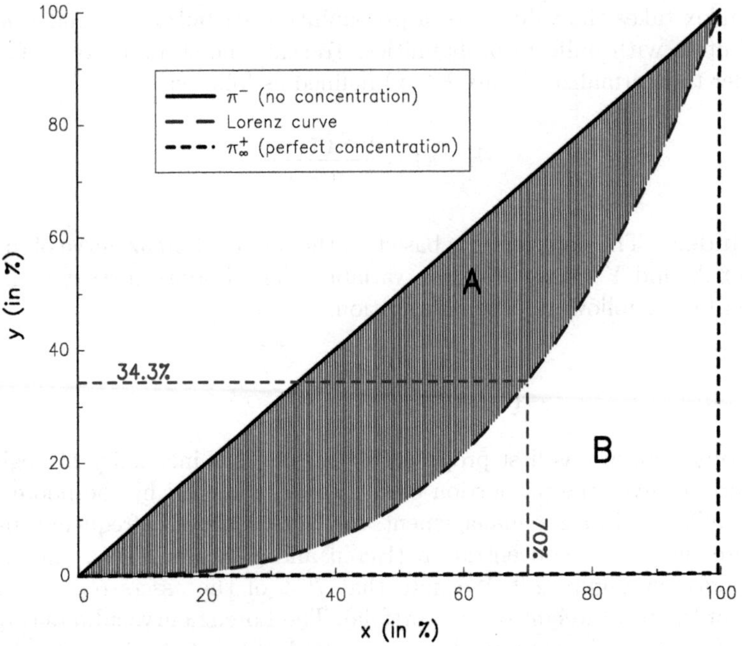

FIGURE 2.9: Geometry of the Lorenz curve

Shannon entropy The Shannon entropy is defined as follows:

$$\mathcal{I}(\pi) = -\sum_{i=1}^{n} \pi_i \ln \pi_i$$

The diversity index corresponds to the statistic $\mathcal{I}^\star(\pi) = \exp(\mathcal{I}(\pi))$. We have $\mathcal{I}^\star(\pi^-) = n$ and $\mathcal{I}^\star(\pi^+) = 1$.

Remark 36 *Bera and Park (2008) propose to penalize the objective function of mean-variance portfolios by using entropy measures. Their motivation is to obtain a 'well-diversified optimal portfolio'. It is also interesting to note that weight or risk concentration is generally associated with the diversification notion, and that the two concepts are synonymous for many authors.*

2.3.4.3 Difficulty of reconciling the different diversification concepts

In order to illustrate the previous diversification measures, we use Example 19. We compute them for different portfolios: the equally weighted portfolio (EW), the minimum variance portfolio (MV), the equal risk contribution portfolio (ERC) and the most diversified portfolio (MDP), which minimizes the diversification index $\mathcal{D}(x)$.

Example 19 *We consider an investment universe with six assets. The volatilities are 25%, 22%, 14%, 30%, 40% and 30% respectively. We use the following correlation matrix:*

$$C = \begin{pmatrix} 100\% & & & & & \\ 60\% & 100\% & & & & \\ 60\% & 60\% & 100\% & & & \\ 60\% & 60\% & 60\% & 100\% & & \\ 60\% & 60\% & 60\% & 60\% & 100\% & \\ 60\% & 60\% & 60\% & 60\% & 20\% & 100\% \end{pmatrix}$$

The correlation matrix is specific, because the correlation is uniform and equal to 60% for all assets except for the correlation between the fifth and sixth assets which is equal to 20%.

Using the volatility risk measure and imposing the long-only constraint, we obtain the results given in Table 2.22. From the Markowitz point of view, the MV portfolio is the most diversified, because it is the only of these portfolios which belongs to the efficient frontier. If we assume that the diversification index is the right measure, the MDP portfolio is the best. Indeed, the diversification index takes the value of 0.77 for this portfolio, which is the lowest[39]. The EW portfolio dominates the other portfolios in terms of weight concentration, whereas it is the ERC that is optimal if we consider the risk concentration. This simple example shows how difficult it is to define the diversification concept.

TABLE 2.22: Diversification measures of MV, ERC, MDP and EW portfolios

Asset	MV		ERC		MDP		EW	
	x_i	$\mathcal{RC}_i^\star$	x_i	$\mathcal{RC}_i^\star$	x_i	$\mathcal{RC}_i^\star$	x_i	$\mathcal{RC}_i^\star$
1	0.00	0.00	15.70	16.67	0.00	0.00	16.67	16.18
2	3.61	3.61	17.84	16.67	0.00	0.00	16.67	14.08
3	96.39	96.39	28.03	16.67	0.00	0.00	16.67	8.68
4	0.00	0.00	13.08	16.67	0.00	0.00	16.67	19.78
5	0.00	0.00	10.86	16.67	42.86	50.00	16.67	24.43
6	0.00	0.00	14.49	16.67	57.14	50.00	16.67	16.86
$\sigma(x)$	13.99		19.53		26.56		21.39	
$\mathcal{D}(x)$	0.98		0.80		0.77		0.80	
$\mathcal{H}^\star$	0.92	0.92	0.02	0.00	0.41	0.40	0.00	0.02
$\mathcal{G}$	0.82	0.82	0.17	0.00	0.69	0.67	0.00	0.16
$\mathcal{I}^\star$	1.17	1.17	5.71	6.00	1.98	2.00	6.00	5.74

[39]However, the ERC and EW portfolios present a diversification index close to that of the MDP.

2.4 Risk budgeting versus weight budgeting

Let $\{b_1, \ldots, b_b\}$ be a vector of budgets. In a risk budgeting (RB) portfolio, we find the weights such that the risk contributions match the budgets:

$$\mathcal{RC}_i = b_i \mathcal{R}(x)$$

We can compare this portfolio construction with a more simple process which is called weight budgeting (WB). In this case, we assign to each asset a weight that is equal to its budget:

$$x_i = b_i$$

In this section, we compare these two portfolios with a third portfolio corresponding to the minimum risk (MR) portfolio:

$$\begin{aligned} x^\star &= \arg\min \mathcal{R}(x) \\ \text{u.c.} \quad & \mathbf{1}^\top x = 1 \end{aligned}$$

2.4.1 Comparing weight budgeting and risk budgeting portfolios

If we consider the WB, MR and RB portfolios, we have the following relationships:

$$\begin{array}{rl} x_i/b_i = x_j/b_j & \text{(wb)} \\ \partial_{x_i} \mathcal{R}(x) = \partial_{x_j} \mathcal{R}(x) & \text{(mr)} \\ \mathcal{RC}_i/b_i = \mathcal{RC}_j/b_j & \text{(rb)} \end{array}$$

In the minimum risk portfolio, the marginal risk of asset i is equal to the marginal risk of asset j. In the weight budgeting portfolio, the ratio between the weight and the budget is the same for all the assets. The risk budgeting portfolio is a combination of these two properties because we have:

$$\frac{x_i \cdot \partial_{x_i} \mathcal{R}(x)}{b_i} = \frac{x_j \cdot \partial_{x_j} \mathcal{R}(x)}{b_j}$$

Thus, a RB portfolio may be viewed as a portfolio located between the MR and WB portfolios. To elaborate further this point of view, let us consider the optimization problem (2.31). We have:

$$\begin{aligned} x^\star(c) &= \arg\min \mathcal{R}(x) \\ \text{u.c.} \quad & \left\{ \begin{array}{l} \sum_{i=1}^n b_i \ln x_i \geq c \\ \mathbf{1}^\top x = 1 \\ x \geq \mathbf{0} \end{array} \right. \end{aligned}$$

We remark that if $c_1 \leq c_2$, we have $\mathcal{R}(x^\star(c_1)) \leq \mathcal{R}(x^\star(c_2))$ because the constraint $\sum_{i=1}^n b_i \ln x_i - c \geq 0$ is less restrictive with c_1 than with c_2. For instance, if $c = -\infty$, the optimization solution is exactly the minimum risk portfolio $x^\star(-\infty)$ (with positive weights) because the first constraint vanishes. We note that the function $\sum_{i=1}^n b_i \ln x_i$ is bounded and we have $\sum_{i=1}^n b_i \ln x_i \leq \sum_{i=1}^n b_i \ln b_i$. The only solution for $c = \sum_{i=1}^n b_i \ln b_i$ is $x_i^\star = b_i$, that is the weight budgeting portfolio. It follows that the solution for the general problem with $c \in \left[-\infty, \sum_{i=1}^n b_i \ln b_i\right]$ satisfies:

$$\mathcal{R}\left(x^\star(-\infty)\right) \leq \mathcal{R}\left(x^\star(c)\right) \leq \mathcal{R}\left(x^\star\left(\sum_{i=1}^n b_i \ln b_i\right)\right)$$

In other words, we have:

$$\mathcal{R}\left(x_{\mathrm{mr}}\right) \leq \mathcal{R}\left(x^\star(c)\right) \leq \mathcal{R}\left(x_{\mathrm{wb}}\right)$$

However, we have shown on page 109 that there exists a constant $c^\star$ such that $x^\star(c^\star)$ is the RB portfolio. This proves that the following inequality holds:

$$\mathcal{R}\left(x_{\mathrm{mr}}\right) \leq \mathcal{R}\left(x_{\mathrm{rb}}\right) \leq \mathcal{R}\left(x_{\mathrm{wb}}\right) \tag{2.38}$$

This result has been found by Bruder and Roncalli (2012). It is a direct extension of the famous inequality of Maillard *et al.* (2010), who state that the volatility of the ERC portfolio is between the volatility of the minimum variance portfolio and the volatility of the equally weighted portfolio:

$$\sigma\left(x_{\mathrm{mv}}\right) \leq \sigma\left(x_{\mathrm{erc}}\right) \leq \sigma\left(x_{\mathrm{ew}}\right) \tag{2.39}$$

Remark 37 *Inequality (2.39) is a special case of Inequality (2.38) with a volatility risk measure –* $\mathcal{R}(x) = \sigma(x)$ *– and uniform budgets –* $b_i = b_j = 1/n$.

Example 20 *We consider an investment universe of three assets. Their expected returns are equal to* 10%, 5% *and* 5% *whereas their volatilities are equal to* 30%, 20% *and* 15%. *The correlation of asset returns is given by the following matrix:*

$$\rho = \begin{pmatrix} 1.00 & & \\ 0.80 & 1.00 & \\ 0.50 & 0.30 & 1.00 \end{pmatrix}$$

We measure the risk with the Gaussian expected shortfall. The risk budgets are set to 50%, 20% and 30%. Weights and risk contributions for the WB, RB and MR portfolios are reported in Table 2.23. We note that these three portfolios are very different in terms of risk decomposition. For instance, the first asset represents respectively 71.40%, 50.00% and 0% of the WB, RB and MR portfolio risk. We note also that the inequality $\mathcal{R}\left(x_{\mathrm{mr}}\right) \leq \mathcal{R}\left(x_{\mathrm{rb}}\right) \leq \mathcal{R}\left(x_{\mathrm{wb}}\right)$ is verified.

2.4.2 New construction of the minimum variance portfolio

Let us introduce the notations $x_{\mathrm{wb}}(b)$ and $x_{\mathrm{rb}}(b)$ to design the weight and risk budgeting portfolios with b the vector of the budgets. By definition,

TABLE 2.23: Risk decomposition of WB, RB and MR portfolios

Portfolio	Asset	x_i	$\mathcal{MR}_i$	$\mathcal{RC}_i$	$\mathcal{RC}_i^\star$
WB	1	50.00	41.59	20.80	71.40
	2	20.00	24.18	4.84	16.60
	3	30.00	11.65	3.50	12.00
Expected shortfall				29.13	
RB	1	30.65	39.07	11.97	50.00
	2	21.04	22.76	4.79	20.00
	3	48.32	14.87	7.18	30.00
Expected shortfall				23.94	
MR	1	0.00	29.11	0.00	0.00
	2	30.34	18.81	5.71	30.34
	3	69.66	18.81	13.10	69.66
Expected shortfall				18.81	

we have $x_{\text{wb}}(b) = b$. Using Inequation (2.38), it follows that $\sigma(x_{\text{rb}}(b)) \leq \sigma(x_{\text{wb}}(b))$ or equivalently:

$$\sigma(x_{\text{rb}}(b)) \leq \sigma(b)$$

Given a portfolio $x^{(0)}$, we then have a simple process to find a portfolio $x^{(1)}$ with a smaller volatility. It suffices to compute the RB portfolio using the weights of $x^{(0)}$ as the risk budgets[40]:

$$\sigma\left(x^{(1)}\right) = \sigma\left(x_{\text{rb}}\left(x^{(0)}\right)\right) \leq \sigma\left(x^{(0)}\right)$$

Let us now consider the iterative portfolio $x^{(k)}$ where k represents the iteration. This portfolio is defined such that the vector of risk budget $b^{(k)}$ at iteration k corresponds to the weight $x^{(k-1)}$ at iteration $k-1$. We then have:

$$\begin{array}{c}
\sigma\left(x^{(1)}\right) = \sigma\left(x_{\text{rb}}\left(x^{(0)}\right)\right) \leq \sigma\left(x^{(0)}\right) \\
\sigma\left(x^{(2)}\right) = \sigma\left(x_{\text{rb}}\left(x^{(1)}\right)\right) \leq \sigma\left(x^{(1)}\right) \\
\vdots \\
\sigma\left(x^{(k)}\right) = \sigma\left(x_{\text{rb}}\left(x^{(k-1)}\right)\right) \leq \sigma\left(x^{(k-1)}\right)
\end{array}$$

We deduce that:

$$\sigma\left(x^{(k)}\right) \leq \sigma\left(x^{(k-1)}\right) \leq \cdots \leq \sigma\left(x^{(2)}\right) \leq \sigma\left(x^{(1)}\right) \leq \sigma\left(x^{(0)}\right)$$

We finally conclude that:

$$\lim_{k\to\infty} \sigma\left(x^{(k)}\right) = \sigma(x_{\text{mv}}) \tag{2.40}$$

[40]We can also interpret this procedure as a risk minimization program with transaction costs (Darolles *et al.*, 2012).

If the portfolio $x^{(k)}$ admits a limit when $k \to \infty$, it is equal to the minimum variance portfolio (with no short-selling constraint). It follows that the long-only MV portfolio is a RB portfolio when the risk budgets are equal to the weights:

$$b^{(\infty)} = x^{(\infty)} = x_{\mathrm{mv}}$$

Remark 38 *This last property can be found in another way. We know that the long-only MV portfolio satisfies:*

$$\frac{\partial\,\sigma(x)}{\partial\,x_i} = \frac{\partial\,\sigma(x)}{\partial\,x_j} = \zeta \quad \textit{if} \quad x_i > 0 \textit{ and } x_j > 0$$

and:

$$\frac{\partial\,\sigma(x)}{\partial\,x_i} > \frac{\partial\,\sigma(x)}{\partial\,x_j} \quad \textit{if} \quad x_i = 0 \textit{ and } x_j > 0$$

The Euler decomposition implies that:

$$\begin{aligned}\sigma(x) &= \sum_{i=1}^{n} x_i \frac{\partial\,\sigma(x)}{\partial\,x_i} \\ &= \zeta \sum_{i=1}^{n} x_i \\ &= \zeta\end{aligned}$$

It follows that:

$$\begin{aligned}b_i &= \frac{\mathcal{RC}_i}{\sigma(x)} \\ &= \frac{x_i \cdot \partial_{x_i}\,\sigma(x)}{\sigma(x)} \\ &= x_i\end{aligned}$$

We use Example 20 to illustrate that the MV portfolio is a limit portfolio given by Equation (2.40). For each portfolio, we report its weights, its absolute risk contributions and its volatility. Starting with $x^{(0)} = (50\%, 20\%, 30\%)$, we obtain the results given in Table 2.24. At each iteration, the volatility of the portfolio $x^{(k)}$ decreases. Starting with the WB portfolio, we have $\sigma\left(x^{(0)}\right) = 20.87\%$. At the first iteration, the risk budgets are set to the weights of the portfolio $x^{(0)}$. The composition of the portfolio $x^{(1)}$ is $(31.15\%, 21.90\%, 49.96\%)$ and its volatility is $\sigma\left(x^{(1)}\right) = 17.49\%$. These weights become the risk budgets of the portfolio at the second iteration. We then have $x^{(2)} = (18.52\%, 22.81\%, 58.67\%)$. We note that the volatility decreases quickly. For instance, we have $\sigma\left(x^{(2)}\right) = 15.58\%$, $\sigma\left(x^{(3)}\right) = 14.65\%$, $\sigma\left(x^{(4)}\right) = 14.19\%$, etc. The convergence to the MV portfolio is illustrated in Figure 2.10. After only six (resp. ten) iterations, the volatility of the RB portfolio is equal to 13.79% (resp. 13.60%). It is very close to the volatility of the minimum variance portfolio, which is equal to 13.57%.

TABLE 2.24: Weights and risk contributions of the iterative RB portfolio $x^{(k)}$

Portfolio	Asset	x_i	$\mathcal{RC}_i^\star$	Portfolio	Asset	x_i	$\mathcal{RC}_i^\star$
$x^{(0)}$	1	50.00	70.43	$x^{(1)}$	1	31.15	50.00
	2	20.00	15.93		2	21.90	20.00
	3	30.00	13.64		3	46.96	30.00
	Volatility		20.87		Volatility		17.49
$x^{(2)}$	1	18.52	31.15	$x^{(3)}$	1	11.04	18.52
	2	22.81	21.90		2	23.71	22.81
	3	58.67	46.96		3	65.25	58.67
	Volatility		15.58		Volatility		14.65
$x^{(4)}$	1	6.67	11.04	$x^{(5)}$	1	4.07	6.67
	2	24.76	23.71		2	25.86	24.76
	3	68.57	65.25		3	70.07	68.57
	Volatility		14.19		Volatility		13.94
$x^{(6)}$	1	2.49	4.07	x_{mv}	1	0.00	0.00
	2	26.87	25.86		2	30.34	30.34
	3	70.63	70.07		3	69.66	69.66
	Volatility		13.79		Volatility		13.57

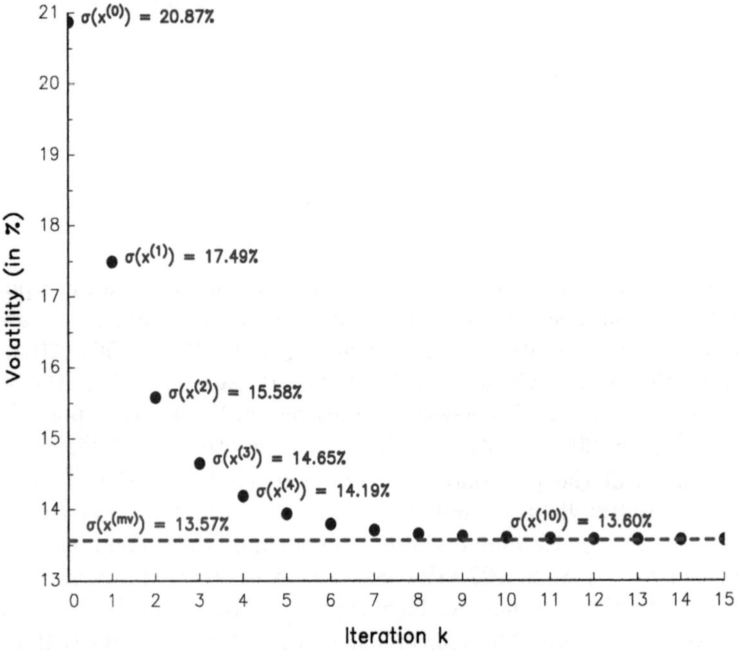

FIGURE 2.10: Convergence of the iterative RB portfolio $x^{(k)}$ to the MV portfolio

2.5 Using risk factors instead of assets

Professionals who use risk parity methods would generally like to control the risk exposures with respect to some specific factors. It is sometimes difficult to achieve this by directly budgeting the risk of the assets. Moreover, we will see below that working directly with assets can be not optimal in some cases. Risk parity methods based on risk factors may therefore be more appropriate.

2.5.1 Pitfalls of the risk budgeting approach based on assets

The risk budgeting approach is sensitive to the universe definition, meaning that the portfolio can change if we consider an investment universe equivalent to the original one. Choueifaty *et al.* (2011) therefore propose a set of desired invariance properties that a portfolio must respect. Among them, it appears that some RB portfolios do not always satisfy the duplication invariance property and the polico invariance property.

2.5.1.1 Duplication invariance property

Let us consider a universe of four assets with equal volatilities and a uniform correlation ρ. The ERC portfolio $x^{(4)}$ corresponds then to the equally weighted portfolio. Let us add to this universe a fifth asset with the same volatility. We assume that this asset is perfectly correlated with the fourth asset. In this case, the composition of the ERC portfolio $x^{(5)}$ depends on the value of the coefficient ρ. For example, if the cross-correlation ρ between the first four assets is zero, the ERC portfolio's composition is $x_i^{(5)} = 22.65\%$ for the first three assets and $x_i^{(5)} = 16.02\%$ for the last two assets.

Such a solution can be confusing to a professional since in reality there are only four and not five assets in our universe[41]. A financial professional would expect that the sum of the exposures to the fourth and fifth assets is equal to the exposure to each of the first three assets, i.e. $x_4^{(5)} + x_5^{(5)} = 25\%$. The weights of assets four and five are therefore equal to the weight of the fourth asset in the four-asset universe divided by two – $x_5^{(5)} = x_4^{(5)} = x_4^{(4)}/2 = 12.5\%$. However, this solution is reached only when the correlation coefficient ρ is at its lowest value[42].

The ERC portfolio is then not invariant if we duplicate an asset. The investment universe is therefore an important factor when considering risk parity portfolios. However, we note that the RB portfolio verifies the duplication invariance property if we split the corresponding risk budget between the two equivalent assets.

[41]The fourth and fifth assets are in fact the same since $\rho_{4,5} = 1$.

[42]ρ is then equal to -33.33%

Let us prove this property in the case of the volatility risk measure. We consider a universe of n assets. We define x the RB portfolio which matches the risk budgets $\{b_1, \ldots, b_n\}$. We duplicate the last asset and we note y a RB portfolio based on the universe of the $n+1$ assets and the risk budgets $\{b'_1, \ldots, b'_{n+1}\}$ such that:

$$\left\{ \begin{array}{ll} b'_i = b_i & \text{if } i < n \\ b'_n + b'_{n+1} = b_n & \end{array} \right. \tag{2.41}$$

We suppose that $\sigma(x) = \sigma(y)$. We have:

$$\sum_{i=1}^{n} x_i \cdot \frac{\partial\,\sigma(x)}{\partial\,x_i} = \sum_{i=1}^{n+1} y_i \cdot \frac{\partial\,\sigma(y)}{\partial\,y_i}$$

Let us try to find a portfolio y such that:

$$x_i \cdot \frac{\partial\,\sigma(x)}{\partial\,x_i} = y_i \cdot \frac{\partial\,\sigma(x)}{\partial\,y_i} \quad \text{if } i < n$$

and:

$$x_n \cdot \frac{\partial\,\sigma(x)}{\partial\,x_n} = y_n \cdot \frac{\partial\,\sigma(y)}{\partial\,y_n} + y_{n+1} \cdot \frac{\partial\,\sigma(y)}{\partial\,y_{n+1}}$$

If $i < n$, the first equation is equivalent to:

$$\begin{aligned} \sum_{j<n} x_i x_j \rho_{i,j} \sigma_i \sigma_j + x_i x_n \rho_{i,n} \sigma_i \sigma_n &= \sum_{j<n} y_i y_j \rho_{i,j} \sigma_i \sigma_j + \\ & \quad y_i y_n \rho_{i,n} \sigma_i \sigma_n + y_i y_{n+1} \rho_{i,n+1} \sigma_i \sigma_{n+1} \\ &= \sum_{j<n} y_i y_j \rho_{i,j} \sigma_i \sigma_j + \\ & \quad y_i (y_n + y_{n+1}) \rho_{i,n} \sigma_i \sigma_n \end{aligned}$$

because $\rho_{i,n+1} = \rho_{i,n}$, $\sigma_{n+1} = \sigma_n$ and $\sigma(x) = \sigma(y)$. For the second equation, we obtain:

$$\begin{aligned} \sum_{j=1}^{n} x_n x_j \rho_{n,j} \sigma_n \sigma_j &= \sum_{j=1}^{n+1} y_n y_j \rho_{n,j} \sigma_n \sigma_j + \sum_{j=1}^{n+1} y_{n+1} y_j \rho_{n+1,j} \sigma_{n+1} \sigma_j \\ &= \sum_{j=1}^{n+1} (y_n + y_{n+1}) y_j \rho_{n,j} \sigma_n \sigma_j \end{aligned}$$

It follows that:

$$\begin{aligned} \sum_{j=1}^{n-1} x_n x_j \rho_{n,j} \sigma_n \sigma_j + x_n^2 \sigma_n^2 &= \sum_{j=1}^{n-1} (y_n + y_{n+1}) y_j \rho_{n,j} \sigma_n \sigma_j + \\ & \quad (y_n + y_{n+1}) y_n \sigma_n^2 + (y_n + y_{n+1}) y_{n+1} \sigma_n^2 \\ &= \sum_{j=1}^{n-1} (y_n + y_{n+1}) y_j \rho_{n,j} \sigma_n \sigma_j + \\ & \quad (y_n + y_{n+1})^2 \sigma_n^2 \end{aligned}$$

because $\rho_{n,n+1} = \rho_{n,n} = 1$. We deduce that the solution satisfies:

$$\begin{cases} y_i = x_i & \text{if } i < n \\ y_n + y_{n+1} = x_n \end{cases}$$

Considering this solution, we easily verify that the statement $\sigma(x) = \sigma(y)$ is true. This proves that the RB portfolio verifies the duplication invariance property if we apply the rule (2.41).

Remark 39 *By considering the systematic rule of equal risk budgets, we do not take into account the structure of the investment universe. Let us consider a universe of multi-asset classes. If the universe includes five equity indices and five bond indices, then the ERC portfolio will be well balanced between equity and bond in terms of risk. Conversely, if the universe includes seven equity indices and three bond indices, the equity risk of the ERC portfolio represents then 70% of the portfolio's total risk, a very unbalanced solution between equity and bond risks. In this case, it is better to define a RB portfolio by adjusting the risk budgets such that the equity risk is equal to the bond risk. Another solution is to adopt the risk parity approach based on risk factors (Roncalli and Weisang, 2012).*

2.5.1.2 Polico invariance property

Choueifaty et al. (2011) define the polico invariance property as follows:

> *"The addition of positive linear combination of assets already belonging to the universe should not impact the portfolio's weights to the original assets".*

For the RB portfolio, this property can be understood as follows. We consider an investment universe of n assets, and we build a RB portfolio x that matches a set of risk budgets $\{b_1, \ldots, b_n\}$. Then, we add another asset which is a linear combination c of the first n assets. With this augmented universe, we define a new RB portfolio y such that:

$$\begin{cases} b'_i + c_i b'_{n+1} = b_i & \text{if } i \leq n \\ \mathbf{1}^\top b' = 1 \end{cases}$$

This portfolio satisfies the polico invariance property if:

$$x_i = y_i + c_i y_{n+1}$$

However, the two previous equations are incompatible[43] meaning that RB portfolios are not polico invariant.

[43] Because they imply that the marginal risk is the same for all the $n + 1$ assets.

2.5.1.3 Impact of the reparametrization on the asset universe

The difficulty with the two previous problems comes from the fact that the assets are not exactly the risk factors which interest the investor. Roncalli and Weisang (2012) also illustrate this problem by showing that a new parameterization of the asset universe implies a new parametrization of the risk budgets.

We consider a set of m primary assets $(\mathcal{A}'_1, \ldots, \mathcal{A}'_m)$ with a covariance matrix Ω. We now define n synthetic assets $(\mathcal{A}_1, \ldots, \mathcal{A}_n)$ which are composed of the primary assets. We denote $W = (w_{i,j})$ the weight matrix such that $w_{i,j}$ is the weight of the primary asset $\mathcal{A}'_j$ in the synthetic asset $\mathcal{A}_i$. Therefore, the synthetic assets can be interpreted as portfolios of the primary assets. For example, $\mathcal{A}'_j$ may represent a stock whereas $\mathcal{A}_i$ may be an index. By construction, we have $\sum_{j=1}^{m} w_{i,j} = 1$. It follows that the covariance matrix of the synthetic assets Σ is equal to $W \Omega W^\top$.

We now consider a portfolio $x = (x_1, \ldots, x_n)$ defined with respect to the synthetic assets. The volatility of this portfolio is then $\sigma(x) = \sqrt{x^\top \Sigma x}$. We deduce that the risk contribution of the synthetic asset i is:

$$\mathcal{RC}_i = x_i \cdot \frac{(\Sigma x)_i}{\sqrt{x^\top \Sigma x}}$$

We can also define the portfolio with respect to the primary assets. In this case, the composition is $y = (y_1, \ldots, y_m)$ where $y_j = \sum_{i=1}^{n} x_i w_{i,j}$. In a matrix form, we have $y = W^\top x$. In a similar way, we may compute the risk contribution of the primary assets j:

$$\mathcal{RC}_j = y_j \cdot \frac{(\Omega y)_j}{\sqrt{y^\top \Omega y}}$$

Example 21 *We have six primary assets. The volatility of these assets is respectively* 20%, 30%, 25%, 15%, 10% *and* 30%. *We assume that the assets are not correlated. We consider three equally weighted synthetic assets with:*

$$W = \begin{pmatrix} 1/4 & 1/4 & 1/4 & 1/4 & & \\ & & 1/4 & 1/4 & 1/4 & 1/4 \\ & & 1/2 & 1/2 & & \end{pmatrix}$$

Let us first consider Portfolio #1 with synthetic weights 36%, 38% and 26%. The risk contributions with respect to the synthetic assets are provided in Table 2.25. Note that this portfolio is very close to the ERC portfolio. It is therefore well-balanced in terms of risk with respect to the synthetic assets. However, if we analyze this portfolio in terms of primary assets, about 80% of the portfolio's risk is then concentrated on the third and fourth primary assets (see Table 2.26). We have here a paradoxical situation. Depending on the analysis, this portfolio is either a well diversified portfolio or a risk concentrated portfolio.

TABLE 2.25: Risk decomposition of Portfolio #1 with respect to the synthetic assets

Asset i	x_i	$\mathcal{MR}_i$	$\mathcal{RC}_i$	$\mathcal{RC}_i^\star$
$\mathcal{A}_1$	36.00	9.44	3.40	33.33
$\mathcal{A}_2$	38.00	8.90	3.38	33.17
$\mathcal{A}_3$	26.00	13.13	3.41	33.50

TABLE 2.26: Risk decomposition of Portfolio #1 with respect to the primary assets

Asset j	y_j	$\mathcal{MR}_j$	$\mathcal{RC}_j$	$\mathcal{RC}_j^\star$
$\mathcal{A}_1'$	9.00	3.53	0.32	3.12
$\mathcal{A}_2'$	9.00	7.95	0.72	7.02
$\mathcal{A}_3'$	31.50	19.31	6.08	59.69
$\mathcal{A}_4'$	31.50	6.95	2.19	21.49
$\mathcal{A}_5'$	9.50	0.93	0.09	0.87
$\mathcal{A}_6'$	9.50	8.39	0.80	7.82

Let us now consider Portfolio #2 with synthetic weights 48%, 50% and 2%. The risk contributions for this portfolio are provided in Table 2.27. In this case, the portfolio is not very well balanced in terms of risk, because the first two assets represent more than 97% of the risk of the portfolio, whereas the risk contribution of the third asset is less than 3%. This portfolio is thus far from the ERC portfolio. However, an analysis in terms of the primary assets shows that this portfolio is less concentrated than the previous one (see Table 2.28). The average risk contribution is 16.67%. Note that primary assets which have a risk contribution above (resp. below) this level in the first portfolio have a risk contribution that decreases (resp. increases) in the second portfolio.

TABLE 2.27: Risk decomposition of Portfolio #2 with respect to the synthetic assets

Asset i	x_i	$\mathcal{MR}_i$	$\mathcal{RC}_i$	$\mathcal{RC}_i^\star$
$\mathcal{A}_1$	48.00	9.84	4.73	49.91
$\mathcal{A}_2$	50.00	9.03	4.51	47.67
$\mathcal{A}_3$	2.00	11.45	0.23	2.42

In Figure 2.11, we have reported the Lorenz curve of the risk contributions of these two portfolios with respect to the primary and synthetic assets. We verify that the second portfolio is less concentrated in terms of primary risk.

TABLE 2.28: Risk decomposition of Portfolio #2 with respect to the primary assets

Asset j	y_j	$\mathcal{MR}_j$	$\mathcal{RC}_j$	$\mathcal{RC}_j^{\star}$
$\mathcal{A}'_1$	12.00	5.07	0.61	6.43
$\mathcal{A}'_2$	12.00	11.41	1.37	14.46
$\mathcal{A}'_3$	25.50	16.84	4.29	45.35
$\mathcal{A}'_4$	25.50	6.06	1.55	16.33
$\mathcal{A}'_5$	12.50	1.32	0.17	1.74
$\mathcal{A}'_6$	12.50	11.88	1.49	15.69

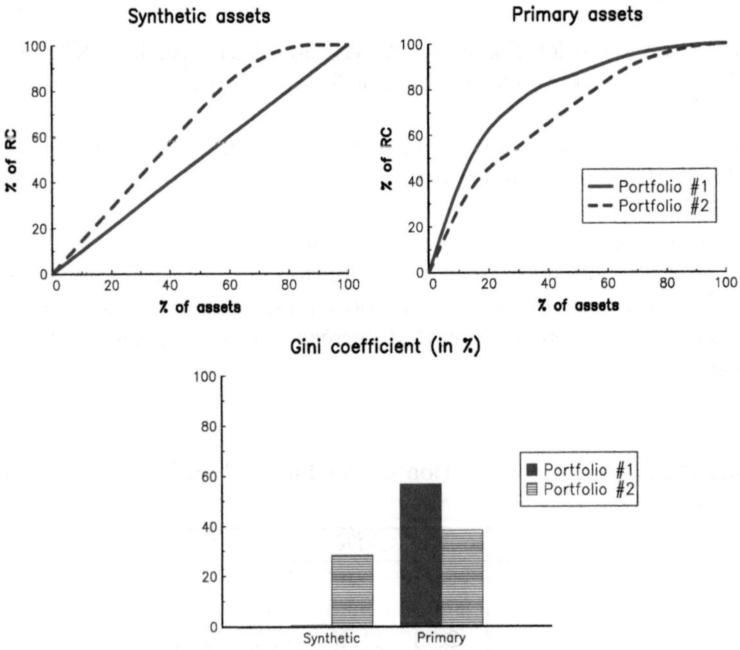

FIGURE 2.11: Lorenz curve of risk contributions

2.5.2 Risk decomposition with respect to the risk factors

The previous section highlights the importance of the risk factors and why it is important to understand the structure of the investment with respect to them. Below, we present the results of Roncalli and Weisang (2012) when we build risk parity portfolios by risk budgeting the risk factors instead of the assets.

We consider a set of n assets $\{\mathcal{A}_1, \ldots, \mathcal{A}_n\}$ and a set of m risk factors $\{\mathcal{F}_1, \ldots, \mathcal{F}_m\}$. Let R_t be the $(n \times 1)$ vector of asset returns at time t, Σ its associated covariance matrix, $\mathcal{F}_t$ the $(m \times 1)$ vector of factor returns at time t and Ω its associated covariance matrix. We assume the following linear factor model:

$$R_t = A\mathcal{F}_t + \varepsilon_t \tag{2.42}$$

where $\mathcal{F}_t$ and ε_t are two uncorrelated random vectors. ε_t is a centered random vector $(n \times 1)$ of covariance D. A is the $(n \times m)$ loadings matrix. Using (2.42), it is easy to deduce the following relationship:

$$\Sigma = A\Omega A^\top + D$$

Meucci (2007, 2009) proposes to decompose the portfolio's asset exposures x by the portfolio's risk factors exposures y in the following way:

$$x = B^+ y + \tilde{B}^+ \tilde{y}$$

where B^+ is the Moore-Penrose inverse[44] of $A^\top$ and $\tilde{B}^+$ is any $n \times (n-m)$ matrix that spans the left nullspace of B^+. $\tilde{y}$ corresponds to $n-m$ residual (or additional) factors that have no economic interpretation[45]. It follows that:

$$\left\{ \begin{array}{l} y = A^\top x \\ \tilde{y} = \tilde{B} x \end{array} \right.$$

where $\tilde{B} = \ker\left(A^\top\right)^\top$. In this case, the marginal risk of assets is related to the marginal risk of factors in the following way:

$$\frac{\partial\,\mathcal{R}(x)}{\partial\,x} = \frac{\partial\,\mathcal{R}(x)}{\partial\,y} B + \frac{\partial\,\mathcal{R}(x)}{\partial\,\tilde{y}} \tilde{B}$$

[44]Let M be a $(r \times c)$ matrix. There are two simple cases where there is an analytical expression of the Moore-Penrose inverse of M:

1. If the columns of M are linearly independent, it implies that $r > c$ and $M^+ = \left(M^\top M\right)^{-1} M^\top$.
2. If the rows of M are linearly independent, it implies that $c > r$ and we have $M^+ = M^\top \left(MM^\top\right)^{-1}$.

In other cases, we need to use a numerical algorithm based on the singular value decomposition to compute M^+.

[45]The residual factors are not determined in a unique way. They are just defined to calculate the risk contributions of the factors y. Therefore, interpreting the risk contributions of these residual factors is not meaningful.

We deduce that the marginal risk of the jth factor exposure is given by:

$$\frac{\partial\,\mathcal{R}(x)}{\partial\,y_j} = \left(A^{+}\frac{\partial\,\mathcal{R}(x)}{\partial\,x^{\top}}\right)_j$$

For the residual factors, we have:

$$\frac{\partial\,\mathcal{R}(x)}{\partial\,\tilde{y}_j} = \left(\tilde{B}\frac{\partial\,\mathcal{R}(x)}{\partial\,x^{\top}}\right)_j$$

Let us note $\mathcal{RC}(\mathcal{F}_j) = y_j \cdot \partial_{y_j}\,\mathcal{R}(x)$ the risk contribution of factor j with respect to risk measure $\mathcal{R}$. We obtain:

$$\mathcal{RC}(\mathcal{F}_j) = \left(A^{\top}x\right)_j \cdot \left(A^{+}\frac{\partial\,\mathcal{R}(x)}{\partial\,x^{\top}}\right)_j$$

and:

$$\mathcal{RC}\left(\tilde{\mathcal{F}}_j\right) = \left(\tilde{B}x\right)_j \cdot \left(\tilde{B}\frac{\partial\,\mathcal{R}(x)}{\partial\,x^{\top}}\right)_j$$

Following Meucci (2007), Roncalli and Weisang (2012) show that these risk contributions satisfy the allocation principle:

$$\mathcal{R}(x) = \sum_{j=1}^{m}\mathcal{RC}(\mathcal{F}_j) + \sum_{j=1}^{n-m}\mathcal{RC}\left(\tilde{\mathcal{F}}_j\right)$$

We also obtain a risk decomposition by factors similar to the Euler decomposition by assets.

Remark 40 *When the risk measure $\mathcal{R}(x)$ is the volatility of the portfolio $\sigma(x) = \sqrt{x^{\top}\Sigma x}$, the risk contribution of the jth factor is:*

$$\mathcal{RC}(\mathcal{F}_j) = \frac{\left(A^{\top}x\right)_j \cdot \left(A^{+}\Sigma x\right)_j}{\sigma(x)}$$

For the residual risk factors $\tilde{\mathcal{F}}_t$, the results become:

$$\mathcal{RC}\left(\tilde{\mathcal{F}}_j\right) = \frac{\left(\tilde{B}x\right)_j \cdot \left(\tilde{B}\Sigma x\right)_j}{\sigma(x)}$$

Example 22 *We consider an investment universe with four assets and three factors. The loadings matrix A is:*

$$A = \begin{pmatrix} 0.9 & 0.0 & 0.5 \\ 1.1 & 0.5 & 0.0 \\ 1.2 & 0.3 & 0.2 \\ 0.8 & 0.1 & 0.7 \end{pmatrix}$$

The three factors are uncorrelated and their volatilities are 20%, 10% *and* 10%. *We assume a diagonal matrix D with specific volatilities* 10%, 15%, 10% *and* 15%.

Using this example, it follows that the corresponding correlation matrix of asset returns is (in %):

$$\rho = \begin{pmatrix} 100.0 & & & \\ 69.0 & 100.0 & & \\ 79.5 & 76.4 & 100.0 & \\ 66.2 & 57.2 & 66.3 & 100.0 \end{pmatrix}$$

and their volatilities are respectively 21.19%, 27.09%, 26.25% and 23.04%. We also obtain that:

$$A^{+} = \begin{pmatrix} 1.260 & -0.383 & 1.037 & -1.196 \\ -3.253 & 2.435 & -1.657 & 2.797 \\ -0.835 & 0.208 & -1.130 & 2.348 \end{pmatrix}$$

and:

$$\tilde{B} = \begin{pmatrix} 0.533 & 0.452 & -0.692 & -0.183 \end{pmatrix}$$

The risk decomposition of the equally weighted portfolio is then given in Tables 2.29 and 2.30. We note that the equally weighted portfolio produces a risk-balanced portfolio in terms of asset's risk contributions, but not in terms of factor's risk contributions. Indeed, the first factor represents more than 80% of the risk. In the next paragraph, we present a method for obtaining portfolios that are more balanced with respect to risk factors.

TABLE 2.29: Risk decomposition of the EW portfolio with respect to the assets

Asset	x_i	$\mathcal{MR}_i$	$\mathcal{RC}_i$	$\mathcal{RC}_i^\star$
1	25.00	18.81	4.70	21.97
2	25.00	23.72	5.93	27.71
3	25.00	24.24	6.06	28.32
4	25.00	18.83	4.71	22.00
Volatility			21.40	

TABLE 2.30: Risk decomposition of the EW portfolio with respect to the risk factors

Factor	y_j	$\mathcal{MR}_j$	$\mathcal{RC}_j$	$\mathcal{RC}_j^\star$
$\mathcal{F}_1$	100.00	17.22	17.22	80.49
$\mathcal{F}_2$	22.50	9.07	2.04	9.53
$\mathcal{F}_3$	35.00	6.06	2.12	9.91
$\tilde{\mathcal{F}}_1$	2.75	0.52	0.01	0.07
Volatility			21.40	

2.5.3 Some illustrations

2.5.3.1 Matching the risk budgets

Suppose that we would like to build a risk budgeting portfolio such that the risk contributions match a set of given risk budgets $\{b_1, \ldots, b_m\}$:

$$\mathcal{RC}(\mathcal{F}_j) = b_j \mathcal{R}(x)$$

Roncalli and Weisang (2012) propose then to solve the following optimization problem:

$$\begin{aligned}(y^\star, \tilde{y}^\star) &= \arg\min \sum_{j=1}^{m} \left(\mathcal{RC}(\mathcal{F}_j) - b_j \mathcal{R}(x)\right)^2 \\ \text{u.c.} \quad & \mathbf{1}^\top \left(B^+ y + \tilde{B}^+ \tilde{y}\right) = 1\end{aligned}$$

If there is a solution to this optimization problem and if the objective function is equal to zero at the optimum, it corresponds to the solution to the matching problem[46]. Such a solution is called a risk factor parity (or RFP) portfolio.

Remark 41 *If we impose long-only constraints, we must add the following inequality restrictions:*

$$\mathbf{0} \leq B^+ y + \tilde{B}^+ \tilde{y} \leq \mathbf{1}$$

We consider Example 22. We have seen that the equally weighted portfolio concentrates the risk on the first factor. We would like to build a portfolio with more balanced risk across the factors. For instance, if $b = (49\%, 25\%, 25\%)$, we obtain the results given in Tables 2.31 and 2.32. The composition of the portfolio that matches these risk budgets is then $(15.08\%, 38.38\%, 0.89\%, 45.65\%)$.

TABLE 2.31: Risk decomposition of the RFP portfolio with respect to the risk factors

Factor	y_j	$\mathcal{MR}_j$	$\mathcal{RC}_j$	$\mathcal{RC}_j^\star$
$\mathcal{F}_1$	93.38	11.16	10.42	49.00
$\mathcal{F}_2$	24.02	22.14	5.32	25.00
$\mathcal{F}_3$	39.67	13.41	5.32	25.00
$\tilde{\mathcal{F}}_1$	16.39	1.30	0.21	1.00
Volatility			21.27	

[46] If we prefer to use a $\mathcal{L}_2$ relative norm, we can replace the objective function as follows:

$$(y^\star, \tilde{y}^\star) = \arg\min_y \sum_{j=1}^{m} \sum_{k=1}^{m} \left(\frac{\mathcal{RC}(\mathcal{F}_j)}{b_j} - \frac{\mathcal{RC}(\mathcal{F}_k)}{b_k}\right)^2$$

TABLE 2.32: Risk decomposition of the RFP portfolio with respect to the assets

Asset	x_i	$\mathcal{MR}_i$	$\mathcal{RC}_i$	$\mathcal{RC}_i^\star$
1	15.08	17.44	2.63	12.36
2	38.38	23.94	9.19	43.18
3	0.89	21.82	0.20	0.92
4	45.65	20.29	9.26	43.54
Volatility			21.27	

Remark 42 *The existence of a solution is not guaranteed in particular if we consider long-only portfolios and if we allocate significant risk budgets to small factors. For instance, if we use the three factors (level, slope and convexity) of the yield curve, it is highly unlikely that there are long-only portfolios of bonds that have high risk contributions on the slope or the convexity factors*[47].

2.5.3.2 Minimizing the risk concentration between the risk factors

We now consider the following problem:

$$\mathcal{RC}(\mathcal{F}_j) \simeq \mathcal{RC}(\mathcal{F}_k)$$

The idea is to find a portfolio that is very well balanced in terms of risk contributions with respect to the common factors. A first idea is to set $b_j = b_k$ and to use the previous framework. Another idea is to minimize the concentration between the risk contributions $(\mathcal{RC}(\mathcal{F}_1), \ldots, \mathcal{RC}(\mathcal{F}_m))$. In this case, we can use the concentration indices given in Section 2.3.4.2 on page 126.

TABLE 2.33: Risk decomposition of the balanced RFP portfolio with respect to the risk factors

Factor	y_j	$\mathcal{MR}_j$	$\mathcal{RC}_j$	$\mathcal{RC}_j^\star$
$\mathcal{F}_1$	91.97	7.91	7.28	33.26
$\mathcal{F}_2$	25.78	28.23	7.28	33.26
$\mathcal{F}_3$	42.22	17.24	7.28	33.26
$\breve{\mathcal{F}}_1$	6.74	0.70	0.05	0.21
Volatility			21.88	

We continue Example 22 by minimizing the risk concentration between the three risk factors. The results are given in Tables 2.33 and 2.34. We have $\mathcal{H}^\star = 0$, $\mathcal{G} = 0$ and $\mathcal{I}^\star = 3$. In this case, the three criteria (Herfindahl, Gini and Shannon entropy) are equivalent. If we impose some constraints, the optimal portfolios will differ. For instance, if we assume that the weights are larger

[47] See Section 4.3.1 on page 216 for further details.

TABLE 2.34: Risk decomposition of the balanced RFP portfolio with respect to the assets

Asset	x_i	$\mathcal{MR}_i$	$\mathcal{RC}_i$	$\mathcal{RC}_i^\star$
1	0.30	16.11	0.05	0.22
2	39.37	23.13	9.11	41.63
3	0.31	20.93	0.07	0.30
4	60.01	21.09	12.66	57.85
Volatility			21.88	

than 10%, we obtain the optimal portfolios given in Table 2.35. In this case, the three optimization problems are not equivalent.

TABLE 2.35: Balanced RFP portfolios with $x_i \geq 10\%$

Criterion	$\mathcal{H}(x)$	$\mathcal{G}(x)$	$\mathcal{I}(x)$
x_1	10.00	10.00	10.00
x_2	22.08	18.24	24.91
x_3	10.00	10.00	10.00
x_4	57.92	61.76	55.09
$\mathcal{H}^\star$	0.0436	0.0490	0.0453
$\mathcal{G}$	0.1570	0.1476	0.1639
$\mathcal{I}^\star$	2.8636	2.8416	2.8643

2.5.3.3 Solving the duplication and polico invariance properties

Let $\Sigma^{(n)}$ be the covariance matrix of n assets. We consider the RB portfolio $x^{(n)}$ corresponding to the risk budgets $b^{(n)}$. We now suppose that we duplicate the last asset. In this case, the covariance matrix of the $n+1$ assets is:

$$\Sigma^{(n+1)} = \begin{pmatrix} \Sigma^{(n)} & \Sigma^{(n)}\mathbf{e}_n \\ \mathbf{e}_n^\top \Sigma^{(n)} & \Sigma_{n,n}^{(n)} \end{pmatrix}$$

We associate the factor model with $\Omega = \Sigma^{(n)}$, $D = \mathbf{0}$ and:

$$A = \begin{pmatrix} I_n \\ \mathbf{e}_n^\top \end{pmatrix}$$

We consider portfolio $x^{(n+1)}$ such that the risk contribution of the factors match the risk budgets $b^{(n)}$. It is then easy to show that the matching portfolio verifies $y^\star = x^{(n)}$ in terms of factor weights. We then have $x_i^{(n+1)} = x_i^{(n)}$ if $i < n$ and $x_n^{(n+1)} + x_{n+1}^{(n+1)} = x_n^{(n)}$. This result shows that a RB portfolio verifies the duplication invariance property if the risk budgets are expressed with respect to factors and not to assets.

We now introduce an asset $n+1$ which is a linear (normalized) combination c of the first n assets. In this case, the covariance matrix of the $n+1$ assets is:

$$\Sigma^{(n+1)} = \begin{pmatrix} \Sigma^{(n)} & \Sigma^{(n)}c \\ c^\top \Sigma^{(n)} & c^\top \Sigma^{(n)} c \end{pmatrix}$$

We associate the factor model with $\Omega = \Sigma^{(n)}$, $D = \mathbf{0}$ and:

$$A = \begin{pmatrix} I_n \\ c^\top \end{pmatrix}$$

We consider the portfolio $x^{(n+1)}$ such that the risk contribution of the factors match the risk budgets $b^{(n)}$. It is then easy to show that the factor weights of the matching portfolio satisfy:

$$y^\star = x^{(n)}$$

We then have $A^\top x^{(n+1)} = x^{(n)}$, which implies that $x_i^{(n)} = x_i^{(n+1)} + c_i x_{n+1}^{(n+1)}$ if $i \leq n$. This result shows that a RB portfolio (and so an ERC portfolio) verifies the polico invariance property if the risk budgets are expressed according to factors and not to assets.

Remark 43 *We then notice that the duplication invariance property is a special case of the polico invariance property with* $c = \mathbf{e}_n$.

Part II

Applications of the Risk Parity Approach

This part is dedicated to the different applications of the risk parity approach. In the third chapter, we consider the topic of risk-based indexation for equities. In particular, we compare the ERC indexation with respect to other heuristic approaches such as the minimum variance portfolio or the most diversified portfolio. In the fourth chapter, we show how the risk budgeting approach can be used to better manage interest rate and credit risk in bond portfolios. The fifth chapter considers the application of risk parity techniques to alternative investments, more especially commodities and hedge funds. Finally, we present how risk parity has become a successful strategy to manage multi-asset portfolios.

Chapter 3

Risk-Based Indexation

Capitalization-weighted indexation (CW) is the most common way to gain access to broad equity market performance. It is often backed by results of modern portfolio theory. In the CAPM framework, the optimal investment strategy is to hold the market portfolio, which corresponds to the capitalization-weighted portfolio under certain assumptions. Because of this presumed efficiency, capitalization-weighted indices play a central role in the investment industry. First, they are a convenient investment solution. The growth of index funds and more recently exchange traded funds (ETF) illustrates perfectly the big interest of investors and passive management represents now a large part of the asset management industry. Second, they represent a benchmark for active management. Generating excess return with respect to capitalization-weighted indices is therefore a challenge for all active managers of equity portfolios.

However, many empirical studies have shown that capitalization-weighted indexation is not generally the optimal portfolio. One reason is that the CAPM assumptions do not necessarily hold. Other criticisms concern the momentum bias, the growth bias and the lack of risk diversification that may affect a CW index. In this context, alternative-weighted (AW) indexation has recently prompted great interest from both academic researchers and market professionals. An alternative-weighted index is defined as an index in which assets are weighted in a different way than those based on market capitalization. Alternative-weighted indexation can be split into two families: fundamental indexation and risk-based indexation. Fundamental indexation defines the weights as a function of economic metrics such as dividends or earnings whereas risk-based indexation defines the weights as a function of individual and common risks (Lee, 2011). In view of the success of these new indexation forms, the asset management has also created a new category of passive management products called smart indexing or smart beta.

In this chapter, we review the rationale behind the capitalization-weighted indexation. We then focus on the alternative-weighted indexation based on the risk parity approach and compare it with other risk-based approaches such as the equally weighted portfolio, the minimum variance portfolio and the most diversified portfolio. We also consider an application with the universe of the Eurostoxx 50 index. The results suggest that the risk parity indexation is a good alternative to capitalization-weighted indexation.

3.1 Capitalization-weighted indexation

In a first section, we consider the theory of equity indices built on market capitalization. We present then the general construction of an equity index. The last section lists the pros and cons of such indexation.

3.1.1 Theory support

The separation theorem states that the tangency portfolio is the unique risky portfolio owned by investors. In 1964, William Sharpe developed the CAPM in order to explain the prices of market securities when investor expectations are homogeneous and the rate of interest is the same for all the investors. The main implication of the CAPM theory is that the market-capitalization portfolio is the tangency portfolio if the market is at equilibrium. This is the reason why we now use the two terms 'tangency portfolio' and 'market portfolio' interchangeably.

Fama (1965) develops the efficient market hypothesis (EMH), which asserts that prices reflect all the available information. As a consequence, one cannot beat the market and consistently achieve excess returns with respect to average market returns. This theory questions the active management industry. The first studies on mutual funds performance are already contradictory. Sharpe (1966) uses the Sharpe ratio to measure fund performances. He ranks the funds over two periods 1944-1953 and 1954-1963, and finds a positive relationship between these two ranking periods. On the contrary, Jensen (1968) measures the performance of 115 mutual funds in the 1945-1964 period using the Jensen alpha measure, and he demonstrates that they do not outperform a buy-the-market strategy. He concludes that there is very little evidence that any individual fund is able to do significantly better than that which is expected from mere random chance.

The seminal work of Michael Jensen had a great influence on the development of passive management. If there is no alpha in mutual funds and if the tangency portfolio is the market portfolio, it implies that an efficient investment consists in buying the market. These results gave the idea to John McQuown to develop the first institutional index fund at Wells Fargo in the early 1970s (Bernstein, 2007). This was the beginning of market-capitalization index funds. The boom of passive management since the 1990s, the success of BGI, Vanguard and State Street, and the incredible development of exchange traded funds (or ETF) these last years have led to a common representation of the equity market shared by all investors. In this context, market-capitalization indices are an appropriate investment solution and represent a benchmark for active management that is difficult to beat[1].

[1]Since the work of Carhart (1997), it is largely admitted that actively managed equity

3.1.2 Constructing and replicating an equity index

Let us consider an index composed of n stocks. Let $P_{i,t}$ be the price of the ith stock and $R_{i,t}$ be the corresponding return between times $t-1$ and t:

$$R_{i,t} = \frac{P_{i,t}}{P_{i,t-1}} - 1$$

The value of the index B_t at time t is defined by:

$$B_t = B_{t-1} \sum_{i=1}^{n} w_{i,t} \left(1 + R_{i,t}\right)$$

where $w_{i,t}$ is the weight of the ith stock in the index satisfying $\sum_{i=1}^{n} w_{i,t} = 1$. The computation of the index value B_t is done at the closing time t, but also in an intraday basis. However, this computation is purely theoretical. In order to replicate this index, we must build a hedging strategy that consists in investing in stocks. Let S_t be the value of the strategy (or the index fund). We have:

$$S_t = \sum_{i=1}^{n} n_{i,t} P_{i,t}$$

where $n_{i,t}$ is the number of stock i held between $t-1$ and t. We define the tracking error as the difference between the return of the strategy and the return of the index:

$$e_t \left(S \mid B\right) = R_{S,t} - R_{B,t}$$

The quality of the replication process is generally measured by the volatility $\sigma \left(e_t \left(S \mid B\right)\right)$ of the tracking error. We may distinguish several cases:

1. We may have an index fund with low tracking error volatility (less than 10 bps). This can be achieved by a pure physical replication (by buying all of the components with the appropriate weights each time) or by a synthetic replication (i.e. entering into a swap agreement with an investment bank).

2. We may have an index fund with moderate tracking error volatility (between 10 bps and 50 bps). For example, this may be the case with an index fund based on sampling techniques.

3. An index fund with higher tracking error volatility (larger than 50 bps) corresponds either to some universes presenting liquidity problems or to enhanced index funds as a part of active management.

---

funds do not produce alpha on average and that performance persistence is not verified, meaning that funds with strong performance over a past period do not continue to overperform in future periods.

It is also important to note the difference between an investable index and a non-investable index. The frontier between these two categories is not precise. From a theoretical point of view, an investable index may be replicated with a tracking error volatility close to zero. For a non-investable index, it is impossible to replicate it perfectly. For instance, stock indices of major market places are investable. This is the case with the S&P 500, DAX, CAC and Nikkei indices. This is not the case with private equity indices, some small-cap indices and for certain market places where it is difficult to invest (i.e. the Middle East). Interesting examples are global stock indices, such as the MSCI World index or the DJ Islamic Market index. They contain many stocks (more than 2000 for the two cited indices) and cover a variety of countries.

By definition, in a capitalization-weighted index, the weights are given by:

$$w_{i,t} = \frac{N_{i,t}P_{i,t}}{\sum_{j=1}^{n} N_{j,t}P_{j,t}}$$

where $N_{i,t}$ is the number of shares outstanding for the ith stock. We note that $C_{i,t} = N_{i,t}P_{i,t}$ is the market capitalization of the ith stock. The weight $w_{i,t}$ then corresponds to the ratio of the market capitalization $C_{i,t}$ of the ith stock with respect to the market capitalization of the index $\sum_{j=1}^{n} C_{j,t}$. Generally, the number of shares is constant $N_{i,t} = N_{i,t-1}$ or changes at a low frequency. We also have:

$$\begin{aligned} w_{i,t} &= \frac{N_{i,t}P_{i,t}}{\sum_{j=1}^{n} N_{j,t}P_{j,t}} \\ &= \frac{N_{i,t-1}P_{i,t}}{\sum_{j=1}^{n} N_{j,t-1}P_{j,t}} \\ &\neq w_{i,t-1} \end{aligned}$$

Regardless of whether the number of shares is constant, the weights of CW indices move every day because of the price effect, giving us:

$$\begin{aligned} w_{i,t} \geq w_{i,t-1} &\Leftrightarrow \frac{C_{i,t}}{\sum_{j=1}^{n} C_{j,t}} \geq \frac{C_{i,t-1}}{\sum_{j=1}^{n} C_{j,t-1}} \\ &\Leftrightarrow \frac{C_{i,t}}{C_{i,t-1}} \geq \frac{\sum_{j=1}^{n} C_{j,t}}{\sum_{j=1}^{n} C_{j,t-1}} \\ &\Leftrightarrow R_{i,t} \geq R_{B,t} \end{aligned}$$

3.1.3 Pros and cons of CW indices

There are strong arguments in favor of capitalization-weighted indices. They concern the representativeness, the efficiency and the track record of such portfolio construction. However, since the 1990s, some criticisms have been expressed regarding these indices. The main drawbacks concern the CAPM theory, which is empirically difficult to test and verify (Roll, 1977) and the momentum bias inherent to such indices.

By construction, the capitalization-weighted index is the most representative portfolio of its market, because it defines the equilibrium of supply and demand of this market. It corresponds then to the aggregated exposure of all the market participants. Moreover, it is the only portfolio which is compatible with the efficient market hypothesis. Another advantage is the simplicity and the objectivity of the index construction. The underlying assets are tradable securities which are sold and bought in an exchange (i.e. a centralized and organized financial market). In this case, stock prices are easily available every time and there is no problem in precisely defining the index rules. These arguments of representativeness and objectivity have naturally led the active management industry to adopt capitalization-weighted indices as the benchmarks of mutual funds.

Another strong argument concerns the cost structure efficiency. The portfolio of the hedging strategy does not change if the structure of the market remains the same (or $N_{i,t} = N_{i,t-1}$). We verify that:

$$n_{i,t} = n_{i,t-1}$$

We do not need to rebalance the hedging portfolio because of the relationship:

$$n_{i,t}P_{i,t} \propto w_{i,t}P_{i,t}$$

This property is one of the main benefits of CW indices and implies low trading costs. In fact, the structure of the market changes, because of capital increases, IPOs and universe modifications. However, a CW index remains the most efficient investment in terms of management simplicity, turnover and transaction costs.

The last but not least argument is the record of passive versus active management (Malkiel, 2003). The first studies in the 1980s based on Jensen's alpha conclude on the persistence of performance on US equity funds (Lehmann and Modest, 1987; Grinblatt and Titman, 1989). The positive persistence in mutual fund performance is also supported by the academic studies in the first half of the 1990s (Grinblatt and Titman, 1992; Hendricks *et al.*, 1993; Goetzmann and Ibbotson, 1994). In a key paper, Carhart (1997) uses an extension of the Fama-French three-factor model. Over the years, this model has become a standard model in most studies on equity funds. Even if there are some results in favor of one-year persistence, there is a fundamental break between Carhart's conclusion and previous ones:

> *"I demonstrate that common factors in stock returns and investment expenses almost completely explain persistence in equity mutual funds' mean and risk-adjusted returns. Hendricks et al. (1993) 'hot hands' result is mostly driven by the one-year momentum effect of Jegadeesh and Titman (1993) [...] The only significant persistence not explained is concentrated in strong underperformance*

by the worst-return mutual funds. The results do not support the existence of skilled or informed mutual fund portfolio managers."

Today, this point of view is largely shared by sophisticated investors such as institutionals and pension funds. Moreover, it has been confirmed by a large body of academic literature (Barras *et al.*, 2010). It is also interesting to note that the oldest funds are passive funds, not active funds. In this context, it is easy to explain the faster growth of passive management with respect to active management.

On the other hand, we can classify the criticisms against CW indices in two categories. The first one concerns the rejection of the CAPM theory, whereas the portfolio construction itself constitutes the second category. We recall that the CAPM implies that:

$$\mathbb{E}[R_i] - r = \beta_i (\mathbb{E}[R_m] - r)$$

where R_i, r and R_m are respectively the return of asset i, the risk-free rate and the return of the market portfolio. This model is also called the single-factor pricing model, because the stock return is entirely explained by one risk factor represented by the market[2]. According to Fama and French (2004), this model is the centerpiece of MBA investment courses and is widely used in finance by practitioners despite a large body of evidence in the academic literature of the invalidation of this model:

> "*The attraction of the CAPM is that it offers powerful and intuitively pleasing predictions about how to measure risk and the relation between expected return and risk. Unfortunately, the empirical record of the model is poor – poor enough to invalidate the way it is used in applications*" (Fama and French, 2004).

Fama and French (1992) study several factors explaining average returns (size, E/P, leverage and book-to-market equity) and extend in 1993 their empirical work and propose a three-factor model:

$$\mathbb{E}[R_i] - r = \beta_i^m (\mathbb{E}[R_m] - r) + \beta_i^{\text{sml}} \mathbb{E}[R_{\text{smb}}] + \beta_i^{\text{hml}} \mathbb{E}[R_{\text{hml}}]$$

where R_{smb} is the return of small stocks minus the return of large stocks and R_{hml} is the return of stocks with high book-to-market values minus the return of stocks with low book-to-market values. Attempts to save the CAPM have been fruitless (Black, 1993; Kothari *et al.*, 1995). This model is supported by a substantial body of empirical literature and is today the standard model in the asset management industry. One implication of the Fama-French model is that the capitalization-weighted portfolio is not efficient:

---

[2]The beta coefficient β_i is the measure of the systematic risk with respect to the market portfolio.

> *"[...] because of the empirical failings of the CAPM, even passively managed stock portfolios produce abnormal returns if their investment strategies involve tilts toward CAPM problems (Elton et al., 1993). For example, funds that concentrate on low beta stocks, small stocks or value stocks will tend to produce positive abnormal returns relative to the predictions of the Sharpe-Lintner CAPM, even when the fund managers have no special talent for picking winners"* (Fama and French, 2004).

The portfolio construction of the capitalization-weighted index is the second criticism. By construction, a CW index is a trend-following strategy, meaning that it incorporates momentum bias which leads to bubble exposure risk as the weight of the best performers increases and the weight of the worst performers decreases. This implies that a CW index generally contains a growth bias, because high-valuation multiple stocks weigh more than low valuation multiple stocks with equivalent realized earnings. One consequence is that the index may suffer from a lack of risk diversification. It can then be exposed to a specific risk and a high drawdown risk. The most famous example to illustrate these points is the dot-com bubble. Another important related issue is the concentration of underlying portfolios. In Table 3.1, we present the Gini coefficient and the values of the Lorenz curve $\mathbb{L}(x)$ for several equity indices (see also Figure 3.1). We note some differences. Indeed, the Gini coefficient of the Stoxx 50 is 30.8% in terms of weight and 26.3% in terms of risk contribution. If we consider the Lorenz curve, we arrive at the following conclusions: 10% (respectively 25% and 50%) of stocks represent 24% (respectively 48% and 71%) of cumulated weights and 19% (respectively 40% and 69%) of cumulated risks. If we consider the IBEX index, we obtain different numbers. For instance, the Gini coefficient is 64.9% in terms of weight and 68.3% in terms of risk contribution. The IBEX index is therefore more concentrated than the Stoxx 50 index both in terms of exposure and risk.

3.2 Alternative-weighted indexation

An alternative-weighted index is defined as an index in which assets are weighted in a different way than those based on market capitalization. The goal is then to improve the risk-return profile of capitalization-weighted indices. We generally distinguish two families of alternative-weighted indexation: fundamental indexation and risk-based indexation. Fundamental indexation considers that market capitalizations are not the right statistics to measure the economic size of a company. It then defines the weights as a function of economic metrics such as sales, dividends or earnings. In the case of risk-based indexation, the weights are defined as a function of individual and common

TABLE 3.1: Weight and risk concentration of several equity indices (June 29, 2012)

Ticker	Weights				Risk contributions			
	$\mathcal{G}(x)$	$\mathbb{L}(x)$			$\mathcal{G}(x)$	$\mathbb{L}(x)$		
		10%	25%	50%		10%	25%	50%
SX5P	30.8	24.1	48.1	71.3	26.3	19.0	40.4	68.6
SX5E	31.2	23.0	46.5	72.1	31.2	20.5	44.7	73.3
INDU	33.2	23.0	45.0	73.5	35.8	25.0	49.6	75.9
BEL20	39.1	25.8	49.4	79.1	45.1	25.6	56.8	82.5
DAX	44.0	27.5	56.0	81.8	47.3	27.2	59.8	84.8
CAC	47.4	34.3	58.3	82.4	44.1	31.9	57.3	79.7
HSCEI	54.8	39.7	69.3	85.9	53.8	36.5	67.2	85.9
SMI	58.1	44.2	70.0	87.8	49.1	30.3	60.2	85.1
NKY	60.2	47.9	70.4	87.7	61.4	49.6	70.9	88.1
UKX	60.8	47.5	73.1	88.6	60.4	46.1	72.8	88.7
SXXE	61.7	49.2	73.5	88.7	63.9	51.6	75.3	90.1
SPX	61.8	52.1	72.0	87.8	59.3	48.7	69.9	86.7
MEXBOL	64.6	48.2	75.1	91.8	65.9	45.7	78.6	92.9
IBEX	64.9	51.7	77.3	90.2	68.3	58.2	80.3	91.4
NDX	66.3	58.6	77.0	89.2	64.6	56.9	74.9	88.6
KOSPI	86.5	80.6	93.9	98.0	89.3	85.1	95.8	98.8

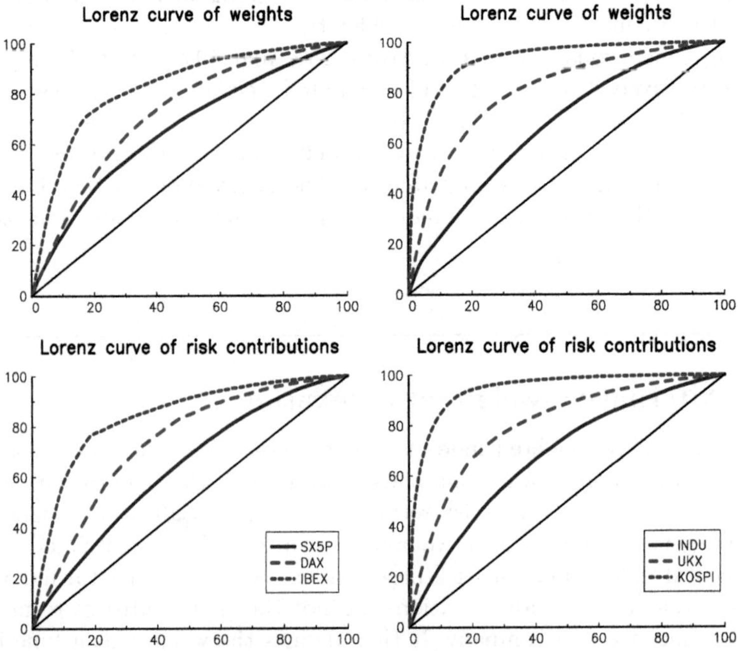

FIGURE 3.1: Lorenz curve of several equity indices (June 29, 2012)

risks. The underlying idea is to build a more diversified portfolio than the capitalization-weighted portfolio in order to better capture the equity risk premium.

3.2.1 Desirable properties of AW indices

We list here the necessary properties of an appropriate alternative-weighted index. In what follows, the CW index is considered the reference index.

- The rules for constructing the AW index are clearly defined, and the computation of the index values may be performed by several third-parties.
- The universe of the AW index is included within the universe of the CW index:

  $$\mathcal{U}_{\mathrm{AW}} \subseteq \mathcal{U}_{\mathrm{CW}}$$

  This conveys that the AW index contains only assets belonging to the corresponding reference index. This is a key property in order to qualify the AW index as a passive strategy and to minimize some style biases between the AW and CW indices.
- In the long-term, the AW index must perform better than the CW index, and/or the volatility of the AW index must be lower than that of the CW index. With respect to the CW index, the AW index must obviously be characterized as an alpha index or a beta index.
- The correlation between the performance of the AW index and the performance of the CW index is strictly different from one. The lower the correlation, the higher the interest of the AW index. This fourth point is the main rationale for the AW index. An investor who prefers to invest in both AW and CW indices does not necessarily seek better performance, but wants to diversify the risk in order to achieve a better risk-return profile.

The first point is of course central and raises the question of what constitutes an index. With the development of proprietary and custom indices by asset managers and investment banks, it is difficult to provide a precise answer. We can say that an index is a common good for the different investors. This means that there is no ambiguity in the index rules implying that the computation of the weights can be checked by the investor. In particular, this excludes discretionary rules or mathematical models which are not fully described. The second point concerns benchmarking. It is important to maintain a strong link with the corresponding CW index. If this is not the case, the behavior of the AW index may not be challenged by a realistic benchmark. This point is related to the last two points. The interest of AW indices is to provide risk-return profiles that are better than or complementary to those of CW indices.

An example of AW indexation is price-weighted indexation. We saw previously that the dynamic of a capitalized-weighted index has two components: a share component and a price component. Generally, the share component is very stable over time, whereas the price component changes every day, meaning that the high-frequency dynamic of the CW index is driven entirely by the price dynamics. Price-weighted (PW) indexation utilizes this idea. The index value is then equal to:

$$B_t = B_{t-1} \frac{\sum_{i=1}^{n} P_{i,t}}{\sum_{i=1}^{n} P_{i,t-1}}$$

Two well-known PW indices are the Nikkei index and the Dow Jones Industrial Average index.

3.2.2 Fundamental indexation

In fundamental indexation, we define the weights as a function of fundamental (or economic) metrics of the enterprise size. A basic example of economic measure is the dividend yield. For instance, in November 2003, Dow Jones launched the DJ US Select Dividend index followed by other dividend indices by country and region. In the same year, Wood and Evans (2003) considered earnings to define an alternative-weighted S&P 100 index and Barclays Global Investors (BGI) was the first asset manager to propose a fundamental-weighted ETF based on Dow Jones indices (Siracusano, 2007). However, the breakthrough in fundamental indexation appears with the publication of Arnott *et al.* (2005). These authors propose to build indices by combining different fundamental metrics:

> "*We constructed indices that use gross revenue, equity book value, gross sales, gross dividends, cash flow, and total employment as weights. If capitalization is a 'Wall Street' definition of the size of an enterprise, these characteristics are clearly 'Main Street' measures. When a merger is announced, the* Wall Street Journal *may cite the combined capitalization but the* New York Post *will focus on the combined sales or total employment.*"

They provide a simulation for the composite index and the different subindices since 1964. Their results are the promise of potentially superior returns:

> "*We show that the fundamentals-weighted, noncapitalization-based indices consistently provide higher returns and lower risks than the traditional cap-weighted equity market indices while retaining many of the benefits of traditional indexing*" (Arnott *et al.*, 2005).

Several research studies have followed after the publication of this article. Most of them find evidence that fundamental indices can outperform capitalization-weighted indices (Estrada, 2008; Hemminki and Puttonen, 2008; Walkshäusl and Lobe, 2010; Hsieh *et al.*, 2012). A theory has also been proposed to explain the superiority of the 'fundamental' portfolio on the market portfolio. This theory based on the 'noisy market hypothesis' states that market prices of stocks deviate from fair values and that fundamental weights are less biased estimators of fair value weights than market weights[3] (Hsu, 2006; Arnott and Hsu, 2008). However, fundamental indexation has also been criticized (Perold, 2007; Kaplan, 2008; Blitz and Swinkels, 2008). The construction of these fundamental indices is not always obvious and some practitioners associate them with a stock-picking value strategy packaged in a passive strategy, meaning that fundamental indices capture the value or size premia of the Fama-French model. Moreover, Perold (2007) and Kaplan (2008) show that the noisy market hypothesis does not imply that the 'fundamental' portfolio is more mean-variance efficient than the market portfolio:

> "*Proponents of fundamental indexation claim that the noisy-market hypothesis implies that fundamental weighting must be superior to market-cap weighting. Perold (2007) and this article demonstrate, in different ways, that these proponents' reasoning is flawed. [...] The proponents of fundamental indexation, rather than having a clear theory on which to base their claims, have only a conjecture that market valuation errors are more variable than fair value multiples. They may have been correct over long historical periods, as the successful backtests of their strategies seem to demonstrate, but they should be much more modest in their claims. In particular, they could argue that it is better to introduce a value bias into a portfolio by using their weighting scheme than by excluding low-yield stocks*" (Kaplan, 2008).

Despite these criticisms, the development of fundamental indexation is strong, with many successes in the investment industry. For instance, WisdomTree, an ETF provider specializing in dividends and earnings indexation, manage \$15 billion in assets under management at June 30, 2012 whereas about \$100 billion in assets are managed worldwide using investment strategies developed by Research Affiliates[4]. In Figure 3.2, we have reported the RAFI® index computed by FTSE and sponsored by Research Affiliates. We show the impressiveness of the performance since 2000 and we note that this fundamental index outperforms the S&P 500 index. Despite the lack of any

---

[3]In this theory, the error between the market price and the fair value of the stock is modeled as a mean-reverting process. In this case, the capitalized-weighted portfolio is a suboptimal portfolio, because more weights are allocated to over-valued stocks and less weight to under-valued stocks.

[4]The firm was founded by Robert Arnott in 2002.

significant over-performance in 2007 and 2008, this index performed solidly in 2009.

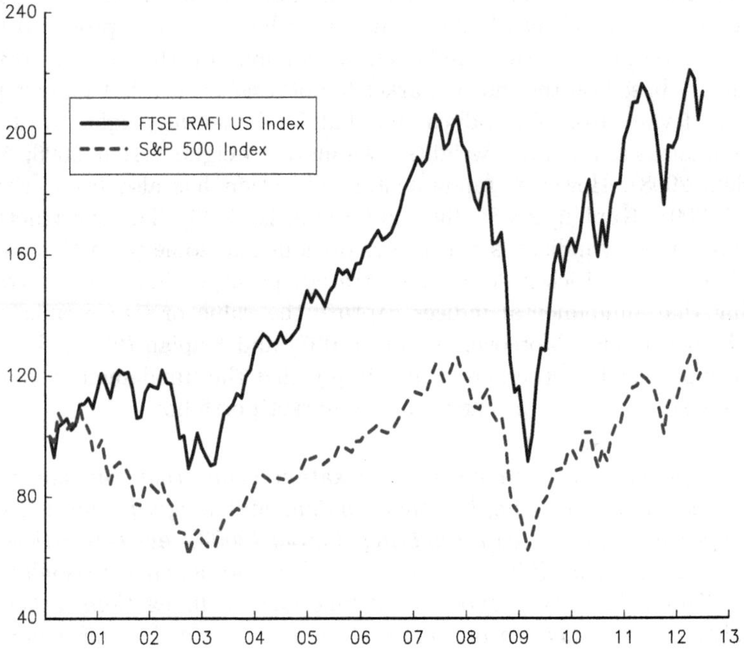

FIGURE 3.2: Performance of the RAFI index since January 2000

3.2.3 Risk-based indexation

While the goal of fundamental indices is to deliver alpha, risk-based indices are more promising for diversification. In a sense, the difference between the two methods comes from the way of modifying the risk-adjusted return ratio. In other words, in the case of fundamental indices, one expects to have superior returns, and to create alpha with respect to the CW index. In the case of risk-based indices, one expects to decrease the portfolio's risk in either absolute or relative value. The AW index may have a smaller risk than the CW index or the combination of the AW index and the CW index may produce lower risk because they are not perfectly correlated.

Consider a portfolio with $(1 - w)\,\%$ of the CW index and $w\%$ of the AW index. We denote the volatility of the returns of the two indices as σ_{CW} and σ_{AW} and the correlation between them as ρ. If we assume that the two indices have the same expected return, then we can compute the optimal portfolio. It appears that w is equal to 0 if and only if we have $\sigma_{\mathrm{CW}} < \rho\sigma_{\mathrm{AW}}$. As such, financial theory explains that if the two indices have the same expected return, it is better to invest in the two indices rather than only in the CW index, except if the volatility of the CW index is particularly low with respect to the volatility of the AW index. In Figure 3.3, we have provided some examples with $\sigma_{CW} = 18\%$. In the first case ($\sigma_{AW} = 20\%$ and $\rho = 95\%$), there is no interest in diversifying the portfolio, because the inequality $\sigma_{CW} > \rho\sigma_{AW}$ does not hold. In other illustrative cases, the investor is particularly interested in diversification if the correlation is low.

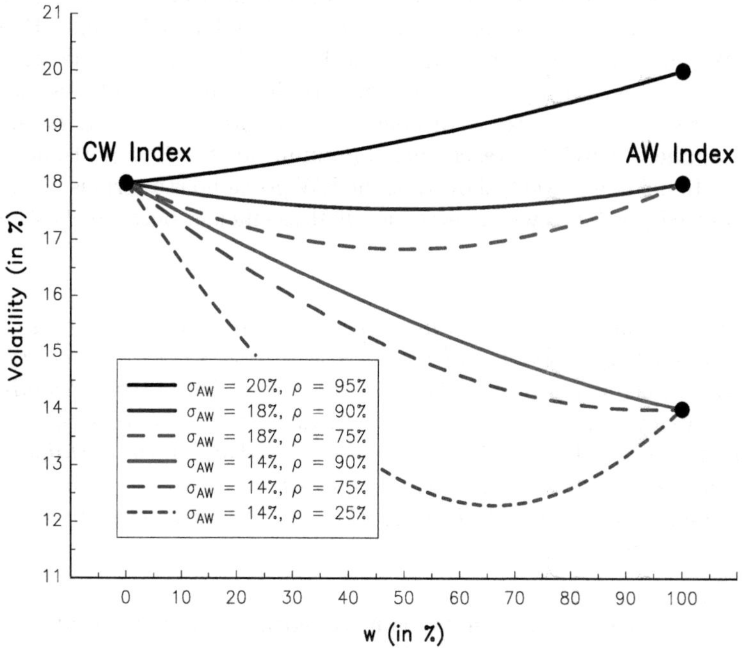

FIGURE 3.3: Illustration of the diversification effect of AW indices

In this section, we focus on risk-based indexation methods based on the following portfolio constructions: the equally weighted (EW) portfolio, the minimum variance (MV) portfolio, the most diversified portfolio (MDP) and the equally weighted risk contribution (ERC) portfolio. These indexation methods are already implemented in the investment industry. Nonetheless, we should note that other solutions, which are more complex, also exist such as maximum entropy portfolios (Bera and Park, 2008).

3.2.3.1 The equally weighted portfolio

The idea of the equally weighted or '$1/n$' portfolio is to define a portfolio independently from the estimated statistics and properties of stocks (Bernartzi and Thaler, 2001; Windcliff and Boyle, 2004). If we assume that it is impossible to predict return and risk, then attributing an equal weight to all of the portfolio components constitutes a natural choice. The structure of the portfolio depends only on the number n of stocks because the weights are equal and uniform:

$$x_i = x_j = \frac{1}{n}$$

This type of indexation is easy to understand, thanks to the uncomplicated rules of construction. It corresponds to a contrarian strategy with a take-profit scheme, because if one stock has a substantial return between two rebalancing dates, its weight will reset to $1/n$ at the next rebalancing date. However, in a CW index, the opposite is true: the larger the return of the stock, the greater its weight. An appealing property is that the EW index is the least concentrated portfolio in terms of weights. Therefore, the Herfindahl and Gini indices applied to weights reach their minimum for the EW portfolio. Moreover, DeMiguel *et al.* (2009) show that the EW portfolio is a serious benchmark for optimized portfolios which are generally less efficient because of estimation errors:

> "*We evaluate the out-of-sample performance of the sample-based mean-variance model, and its extensions designed to reduce estimation error, relative to the naive $1/n$ portfolio. Of the 14 models we evaluate across seven empirical datasets, none is consistently better than the $1/n$ rule in terms of Sharpe ratio, certainty-equivalent return, or turnover, which indicates that, out of sample, the gain from optimal diversification is more than offset by estimation error.*"

The main weakness is then its main strength: it does not consider individual risks and correlations between these risks, which implies that it is difficult to locate this portfolio in a mean-variance framework. From a theoretical point of view, the EW portfolio coincides with the efficient portfolio if the expected returns and volatilities of stocks are assumed to be equal and correlation is uniform. If we are not far from these assumptions, we may consider the EW portfolio to be similar to the efficient portfolio.

3.2.3.2 The minimum variance portfolio

We have seen previously that the minimum variance (MV) portfolio corresponds to the following optimization program:

$$\begin{aligned} x^\star &= \arg\min \frac{1}{2} x^\top \Sigma x \\ & \text{u.c.} \quad \mathbf{1}^\top x = 1 \end{aligned}$$

The solution is:

$$x^\star = \frac{\Sigma^{-1}\mathbf{1}}{\mathbf{1}^\top \Sigma^{-1}\mathbf{1}} \tag{3.1}$$

If we assume that the correlation matrix is constant – $C = C_n(\rho)$, the optimal weights are given by the following relationship:

$$x_i \propto \frac{((n-1)\rho + 1)}{\sigma_i^2} - \frac{\rho}{\sigma_i} \sum_{j=1}^{n} \frac{1}{\sigma_j}$$

In particular, if the assets are uncorrelated ($\rho = 0$), we obtain:

$$x_i = \frac{\sigma_i^{-2}}{\sum_{j=1}^{n} \sigma_j^{-2}}$$

In this case, we observe an inverse relationship between the weights and the variance of the assets. The MV portfolio is the only portfolio located on the efficient frontier that is not dependent on the expected returns hypothesis (see Figure 3.4). It is also the tangency portfolio if and only if expected returns are equal for all stocks[5]. If the correlation is uniform, the weights may be negative and we must impose the long-only constraint $\mathbf{0} \leq x \leq \mathbf{1}$.

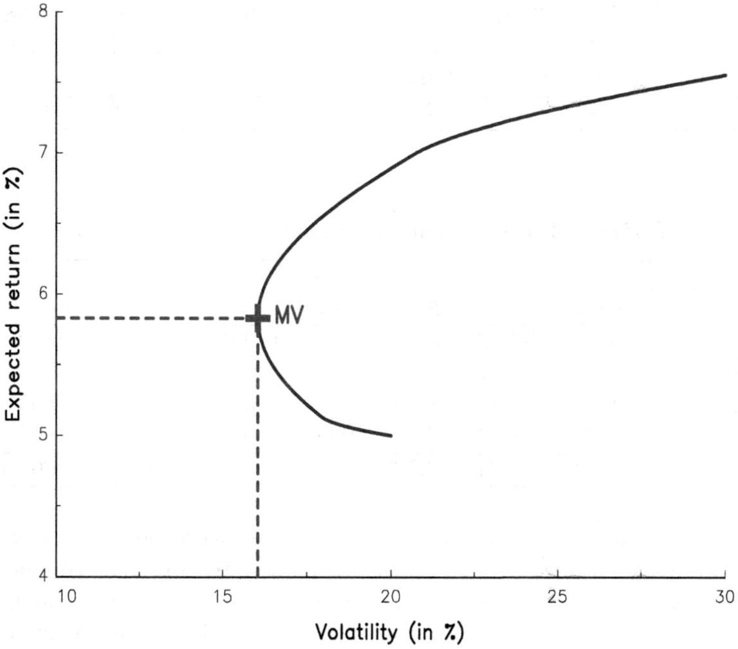

FIGURE 3.4: Location of the minimum variance portfolio in the efficient frontier

[5]In the case of no short-selling constraint, we may show that every optimal portfolio located in the efficient frontier contains a proportion of the MV portfolio. However, it does not hold when we impose a no-short-selling constraint.

Example 23 *We consider a universe of four assets. Their volatilities are respectively equal to* 4%, 6%, 8% *and* 10%. *We assume also that the correlation matrix* C *is uniform and is equal to* $C_n(\rho)$.

We have reported the composition (in %) of the MV portfolio in Tables 3.2 and 3.3 with respect to different values of ρ. We note that the long-only portfolio behaves differently than the long-short portfolio when the correlation increases. If we impose that the weights are positive, we observe that the volatility increases with the correlation. In the case of the long-short portfolio, the volatility increases in a first time and then decreases when the correlation is high[6]. In fact, when the correlation is high, we may always hedge long exposures in low-volatility assets by short exposures in high-volatility assets.

TABLE 3.2: Unconstrained minimum variance portfolios

Asset	ρ						
	−20%	0%	20%	50%	70%	90%	99%
1	44.35	53.92	65.88	90.65	114.60	149.07	170.07
2	25.25	23.97	22.36	19.04	15.83	11.20	8.38
3	17.32	13.48	8.69	−1.24	−10.84	−24.67	−33.09
4	13.08	8.63	3.07	−8.44	−19.58	−35.61	−45.37
$\sigma(x^\star)$	1.93	2.94	3.52	3.86	3.62	2.52	0.87

TABLE 3.3: Long-only minimum variance portfolios

Asset	ρ						
	−20%	0%	20%	50%	70%	90%	99%
1	44.35	53.92	65.88	85.71	100.00	100.00	100.00
2	25.25	23.97	22.36	14.29	0.00	0.00	0.00
3	17.32	13.48	8.69	0.00	0.00	0.00	0.00
4	13.08	8.63	3.07	0.00	0.00	0.00	0.00
$\sigma(x^\star)$	1.93	2.94	3.52	3.93	4.00	4.00	4.00

There is a substantial body of academic literature on the performance of the minimum variance portfolio that generally reports that this portfolio presents some good out-of-sample performance in a complete economic cycle (Haugen and Baker, 1991; Clarke *et al.*, 2006). The main explanation is an anomaly called the low volatility effect (Ang *et al.* 2006; Blitz and van Vliet, 2007), which states that stocks with high idiosyncratic volatility are less rewarded than stocks with low idiosyncratic volatility in a risk-return framework[7].

[6] In particular, the volatility is equal to zero if the correlation is equal to one.

[7] This result can be interpreted as a new formulation of the low beta effect (Black *et al.*, 1972), which states that low beta stocks produce positive alpha.

To illustrate that the minimum variance portfolio is linked to the low volatility effect, we follow the analysis of Scherer (2011). In the CAPM, the covariance matrix Σ can be decomposed as:

$$\Sigma = \beta\beta^\top \sigma_m^2 + D$$

where $\beta = (\beta_1, \ldots, \beta_n)$ is the vector of betas, σ_m^2 is the variance of the market portfolio and $D = \operatorname{diag}\left(\tilde{\sigma}_1^2, \ldots, \tilde{\sigma}_n^2\right)$ is the diagonal matrix of specific variances. Using the Sherman-Morrison-Woodbury formula[8], we deduce that the inverse of the covariance matrix is:

$$\Sigma^{-1} = D^{-1} - \frac{\sigma_m^2}{1 + \sigma_m^2 \kappa} \tilde{\beta}\tilde{\beta}^\top$$

with $\tilde{\beta}_i = \beta_i / \tilde{\sigma}_i^2$ and $\kappa = \tilde{\beta}^\top \beta$. Solution (3.1) becomes:

$$x^\star = \sigma^2\left(x^\star\right)\left(D^{-1}\mathbf{1} - \frac{\sigma_m^2}{1 + \sigma_m^2 \kappa} \tilde{\beta}\tilde{\beta}^\top \mathbf{1}\right)$$

Using this new expression, Scherer (2011) shows that:

$$x_i^\star = \frac{\sigma^2\left(x^\star\right)}{\tilde{\sigma}_i^2}\left(1 - \frac{\beta_i}{\beta^\star}\right) \tag{3.2}$$

with:

$$\beta^\star = \frac{1 + \sigma_m^2 \kappa}{\sigma_m^2 \tilde{\beta}^\top \mathbf{1}}$$

If we consider this formula, we note that the minimum variance portfolio is exposed to stocks with low volatility and low beta. More precisely, if asset i has a beta β_i smaller than $\beta^\star$, the weight of this asset is positive ($x_i^\star > 0$). If $\beta_i > \beta^\star$, then $x_i^\star < 0$. Clarke *et al.* (2010) extend Formula (3.2) to the long-only case with the threshold $\beta^\star$ defined as follows:

$$\beta^\star = \frac{1 + \sigma_m^2 \sum_{\beta_i < \beta^\star} \tilde{\beta}_i \beta_i}{\sigma_m^2 \sum_{\beta_i < \beta^\star} \tilde{\beta}_i}$$

In this case, if $\beta_i > \beta^\star$, $x_i^\star = 0$.

Remark 44 *We can also hope that the minimum variance portfolio presents a low correlation with respect to the capitalization-weighted portfolio. The minimum variance portfolio is then a serious candidate to build alternative-weighted indices*[9].

---

[8]Suppose u and v are two vectors and A is an invertible square matrix. It follows that:

$$\left(A + uv^\top\right)^{-1} = A^{-1} - \frac{1}{1 + v^\top A^{-1} u} A^{-1} u v^\top A^{-1}$$

The expression of Σ^{-1} is then obtained with $A = D$ and $u = v = \sigma_m \beta$.

[9]Notwithstanding these advantages, the MV portfolio poses a serious drawback: diversification of volatility, but not diversification of weight. This implies that the portfolio is concentrated in relatively few stocks.

Example 24 *We consider an investment universe of five assets. Their beta is respectively equal to* 0.9, 0.8, 1.2, 0.7 *and* 1.3 *whereas their specific volatility is* 4%, 12%, 5%, 8% *and* 5%. *We also assume that the market portfolio volatility is equal to* 25%.

Using these figures, we have computed the composition of the minimum variance portfolio. The results are reported in Table 3.4. The second column corresponds to the beta β_i, the value taken by $\tilde{\beta}_i$ is given in the third column and the fourth and fifth columns contain the weights (in %) of the unconstrained and long-only MV portfolio. In the case of the unconstrained portfolio, we have $\kappa = 1879.26$ and $\beta^\star = 1.0972$. We deduce then that long exposures concern the first, second and fourth assets whereas the short exposures concern the third and fifth assets. For the long-only portfolio, we obtain $\kappa = 121.01$ and $\beta^\star = 0.8307$. This implies that only the second and fourth assets are represented in the long-only MV portfolio.

TABLE 3.4: Composition of the MV portfolio

Asset	β_i	$\tilde{\beta}_i$	$x_i^\star$	
			Unconstrained	Long-only
1	0.90	562.50	147.33	0.00
2	0.80	55.56	24.67	9.45
3	1.20	480.00	−49.19	0.00
4	0.70	109.37	74.20	90.55
5	1.30	520.00	−97.01	0.00
Volatility			11.45	19.19

3.2.3.3 The most diversified portfolio

Choueifaty and Coignard (2008) introduce the concept of diversification ratio, which corresponds to the following expression:

$$\mathcal{DR}(x) = \frac{\sum_{i=1}^n x_i \sigma_i}{\sigma(x)} = \frac{x^\top \sigma}{\sqrt{x^\top \Sigma x}} \tag{3.3}$$

$\mathcal{DR}(x)$ is the ratio between the weighted average volatility and the portfolio volatility. The numerator is then the portfolio's volatility when we do not take into account the diversification effect induced by the correlations[10]. By construction, the diversification ratio of a portfolio fully invested in one asset is equal to one:

$$\mathcal{DR}(e_i) = 1$$

whereas it is larger than one in the general case:

$$\mathcal{DR}(x) \geq 1$$

[10]The diversification ratio is then the inverse of the diversification index, which is defined on page 125.

The MDP is then defined as the portfolio which maximizes the diversification ratio or equivalently its logarithm:

$$\begin{aligned} x^\star &= \arg\max \ln \mathcal{DR}(x) \\ \text{u.c.} & \quad \left\{ \begin{array}{l} \mathbf{1}^\top x = 1 \\ \mathbf{0} \leq x \leq \mathbf{1} \end{array} \right. \end{aligned} \tag{3.4}$$

The second constraint $\mathbf{0} \leq x \leq \mathbf{1}$ is added when we consider long-only portfolios. The associated Lagrange function is then:

$$\mathcal{L}(x; \lambda_0, \lambda) = \ln\left(x^\top \sigma\right) - \frac{1}{2}\ln\left(x^\top \Sigma x\right) + \lambda_0\left(\mathbf{1}^\top x - 1\right) + \lambda^\top x$$

with $\lambda_0 \in \mathbb{R}$ and $\lambda \in \mathbb{R}^n$. The solution $x^\star$ verifies also the first-order condition:

$$\frac{\partial\, \mathcal{L}(x; \lambda_0, \lambda)}{\partial\, x} = \frac{\sigma}{x^\top \sigma} - \frac{\Sigma x}{x^\top \Sigma x} + \lambda_0 \mathbf{1} + \lambda = 0$$

whereas the Kuhn-Tucker conditions are $\min(\lambda_i, x_i) = 0$ for $i = 1, \ldots, n$. We note that the constraint $\mathbf{1}^\top x - 1$ can always be matched because scaling the portfolio does not change the diversification ratio[11]. Choueifaty *et al.* (2011) deduce that the MDP satisfies:

$$\begin{aligned} \Sigma x^\star &= \frac{\sigma^2(x^\star)}{x^{\star\top}\sigma}\sigma + \lambda\sigma^2(x^\star) \\ &= \frac{\sigma(x^\star)}{\mathcal{DR}(x^\star)}\sigma + \lambda\sigma^2(x^\star) \end{aligned}$$

The previous analysis remains valid if we do not consider the long-only constraint. In this case, we have $\lambda = \mathbf{0}$.

The correlation between a portfolio x and the MDP $x^\star$ is given by:

$$\begin{aligned} \rho(x, x^\star) &= \frac{x^\top \Sigma x^\star}{\sigma(x)\,\sigma(x^\star)} \\ &= \frac{1}{\sigma(x)\,\mathcal{DR}(x^\star)} x^\top \sigma + \frac{\sigma(x^\star)}{\sigma(x)} x^\top \lambda \\ &= \frac{\mathcal{DR}(x)}{\mathcal{DR}(x^\star)} + \frac{\sigma(x^\star)}{\sigma(x)} x^\top \lambda \end{aligned}$$

If $x^\star$ is the long-only MDP, we obtain[12]:

$$\rho(x, x^\star) \geq \frac{\mathcal{DR}(x)}{\mathcal{DR}(x^\star)} \tag{3.5}$$

[11]We have $\mathcal{DR}(c \cdot x) = \mathcal{DR}(x)$ with c a constant.

[12]Because $\lambda \geq \mathbf{0}$.

whereas we have for the unconstrained MDP:

$$\rho(x, x^\star) = \frac{\mathcal{DR}(x)}{\mathcal{DR}(x^\star)}$$

Result (3.5) is called the 'core property' of the MDP by Choueifaty *et al.* (2011):

> "*The long-only MDP is the long-only portfolio such that the correlation between any other long-only portfolio and itself is greater than or equal to the ratio of their diversification ratios.*"

If we now consider the correlation between asset i and the MDP, we have:

$$\begin{aligned}\rho(e_i, x^\star) &= \frac{\mathcal{DR}(e_i)}{\mathcal{DR}(x^\star)} + \frac{\sigma(x^\star)}{\sigma(e_i)} e_i^\top \lambda \\ &= \frac{1}{\mathcal{DR}(x^\star)} + \frac{\sigma(x^\star)}{\sigma_i} \lambda_i\end{aligned}$$

We deduce that[13]:

$$\rho(e_i, x^\star) = \frac{1}{\mathcal{DR}(x^\star)} \quad \text{if} \quad x_i^\star > 0$$

and:

$$\rho(e_i, x^\star) \geq \frac{1}{\mathcal{DR}(x^\star)} \quad \text{if} \quad x_i^\star = 0$$

Choueifaty *et al.* (2011) conclude that:

> "*Any stock not held by the MDP is more correlated to the MDP than any of the stocks that belong to it. Furthermore, all stocks belonging to the MDP have the same correlation to it. [...] This property illustrates that all assets in the universe are effectively represented in the MDP, even if the portfolio does not physically hold them. [...] This is consistent with the notion that the most diversified portfolio is the un-diversifiable portfolio.*"

In the case when the long-only constraint is omitted, we have $\rho(e_i, x^\star) = \rho(e_j, x^\star)$ meaning that the correlation with the MDP is the same for all the assets.

Example 25 *We consider an investment universe of four assets. Their volatilities are equal to 20%, 10%, 20% and 25%. The correlation of asset returns is given by the following matrix:*

$$\rho = \begin{pmatrix} 1.00 & & & \\ 0.80 & 1.00 & & \\ 0.40 & 0.30 & 1.00 & \\ 0.50 & 0.10 & -0.10 & 1.00 \end{pmatrix}$$

[13]Because $\lambda_i = 0$ if $x_i^\star > 0$ and $\lambda_i > 0$ if $x_i^\star = 0$.

The results are given in Table 3.5. We verify that the correlation with the unconstrained MDP is the same for all the assets. For the long-only MDP, this property is only verified for the assets that are represented in the portfolio.

TABLE 3.5: Composition of the MDP

	Unconstrained		Long-only	
Asset	$x_i^\star$	$\rho(e_i, x^\star)$	$x_i^\star$	$\rho(e_i, x^\star)$
1	−18.15	61.10	0.00	73.20
2	61.21	61.10	41.70	62.40
3	29.89	61.10	30.71	62.40
4	27.05	61.10	27.60	62.40
$\sigma(x^\star)$	9.31		10.74	
$\mathcal{DR}(x^\star)$	1.64		1.60	

Choueifaty and Coignard (2008) find another interesting property concerning the optimality of the MDP. If all the assets have the same Sharpe ratio:

$$\frac{\mu_i - r}{\sigma_i} = \text{SR}$$

the diversification ratio of portfolio x is proportional to its Sharpe ratio:

$$\begin{aligned}\mathcal{DR}(x) &= \frac{1}{\text{SR}} \frac{\sum_{i=1}^n x_i (\mu_i - r)}{\sigma(x)} \\ &= \frac{1}{\text{SR}} \frac{x^\top \mu - r}{\sigma(x)} \\ &= \frac{\text{SR}(x \mid r)}{\text{SR}}\end{aligned}$$

Maximizing the diversification ratio is then equivalent to maximizing the Sharpe ratio. Therefore, the MDP is the tangency portfolio. This is why the MDP is synonymous with the maximum Sharpe ratio (or MSR) portfolio presented by Martellini (2008).

Remark 45 *Amenc* et al. *(2010) extend the concept of the MSR portfolio by assuming that the risk premium of one asset is related to its downside risk. For instance, they propose to replace the volatility σ_i by the square root of the semi-variance below zero (sometimes called the semi-deviation ς_i):*

$$\varsigma_i = \sqrt{\mathbb{E}\left[\min(0, R)\right]^2}$$

In this case, the objective function of the MSR portfolio becomes:

$$x^\star = \arg\max \frac{x^\top \varsigma}{\sqrt{x^\top \Sigma x}}$$

Using the CAPM framework presented on page 167, Clarke *et al.* (2012) find a similar expression to the one obtained for the minimum variance. Thus, the weights of the MDP are equal to:

$$x_i^\star = \mathcal{DR}(x^\star) \frac{\sigma_i \sigma(x^\star)}{\tilde{\sigma}_i^2} \left(1 - \frac{\rho_{i,m}}{\rho^\star}\right)$$

where $\sigma_i = \sqrt{\beta_i^2 \sigma_m^2 + \tilde{\sigma}_i^2}$ is the volatility of asset i, $\rho_{i,m} = \beta_i \sigma_m / \sigma_i$ is the correlation between asset i and the market portfolio and $\rho^\star$ is the threshold correlation given by this formula:

$$\rho^\star = \left(1 + \sum_{i=1}^n \frac{\rho_{i,m}^2}{1 - \rho_{i,m}^2}\right) \Big/ \left(\sum_{i=1}^n \frac{\rho_{i,m}}{1 - \rho_{i,m}^2}\right)$$

The weights are then strictly positive if $\rho_{i,m} < \rho^\star$. Clarke *et al.* (2012) also note that the MDP tends to be less concentrated than the minimum variance portfolio, because it is (approximately) sensitive to the inverse of the idiosyncratic volatility[14] $\tilde{\sigma}_i$ whereas the minimum variance is sensitive to the inverse of the idiosyncratic variance $\tilde{\sigma}_i^2$.

3.2.3.4 The ERC portfolio

The last construction is the ERC portfolio, which has been extensively presented in the second chapter. We simply remind that the ERC portfolio corresponds to the portfolio in which the risk contribution from each stock is made equal. It is the simplest risk budgeting rule. If we assume that volatilities and correlations can be reasonably forecast but that it is impossible to predict asset returns, then attributing an equal budget of risk to all of the portfolio components seems natural.

The main advantages of the ERC allocation are the following:

1. It defines a portfolio that is well diversified in terms of risk and weights.

2. Like the three previous heuristic methods, it does not depend on any expected returns hypothesis.

3. It is less sensitive to small changes in the covariance matrix than the MV or MDP portfolios (Demey *et al.*, 2010).

Like the equally weighted portfolio, it is difficult to locate the ERC portfolio on the mean-variance framework, but it corresponds to the optimal portfolio when the correlation is uniform and the assets have the same Sharpe ratio. This is the reason why the ERC portfolio coincides with the MDP when the correlation is uniform.

[14]The right quantity is $\sigma_i / \tilde{\sigma}_i^2$.

Like the minimum variance and most diversified portfolios, Clarke *et al.* (2012) find the following useful decomposition[15]:

$$x_i^\star = \frac{\sigma^2(x^\star)}{\tilde{\sigma}_i^2}\left(\sqrt{\frac{\beta_i^2}{\beta^{\star 2}} + \frac{\tilde{\sigma}_i^2}{n\sigma^2(x^\star)}} - \frac{\beta_i}{\beta^\star}\right)$$

where:

$$\beta^\star = \frac{2\sigma^2(x^\star)}{\beta(x^\star)\,\sigma_m^2}$$

The authors show then that weights decline with asset beta and that idiosyncratic risks have little impact on the magnitude of the weights. However, contrary to the previous decompositions this one is more difficult to interpret.

3.2.3.5 Comparison of the risk-based allocation approaches

In what follows, we restrict our analysis to long-only portfolios, because weights are necessarily positive in an alternative-weighted index.

Some properties Despite being based on different approaches, the four methods present similarities. Demey *at al.* (2010) compare them in terms of weights and marginal volatilities:

$$x_i = x_j \qquad \text{(EW)}$$

$$\frac{\partial\,\sigma(x)}{\partial\,x_i} = \frac{\partial\,\sigma(x)}{\partial\,x_j} \qquad \text{(MV)}$$

$$\frac{1}{\sigma_i}\cdot\frac{\partial\,\sigma(x)}{\partial\,x_i} = \frac{1}{\sigma_j}\cdot\frac{\partial\,\sigma(x)}{\partial\,x_j} \qquad \text{(MDP)}$$

$$x_i\cdot\frac{\partial\,\sigma(x)}{\partial\,x_i} = x_j\cdot\frac{\partial\,\sigma(x)}{\partial\,x_j} \qquad \text{(ERC)}$$

The weights are equal in the EW portfolio whereas the marginal volatilities are equal in the MV portfolio. In the case of the ERC portfolio, this is the products of the weight times the marginal risk which are equal. For the MDP portfolio, the equality is on the marginal risk divided by the volatility[16] (this measure

[15] See also Jurczenko *et al.* (2013) for another similar decomposition.

[16] For the unconstrained MDP portfolio, we recall that the first-order condition is given by:

$$\frac{\partial\,\mathcal{L}(x;\lambda_0,\lambda)}{\partial\,x_i} = \frac{\sigma_i}{x^\top\sigma} - \frac{(\Sigma x)_i}{x^\top\Sigma x} = 0$$

It follows also that the scaled marginal volatility is equal to the inverse of the diversification ratio of the MDP:

$$\begin{aligned}\frac{1}{\sigma_i}\cdot\frac{\partial\,\sigma(x)}{\partial\,x_i} &= \frac{1}{\sigma_i}\cdot\frac{(\Sigma x)_i}{\sqrt{x^\top\Sigma x}} \\ &= \frac{1}{\mathcal{DR}(x)}\end{aligned}$$

may be interpreted as relative or scaled marginal volatility). We note that the equalities are verified in the case of the MV or the MDP/MSR portfolios only for assets with non-zero weight.

Another important result is that the volatility of the MV, ERC and EW portfolios may be ranked in the following order (Maillard *et al.*, 2010):

$$\sigma_{\mathrm{mv}} \leq \sigma_{\mathrm{erc}} \leq \sigma_{\mathrm{ew}}$$

The ERC portfolio may then be viewed as a portfolio between the MV and the EW portfolios. Of course, in the case of the MDP, we also have $\sigma_{\mathrm{mv}} \leq \sigma_{\mathrm{mdp}}$, but a comparison with the ERC and EW portfolios is not possible. The volatility of the MDP may be either greater or lower than the volatility of the ERC and EW portfolios.

We can classify the risk-based indexations in two families:

$$\forall i : x_i > 0 \qquad \text{(EW/ERC)}$$
$$\exists i : x_i = 0 \qquad \text{(MV/MDP)}$$

In the first one, all the assets are represented in the portfolio, because the weights are strictly positive. In the second family, the portfolio cannot be invested in some assets. It is then obvious the MV and MDP portfolios take more bets than the EW and ERC portfolios. This is the reason why EW and ERC indices will present smaller tracking error volatility with respect to the CW index than the MV and MDP indices.

It is important to mention that the ERC and MDP portfolios coincide when the correlation is uniform across assets returns (Maillard *et al.*, 2010). In this case, the weight x_i of the ith stock is inversely proportional to its volatility σ_i. The MDP portfolio corresponds to the MV portfolio when the individual volatilities σ_i are equal (Choueifaty and Coignard, 2008). Curiously, the ERC and MV portfolios are the same when the correlation is uniform and is equal to the lower bound $\rho = -1/(n-1)$, i.e. when diversification from correlation is maximum (Maillard *et al.*, 2010).

Some examples We illustrate here the properties of these four portfolios by looking at several numerical examples, for a better understanding of the characteristics of these four risk-based portfolios and the issues faced in practice when they are implemented. To compare them, we consider the following criteria:

- $\mathcal{H}^{\star}(x)$ is the modified Herfindahl index of the weights. It is optimal and takes the value 0 for the equally weighted portfolio.
- $\sigma(x)$ is the portfolio's volatility. It takes the minimum value for the minimum variance portfolio.
- $\mathcal{DR}(x)$ is the diversification ratio. The maximum value is reached for the MDP.

- The risk concentration is measured using the Herfindahl index $\mathcal{H}^{\star}(\mathcal{RC})$ applied to risk contributions. For the ERC portfolio, there is no risk concentration and $\mathcal{H}^{\star}(\mathcal{RC})$ takes the value 0.

For each portfolio, we also indicate the weights and the corresponding risk contributions, which are the (ex-ante) performance contributions in an equilibrium framework. All the statistics are expressed in % except the diversification ratio.

Example 26 *We consider an investment universe with four assets. We assume that the volatility σ_i is the same and equal to 20% for all four assets. The correlation matrix C is equal to:*

$$C = \begin{pmatrix} 100\% & & & \\ 80\% & 100\% & & \\ 0\% & 0\% & 100\% & \\ 0\% & 0\% & -50\% & 100\% \end{pmatrix}$$

The results are reported in Table 3.6. We verify that because the volatility of the assets is the same, the MDP is equal to the MV portfolio. We also check that the marginal risks $\mathcal{MR}_i$ are equal for the MV portfolio[17] whereas it is the risk contributions $\mathcal{RC}_i$ that are equal for the ERC portfolio. We note that the MV and the ERC portfolios are similar in terms of weights, but the weight concentration is smaller for the ERC portfolio than for the MV portfolio. In terms of risk concentration, the EW, MV and MDP portfolios are equivalent.

TABLE 3.6: Weights and risk contributions (Example 26)

Asset	EW		MV		MDP		ERC	
	x_i	$\mathcal{RC}_i$	x_i	$\mathcal{RC}_i$	x_i	$\mathcal{RC}_i$	x_i	$\mathcal{RC}_i$
1	25.00	4.20	10.87	0.96	10.87	0.96	17.26	2.32
2	25.00	4.20	10.87	0.96	10.87	0.96	17.26	2.32
3	25.00	1.17	39.13	3.46	39.13	3.46	32.74	2.32
4	25.00	1.17	39.13	3.46	39.13	3.46	32.74	2.32
$\mathcal{H}^{\star}(x)$	0.00		10.65		10.65		3.20	
$\sigma(x)$	10.72		8.85		8.85		9.26	
$\mathcal{DR}(x)$	1.87		2.26		2.26		2.16	
$\mathcal{H}^{\star}(\mathcal{RC})$	10.65		10.65		10.65		0.00	

Remark 46 *For the minimum variance portfolio, the risk contributions are proportional to the weights because the marginal volatilities are equal for all*

[17] We have $\mathcal{MR}_i = 8.8\%$ for the four assets.

assets with positive weights[18]*:*

$$\begin{aligned}\mathcal{RC}_i &= x_i \cdot \partial_{x_i} \sigma(x) \\ &\propto x_i\end{aligned}$$

This implies $\mathcal{H}^\star(x) = \mathcal{H}^\star(\mathcal{RC})$ meaning that the weight concentration is equal to the risk concentration in the case of the minimum variance portfolio.

Example 27 *We modify the previous example by introducing differences in volatilities. They are 10%, 20%, 30% and 40% respectively. The correlation matrix remains the same as in Example 26.*

If we consider the results in Table 3.7, we note that the MV portfolio is concentrated in the first asset because of its low level of volatility. The weight of the second asset is 0%, because although its volatility is smaller than that of the third and fourth asset, its correlation with the first asset is high. We verify that the values of the marginal risk $\mathcal{MR}_i$ in the MV portfolio are equal for assets with non-zero weights (they are equal to 8.6%). The ERC and MDP portfolios produce more balanced portfolios in terms of weights. In this example, we check the inequalities $\sigma_{\text{mv}} \le \sigma_{\text{erc}} \le \sigma_{\text{ew}}$. Unlike the previous example, however, the volatility of the MDP portfolio is now higher than the volatility of the ERC portfolio. We also note that the ranking between the EW, MV and ERC portfolios with respect to the diversification ratio is reversed in this example compared to Example 26.

TABLE 3.7: Weights and risk contributions (Example 27)

Asset	EW		MV		MDP		ERC	
	x_i	$\mathcal{RC}_i$	x_i	$\mathcal{RC}_i$	x_i	$\mathcal{RC}_i$	x_i	$\mathcal{RC}_i$
1	25.00	1.41	74.48	6.43	27.78	1.23	38.36	2.57
2	25.00	3.04	0.00	0.00	13.89	1.23	19.18	2.57
3	25.00	1.63	15.17	1.31	33.33	4.42	24.26	2.57
4	25.00	5.43	10.34	0.89	25.00	4.42	18.20	2.57
$\mathcal{H}^\star(x)$	0.00		45.13		2.68		3.46	
$\sigma(x)$	11.51		8.63		11.30		10.29	
$\mathcal{DR}(x)$	2.17		1.87		2.26		2.16	
$\mathcal{H}^\star(\mathcal{RC})$	10.31		45.13		10.65		0.00	

Example 28 *We now reverse the volatilities of Example 27. They are now equal to 40%, 30%, 20% and 10%.*

The results are reported in Table 3.8. The weights of the MV, ERC and MDP portfolios are similar. The volatilities of the corresponding portfolios

[18] If $x_i = 0$, $\mathcal{RC}_i = 0$. Then the proportionality between weights and risk contributions remains valid in the general case.

are comparable, but the EW portfolio has a high volatility with respect to the three other portfolios. This can be explained by the fact that the first and second assets are more volatile and highly correlated, meaning that diversification effects are low. We also note that the MV and MDP portfolios present some weight concentrations whereas the MV and EW portfolios present some risk concentrations. For the EW portfolio, we observe that the risk contribution of the fourth asset is exactly equal to 0. This is due to the fact that this asset has a very low volatility and is not positively correlated with the other assets. If we consider portfolios with a weight in the fourth asset below 25%, the risk contribution may even be negative.

TABLE 3.8: Weights and risk contributions (Example 28)

	EW		MV		MDP		ERC	
Asset	x_i	$\mathcal{RC}_i$	x_i	$\mathcal{RC}_i$	x_i	$\mathcal{RC}_i$	x_i	$\mathcal{RC}_i$
1	25.00	9.32	0.00	0.00	4.18	0.74	7.29	1.96
2	25.00	6.77	4.55	0.29	5.57	0.74	9.72	1.96
3	25.00	1.09	27.27	1.74	30.08	2.66	27.66	1.96
4	25.00	0.00	68.18	4.36	60.17	2.66	55.33	1.96
$\mathcal{H}^\star(x)$	0.00		38.84		27.65		19.65	
$\sigma(x)$	17.18		6.40		6.80		7.82	
$\mathcal{DR}(x)$	1.46		2.13		2.26		2.16	
$\mathcal{H}^\star(\mathcal{RC})$	27.13		38.84		10.65		0.00	

Example 29 *We consider an investment universe of four assets. The volatility is respectively equal to* 15%, 30%, 45% *and* 60% *whereas the correlation matrix* C *is equal to:*

$$C = \begin{pmatrix} 100\% & & & \\ 10\% & 100\% & & \\ 30\% & 30\% & 100\% & \\ 40\% & 20\% & -50\% & 100\% \end{pmatrix}$$

This example is very interesting because the MDP is sometimes assimilated to the minimum variance portfolio. The results in Table 3.9 show that the compositions of the MDP and MV portfolios may be completely different. The MV portfolio selects the first and the second assets whereas the MDP selects the third and fourth assets. In this case, the volatility of the MDP is about twice the volatility of the MV portfolio. This example illustrates a drawback of the diversification ratio, because it focuses primarily on correlations and less on volatilities. For instance, if we multiply the volatilities of the third and fourth assets by a factor, the composition of the MDP does not change. Indeed, the weights of the MDP when the volatility of the assets is respectively 15%, 30%, 450% and 600% remain the same, meaning that the volatility of the MDP becomes 257.1% whereas the volatility of the MV remains at 13.92%!

TABLE 3.9: Weights and risk contributions (Example 29)

	EW		MV		MDP		ERC	
Asset	x_i	$\mathcal{RC}_i$	x_i	$\mathcal{RC}_i$	x_i	$\mathcal{RC}_i$	x_i	$\mathcal{RC}_i$
1	25.00	2.52	82.61	11.50	0.00	0.00	40.53	4.52
2	25.00	5.19	17.39	2.42	0.00	0.00	22.46	4.52
3	25.00	3.89	0.00	0.00	57.14	12.86	21.12	4.52
4	25.00	9.01	0.00	0.00	42.86	12.86	15.88	4.52
$\mathcal{H}^\star(x)$	0.00		61.69		34.69		4.61	
$\sigma(x)$	20.61		13.92		25.71		18.06	
$\mathcal{DR}(x)$	1.82		1.27		2.00		1.76	
$\mathcal{H}^\star(\mathcal{RC})$	7.33		61.69		33.33		0.00	

Example 30 *Now we consider an example with six assets. The volatilities are 25%, 20%, 15%, 18%, 30% and 20% respectively. We use the following correlation matrix:*

$$C = \begin{pmatrix} 100\% & & & & & \\ 20\% & 100\% & & & & \\ 60\% & 60\% & 100\% & & & \\ 60\% & 60\% & 60\% & 100\% & & \\ 60\% & 60\% & 60\% & 60\% & 100\% & \\ 60\% & 60\% & 60\% & 60\% & 60\% & 100\% \end{pmatrix}$$

The correlation matrix is specific, because the correlation is uniform and equal to 60% for all assets except for the correlation between the first and second assets which is equal to 20%. The results are surprising. Whereas the MV portfolio concentrates the weights in the third and fourth assets, the MDP concentrates the weights in the first and second assets. The ERC portfolio overweighs the third asset, but the weights are close to the EW portfolio. In this example, it is the MDP portfolio that has the largest volatility. It is interesting to note that in this case, the MV portfolio is sensitive to the *specific volatility risk* whereas the MDP portfolio is sensitive to the *specific correlation risk*.

Example 31 *To illustrate how the MV and MDP portfolios are sensitive to specific risks, we consider a universe of n assets with volatility equal to 20%. The structure of the correlation matrix is the following:*

$$C = \begin{pmatrix} 100\% & & & & \\ \rho_{1,2} & 100\% & & & \\ 0 & \rho & 100\% & & \\ \vdots & \vdots & \ddots & 100\% & \\ 0 & \rho & \cdots & \rho & 100\% \end{pmatrix}$$

TABLE 3.10: Weights and risk contributions (Example 30)

	EW		MV		MDP		ERC	
Asset	x_i	$\mathcal{RC}_i$	x_i	$\mathcal{RC}_i$	x_i	$\mathcal{RC}_i$	x_i	$\mathcal{RC}_i$
1	16.67	3.19	0.00	0.00	44.44	8.61	14.51	2.72
2	16.67	2.42	6.11	0.88	55.56	8.61	18.14	2.72
3	16.67	2.01	65.16	9.33	0.00	0.00	21.84	2.72
4	16.67	2.45	22.62	3.24	0.00	0.00	18.20	2.72
5	16.67	4.32	0.00	0.00	0.00	0.00	10.92	2.72
6	16.67	2.75	6.11	0.88	0.00	0.00	16.38	2.72
$\mathcal{H}^{\star}(x)$	0.00		37.99		40.74		0.83	
$\sigma(x)$	17.14		14.33		17.21		16.31	
$\mathcal{DR}(x)$	1.24		1.14		1.29		1.25	
$\mathcal{H}^{\star}(\mathcal{RC})$	1.36		37.99		40.00		0.00	

The first asset is only correlated with the second asset. The correlation of the second asset to the nth asset is uniform and equal to ρ. The correlation of the first two assets is set to $\rho_{1,2}$. The correlation matrix is more specific, because it is similar to a constant-correlation matrix except for one asset. In Figure 3.5, we report the sum $x_1 + x_2$ with respect to the number n of assets by considering several values of ρ and $\rho_{1,2}$. It is interesting to note the significant difference between the MV and MDP portfolios[19] on one side and the ERC and EW portfolios on the other side. The sum $x_1 + x_2$ decreases faster for the ERC and EW portfolio whereas the decrease is slow for the MV and MDP portfolios. For example, if $\rho = 70\%$ and $\rho_{1,2} = 20\%$, the sum x_1+x_2 is equal to 58.6% (MV and MDP), 54.3% (ERC) and 50% (EW) respectively if n is equal to 4. If n is now set to 50, the sum $x_1 + x_2$ becomes 54.1% (MV and MDP), 18.1% (ERC) and 4% (EW) respectively. The number of assets only marginally impacts MV and MDP weights because they rely solely on the covariance matrix. They concentrate their weights in the relatively least volatile or least correlated assets. The ERC portfolio, with its implicit diversification constraint, naturally dilutes weights among components as the number of assets increases.

Example 32 *We assume that asset returns follow the one-factor CAPM model. The idiosyncratic volatility $\tilde{\sigma}_i$ is set to 5% for all the assets whereas the volatility of the market portfolio σ_m is equal to 25%.*

In Figure 3.6, we have reported the composition of the different portfolios when the number of assets is equal to 50. The values of beta β_i have been simulated uniformly between -1 and 3. We retrieve the main results of Clarke *et al.* (2012).

[19]They are equal because the volatilities are the same.

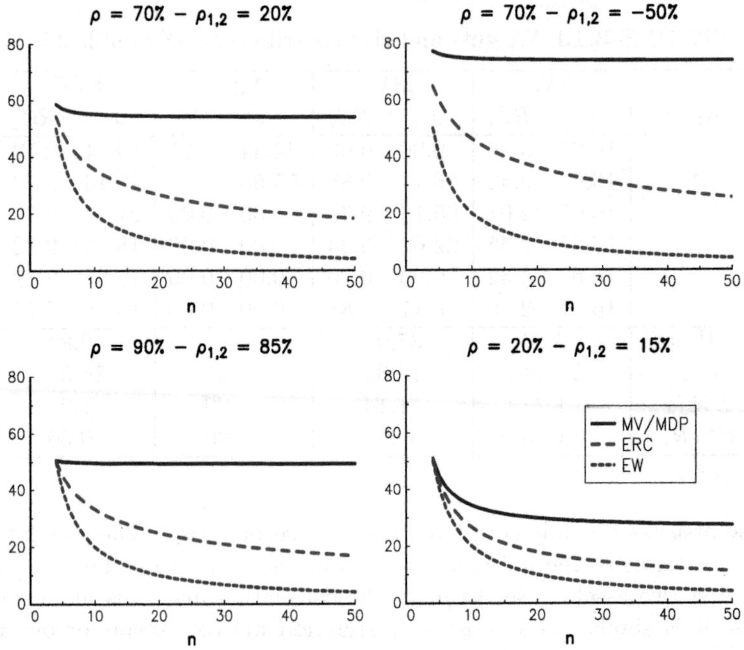

FIGURE 3.5: Weight of the first two assets in AW portfolios (Example 31)

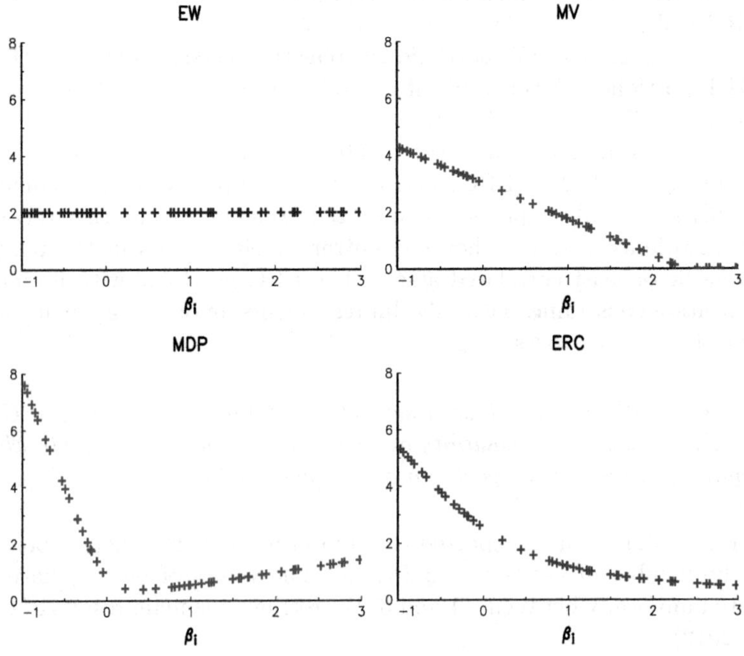

FIGURE 3.6: Weight with respect to the asset beta β_i (Example 32)

3.3 Some illustrations

In this section, we update a simulation performed by Demey *et al.* (2010) concerning backtesting of the four risk-based indexations based on the Euro Stoxx 50 universe. We also discuss some practical issues regarding the turnover, the concentration and the stability of some risk-based strategies. We then compare our results with other existing empirical works, and we investigate how risk-based portfolios are related to factors of the Fama-French model.

3.3.1 Simulation of risk-based indices

We consider the universe of the Euro Stoxx 50 index from December 31, 1992, to September 28, 2012. Even if the results are sensitive to the universe, we believe that we can draw some interesting conclusions from this specific example. For our purpose, we build risk-based indices by using the following characteristics:

- Every month, we consider only the stocks belonging to the Euro Stoxx 50 index.
- We compute the empirical covariance matrix using daily returns and a one-year rolling window.
- The portfolio is rebalanced on the first trading day of the next month.
- The risk-based index is computed daily as a price index.

In Table 3.11, we report the main statistics by considering the Euro Stoxx 50 index as the benchmark[20]. We first note that the four risk-based indices outperform the Euro Stoxx 50 index. However, we must keep in mind that we are not taking trading costs into account, an issue discussed later. Comparing risk-based indexations with the capitalized-weighted indexation, we observe that they present lower volatility or maximum drawdown except for the EW portfolio. This is not true for skewness and kurtosis measures. Finally, we report the return correlation and the beta of these alternative-weighted indices with respect to the Euro Stoxx 50 index. The least correlated is the MV portfolio followed by the MDP.

Simulated performances by calendar year are reported in Table 3.12. We can make two remarks. First, the performance of risk-based indexations relative to the capitalization-weighted index depends on market regimes. Indeed,

[20]They are the yearly return $\mu(x)$ (in %), the volatility $\sigma(x)$ (in %), the Sharpe ratio $\mathrm{SR}(x \mid r)$, the volatility of the tracking error $\sigma(x \mid b)$ (in %), the information ratio $\mathrm{IR}(x \mid r)$, the maximum drawdown $\mathcal{MDD}(x)$ (in %), the daily and monthly skewness $\gamma_1(x)$ and $\gamma_1^\star(x)$, the daily and monthly excess kurtosis $\gamma_2(x)$ and $\gamma_2^\star(x)$, the correlation with the benchmark $\rho(x \mid b)$ (in %) and the beta $\beta(x \mid b)$.

TABLE 3.11: Main statistics of AW indexations (Jan. 1993 – Sep. 2012)

	CW	EW	MV	MDP	ERC
$\mu(x)$	4.47	6.92	7.36	10.15	8.13
$\sigma(x)$	22.86	23.05	17.57	20.12	21.13
$\mathrm{SR}(x \mid r)$	0.05	0.16	0.23	0.34	0.23
$\sigma(x \mid b)$		4.18	14.85	12.79	5.65
$\mathrm{IR}(x \mid b)$		0.56	0.19	0.42	0.62
$\mathcal{MDD}(x)$	−66.88	−61.67	−56.04	−50.21	−56.85
$\gamma_1(x)$	0.10	0.11	1.83	2.91	0.23
$\gamma_2(x)$	5.28	6.06	49.88	74.13	7.13
$\gamma_1^\star(x)$	−0.46	−0.41	−1.00	−0.54	−0.50
$\gamma_2^\star(x)$	0.63	1.33	2.21	0.97	1.09
$\rho(x \mid b)$		98.35	76.03	83.02	97.00
$\beta(x \mid b)$		0.99	0.58	0.73	0.90

TABLE 3.12: Simulated performance of AW portfolios by year (in %)

Year	CW	EW	MV	MDP	ERC
1993	38.7	44.5	38.3	45.5	43.5
1994	−7.9	−2.6	−3.9	7.7	−1.9
1995	14.1	13.2	16.8	19.2	14.7
1996	22.8	30.3	28.1	34.9	30.0
1997	36.8	44.5	38.8	45.7	44.9
1998	32.0	34.2	47.5	50.6	35.1
1999	46.7	41.0	20.8	25.6	35.9
2000	−2.7	2.4	4.3	3.8	5.5
2001	−20.2	−17.6	−11.5	−10.7	−13.8
2002	−37.3	−34.8	−34.7	−29.2	−32.9
2003	15.7	23.3	4.3	25.0	18.8
2004	6.9	8.0	15.6	8.3	10.0
2005	21.3	20.4	16.9	16.1	20.0
2006	15.1	18.3	17.1	16.5	18.6
2007	6.8	5.2	−2.3	−2.5	5.0
2008	−44.4	−44.3	−15.7	−20.1	−36.3
2009	21.1	29.4	−5.2	16.9	25.5
2010	−5.8	−3.3	−5.8	0.0	−2.5
2011	−17.1	−18.0	1.6	−15.5	−15.3
2012	5.9	5.4	11.1	5.1	6.6

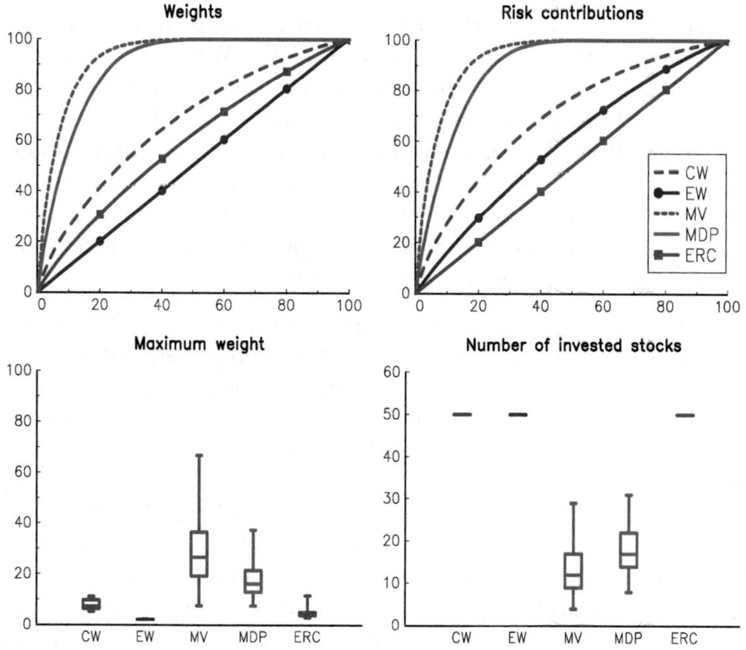

FIGURE 3.7: Concentration statistics of AW portfolios

they outperform the CW index in bear market periods (1994, 2000-2002, 2008 and 2011). However, they may largely underperform in bull market periods (for example in 1999). Second, we also note that their behavior is not homogeneous over time. This is particularly true if we compare the year 2009 with the years 2005 and 1996. These years present similar performance in terms of the CW index, but not in terms of the AW index.

3.3.2 Practical issues of risk-based indexation

Figure 3.7 presents some statistics regarding concentration. The top-left panel corresponds to the average of all the Lorenz curves of weights. The MV and MDP portfolios are more concentrated than the CW index, but this is not the case for the ERC portfolio. Of course, the EW portfolio appears to be the least concentrated. If we build the Lorenz curve on risk contributions (top-right panel), we obtain the same conclusions: the MV and MDP portfolios are the most concentrated indices. The bottom-left panel represents the box plot[21] of the maximum weight for all the 238 monthly rebalancing dates. We observe that the MV and MDP portfolios may be highly concentrated in one stock. Indeed, the maximum weight in the MV (resp. MDP) index reaches a top at

[21] In the box plot, we indicate the statistics of maximum, minimum, median and the 25th and 75th percentiles.

66.8% (resp. 37.2%). Finally, the last panel corresponds to the box plot of the number of invested stocks. In the case of the CW, EW and ERC indices, there are always 50 stocks present in the portfolio. For the MV and MDP portfolios, the median is 12 and 17 stocks, and the maximum number is never higher than 30 stocks. It is obvious that such portfolios are not appropriate from a professional point of view, because of two main reasons. First, an investor may be not confident with such concentrated portfolios, in particular when he considers passive management. Second, optimized concentrated portfolios may induce some stability issues in terms of turnover.

These observations imply that we must introduce some constraints in the MV and MDP indexations if we want to obtain an investment strategy that makes sense. This is the reason why we simulate the two strategies by adding an upper bounds of 10% or 5%. In Table 3.13, we have computed the annualized turnover based on the monthly weights. In the case of the one-year estimated covariance matrix (which is our default case), the turnover is equal respectively to 19%, 63%, 330% and 343% for the EW[22], ERC, MV and MDP indices. The turnover of the MV and MDP portfolios is relatively high. However, by adding an upper bound, we reduce this turnover. Indeed, it becomes 161% and 197% with a 5% upper bound. Another way to reduce the turnover is to use a longer window for the covariance matrix estimation. For instance, if we use a three-year rolling window, the turnover is divided by two.

TABLE 3.13: Annualized monthly turnover of AW portfolios (in %)

Lag	EW	ERC	MV			MDP		
				10%	5%		10%	5%
1M	19	551	1765	1401	969	1779	1571	1098
2M	19	290	1234	911	620	1313	1061	718
3M	20	195	909	681	462	985	806	539
1Y	19	63	330	248	161	343	290	197
2Y	19	43	202	152	101	223	196	127
3Y	18	34	149	113	73	164	147	95

Remark 47 *High turnover does not necessarily imply high transaction costs. For instance, in the case of the Euro Stoxx 50 universe, trading costs are about 2 bps. So, turnover of 300% implies transaction costs lower than 10 bps. However, turnover is clearly a negative factor when we consider larger universes with less liquid stocks. In this case, turnover of more than 100% is difficult for investors to accept.*

[22]We notice that the turnover is not equal to zero for the EW portfolio because of the entry/exit in the universe.

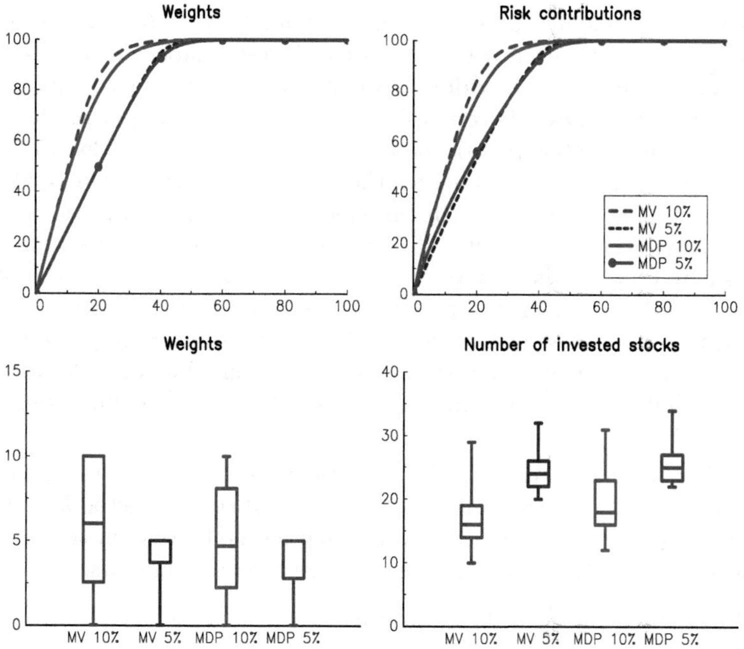

FIGURE 3.8: Concentration statistics of constrained MV and MDP indexations

TABLE 3.14: Main statistics of constrained MV and MDP indexations (Jan. 1993 – Sep. 2012)

		MV			MDP	
		10%	5%		10%	5%
$\mu(x)$	7.36	8.49	8.57	10.15	9.72	9.32
$\sigma(x)$	17.57	17.34	17.79	20.12	19.03	19.52
$\text{SR}(x \mid r)$	0.23	0.30	0.30	0.34	0.34	0.31
$\sigma(x \mid b)$	14.85	11.67	8.91	12.79	9.40	7.52
$\text{IR}(x \mid b)$	0.19	0.33	0.44	0.42	0.53	0.62
$\mathcal{MDD}(x)$	−56.04	−55.10	−50.35	−50.21	−49.73	−51.04
$\gamma_1(x)$	1.83	0.79	0.14	2.91	0.11	0.02
$\gamma_2(x)$	49.88	17.32	7.00	74.13	5.50	5.43
$\gamma_1^\star(x)$	−1.00	−0.71	−0.67	−0.54	−0.42	−0.52
$\gamma_2^\star(x)$	2.21	1.25	0.95	0.97	0.64	1.06
$\rho(x \mid b)$	76.03	86.65	93.41	83.02	91.52	94.91
$\beta(x \mid b)$	0.58	0.66	0.73	0.73	0.76	0.81

In Figure 3.8, we have reported the concentration statistics of these indexations when we take into account an upper bound. We note that the concentration remains high, even if the number of invested stocks increases[23]. The reason for this is that, by adding a weight constraint, the weight of many invested stocks reaches this constraint. Let us analyze the bottom-left panel which represents the box plot of the weights of all invested stocks. 75% of the invested stocks have a weight larger than 4% in the MV portfolio when we impose an upper bound of 5%. In this case, the MV method is more a stock picking process than an allocation process. However, we note that adding constraints may improve the results[24] (see Table 3.14). For instance, the Sharpe ratio of the MV strategy is better with weight constraints than without them. These constraints allow decreasing the tracking error volatility. Concerning the performance, we observe a curious phenomenon. Whereas adding an upper bound improves the performance of the MV index, it reduces the return of the MDP index.

Another important issue is the statistical method used to estimate the covariance matrix. On page 49, we have reviewed several methods to obtain a more robust estimator than the traditional method based on the empirical covariance matrix. In Table 3.15, we compare then 5 estimators:

- EMP corresponds to the empirical covariance matrix.
- CCM is the estimator based on the constant correlation matrix.
- RMT denoises the empirical covariance matrix using the random matrix theory, as suggested by Laloux *et al.* (1999).
- CCS is the original shrinkage estimator with a uniform correlation proposed by Ledoit and Wolf (2003).
- FSE combines the general shrinkage estimator based on factors and developed by Ledoit and Wolf (2004), and the denoised PCA factors.

For each estimator and each risk-based method, we report the yearly turnover $\tau(x)$ (expressed in %), the average Gini coefficient of weights $\mathcal{G}(x)$ and risk contributions $\mathcal{G}(\mathcal{RC})$, and the information ratio $\text{IR}(x \mid b)$. All estimators use a window lag of one year. We note that using a more robust covariance matrix estimate generally allows reducing the turnover. The impact on the information ratio is less obvious. Indeed, robust estimators improve the performance of the MDP portfolio, but this is not necessarily the case of the MV portfolio. Furthermore, we also observe that, except for the CCM estimator applied to the MDP method[25], they are not sufficient to solve the concentration issue.

[23]The average number is 18 and 25 stocks for the MV and MDP portfolios with the 5% upper bound constraint.

[24]This result may be explained by the shrinkage effect of weight constraints (Jagannathan and Ma, 2003).

[25]This result was foreseeable because the MDP corresponds to the ERC portfolio when the correlation is uniform.

TABLE 3.15: Influence of the covariance estimator

	Statistic	ERC	MV			MDP		
				10%	5%		10%	5%
EMP	$\tau(x)$	63	330	248	161	343	290	197
	$\mathcal{G}(x)$	0.18	0.86	0.77	0.59	0.79	0.74	0.59
	$\mathcal{G}(\mathcal{RC})$	0.01	0.86	0.77	0.61	0.78	0.74	0.62
	$\text{IR}(x \mid b)$	0.62	0.19	0.33	0.44	0.42	0.45	0.41
CCM	$\tau(x)$	47	289	210	132	47	47	47
	$\mathcal{G}(x)$	0.14	0.87	0.77	0.59	0.14	0.14	0.14
	$\mathcal{G}(\mathcal{RC})$	0.01	0.87	0.76	0.60	0.01	0.01	0.01
	$\text{IR}(x \mid b)$	0.60	0.08	0.26	0.32	0.60	0.60	0.60
RMT	$\tau(x)$	64	262	198	144	260	228	168
	$\mathcal{G}(x)$	0.19	0.86	0.77	0.59	0.77	0.74	0.58
	$\mathcal{G}(\mathcal{RC})$	0.01	0.86	0.77	0.61	0.76	0.74	0.61
	$\text{IR}(x \mid b)$	0.65	0.23	0.32	0.46	0.54	0.54	0.55
CCS	$\tau(x)$	57	314	233	148	304	266	194
	$\mathcal{G}(x)$	0.16	0.86	0.77	0.59	0.74	0.71	0.57
	$\mathcal{G}(\mathcal{RC})$	0.01	0.86	0.77	0.60	0.73	0.70	0.59
	$\text{IR}(x \mid b)$	0.68	0.16	0.32	0.39	0.46	0.49	0.43
FSE	$\tau(x)$	63	306	231	155	309	267	187
	$\mathcal{G}(x)$	0.18	0.86	0.77	0.59	0.77	0.74	0.58
	$\mathcal{G}(\mathcal{RC})$	0.01	0.86	0.77	0.61	0.76	0.73	0.62
	$\text{IR}(x \mid b)$	0.62	0.21	0.32	0.43	0.48	0.47	0.42

3.3.3 Findings of other empirical works

3.3.3.1 What is the best alternative-weighted indexation?

There are several empirical works that compare the different alternative-weighted indexation methods. We can cite Arnott *et al.* (2010) (hereafter AKMS), Chow *et al.* (2011) (hereafter CHKL), Choueifaty *et al.* (2011) (hereafter CFR), Carvalho *et al.* (2012) (hereafter CLM), NBIM (2012), Lohre *et al.* (2012) (hereafter LNZ). In the following table, we indicate the AW indexations tested by the different researchers:

Authors	EW	MV	MDP	ERC	MSR	RP	FW
AKMS	✓	✓					✓
CHKL	✓	✓	✓		✓		✓
CFR	✓	✓	✓	✓		✓	
CLM	✓	✓	✓	✓		✓	
NBIM	✓	✓	✓	✓		✓	✓
LNZ	✓	✓	✓	✓			

MSR stands for a new version of the maximum Sharpe ratio portfolio where the expected return of assets is proportional to their (downside) semi-volatility[26] (Amenc *et al.*, 2010), RP is an ERC portfolio with a constant correlation matrix[27] while FW corresponds to the fundamental-weighted indexation. One difficulty when comparing these different studies is the heterogeneity of the portfolio construction. For instance, most studies use an annual frequency to rebalance the portfolio, whereas others use semi-annual, quarterly or monthly frequencies. If we consider the estimated covariance, we encounter the same heterogeneity problem. Some studies use monthly returns, other weekly returns, some consider shrinkage methods, etc. Another difference comes from the constraints specified for the portfolio optimization. Indeed, they may use upper and lower bounds, which may differ substantially from one study to another.

Authors	Study period	Universe	Frequency	Window
AKMS	1993-2009	World DC	annually?	
CHKL	1964-2009	US	annually	five-year
	1987-2009	World	annually	five-year
CFR	1999-2010	MSCI World	semi-annually	
CLM	1997-2010	MSCI World	quarterly	two-year
NBIM	1999-2011	FTSE World	annually	five-year
LNZ	1989-2011	S&P 500	monthly	five-year

In this context, it is difficult to draw conclusions as to the superiority of one methodology with respect to the other portfolio constructions. For instance, the minimum variance portfolio is certainly one of the best methods if we consider the Sharpe ratio, but it is one of the weakest methods if we consider the information ratio. However, we observe that alternative-weighted indexations generally outperform the capitalization-weighted indexation. Of course, we must be careful with such statements because there is always a gap between simulated backtests and real life. As noted by Chow *et al.* (2011), "*the excess turnover, reduced portfolio liquidity, and decreased investment capacity, in addition to the fees and expenses associated with managing a more complex index portfolio strategy, may erode much of the anticipated performance advantage.*"

Remark 48 *It is pointless to think that one alternative-weighted indexation is superior to other methods, because the answer depends on the investor's point of view. For instance, some investors think that high tracking error volatility is not an acceptable risk for their passive management portfolios. Conversely,*

[26] Let X be a centered random variable. The semi-variance of X is defined by:

$$\operatorname{sv}(X) = \mathbb{E}\left[\min(0, X)^2\right]$$

It is then the variance of the negative part of X. The semi-volatility is simply the square root of the semi-variance. In finance, it corresponds to the volatility of negative returns.

[27] The weights are then proportional to the volatilities. This weighting scheme is the original risk parity portfolio.

others are precisely attracted by smart indices[28]*, which are far removed from capitalized-weighted indices. Today, the question of large institutional investors is not what is the best smart index, but how to allocate between the different smart indices, including capitalized-weighted indices.*

3.3.3.2 Style analysis of alternative-weighted indexation

If we consider the style analysis of the different alternative-weighted indexations, these studies share the same point of view. They are all exposed to the four-factor Fama-French-Carhart model. Regression analysis shows that these factors explain a large part of the outperformance of these alternative-weighted indexations with high R^2 and significant regression coefficients. More surprising, the intercept of these linear regressions is not statistically significant, meaning that the principal reason for the outperformance is the exposure of these indexations to the Fama-French-Carhart factors. Chow *et al.* (2011) concludes then:

> "*Using the Carhart four-factor model, we identified the sources of outperformance as exposure to the value and size factors, with risk-adjusted alpha not statistically different from zero. [...] This finding leads us to conclude that, despite the unique investment insights and technological sophistication claimed by the purveyors of these strategies, the performances are directly related to a strategy of naive equal weighting, which produces outperformance by tilting toward value and size factors. Nonetheless, the alternative betas represent an efficient and potentially low-cost way to access the value and size premiums because traditional style indices tend to have negative Fama-French alpha and direct replication of Fama-French factors is often impractical and costly.*"

In a sense, smart indexations may be viewed as another way to capture the systematic risk premia (market, value, small caps and momentum) in a systematic way. However, contrary to active management, the excess performance of these investments is linked to their beta exposures and is not considered as pure alpha.

[28] We recall that smart indexation (or smart beta) is another term to design alternative-weighted indexation.

Chapter 4

Application to Bond Portfolios

Bond indexing is generally based on the classical market-capitalization weighting scheme. However, with the recent development of the European debt crisis traditional index bond management has been severely called into question. More generally, the management of bond portfolios has suffered since the credit crisis of 2008, because it has been driven these last years by the search of yield with little consideration for the management of credit risk.

The bond risk premium has been high since the beginning of the 2000s. This is explained by the continuous fall of interest rates during this period, due to the decrease of the inflation risk. Today, the situation is very different because the interest rates are close to zero. The major risk is a new inflation regime and an increase of nominal interest rates. In this case, it is uncertain whether the performance of the bond market over the next ten years will be the same as the performance observed these past ten years. In this context, the buy-and-hold strategy is not appropriate and risk management will gain in importance in the future.

In this chapter, we show how the risk parity approach is interesting for the measurement and management of the risk of a bond portfolio. In the first section, we discuss some issues concerning the risk management of bond portfolios. In the second section, we define different risks of bond portfolios and show how to measure them. In particular, we present the analytics of the yield curve and the use of the term structure of interest rates to define risk factors. We also extend this analysis by incorporating the default risk. Finally, we apply the risk budgeting framework to two cases: the management of yield curve risk factors and the management of the sovereign credit risk.

4.1 Some issues in bond management

4.1.1 Debt-weighted indexation

Traditionally, bond indices have been constructed according to the methodology of weighting by market capitalization. This means that each issuer in the index is given a weight proportional to its level of outstanding debt[1]. The simplicity of this approach and the recognition of a capitalization-weighted index as the market portfolio have contributed to the success of the methodology. Yet, intuitively, it is easy to note a basic flaw in this allocation scheme, since it gives higher weights to the most indebted issuers, regardless of their capacity to service their debt. An issuer facing financial hardship and trapped in a debt spiral to remain solvent would see its index weight increase until the whole mechanism collapses and an exclusion from the index occurs. Depending on the index, exclusion can be triggered by specific events, such as a downgrade or, in the worst case, a default.

In order to circumvent this problem, index providers prefer to use the debt market price instead of the debt notional. However, bonds are generally traded in an over-the-counter market and not in an exchange. Contrary to the equity market, the notion of closing prices is not adequate for the bond market. In this case, the mark-to-market price of one bond may differ from one index provider to another. Another implication of this OTC market is that access to these prices is difficult. This explains why most index providers are investment banks, such as Barclays, Citigroup and JP Morgan. In this context, the question of the price adjustment process is essential (Katz, 1974). However, the academic literature on this subject is scarce (Brennan and Schwartz, 1982; Hotchkiss and Ronen, 2002). It is then difficult to conclude that the bond market is efficient and incorporates all the information as quickly as the equity market. This question is particularly interesting if we consider sovereign debt. The behavior of the Euro bond market after the default of Lehman Brothers shows that credit risk was not fully incorporated into bond prices.

The case of Greece illustrates the two previous points. Since the beginning of the financial crisis, Greece has struggled with a high debt and refinancing burden. As the country relied more and more on borrowing, the weight of Greek debt increased in European bond indices until the point where Greece was downgraded and excluded from some of the key indices. Passive investors then had no choice but to sell their distressed bonds into a depressed market, leading to significant losses. From the perspective of efficient markets theory, such a risk could be acceptable if it is compensated by an additional return, resulting in similar risk-adjusted returns from the debt of different countries. However, this does not seem to be verified in practice, as Robert Arnott has pointed out (2010). These drawbacks have led Toloui (2010) to write a famous article entitled "*Time To Rethink Bond Indexes?*".

However, the construction problem of bond indices is not limited to sovereign bond indices. It is also present in corporate bond, high yield or global aggregate indexations (Reilly *et al.*, 1992). For instance, Goltz and Campani (2011) review corporate investment-grade bond indices for the US and euro-denominated markets. They analyze the heterogeneity of these indices and conclude as follows:

[1]In the rest of the chapter, 'capitalization-based' and 'debt-based' (or 'capitalization-weighted' and 'debt-weighted') are used interchangeably to refer to such a weighting scheme.

"The duration, yield, and time to maturity of these indices fluctuate persistently, and the fluctuations are even greater in the two indices with the smallest number of bonds, that is, the most investable ones. In short, the more investable the index is meant to be, the less reliable (the less stable) it is. For an investor, stable risk exposures are important so that allocation decisions are not compromised by uncontrolled fluctuations of risk exposures in the building blocks chosen to implement such decisions."

4.1.2 Yield versus risk

Another problem with bond portfolio management is that it has been driven these last years by the search of yield with little consideration for risk management, in particular concerning the credit risk. Most investors have forgotten that debt means risk and sometimes default. Some years ago, fund managers explained that they prefer to invest in Italian bonds than in German bonds because there was an arbitrage opportunity. However, in reality it was not an arbitrage position, but a carry position of credit risk.

There is a widespread belief in asset management and among institutional investors that it is easier to create alpha in bond management than in equity management. This explains why most institutional investments in equities are carried out with passive management whereas bond portfolios are generally actively managed. This question has been studied by academics since the beginning of the 1990s, even if the number of academic studies on bond funds is small compared to the number of studies on equity funds. Similarly to the results on equity funds, Blake *et al.* (1993) and Detzler (1999) show that bond funds underperform their benchmarks on average and they conclude that the underperformance is approximately equal to the management fees. Conclusions are less clear regarding the persistence of performance. Blake *et al.* (1993) do not find any persistence, and suggest that the performance differences between actively managed bond funds can be explained either by differences in the duration or by differences in the credit risk premia. In a more recent study, Huij and Derwall (2008) find different results. Using a large universe of bond funds from 1990 to 2003, they measure the funds' performance using a three-factor model[2]. They find that funds with strong (resp. weak) performance over a past period continue to outperform (resp. underperform) in future periods. Since the crisis of 2008, the performance of active management is even more difficult to investigate, because this last period is characterized by large changes in economic conditions that have a large impact on the funds' relative performances[3]. For instance, diversified

[2]The three factors are an investment-grade index, a high yield index, and a mortgage bonds index.

[3]The SPIVA (Standard & Poor's Indices Versus Active Funds Scorecard) for 2011 shows that 96.7% of active managers on the long end of the treasury yield curve were beaten by their benchmarks.

bond funds with a defensive style (exposure to govies rather than corporate bonds) largely outperformed more aggressive diversified bond funds in 2008, whereas the contrary was the case in 2009.

The fact that sovereign bonds were considered to be a safe asset before the crisis and the increase of the volatility over the past years imply rebalancing of performance and risk. This explains why risk management is becoming increasingly important in the management of bond portfolios.

4.2 Bond portfolio management

In this section, we present the term structure of interest rates. We consider the pricing of bonds and show how to take the default risk into account. Finally, we study different models to measure and manage the credit risk of bond portfolios.

4.2.1 Term structure of interest rates

Let r_t be the instantaneous interest rate. The price of the zero-coupon bond[4] is equal to:

$$B_t(T) = \mathbb{E}\left[e^{-\int_t^T r_s \, \mathrm{d}s} \middle| \mathcal{F}_t \right]$$

where T is the maturity of the bond and $\mathcal{F}_t$ is the adequate filtration associated with the stochastic process r_t. We define the zero-coupon rate (or spot rate) as the following rate:

$$R_t(T) = -\frac{1}{T-t} \ln B_t(T)$$

The set of zero-coupon rates $R_t(T)$ with maturity $T \geq t$ is called the term structure of interest rates or the 'yield curve'. We generally distinguish three generic forms for the yield curve:

1. The yield curve is flat when long-term interest rates are equal to short-term interest rates.

2. The normal configuration corresponds to a growing curve, when long-term interest rates are larger than short-term interest rates.

3. If short-term interest rates are larger than long-term interest rates, we say that the term structure is inverted.

[4] A zero-coupon bond is a bond with no coupon. It does not pay interest during the life of the bond and the notional is generally normalized to one.

Let $F_t(T, m)$ be the forward rate at time t for the period $[T, T+m]$. It verifies the following relationship:

$$B_t(T+m) = e^{-mF_t(T,m)} B_t(T)$$

We deduce that the expression of $F_t(T, m)$ is:

$$F_t(T, m) = -\frac{1}{m} \left(\ln B_t(T+m) - \ln B_t(T) \right)$$

It follows that the instantaneous forward rate is given by this equation:

$$F_t(T) = F_t(T, 0) = -\frac{\partial \ln B_t(T)}{\partial T}$$

Modeling the yield curve consists in defining a model for the stochastic interest rate and characterizing the zero-coupon rate $R_t(T)$. For instance, if we assume that the interest rate r_t is constant and equal to r, we obtain $B_t(T) = e^{-r(T-t)}$. It follows that $R_t(T) = r$ meaning that the yield curve is flat. Another example is the stochastic model of Vasicek (1977), which assumes that the interest rate follows an Ornstein-Uhlenbeck process:

$$\mathrm{d}r_t = a(b - r_t)\,\mathrm{d}t + \sigma\,\mathrm{d}W_t$$

where W_t is a standard Brownian motion. If $a > 0$, this model is a mean-reverting process because r_t tends to return to the equilibrium b. The Vasicek model is very famous because we have a closed-form formula to characterize the zero-coupon rate $R_t(T)$. This model has been extended in two directions:

1. Some models use a more complex stochastic differential equation to characterize r_t (Cox *et al.*, 2005; Chan *et al.*, 1992; Hull and White, 1993).

2. Other models consider factor models to define the behavior of the interest rate (Heath *et al.*, 1992; Duffie and Kan, 1996; Ahn *et al.*, 2002).

Calibrating the yield curve consists in estimating a parametric model or a non-parametric function for the zero-coupon rates $R_t(T)$ using the market prices of zero-coupon bonds. One of the most famous models is the parsimonious functional form proposed by Nelson and Siegel (1987):

$$\begin{aligned} R_t(T) &= \theta_1 + \theta_2 \left(\frac{1 - \exp(-(T-t)/\theta_4)}{(T-t)/\theta_4} \right) + \\ & \quad \theta_3 \left(\frac{1 - \exp(-(T-t)/\theta_4)}{(T-t)/\theta_4} - \exp(-(T-t)/\theta_4) \right) \end{aligned}$$

This is a model with four parameters: θ_1 is a parameter of level, θ_2 is a parameter of rotation, θ_3 controls the shape of the curve and θ_4 permits to localize the break of the curve. We also note that $R_t(t) = \theta_1 + \theta_2$ and $R_t(\infty) = \theta_1$.

Example 33 *We consider the term structure of interest rates generated by the Nelson-Siegel model with* $\theta_1 = 5\%$, $\theta_2 = -5\%$, $\theta_3 = 6\%$ *and* $\theta_4 = 10$.

We deduce that the zero-coupon interest rates from one to five years are respectively equal to 0.52%, 0.99%, 1.42%, 1.80% and 2.15%. In Figure 4.1, we have represented the yield curve and the term structure of forward interest rates.

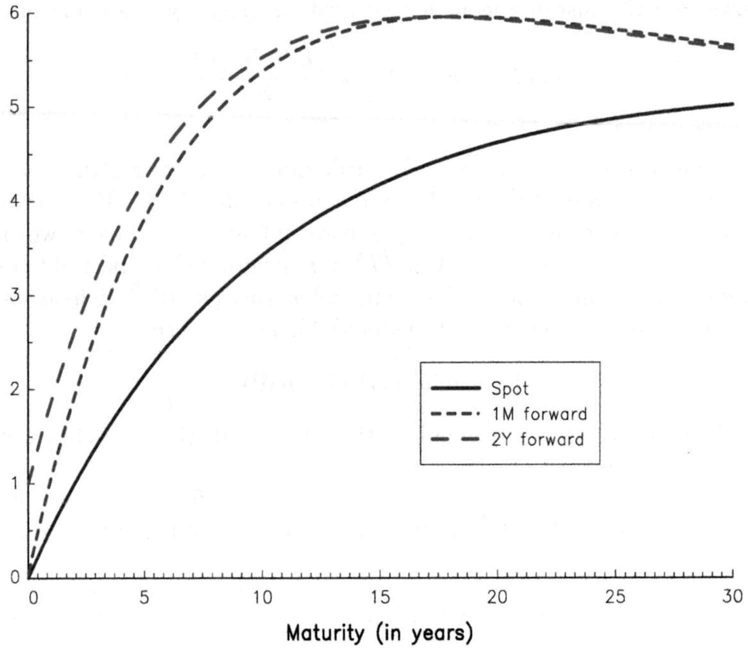

FIGURE 4.1: Term structure of spot and forward interest rates (in %)

Remark 49 *Another representation of the yield curve has been formulated by Litterman and Scheinkman (1991), who propose to characterize the factors using the PCA approach. For instance, the first three factors of the US yield curve*[5] *are represented in Figure 4.2. For each maturity, we provide its sensitivity with respect to the factor. We retrieve the three factors of Litterman and Scheinkman, which are a factor of general level, a slope factor and a convexity (or curvature) factor.*

---

[5]The PCA analysis is done on the daily zero-coupon interest rates computed by Datastream for the period January 2003 – June 2012.

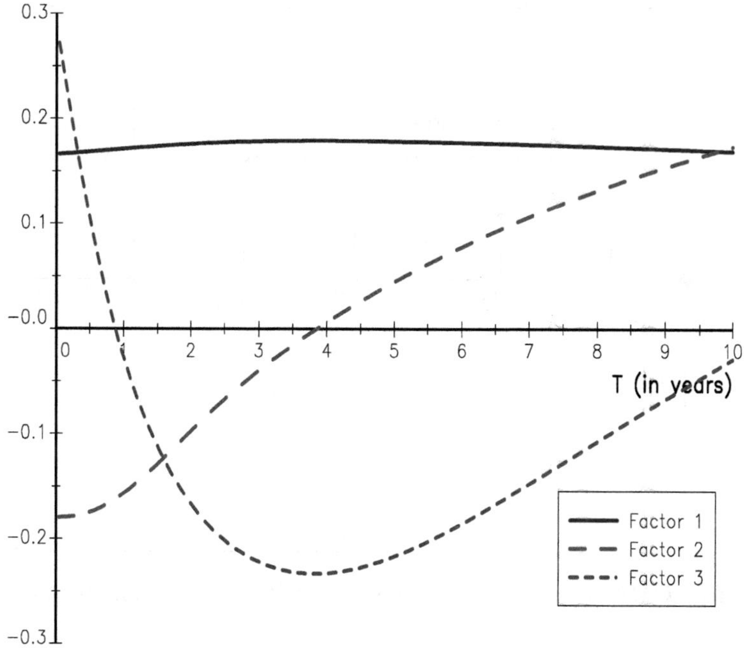

FIGURE 4.2: PCA factors of the US yield curve (Jan. 2003 – Jun. 2012)

4.2.2 Pricing of bonds

4.2.2.1 Without default risk

We consider that the bond pays coupons $C(t_m)$ with fixing dates t_m and the notional N (or the par value) at maturity T. We have reported an example of a cash flows scheme in Figure 4.3. Knowing the yield curve, the price of the bond at date 0 is the sum of the present values of all expected coupon payments and the par value:

$$P_0 = \sum_{m=1}^{M} C(t_m) B_0(t_m) + N B_0(T)$$

When the valuation date is not the issuance date, the previous formula remains valid if we take into account the accrued interests[6]. The term structure

[6] When $t > 0$, the buyer of the bond has the benefit of the next coupon. The price of the bond then satisfies:

$$P_t + AC_t = \sum_{t_m \geq t} C(t_m) B_t(t_m) + N B_t(T)$$

Here, AC_t is the accrued coupon:

$$AC_t = C(t_c) \cdot \frac{t - t_c}{365}$$

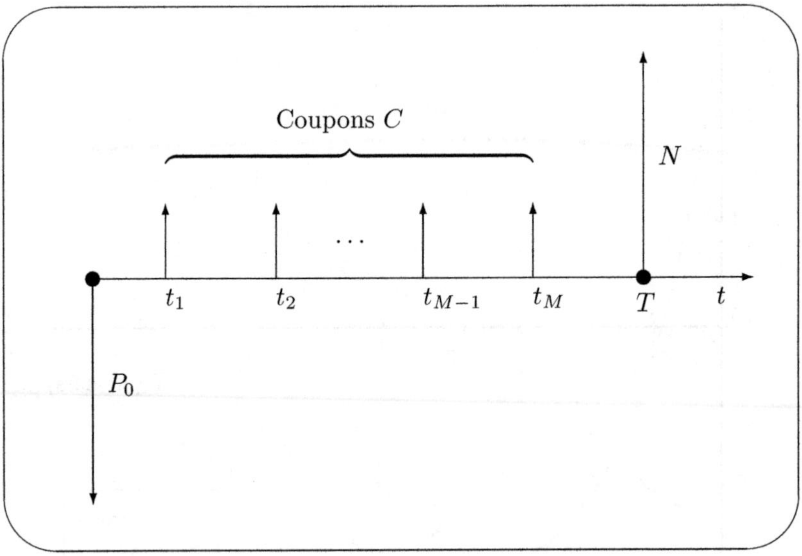

FIGURE 4.3: Cash flows of a bond with a fixed coupon rate

of interest rates impacts the bond price. We generally distinguish three movements:

1. The movement of level corresponds to a parallel shift of interest rates.
2. A twist in the slope of the yield curve indicates how the spread between long and short interest rates moves.
3. A change in the curvature of the yield curve affects the convexity of the term structure.

All these movements are illustrated using Example 33 in Figure 4.4.

The yield to maturity $r^\star$ of a bond is the discount rate which returns its market price:

$$\sum_{t_m \geq t} C(t_m) e^{-(t_m - t)r^\star} + N e^{-(T-t)r^\star} = P_t + AC_t$$

We also define the sensitivity[7] S of the bond price as the derivative of the clean price P_t with respect to the yield to maturity $r^\star$:

$$\begin{aligned} S &= \frac{\partial P_t}{\partial r^\star} \\ &= -\sum_{t_m \geq t} (t_m - t) C(t_m) e^{-(t_m - t)r^\star} - (T - t) N e^{-(T-t)r^\star} \end{aligned}$$

and t_c is the last coupon payment date with $c = \{m : t_{m+1} > t, t_m \leq t\}$. $P_t + AC_t$ is called the *dirty price* whereas P_t refers to the *clean price.*

[7]This sensitivity is also called the \$-duration or DV01.

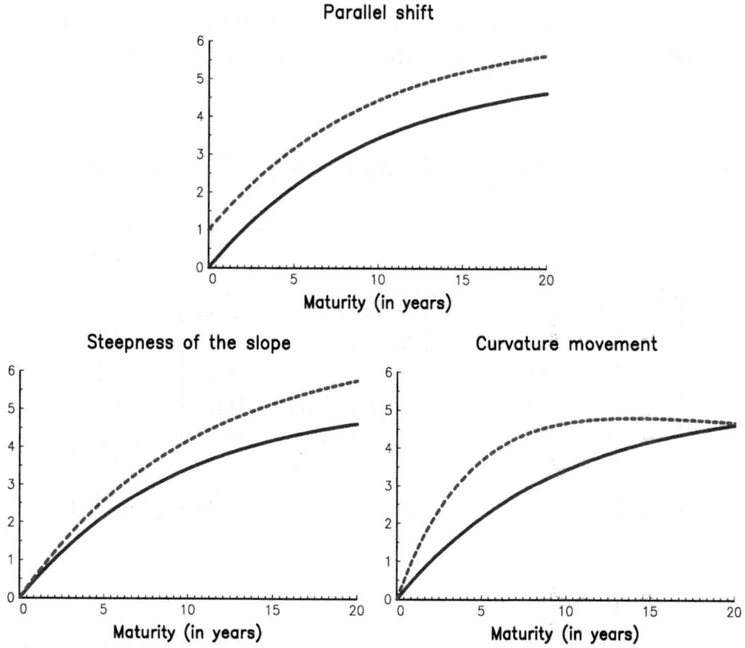

FIGURE 4.4: Movements of the yield curve

It indicates how the P&L of a long position in the bond moves when the yield to maturity changes:

$$\mathrm{PnL} \simeq S \cdot \Delta r^\star$$

Because $S < 0$, the bond price is a decreasing function with respect to interest rates. This implies that an increase of interest rates reduces the value of the bond portfolio.

TABLE 4.1: Price, yield to maturity and sensitivity of bonds

T	$R_0(T)$	$B_0(T)$	P_0	$r^\star$	S
1	0.52	99.48	104.45	0.52	−104.45
2	0.99	98.03	107.91	0.98	−210.86
3	1.42	95.83	110.50	1.39	−316.77
4	1.80	93.04	112.36	1.76	−420.32
5	2.15	89.82	113.63	2.08	−520.16

Using the yield curve defined in Example 33, we report in Table 4.1 the price of a bond with maturity T (expressed in years) with a 5% annual coupon. We also indicate the yield to maturity $r^\star$ (in %) and the corresponding sensitivity S. Let $\breve{P}_0$ (resp. $\hat{P}_0$) be the bond price by taking into account a parallel shift $\Delta r^\star$ (in bps) directly on the zero-coupon rates (resp. on the yield to maturity). The results are given in Table 4.2 in the case of the bond with a five-year maturity. We verify that the computation based on the sensitivity provides a good approximation.

TABLE 4.2: Impact of a parallel shift of the yield curve on the bond with five-year maturity

$\Delta r^{\star}$ (in bps)	$\breve{P}_0$	ΔP	$\hat{P}_0$	ΔP	$S \cdot \Delta r^{\star}$
−50	116.26	2.63	116.26	2.63	2.60
−30	115.20	1.57	115.20	1.57	1.56
−10	114.15	0.52	114.15	0.52	0.52
0	113.63	0.00	113.63	0.00	0.00
10	113.11	−0.52	113.11	−0.52	−0.52
30	112.08	−1.55	112.08	−1.55	−1.56
50	111.06	−2.57	111.06	−2.57	−2.60

4.2.2.2 With default risk

In the previous section, we implicitly assumed that there is no default risk. If the issuer defaults at time τ before the bond maturity T, some coupons and the notional are not paid. In Figure 4.5, we provide an example of cash flows. In this case, the buyer of the bond recovers part of the notional. We note $\Re$ the corresponding recovery rate. It is then easy to show that the price of the bond becomes:

$$\begin{aligned} P_t + AC_t &= \sum_{t_m \geq t} C(t_m)\, B_t(t_m)\, \mathbf{S}_t(t_m) + N B_t(T)\, \mathbf{S}_t(T) + \\ &\quad N\Re \int_t^T B_t(u)\, f_t(u)\, \mathrm{d}u \end{aligned} \qquad (4.1)$$

where $\mathbf{S}_t(u)$ is the survival function at time u and $f_t(u)$ the associated density function[8].

Remark 50 *If we consider an exponential default time with parameter* λ – $\tau \sim \mathcal{E}(\lambda)$, *we have:*

$$\begin{aligned} P_t + AC_t &= \sum_{t_m \geq t} C(t_m)\, B_t(t_m)\, e^{-\lambda(t_m - t)} + N B_t(T)\, e^{-\lambda(T-t)} + \\ &\quad \lambda N\Re \int_t^T B_t(u)\, e^{-\lambda(u-t)}\, \mathrm{d}u \end{aligned}$$

[8] We have:

$$\mathbf{S}_t(u) = \mathbb{E}\left[\mathbb{1}\{\tau > u \mid \tau > t\}\right] = \Pr\{\tau > u \mid \tau > t\}$$

The density function is then given by $f_t(u) = -\partial_u \mathbf{S}_t(u)$.

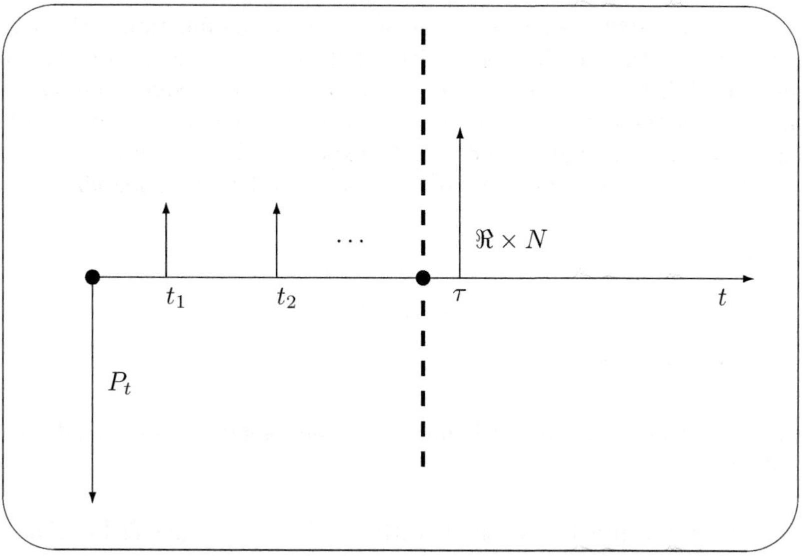

FIGURE 4.5: Cash flows of a bond with default risk

If we assume a flat yield curve – $R_t(u) = r$, we obtain:

$$\begin{aligned} P_t + AC_t &= \sum_{t_m \geq t} C(t_m) e^{-(r+\lambda)(t_m - t)} + N e^{-(r+\lambda)(T-t)} + \\ & \frac{\lambda}{r+\lambda} N \Re \left(1 - e^{-(r+\lambda)(T-t)}\right) \end{aligned}$$

Example 34 *We consider a bond with ten-year maturity. The notional is 100 dollars whereas the annual coupon rate is equal to 4.5%.*

If we consider that $r = 0$, the price of the non-risky bond is 145 dollars. With $r = 5\%$, the price becomes 95.19 dollars. Let us now take into account the default risk. We assume that the recovery rate $\Re$ is 40%. If $\lambda = 2\%$ (resp. 10%), the price of the risky bond is 86.65 dollars (resp. 64.63 dollars). If the yield curve is not flat, we must use the general formula (4.1) to compute the price of the bond. In this case, the integral is evaluated with a numerical integration procedure, typically a Gauss-Legendre quadrature. For instance, if we consider the yield curve defined in Example 33, the bond price is equal to 110.13 if there is no risk, 99.91 if $\lambda = 2\%$ and 73.34 if $\lambda = 10\%$.

The yield to maturity of the defaultable bond is computed exactly in the same way as without default risk. The credit spread $\mathfrak{s}$ is then defined as the difference of the yield to maturity with default risk $\tilde{r}^\star$ and the yield to maturity without default risk $r^\star$:

$$\mathfrak{s} = \tilde{r}^\star - r^\star$$

This spread is a credit risk measure and is an increasing function of the default risk. Reconsider the simple model with a flat yield curve and an exponential default time[9]. If the recovery rate $\mathfrak{R}$ is equal to zero, we deduce that the yield to maturity of the defaultable bond is $\tilde{r}^{\star} = r + \lambda$. It follows that the credit spread is equal to the parameter λ of the exponential distribution. Moreover, if λ is relatively small (less than 20%), the annual default probability is:

$$\begin{aligned} p &= S_t(t+1) \\ &= 1 - e^{-\lambda} \\ &\simeq \lambda \end{aligned}$$

In this case, the credit spread is approximately equal to the annual default probability ($\mathfrak{s} \simeq p$).

If we reuse our previous example with the yield curve specified in Example 33, we obtain the results reported in Table 4.3. For instance, the yield to maturity of the bond is equal to 3.24% without default risk. If λ and $\mathfrak{R}$ are set to 200 bps and 0%, the yield to maturity becomes 5.22% which implies a credit spread of 198.1 bps. If the recovery rate is higher, the credit spread decreases. Indeed, with λ equal to 200 bps, the credit spread is equal to 117.1 bps if $\mathfrak{R} = 40\%$ and only 41.7 bps if $\mathfrak{R} = 80\%$.

TABLE 4.3: Computation of the credit spread $\mathfrak{s}$

$\mathfrak{R}$ (in %)	λ (in bps)	PD (in bps)	P_t (in \$)	$\tilde{r}^{\star}$ (in %)	$\mathfrak{s}$ (in bps)
0	0	0.0	110.1	3.24	0.0
	10	10.0	109.2	3.34	9.9
	200	198.0	93.5	5.22	198.1
	1000	951.6	50.4	13.13	988.9
40	0	0.0	110.1	3.24	0.0
	10	10.0	109.6	3.30	6.0
	200	198.0	99.9	4.41	117.1
	1000	951.6	73.3	8.23	498.8
80	0	0.0	110.1	3.24	0.0
	10	10.0	109.9	3.26	2.2
	200	198.0	106.4	3.66	41.7
	1000	951.6	96.3	4.85	161.4

[9] See Remark 50 on page 200.

4.2.3 Risk management of bond portfolios

Let us consider a portfolio of n bonds. The value of this portfolio is given by:

$$V_t = \sum_{i=1}^{n} \varpi_i \left(P_t^{(i)} + AC_t^{(i)} \right)$$

where ϖ_i is the number of bonds i held in the portfolio[10]. We deduce that:

$$\begin{aligned} V_t &= \sum_{i=1}^{n} \varpi_i \sum_{t_m \geq t} C^{(i)}(t_m) B_t(t_m) \mathbf{S}_t^{(i)}(t_m) + \sum_{i=1}^{n} \varpi_i N_i B_t(T_i) \mathbf{S}_t^{(i)}(T_i) + \\ & \sum_{i=1}^{n} \varpi_i N_i \Re_i \int_t^{T_i} B_t(u) f_t^{(i)}(u)\, du \end{aligned}$$

where $\{t_m \geq t\}$ is the set of all fixing dates[11]. We must distinguish two cases:

1. If there is no default risk, we have:

$$V_t = \sum_{i=1}^{n} \sum_{t_m \geq t} \varpi_i C^{(i)}(t_m) B_t(t_m) + \sum_{i=1}^{n} \varpi_i N_i B_t(T_i)$$

   The bond portfolio can then be viewed as a meta-bond with a complex cash flows scheme.

2. In the presence of default risk, measuring the risk is more complicated because of the dependency of the default times $(\tau_1, \ldots, \tau_n)$.

The risk management of bond portfolios is then similar to the risk management of equity portfolios. The main difference comes from the specification of factors that are very specific to fixed-income instruments. Whereas stock returns generally represent the set of factors in an equity portfolio, it is the yield curve that defines the factors in a bond portfolio. Moreover, taking credit risk into consideration makes the risk management of such a portfolio slightly more complex.

[10] We use the exponent i to distinguish the different bonds.

[11] For instance, if we consider a bond with $C(1) = C(2) = 5$ and another bond with $C(0.5) = C(1.5) = 3$, we have $t_m \in \{0.5, 1, 1.5, 2\}$. For the first bond, we deduce that $C^{(1)}(0.5) = 0$, $C^{(1)}(1) = 5$, $C^{(1)}(1.5) = 0$ and $C^{(1)}(2) = 5$ whereas we have $C^{(2)}(0.5) = 3$, $C^{(2)}(1) = 0$, $C^{(2)}(1.5) = 3$ and $C^{(2)}(2) = 0$ for the second bond.

4.2.3.1 Using the yield curve as risk factors

If there is no default risk, we deduce that the P&L Π between t and $t+h$ is:

$$\begin{aligned}\Pi &= V_{t+h} - V_t \\ &= \sum_{i=1}^{n} \sum_{t_m \geq t} \varpi_i C^{(i)}(t_m)\left(B_{t+h}(t_m) - B_t(t_m)\right) + \\ &\quad \sum_{i=1}^{n} \varpi_i N_i \left(B_{t+h}(T_i) - B_t(T_i)\right)\end{aligned}$$

If we interpret the notional of a bond as a coupon[12], another expression of the P&L is:

$$\Pi = \sum_{t_m \in \mathcal{T}} \sum_{i=1}^{n} \varpi_i C^{(i)}(t_m)\left(B_{t+h}(t_m) - B_t(t_m)\right)$$

where $\mathcal{T}$ is the set of all fixing dates and maturities[13]. Using the previous expression, we can easily calculate Π for different historical or simulated scenarios by considering the zero-coupon interest rates as factors. Computing the historical value-at-risk or the expected shortfall is then straightforward. If we prefer an analytical value-at-risk, we note that:

$$\begin{aligned}D_t(u) &= \lim_{\Delta R_t(u) \to 0} -\frac{\Delta B_t(u)}{\Delta R_t(u)} \Big/ B_t(u) \\ &= -\frac{\partial \ln B_t(u)}{\partial R_t(u)} \\ &= u - t\end{aligned}$$

We deduce that:

$$\Pi \simeq \sum_{t_m \in \mathcal{T}} \delta(t_m) \Delta_h R_t(t_m)$$

with:

$$\delta(u) = -D_t(u) B_t(u) \sum_{i=1}^{n} \varpi_i C^{(i)}(u)$$

$\delta(u)$ is then the exposure of the P&L to the factor $\Delta_h R_t(u)$. If we assume that the variations of zero-coupon interest rates are distributed according to the Gaussian distribution $\mathcal{N}(\mu, \Sigma)$, we deduce that the value-at-risk is equal to:

$$\mathrm{VaR}_\alpha(L) = -\delta^\top \mu + \Phi^{-1}(\alpha)\sqrt{\delta^\top \Sigma \delta} \tag{4.2}$$

[12] In this case, we have $C^{(i)}(T_i) = N_i$.

[13] It implies that $\{T_1, \ldots, T_n\} \subset \mathcal{T}$.

where δ is the vector defined by the values $\delta(t_m)$. We then obtain a value-at-risk formula by considering the exposures $\delta(t_m)$ to the zero-coupon interest rates. We note that another expression of the P&L is:

$$\Pi \simeq \sum_{i=1}^{n} \varpi_i Z_i$$

with:

$$Z_i = \sum_{t_m \in \mathcal{T}} -D_t(t_m) B_t(t_m) C^{(i)}(t_m) \Delta_h R_t(t_m)$$

Z_i is then the P&L of the bond i. Let $Z = (Z_1, \ldots, Z_n)$ be the P&L vector of the n bonds. We deduce that:

$$Z \sim \mathcal{N}\left(A\mu, A\Sigma A^\top\right)$$

where A is the matrix $(A_{i,j})$ with the index j corresponding to the jth fixing date t_m and:

$$A_{i,j} = -D_t(t_m) B_t(t_m) C^{(i)}(t_m)$$

Another expression of the value-at-risk is then:

$$\text{VaR}_\alpha(L) = -\varpi^\top A\mu + \Phi^{-1}(\alpha)\sqrt{\varpi^\top A\Sigma A^\top \varpi} \tag{4.3}$$

We then obtain a value-at-risk formula by considering the exposures ϖ_i to the bonds.

Example 35 *We consider an exposure of 1 MUSD on a bond at June 29, 2012. The notional of the bond is 100 whereas the coupons are equal to 5. The remaining maturity is five years and the fixing dates are at the end of June.*

In Figure 4.6, we have reported the historical evolution of the US zero-coupon interest rates[14] for the period June 2010 – June 2012. At the end of June 2012, we have $R_t(1) = 0.4930$, $R_t(2) = 0.5441$, $R_t(3) = 0.6256$, $R_t(4) = 0.7733$ and $R_t(5) = 0.9598$. This implies that the price of the bond is equal to 119.76 (see the computational details given in Table 4.4). We deduce that the portfolio contains 8 350 shares of the bond. We now compute some risk measures with a 99% confidence level and a one-day holding period in Table 4.5. G1 is the Gaussian value-at-risk given by formula (4.2), G2 is similar to G1 except that we ignore the mean effect ($\mu = \mathbf{0}$) and ES is the expected shortfall. H1, H2 and H3 correspond to the historical value-at-risk computed with different methods: the order statistic for H1, the uniform window smoothing approach (see Remark 23 on page 88) and the covariance principle (see Equation (2.17) on page 92). For each risk measure, we indicate

[14]These data come from the Datastream database. The zero-coupon interest rate of maturity yy years and mm months corresponds to the variable USyyYmm.

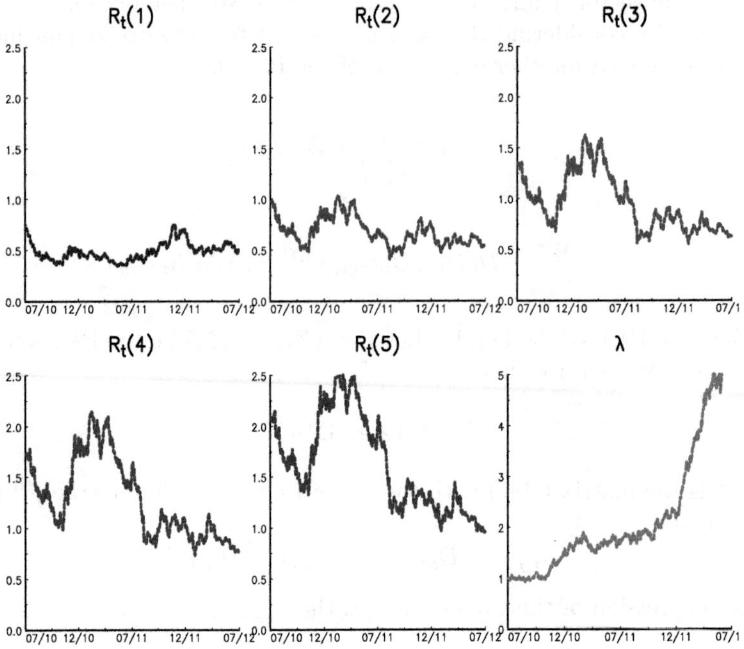

FIGURE 4.6: Evolution of the zero-coupon interest rates and the intensity (June 2010 – June 2012)

TABLE 4.4: Pricing of the bond

t_m	$C(t_m)$	$R_t(t_m)$	$B_t(t_m)$	$\delta(t_m)$
1	5	0.493	0.995	−4.975
2	5	0.544	0.989	−9.892
3	5	0.626	0.981	−14.721
4	5	0.773	0.970	−19.391
5	105	0.960	0.953	−500.400

TABLE 4.5: Risk measure and decomposition of the bond exposure

t_m	G1	G2	H1	H2	H3	ES
1	6.3	6.5	−1.6	7.2	6.4	15.1
2	41.6	42.3	29.4	50.1	46.7	81.7
3	100.7	102.4	102.4	121.7	114.2	172.6
4	169.8	172.8	195.1	206.6	193.0	271.0
5	4941.7	5032.2	5658.7	5928.3	5623.6	7399.3
$\mathcal{R}$	5260.2	5356.3	5984.0	6313.9	5984.0	7939.6

its value and also the corresponding risk decomposition[15] with respect to the 5 zero-coupon interest rates. We note that Gaussian value-at-risk measures give smaller values than historical value-at-risk. We also observe that most of the risk is concentrated in the five-year zero-coupon interest rate. A curious figure is the negative risk contribution of the one-year zero-coupon interest-rate when we consider the classical historical value-at-risk H1. We explain this result by the fact that we only consider one observation to estimate it. By using the window smoothing approach or the covariance principle, this risk contribution becomes positive, which is more realistic.

Example 36 *We consider a portfolio of three bonds with the following characteristics at June 29, 2012:*

Bond	*Notional*	*Coupon*	*Maturity*	*Fixing dates*
#1	100	5	5.0	*End of June*
#2	100	2	4.5	*End of December*
#3	100	3	2.0	*End of June*

The composition of the portfolio is $\varpi_1 = 3\,340$, $\varpi_2 = 3\,193$ *and* $\varpi_3 = 1\,907$.

Using the historical data, we find that the price of the three bonds is respectively \$119.76, \$125.25 and \$104.87. We deduce that the relative weights are 40%, 40% and 20%. In Table 4.6, we have reported the value taken by the different risk measures. As in the previous example, we have also given the risk contributions with respect to the maturity buckets. However, we can also allocate the risk with respect to the three bonds[16]. For instance, the first (resp. the second and the third) bond represents 49.5% (resp. 45.7% and 4.8%) of the risk if we consider the value-at-risk G1. We note that the third bond has a smaller relative risk contribution with respect to its weight of 20% in the portfolio. Indeed, it is between 2.9% and 4.8% for the value-at-risk, but it increases significantly for the expected shortfall (it is equal to 5.9%).

Remark 51 *We have computed the risk contribution with respect to two dimensions: zero-coupon interest rates and bonds. However, we can use other dimensions. For instance, we can allocate the risk with respect to the PCA factors. In the case of the previous example, the results are given in Table 4.7. 98.9% of the Gaussian value-at-risk is explained by the first PCA factor or*

[15]For H1, H2, H3 and ES, formulas for computing the risk contribution are given in the second chapter. For G1 and G2, we have:

$$\mathcal{RC}(t_m) = \delta(t_m) \cdot \left(-\mu + \Phi^{-1}(\alpha) \frac{\Sigma\delta}{\sqrt{\delta^\top \Sigma \delta}}\right)_m$$

[16]For the Gaussian value-at-risk, we have:

$$\mathcal{RC}_i = \varpi_i \cdot \left(-A\mu + \Phi^{-1}(\alpha) \frac{A\Sigma A^\top \varpi}{\sqrt{\varpi^\top A \Sigma A^\top \varpi}}\right)_i$$

TABLE 4.6: Risk allocation of the bond portfolio

	G1	G2	H1	H2	H3	ES
$\mathcal{R}$	4244.6	4321.0	4749.2	5114.1	4749.2	6552.3
	Risk contribution with respect to $R_t(t_m)$					
0.5	0.2	0.2	−0.3	0.1	0.1	0.3
1.0	3.8	3.9	−0.9	3.8	3.8	8.1
1.5	3.2	3.2	1.0	3.5	3.4	6.4
2.0	221.3	224.8	150.1	255.6	244.6	417.0
2.5	10.9	11.1	9.5	12.7	12.2	19.3
3.0	41.1	41.8	41.0	48.7	45.9	69.0
3.5	20.9	21.2	22.8	25.0	23.3	34.0
4.0	68.4	69.6	78.0	82.6	76.5	108.4
4.5	1904.4	1938.5	2184.5	2310.6	2132.3	2930.2
5.0	1970.5	2006.7	2263.5	2371.3	2207.2	2959.7
sum	4244.6	4321.0	4749.2	5114.1	4749.2	6552.3
	Risk contribution with respect to ϖ_i					
#1	2100.2	2138.6	2393.6	2525.5	2351.6	3175.8
#2	1939.5	1974.2	2217.5	2351.9	2171.3	2990.1
#3	205.0	208.2	138.2	236.6	226.4	386.4
sum	4244.6	4321.0	4749.2	5114.1	4749.2	6552.3
	Risk contribution with respect to ϖ_i (in %)					
#1	49.5	49.5	50.4	49.4	49.5	48.5
#2	45.7	45.7	46.7	46.0	45.7	45.6
#3	4.8	4.8	2.9	4.6	4.8	5.9

TABLE 4.7: Risk decomposition of the bond portfolio with respect to the PCA factors

Factor	G1	G1	H1	H2	H3	ES
1	4198.4	4274.7	4406.2	4974.6	4406.2	6718.0
2	39.0	39.0	475.3	120.6	475.3	−139.8
3	6.0	6.0	−66.1	25.2	−66.1	−42.8
4	1.1	1.1	−57.5	−11.3	−57.5	25.2
5	0.1	0.1	−0.4	2.9	−0.4	−4.3
6	0.0	0.0	0.1	−0.1	0.1	0.0
7	0.0	0.0	−5.8	−2.3	−5.8	−2.2
8	0.0	0.0	0.9	2.9	0.9	−0.6
9	0.0	0.0	0.2	0.2	0.2	0.2
10	0.0	0.0	−3.5	1.4	−3.5	−1.3
sum	4244.6	4321.0	4749.2	5114.1	4749.2	6552.3

the level factor, whereas the slope and convexity factors explain only 0.9% *and* 0.1% *of the portfolio risk. For the historical value-at-risk, the part of the level factor is smaller and is equal to* 92.8%. *We note that the slope and convexity factors may have a negative risk contribution. This is the case for the expected shortfall where their risk contributions are* −2.1% *and* −0.7%.

4.2.3.2 Taking into account the default risk

By considering the default risk, the expression of the P&L between t and $t+h$ depends on the survival function $\mathbf{S}_t^{(i)}(t_m)$ of the ith issuer[17]. The risk factors then correspond to the yield curve and to the default times of the different issuers. However, it is extremely difficult to decompose the P&L as a linear function of these risk factors. This is why we cannot propose an analytical formula of the value-at-risk, contrary to the previous section. However, historical or Monte Carlo value-at-risk measures are easy to implement. The only difficulty lies in defining the factors of the default risk.

We consider again Example 35, but we assume that there is a credit risk in the bond price. For this purpose, we suppose that λ and $\Re$ are equal to 5.55% and 50% at the end of June 2012. The historical evolution of zero-coupon interest rates and the default parameter λ are those given in Figure 4.6. We recall that the historical value-at-risk is equal to \$5 984 for an exposure of 1 MUSD on the non-defaultable bond. If we take into account the default risk[18], this value-at-risk becomes \$10 658, a significant increase of more than 75%. If we add the recovery risk by considering that the recovery rate is distributed uniformly between 45% and 55%, we obtain a value-at-risk equal to \$16 436. In Figure 4.7, we report the estimated density of the daily loss for the different historical scenarios. We note the substantial impacts of default and recovery risks.

The computation of value-at-risk and expected shortfall for a bond portfolio is no more complicated than for a bond if we use historical scenarios. However, it becomes somewhat tricky when we consider Monte Carlo simulations. The problem comes from the specification of multivariate default times (and potentially multivariate recovery rates). For this purpose,

[17]The expression of the P&L between t and $t+h$ becomes:

$$\begin{aligned}\Pi &= \sum_{i=1}^{n}\sum_{t_m\geq t}\varpi_i C^{(i)}(t_m)\left(B_{t+h}(t_m)\mathbf{S}_{t+h}^{(i)}(t_m)-B_t(t_m)\mathbf{S}_t^{(i)}(t_m)\right)+\\ &\quad \sum_{i=1}^{n}\varpi_i N_i\left(B_{t+h}(T_i)\mathbf{S}_{t+h}^{(i)}(T_i)-B_t(T_i)\mathbf{S}_t^{(i)}(T_i)\right)+\\ &\quad \sum_{i=1}^{n}\varpi_i N_i\Re_i\int_t^{T_i}\left(B_{t+h}(u)f_{t+h}^{(i)}(u)-B_t(u)f_t^{(i)}(u)\right)\,\mathrm{d}u\end{aligned}$$

[18]For each historical scenario, we price the bond by applying a shock on the default parameter λ. This shock corresponds to the relative variation between two consecutive trading dates.

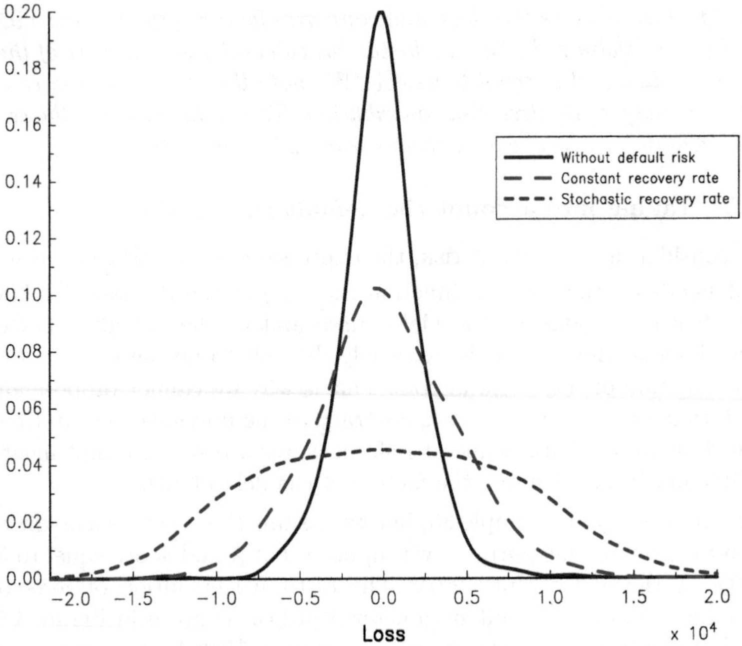

FIGURE 4.7: Loss distribution of the bond portfolio with and without default risk

we use the copula functions presented in Appendix A.2 on page 308. Let $\mathbf{S}(t_1, \ldots, t_n) = \Pr\{\tau_1 > t_1, \ldots, \tau_n > t_n\}$ be a multivariate survival function. We can always decompose $\mathbf{S}$ into the margins $\mathbf{S}_i$ and the survival copula $\mathbf{C}$:

$$\mathbf{S}(t_1, \ldots, t_n) = \mathbf{C}(\mathbf{S}_1(t_1), \ldots, \mathbf{S}_n(t_n))$$

With this representation, simulating default times is straightforward. The algorithm is the following:

1. We simulate a vector of random variates $(u_1, \ldots, u_n)$ from the copula $\mathbf{C}$.

2. The simulated default times $(\tau_1, \ldots, \tau_n)$ are given by:

$$\tau_i = \mathbf{S}_i^{-1}(u_i)$$

If we only consider the credit risk, and no other risks such as the interest rate risk, we may obtain analytical formulas with the copula approach. We generally define the credit loss of a portfolio in the following way[19]:

$$L = \sum_{i=1}^{n} x_i G_i D_i \tag{4.4}$$

[19]In this equation, G_i and D_i are two random variables whereas x_i is a scalar.

where x_i, G_i and D_i are the weight, the loss given default[20] and the default indicator of the ith issuer. Let T_i be the maturity of the ith credit. The default indicator is related to the default time by the following relationship:

$$D_i = \begin{cases} 1 & \text{if } \tau_i \leq T_i \\ 0 & \text{otherwise} \end{cases}$$

We note that the loss definition (4.4) corresponds to instruments with face value amount returned at maturity. We consider the following assumptions:

1. The default time τ_i depends on a set of risk factors Y with distribution $\mathbf{H}$. We note $p_i(y)$ (resp. p_i) the conditional (resp. unconditional) default probability when $Y = y$.

2. The default time τ_i is independent of the loss given default G_i.

3. The credit portfolio is infinitely fine-grained meaning that:

$$\lim_{n \to \infty} \sum_{i=1}^{n} x_i^2 = 0$$

Conditionally to the risk factors y, the default indicator $D_i = \mathbb{1}\{\tau_i \leq T_i\}$ is a Bernoulli random variable with parameter $p_i(y)$. It follows that:

$$\mathbb{E}[L \mid Y = y] = \sum_{i=1}^{n} x_i \mu(G_i) p_i(y)$$

where $\mu(G_i) = \mathbb{E}[G_i]$ and:

$$\sigma^2(L \mid Y = y) = \sum_{i=1}^{n} x_i^2 \left(\mu^2(G_i) p_i(y)(1 - p_i(y)) + \sigma^2(G_i) p_i(y)\right)$$

If the portfolio is infinitely fine-grained, we have $\sigma^2(L \mid Y = y) = 0$ (Wilde, 2001). Conditionally to the risk factors, the loss of an infinitely fine-grained portfolio is then equal to its conditional expected loss:

$$\begin{aligned} g(y) &= \mathbb{E}[L \mid Y = y] \\ &= \sum_{i=1}^{n} x_i \mu(G_i) p_i(y) \end{aligned}$$

Let $\mathbf{F}$ be the unconditional loss distribution. We have:

$$\begin{aligned} \mathbf{F}(\ell) &= \Pr\{L \leq \ell\} \\ &= \Pr\{g(Y) \leq \ell\} \\ &= \int \cdots \int \mathbb{1}\{g(y) \leq \ell\} \, \mathrm{d}\mathbf{H}(y) \end{aligned}$$

[20]The loss given default can be viewed as the complement of the recovery rate: $G_i = 1 - \Re_i$.

This expression is very useful to evaluate $\mathbf{F}(\ell)$ by Monte Carlo simulations. However, it does not permit to find an analytical expression of the portfolio value-at-risk $\mathrm{VaR}_\alpha(x)$ or expected shortfall $\mathrm{ES}_\alpha(x)$.

Let us consider the case with only one risk factor. The definition of the value-at-risk implies that $\Pr\{g(Y) \leq \mathrm{VaR}_\alpha(x)\} = \alpha$. If $g(y)$ is a non-decreasing function, we have $\Pr\left\{Y \leq g^{-1}(\mathrm{VaR}_\alpha(x))\right\} = \alpha$. We deduce that $g^{-1}(\mathrm{VaR}_\alpha(x)) = \mathbf{H}^{-1}(\alpha)$. The expression of the value-at-risk is then[21]:

$$\begin{aligned} \mathrm{VaR}_\alpha(x) &= g\left(\mathbf{H}^{-1}(\alpha)\right) \\ &= \sum_{i=1}^{n} x_i \mu(G_i)\, p_i\left(\mathbf{H}^{-1}(\alpha)\right) \end{aligned}$$

This formula is very interesting because this value-at-risk satisfies the Euler allocation principle. We deduce that the risk contribution $\mathcal{RC}_i$ of the ith credit is:

$$\mathcal{RC}_i = x_i \mu(G_i)\, p_i\left(\mathbf{H}^{-1}(\alpha)\right)$$

If the risk measure is the expected shortfall, we have[22]:

$$\begin{aligned} \mathrm{ES}_\alpha(x) &= \mathbf{E}\left[L \mid L \geq \mathrm{VaR}_\alpha(x)\right] \\ &= \mathbf{E}\left[L \mid g(Y) \geq \mathrm{VaR}_\alpha(x)\right] \\ &= \mathbf{E}\left[L \mid Y \geq g^{-1}(\mathrm{VaR}_\alpha(x))\right] \\ &= \mathbf{E}\left[\sum_{i=1}^{n} x_i \mu(C_i)\, p_i(Y) \mid Y \geq \mathbf{H}^{-1}(\alpha)\right] \\ &= \sum_{i=1}^{n} x_i \mu(G_i)\, \mathbf{E}\left[p_i(Y) \mid Y \geq \mathbf{H}^{-1}(\alpha)\right] \end{aligned}$$

The risk contribution of the ith credit with respect to the expected shortfall is then:

$$\mathcal{RC}_i = x_i \mu(G_i)\, \mathbf{E}\left[p_i(Y) \mid Y \geq \mathbf{H}^{-1}(\alpha)\right]$$

To obtain a comprehensive analytical formula, we generally refer to the seminal framework of Merton (1974). In this credit model, the default of the firm occurs when the value of its assets Z_i falls below a certain threshold B_i (also called the default barrier):

$$D_i = 1 \Leftrightarrow Z_i \leq B_i$$

---

[21] If $g(y)$ is a non-increasing function, the expression of the value-at-risk becomes:

$$\mathrm{VaR}_\alpha = \sum_{i=1}^{n} x_i \mu(G_i)\, p_i\left(\mathbf{H}^{-1}(1-\alpha)\right)$$

[22] If $g(y)$ is a non-increasing function, the condition becomes $Y \leq \mathbf{H}^{-1}(1-\alpha)$.

Vasicek (2002) proposes to link the asset value Z_i to one common risk factor Y and an idiosyncratic risk ε_i:

$$Z_i = \sqrt{\rho} Y + \sqrt{1-\rho}\varepsilon_i$$

By assuming that Y and ε_i are independent Gaussian random variables with distribution $\mathcal{N}(0,1)$, it follows that Z_i has a $\mathcal{N}(0,1)$ distribution too. Because the default probability p_i satisfies:

$$\begin{aligned} p_i &= \Pr\{D_i = 1\} \\ &= \Pr\{Z_i \leq B_i\} \\ &= \Phi(B_i) \end{aligned}$$

we conclude that the default barrier B_i is equal to $\Phi^{-1}(p_i)$. It follows that the conditional default probability is[23]:

$$\begin{aligned} p_i(y) &= \Pr\{D_i = 1 \mid Y = y\} \\ &= \Pr\{Z_i < B_i \mid Y = y\} \\ &= \Pr\left\{\sqrt{\rho} y + \sqrt{1-\rho}\varepsilon_i < B_i\right\} \\ &= \Pr\left\{\varepsilon_i < \frac{B_i - \sqrt{\rho} y}{\sqrt{1-\rho}}\right\} \\ &= \Phi\left(\frac{\Phi^{-1}(p_i) - \sqrt{\rho} y}{\sqrt{1-\rho}}\right) \end{aligned}$$

It is easy to show that the default times are independent conditionally to the risk factor Y. As the function $\sum_{i=1}^{n} x_i \mu(G_i) p_i(y)$ is a non-increasing function[24], we obtain that the risk contribution with the value-at-risk as the risk measure is:

$$\mathcal{RC}_i = x_i \mu(G_i) \Phi\left(\frac{\Phi^{-1}(p_i) + \sqrt{\rho}\Phi^{-1}(\alpha)}{\sqrt{1-\rho}}\right) \tag{4.5}$$

Remark 52 *If we replace in Equation (4.5) the parameters x_i, $\mu(G_i)$ and p_i by the parameters* EAD_i *(exposure at default),* LGD_i *(loss given default)*

---

[23]This factor model has a very simple copula representation. We note that the probability distribution of the asset value vector is $\mathcal{N}(\mathbf{0}, \Sigma)$ where Σ is equal to the constant correlation matrix $C_n(\rho)$. We can decompose the multivariate survival function as follows:

$$\begin{aligned} \mathbf{S}(t_1, \ldots, t_n) &= \Pr\{\tau_1 > t_1, \ldots, \tau_n > t_n\} \\ &= \Pr\left\{Z_1 > \Phi^{-1}(p_1), \ldots, Z_n > \Phi^{-1}(p_n)\right\} \\ &= \mathbf{C}(1-p_1, \ldots, 1-p_I; \Sigma) \\ &= \mathbf{C}(\mathbf{S}_1(t_1), \ldots, \mathbf{S}_n(t_n); \Sigma) \end{aligned}$$

The survival copula is then normal with the parameter matrix $C_n(\rho)$.

[24]We assume that the weights x_i are positive. Moreover, we have $\mu(G_i) \geq 0$ and $\partial_y p_i(y) \leq 0$.

and PD_i *(probability of default), we obtain exactly the internal-rating based (or IRB) formula used in the Basle II framework:*

$$\mathcal{RC}_i = \mathrm{EAD}_i \cdot \mathrm{LGD}_i \cdot \Phi\left(\frac{\Phi^{-1}(\mathrm{PD}_i) + \sqrt{\rho}\Phi^{-1}(\alpha)}{\sqrt{1-\rho}}\right)$$

If we prefer to use the expected shortfall, we need to compute the conditional expected default probability[25] $\bar{p} = \mathbf{E}\left[p_i(Y) \mid Y \le \Phi^{-1}(1-\alpha)\right]$:

$$\begin{aligned}
\bar{p} &= \mathbf{E}\left[\Phi\left(\frac{\Phi^{-1}(p_i) - \sqrt{\rho}Y}{\sqrt{1-\rho}}\right) \mid Y \le \Phi^{-1}(1-\alpha)\right] \\
&= \int_{-\infty}^{\Phi^{-1}(1-\alpha)} \Phi\left(\frac{\Phi^{-1}(p_i)}{\sqrt{1-\rho}} + \frac{-\sqrt{\rho}}{\sqrt{1-\rho}}y\right) \frac{\phi(y)}{\Phi(\Phi^{-1}(1-\alpha))}\,\mathrm{d}y \\
&= \frac{1}{1-\alpha}\Phi_2\left(\Phi^{-1}(1-\alpha), \Phi^{-1}(p_i); \sqrt{\rho}\right) \\
&= \frac{1}{1-\alpha}\mathbf{C}\left(1-\alpha, p_i; \sqrt{\rho}\right)
\end{aligned}$$

with $\mathbf{C}$ the normal copula. We finally obtain that the risk contribution has the following expression:

$$\mathcal{RC}_i = \frac{1}{1-\alpha}x_i\mu(G_i)\,\mathbf{C}\left(1-\alpha, p_i; \sqrt{\rho}\right)$$

This result has been obtained by Pykhtin (2004).

Remark 53 *The previous framework to compute credit risk measures has been generalized to non-granular portfolios. For instance, Emmer and Tasche (2005) present an approximated formula of the value-at-risk in the case of the single factor model (Gordy, 2003). In a similar way, Pykhtin (2004) provides some adjustments for the multi-factor model with respect to the infinitely fine-grained portfolio.*

Example 37 *We consider a portfolio with five bonds whose characteristics are given in Table 4.8.* x_i *is the weight of the* i^{th} *bond in the portfolio. We assume that the default times are exponentially distributed. The default probability* p_i *is then equal to* $1 - \exp(-\lambda_i D_i)$ *with* λ_i *the credit spread of the issuer and* D_i *the duration of the bond. We also indicate the expected loss given default* $\mu(G_i)$ *in the last column.*

[25] We use the general result:

$$\int_{-\infty}^{c} \Phi(a+by)\phi(y)\,\mathrm{d}y = \Phi_2\left(c, \frac{a}{\sqrt{1+b^2}}; \frac{-b}{\sqrt{1+b^2}}\right)$$

with $\Phi_2(x,y;\rho)$ the cumulative distribution function of the standardized bivariate normal density with correlation ρ on the space $[-\infty, x] \times [-\infty, y]$ (lower tail).

TABLE 4.8: Characteristics of the bond portfolio

Bond	x_i (in %)	D_i (in years)	λ_i (in bps)	p_i (in %)	$\mu(G_i)$ (in %)
1	20.00	5.00	100	4.88	70.00
2	30.00	6.00	120	6.95	70.00
3	10.00	8.00	85	6.57	50.00
4	15.00	7.00	115	7.73	50.00
5	25.00	5.00	250	11.75	50.00

The computation of the normalized risk contributions are given in Table 4.9 for different parameters (ρ, α). For instance, if $\rho = 30\%$ and $\alpha = 99\%$, the credit value-at-risk of the bond portfolio is equal to 24.9% of the portfolio value. 33.9% comes from the second bond, whereas the third bond only represents 7.8% of the portfolio risk. We verify of course that $\mathrm{ES}_\alpha(x) > \mathrm{VaR}_\alpha(x)$. We also note that the risk decomposition is sensitive to the correlation parameter ρ and to the confidence level α. As α tends to infinity, the risk measures $\mathrm{VaR}_\alpha(x)$ and $\mathrm{ES}_\alpha(x)$ converge to the maximum loss L^+ which is equal to $\sum_{i=1}^n x_i \mu(G_i)$. In this example, L^+ takes the value of 60%.

TABLE 4.9: Normalized risk contributions $\mathcal{RC}_i^\star$ of the bond portfolio (in %)

Bond	(30%, 99%) VaR	(30%, 99%) ES	(50%, 99%) VaR	(50%, 99%) ES	(30%, 99.9%) VaR	(30%, 99.9%) ES
1	18.2	19.0	19.1	20.2	19.8	20.3
2	33.9	34.2	34.4	34.7	34.5	34.6
3	7.8	7.9	8.0	8.1	8.0	8.1
4	12.9	12.9	12.9	12.8	12.8	12.8
5	27.1	26.0	25.6	24.2	24.9	24.1
$\mathcal{R}(x)$	24.9	30.1	36.2	42.8	36.6	40.4

4.3 Some illustrations

In this section, we present two applications using the risk parity approach to manage bond portfolios. The first application concerns the budgeting of the risk factors, which are measured using a principal component analysis of the yield curve. In the second application, we present a method to manage the credit risk of bond portfolios. This method can be viewed as a risk-based indexation. It will allow us to discuss and compare the different forms of alternative-weighted indexation for bonds.

4.3.1 Managing risk factors of the yield curve

We have seen previously that interest rates are influenced by three main factors: the general level of interest rates, the slope of the yield curve and the convexity. We can then analyze the risk of a bond portfolio with respect to these factors. For this purpose, we build them with the principal component analysis and we use the framework of factor risk parity portfolios presented in Section 2.5 on page 135.

Let us consider a strip portfolio of 10 zero-coupon bonds with maturities of one to ten years. The bond valuation is done for the date of June 30, 2012 while the risk measure corresponds to the previous Gaussian value-at-risk G2. Let R_t be the vector of the zero-coupon interest rates $\{R_t(t+1), \ldots, R_t(t+10)\}$. The factor model is specified as follows:

$$\Delta_h R_t = A\mathcal{F}_t + \varepsilon_t$$

with $\Delta_h R_t = R_t - R_{t-h}$, $\Sigma = \operatorname{cov}(\Delta_h R_t)$, $\Omega = \operatorname{cov}(\mathcal{F}_t)$ and $D = \operatorname{cov}(\varepsilon_t)$. We note $\hat{\Sigma}$ the empirical covariance matrix of $\Delta_h R_t$. We use the PCA of the US yield curve (see Figure 4.2 on page 197) and compute the eigendecomposition of $\hat{\Sigma}$:

$$\hat{\Sigma} = V\Lambda V^\top$$

with $V = (v_1, \ldots, v_{10})$ the matrix of eigenvectors, λ_j the eigenvalue[26] associated with the eigenvector v_j and $\Lambda = \operatorname{diag}(\lambda_1, \ldots, \lambda_{10})$. The model with the first three PCA factors $\mathcal{F}_1$, $\mathcal{F}_2$ and $\mathcal{F}_3$ is given by[27]:

$$A = \begin{pmatrix} v_1 & v_2 & v_3 \end{pmatrix}, \quad \Omega = \operatorname{diag}(\lambda_1, \lambda_2, \lambda_3) \quad \text{and} \quad D = \hat{\Sigma} - A\Omega A^\top$$

For each portfolio, we have reported its composition v_i with respect to the nominal value of the long leg[28], the relative risk contribution $\mathcal{RC}_i^\star$ of each zero-coupon bond and the relative risk contribution $\mathcal{RC}^\star(\mathcal{F}_j)$ of the level, slope and convexity factors[29]:

$$\mathcal{RC}^\star(\mathcal{F}_j) = \frac{1}{\mathcal{R}(x)} \left(A^\top x\right)_j \cdot \left(A^+ \frac{\partial\, \mathcal{R}(x)}{\partial\, x}\right)_j$$

with:

$$x_i = -\varpi_i D_t(t+i) B_t(t+i)$$

[26] We assume that the eigenvalues are sorted in descending order: $\lambda_1 \geq \lambda_2 \geq \cdots \geq \lambda_{10}$.

[27] We deduce this result because the PCA factors are $V^\top \Delta_h R_t$ and $V^{-1} = V^\top$.

[28] Let ϖ_i be the number of ith zero-coupon bonds held in the portfolio. The value v_i is then equal to:

$$v_i = \frac{\varpi_i P_t^{(i)}}{\sum_{i=1}^n \max\left(0, \varpi_i P_t^{(i)}\right)}$$

where $P_t^{(i)} = B_t(t+i)$ is the mark-to-market price of the ith zero-coupon bond.

[29] The risk contribution of factor 4+ corresponds to the remaining risk contributions of the other factors: $\mathcal{RC}(\mathcal{F}_{4+}) = \mathcal{R}(x) - \sum_{i=1}^3 \mathcal{RC}(\mathcal{F}_j)$.

Figure 4.8 corresponds to the equally weighted portfolio $\varpi_i = 1$. We note that the risk contribution of the zero-coupon bonds increases with the maturity, because of the duration sensitivity. So, most of the risk is concentrated on the long maturities, implying that the level factor explains more than 99% of the portfolio risk. We then consider three long-short portfolios:

Maturity	1	2	3	4	5	6	7	8	9	10
#1	1	1	1	1	1	1	1	1	1	1
#2	−2	−2	−2	−2	−2	1	1	1	1	1
#3	10	10	10	10	10	−4	−4	−4	−4	−4
#4	53	−8	−7	−6	−5	−4	0	3	3	3

The results are reported in Figures 4.9, 4.10 and 4.11. Portfolio #2 is principally exposed to the level and slope factors. Concerning Portfolio #3, 90% of its risk is explained by the slope of the yield curve, whereas the convexity factor is the main risk of Portfolio #4.

Remark 54 *In the previous examples, slope and convexity risk contributions are significant because we consider long-short portfolios. With long-only portfolios, it is very difficult to be highly exposed to these two factors, because the level factor is the main contributor of their performances. Indeed, with the ten previous zero-coupon bonds, the risk contribution of the slope and convexity factors reaches their maximum at 32.5% and 2.9% in the case of long-only portfolios. However, we obtain very singular portfolios. If we impose a maximum weight of 25%, these figures become 13.9% and 0.5%.*

The traditional way to immunize the portfolio against the level of interest rates is to use barbell strategies. These strategies are particularly interesting when one would like to play a yield curve scenario, typically the steepness of the term structure. A typical barbell portfolio consists in investing in bonds with maturities equal to 2, 5 and 10 years. Let ϖ_i be the share number of the ith bond. We note T_i the respective maturity with $T_1 < T_2 < T_3$. The sensitivity of the bond portfolio is zero if we have:

$$\varpi_1 S_1 + \varpi_2 S_2 + \varpi_3 S_3 = 0$$

with S_i the sensitivity of the ith bond. It is conventional to fix ϖ_2 equal to -1. To find ϖ_1 and ϖ_3, we must impose other constraints. Martellini *et al.* (2003) consider four methods:

1. In a 50/50 barbell portfolio, the sensitivity of two long legs are equal: $\varpi_1 S_1 = \varpi_3 S_3$. We deduce that:

$$\begin{cases} \varpi_1 = -\frac{1}{2}\varpi_2 \frac{S_2}{S_1} \\ \varpi_3 = -\frac{1}{2}\varpi_2 \frac{S_2}{S_3} \end{cases}$$

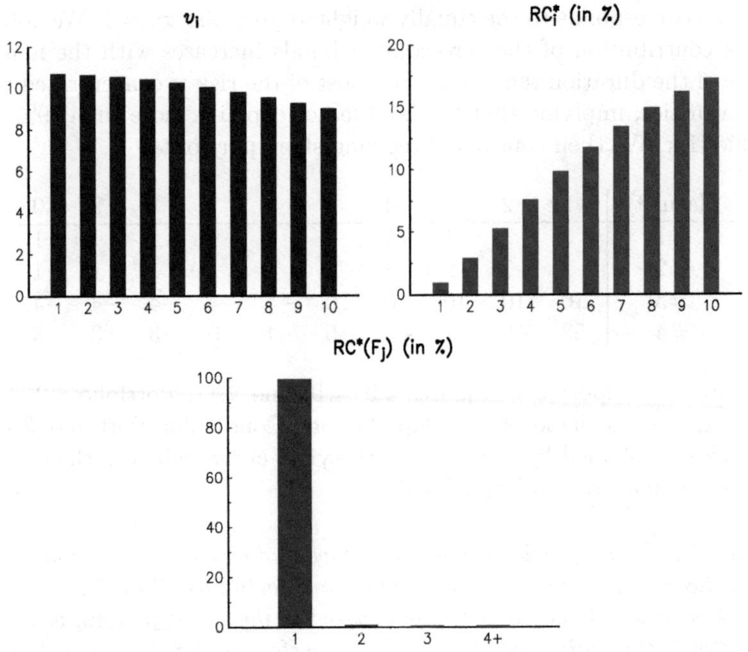

FIGURE 4.8: Risk factor contributions of the EW Portfolio #1

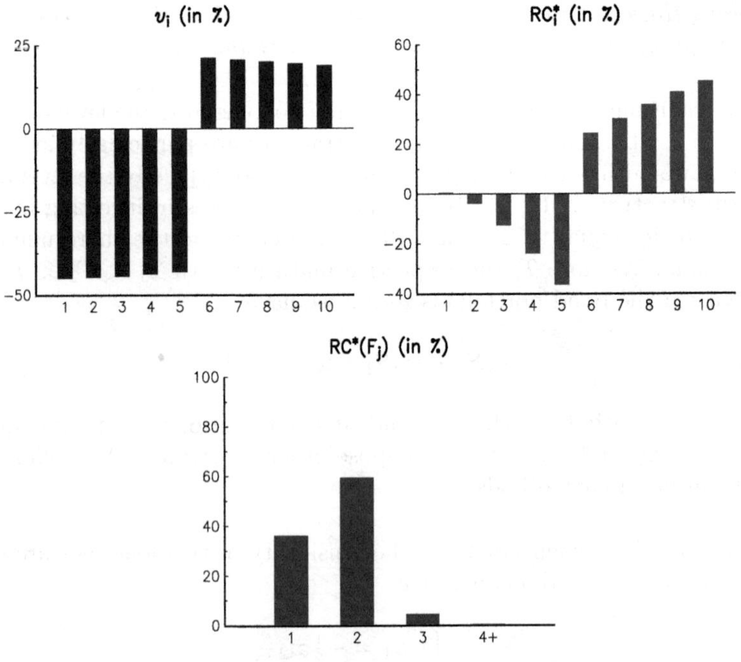

FIGURE 4.9: Risk factor contributions of the long-short Portfolio #2

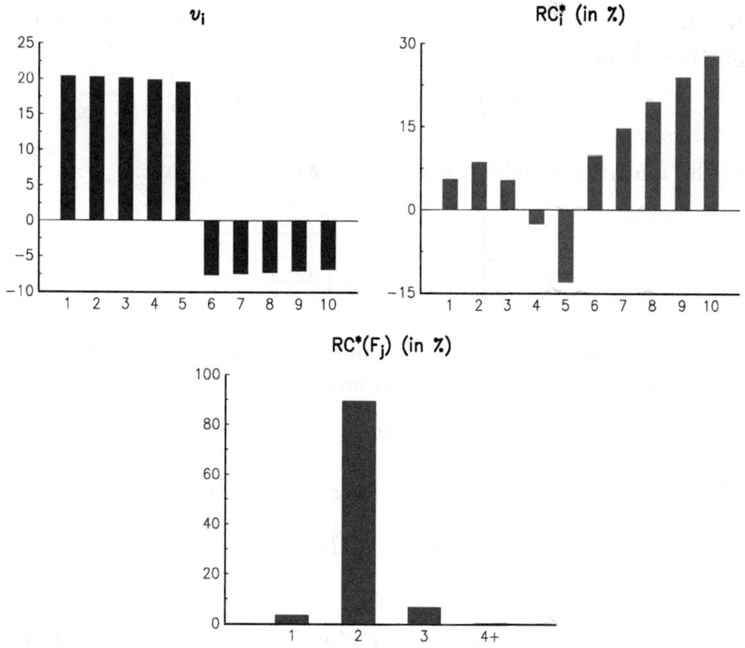

FIGURE 4.10: Risk factor contributions of the long-short Portfolio #3

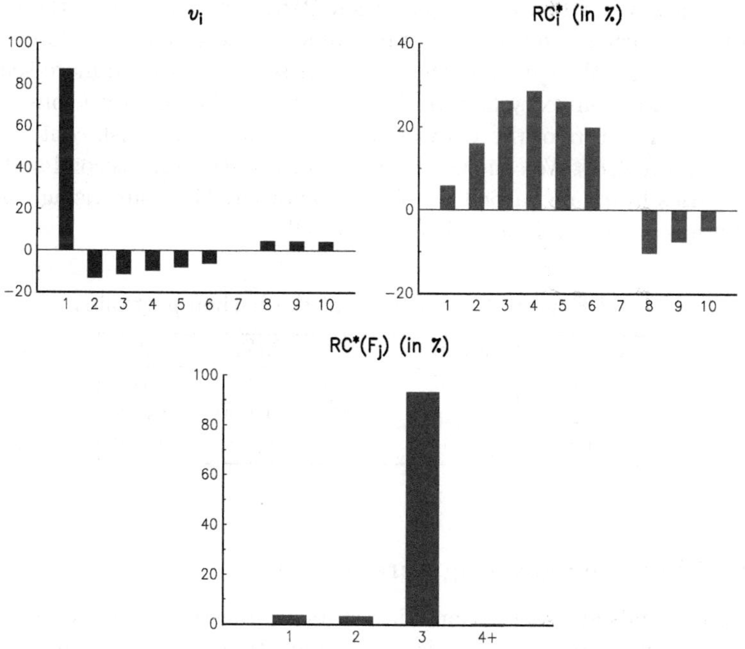

FIGURE 4.11: Risk factor contributions of the long-short Portfolio #4

2. A barbell portfolio is cash-neutral (or CN) if the initial value of the portfolio is zero:

$$\varpi_1 P_t(T_1) + \varpi_2 P_t(T_2) + \varpi_3 P_t(T_3) = 0$$

3. In the maturity-weighted (or MW) portfolio, we have:

$$\begin{cases} \varpi_1 = -\varpi_2 \left(\frac{T_2 - T_1}{T_3 - T_1}\right) \frac{S_2}{S_1} \\ \varpi_3 = -\varpi_2 \left(\frac{T_3 - T_2}{T_3 - T_1}\right) \frac{S_2}{S_3} \end{cases}$$

4. The regression-weighted (or RW) portfolio takes into account the volatility of the two long legs. We then have:

$$\varpi_1 S_1 = \beta \varpi_3 S_3$$

where β is the coefficient of the linear regression:

$$R_t(T_2) - R_t(T_1) = \beta (R_t(T_3) - R_t(T_2)) + \varepsilon_t$$

We consider the Nelson-Siegel yield curve given in Figure 4.1. The three bonds correspond to zero-coupon bonds with maturities 2, 5 and 10 years. Using the previous formulas, we deduce the weights of the different barbell portfolios[30]. We note that the compositions of the four portfolios are very different. However, when we compute their P&L Π by translating the yield to maturity $r^\star$ of each bond by a constant value, we note that it is close to zero (see Figure 4.12). We can then think that these portfolios are more sensitive to the slope and convexity factors than to the level factor. If we consider the previous PCA factors of the US yield curve, we obtain the risk contributions given in Figure 4.13. We conclude that the sensitivity to the second and third factors varies from one barbell portfolio to another. More surprising, the risk contribution of the level factor is not residual!

TABLE 4.10: Composition of the barbell portfolios

Maturity	50/50	CN	MW	RW
2Y	1.145	0.573	0.859	0.763
5Y	−1.000	−1.000	−1.000	−1.000
10Y	0.316	0.474	0.395	0.422

4.3.2 Managing sovereign credit risk

Pension funds and institutional investors are massively invested in bonds, and in particular sovereign bonds. For a long time, the sovereign bonds of

[30] β is set to 50%.

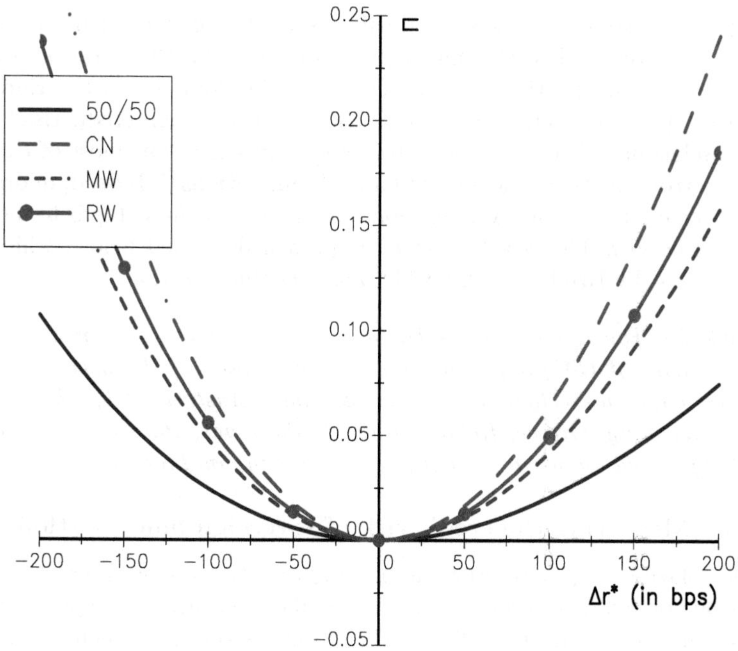

FIGURE 4.12: P&L of the barbell portfolios due to a YTM variation

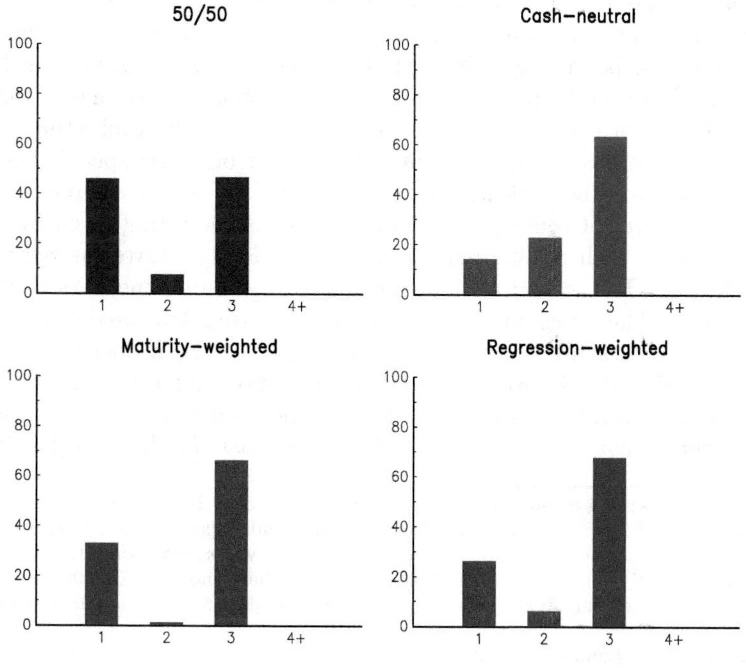

FIGURE 4.13: Risk factor contributions of the barbell portfolios

developed countries were considered safe assets. The management of such assets was then reduced to the management of the yield, the duration and the rating of the bond portfolio. With the recent development of the European debt crisis, portfolio managers have rediscovered sovereign credit risk[31]. Different models have been developed to manage this risk, but most of them do not have strong mathematical and financial foundations[32]. Indeed, in order to manage the credit risk of sovereign bond portfolios, the first step is to measure the risk precisely. The risk budgeting approach described below, which was first proposed by Bruder *et al.* (2011), answers this question.

Remark 55 *The application is based on the European Government Bond Index (EGBI) of Citigroup. This index is composed of the sovereign bonds of eleven countries belonging to the eurozone: Austria, Belgium, Finland, France, Germany, Greece, Ireland, Italy, Netherlands, Portugal and Spain*[33]. *The study period begins in January 2008 and ends in June 2012.*

4.3.2.1 Measuring the credit risk of sovereign bond portfolios

The volatility of prices return, a measure traditionally used for equity risk, cannot be applied to sovereign bonds. Instead of measuring the specific risk of a sovereign country, this volatility will also reflect global movements in interest rate yield curves. Even if we consider a portfolio of sovereign bonds following a single yield curve, the underlying bonds will react differently to interest rate movements, due to their different sensitivities (duration and convexity). For instance, low duration bonds will consistently display lower volatilities than longer duration bonds, regardless of their credit risk. The same conclusion applies when price returns are used to measure bonds' co-movements through a computation of correlations. As our interest lies in discriminating among countries on the basis of their credit risk, and not on issuer-specific features, we must turn to other risk measures. In Table 4.11, we have reported some popular measures of country risk. The first one is the rating given by credit rating agencies such as Standard and Poor's, Moody's Investors Service or Fitch Ratings. The second one corresponds to the Euromoney Country Risk (ECR) index. These two measures are very interesting, but are difficult to use to define a numerical risk measure. For example, the ECR does not measure exactly the default of sovereign debt. Ratings are better for that purpose, but we observe a relative inertia in revising these ratings. Moreover, there is some evidence that rating movements lag market-based indicators (Di Cesare,

[31] However, sovereign default risk has a long history. Recent debt crises, such as those in 1982 (Mexico and Latin America), 1997-1998 (Russia and Asia) and 2001 (Argentina and Turkey), have been concentrated on emerging markets. However, many decades have passed since a 'developed' country defaulted. According to Reinhart and Rogoff (2009), there have been at least 250 sovereign external defaults and 68 defaults on domestic public debt since 1800.

[32] They are more bond picking models.

[33] The ISO codes of these countries are AT, BE, FI, FR, DE, GR, IE, IT, NL, PT and ES.

TABLE 4.11: Some measures of country risk (October 2011)

Country	Rating	ECR		CDS Spread	
		Score	Rank	01/09/11	04/10/11
Austria	AAA	84.01	13	123	186
Belgium	AA+	77.81	19	249	309
Finland	AAA	86.96	8	64	85
France	AAA	80.90	16	163	201
Germany	AAA	84.98	11	76	122
Greece	CCC	52.38	65	2291	5736
Ireland	BBB+	62.33	43	781	726
Italy	A	71.20	30	384	487
Netherlands	AAA	86.67	9	80	117
Portugal	BBB−	61.35	44	957	1167
Spain	AA	66.71	36	376	391
Norway	AAA	93.44	1	44	52
Switzerland	AAA	90.31	3	58	79
Denmark	AAA	89.21	4	100	153
Sweden	AAA	88.74	5	54	66
United Kingdom	AAA	80.22	17	76	102
United States	AA+	82.10	15	52	52
Japan	AA−	74.66	25	102	155

2006). In Table 4.11, we have also reported the credit default swap (or CDS) spreads (in bps) of several countries for September 1 and October 4, 2011. For example, the downgrades of Spain and Italy during the first two weeks of October seem to have been anticipated by the CDS market. More generally, market-based indicators are preferable to measure the credit risk of sovereign bond portfolios. For instance, bond yield spreads (or asset swap spreads) are another good measure, because they act as an incremental return, compensating the investor for the risk associated with the issuer. Ceteris paribus, the higher the yield spread, the higher the credit risk. The difficulty is that this information is not public; indeed only a handful of investment banks have access to it. An alternative to using bond yield spreads is to consider CDS spreads, which correspond to the value investors are willing to pay to insure against an issuer's default risk[34]. They therefore allow us to isolate the sovereign credit risk of an issuer, regardless of its currency, yield curve or maturity[35].

Let $\mathfrak{s}_i(t)$ be the credit spread of the ith country. We assume that it follows

[34]One drawback of this measure is that the CDS market is less liquid than the bond market.

[35]This approach is frequently used by recent studies on this subject (Longstaff *et al.*, 2011).

a general diffusion process[36]:

$$\mathrm{d}\mathfrak{s}_i(t) = \sigma_i^{\mathfrak{s}} \mathfrak{s}_i(t)^{\beta_i}\, \mathrm{d}W_i(t)$$

where $W_i(t)$ is a standard Brownian motion and $\sigma_i^{\mathfrak{s}}$ is a volatility parameter. Moreover, we assume that the Brownian motions are correlated, and we note $\rho_{i,j} = \mathbb{E}[W_i(t)\, W_j(t)]$ the cross-correlation between $W_i(t)$ and $W_j(t)$. The exponent β_i defines the shape of the distribution:

1. If $\beta_i = 0$, we assume that the credit spread is Gaussian and the risk factor corresponds to the absolute variation of the credit spread.

2. If $\beta_i = 1$, we assume that the credit spread is log-normal and the risk factor becomes the relative variation of the credit spread.

Using historical observations of CDS spreads, we can calibrate the parameter β_i by the maximum likelihood method[37]. The results are reported in Table 4.12. We reject the hypothesis $\mathcal{H}_0 : \beta_i = 0$ for all the countries at the confidence level 99%. We also note that the estimates vary within a 0.78/1.22 range and the average is close to one[38]. This is why we consider that the spreads follow a log-normal diffusion, which implies that spreads vary proportionally to their absolute levels. In view of CDS spread data over the last few years, this assumption seems reasonable.

TABLE 4.12: ML estimate of the parameter β_i (Jan. 2008 – Jun. 2012)

Country	AT	BE	FI	FR	DE	GR
$\hat{\beta}_i$	0.953	0.969	0.780	0.806	0.853	1.219
$\hat{\sigma}(\hat{\beta}_i)$ (in %)	1.015	1.410	1.415	1.085	1.638	0.462
Country	IE	IT	NL	PT	ES	AC
$\hat{\beta}_i$	0.786	1.033	0.790	0.911	1.042	0.922
$\hat{\sigma}(\hat{\beta}_i)$ (in %)	0.633	1.697	0.870	0.950	1.776	1.178

Moreover, the assumption $\mathcal{H}_1 : \beta_i = 1$ simplifies the estimation of the

[36]Such a process has already been used in the SABR model (Hagan *et al.*, 2002).

[37]We assume that we observe spreads at some known dates $t_0, \ldots, t_m$. Let $\mathfrak{s}_i(t_j)$ be the observed spread for the ith country at date t_j. The log-likelihood function for the ith country is given by the following formula:

$$\begin{aligned}\ell(\sigma_i^{\mathfrak{s}}, \beta_i) &= -\frac{m}{2}\ln(2\pi) - m\ln\sigma_i^{\mathfrak{s}} - \frac{1}{2}\sum_{j=1}^{m}\ln(t_j - t_{j-1}) - \\ &\quad \beta_i\sum_{j=1}^{m}\ln\mathfrak{s}_i(t_{j-1}) - \frac{1}{2}\sum_{j=1}^{m}\frac{(\mathfrak{s}_i(t_j) - \mathfrak{s}_i(t_{j-1}))^2}{(t_j - t_{j-1})\left(\sigma_i^{\mathfrak{s}}\mathfrak{s}_i^{\beta_i}(t_{j-1})\right)^2}\end{aligned}$$

We estimate the coefficient β_i by maximizing the concentrated log-likelihood function.

[38]The column AC corresponds to the average of all countries.

TABLE 4.13: Spread $\mathfrak{s}_i(t)$ (in bps)

Country	Jan. 08	Jan. 09	Jan. 10	Jan. 11	Jan. 12
Austria	6	128	85	100	190
Belgium	12	83	54	218	316
Finland	6	57	27	33	78
France	7	54	32	101	222
Germany	6	45	27	58	104
Greece	22	228	282	1074	8786
Ireland	15	177	160	615	726
Italy	21	165	108	238	503
Netherlands	6	83	32	63	122
Portugal	18	92	91	500	1093
Spain	20	103	113	350	394
Median	12	92	85	218	316

parameters $\sigma_i^{\mathfrak{s}}$ and $\rho_{i,j}$. Indeed, they may be estimated using the empirical covariance matrix of the CDS spread relative variations. In Tables 4.13 and 4.14, we report the value of the spread and the estimated volatility $\sigma_i^{\mathfrak{s}}$ at the beginning of several years. We note the high increase of the CDS spread level. At the beginning of 2008, the median spread was equal to 12 bps. Five years later, it had multiplied by 25. Concerning the spread volatility, it moved more slowly than the spread level. Indeed, the mean fluctuated between 52% and 93%. More interesting is the dynamics of the correlation. In Figure 4.14, we have represented the average correlation for different groups of countries. We observe some interesting facts. First, the default of Lehman Brothers shifted the global correlation in September 2008[39] and it increased almost continuously until mid-2010. Second, we also observe that the correlation between GIIPS[40] reached its maximum in October 2010, and has been decreasing since this date. Third, the correlation between safe countries (Finland, Germany and the Netherlands) has increased recently illustrating a period of 'flight to quality' (Ilmanen, 2003).

Let us now define the credit risk measure of a bond. We note $B_i(t, D_i)$ the zero-coupon bond price with maturity (or duration) D_i of the country i. If we assume that the recovery rate is zero, we have:

$$\mathrm{d}\ln B_i(t, D_i) = -D_i\,\mathrm{d}r_t - D_i\,\mathrm{d}\mathfrak{s}_i(t)$$

with r_t the risk-free interest rate. If we assume that the credit spread is not correlated with the risk-free interest rate, we deduce that:

$$\sigma^2(\mathrm{d}\ln B_i(t, D_i)) = D_i^2\sigma^2(\mathrm{d}r_t) + D_i^2\sigma^2(\mathrm{d}\mathfrak{s}_i(t))$$

[39]This is particularly obvious for the 'safe' countries such as Finland, Germany and the Netherlands.

[40]GIIPS stands for Greece, Ireland, Italy, Portugal, and Spain.

TABLE 4.14: Estimated values of the volatility σ_i^s (in %)

Country	Jan. 08	Jan. 09	Jan. 10	Jan. 11	Jan. 12
Austria	56.6	96.0	72.4	76.3	69.7
Belgium	65.8	70.2	83.4	73.2	74.1
Finland	103.9	107.6	80.9	61.7	66.7
France	50.2	92.5	97.1	77.5	68.5
Germany	69.2	96.8	76.2	72.3	65.0
Greece	60.4	57.3	64.4	89.3	85.2
Ireland	76.5	97.4	63.2	78.1	52.1
Italy	48.8	65.5	62.8	90.6	74.3
Netherlands	81.7	108.6	78.3	61.2	66.5
Portugal	56.6	64.4	84.2	106.6	54.8
Spain	67.7	63.7	69.2	92.0	72.8
Mean	67.0	83.6	75.6	79.9	68.1

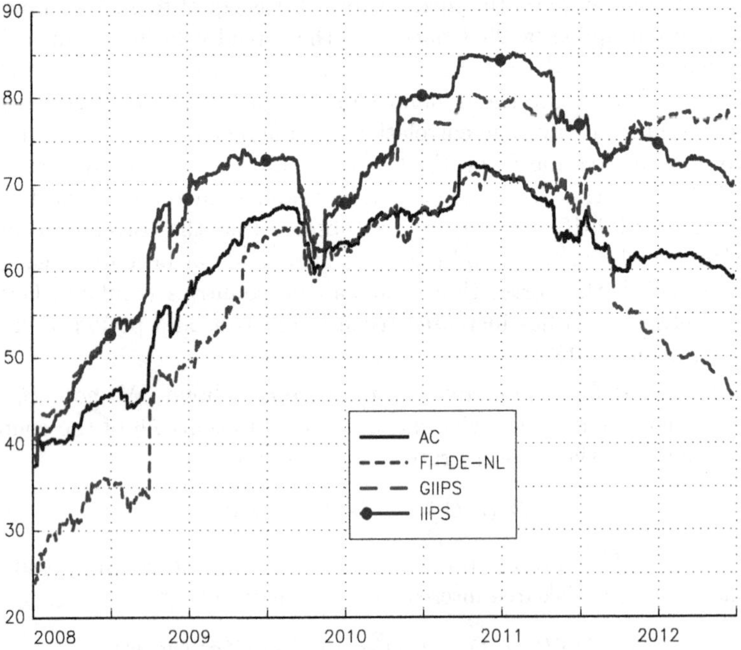

FIGURE 4.14: Average correlation of credit spreads (in %)

The risk of the defaultable bond may be decomposed into two components: an interest-rate risk component and a credit risk component. Because we would like to manage this second risk, we define the credit risk measure for one bond as follows:

$$\begin{aligned} \mathcal{R}(B_i) &= \sqrt{D_i^2 \sigma^2 \left(\mathrm{d}\mathfrak{s}_i(t)\right)} \\ &= D_i \sigma_i^{\mathfrak{s}} \mathfrak{s}_i(t)\,\mathrm{d}t \\ &\propto D_i \sigma_i^{\mathfrak{s}} \mathfrak{s}_i(t) \end{aligned}$$

The credit risk measure $\sigma_i^{\mathfrak{c}} = D_i \sigma_i^{\mathfrak{s}} \mathfrak{s}_i(t)$ of the bond can then be estimated as the product of its duration D_i, the spread volatility $\sigma_i^{\mathfrak{s}}$ and the spread level $\mathfrak{s}_i(t)$. In the case of a bond portfolio with weights $(x_1, \ldots, x_n)$, we obtain[41]:

$$\begin{aligned} \mathcal{R}(x) &= \sigma\left(\sum_{i=1}^{n} -x_i D_i\,\mathrm{d}\mathfrak{s}_i(t)\right) \\ &= \sqrt{\sum_{i=1}^{n}\sum_{j=1}^{n} x_i x_j D_i D_j \sigma_i^{\mathfrak{s}} \sigma_j^{\mathfrak{s}} \mathfrak{s}_i(t)\,\mathfrak{s}_j(t)\,\rho_{i,j}\,\mathrm{d}t} \\ &\propto \sqrt{x^\top \Sigma x} \end{aligned}$$

with $\Sigma_{i,j} = \rho_{i,j}\sigma_i^{\mathfrak{c}}\sigma_j^{\mathfrak{c}}$. We note that $\mathcal{R}(x) = \sqrt{x^\top \Sigma x}$ may be interpreted as the integrated volatility of the CDS basket which would hedge the credit risk of the bond portfolio. $\mathcal{R}(x)$ is then the credit risk measure of the bond portfolio. It has two appealing properties:

1. It is based on few parameters: two 'portfolio' parameters x_i and D_i, and three 'market-based' parameters $\mathfrak{s}_i(t)$, $\sigma_i^{\mathfrak{s}}$ and $\rho_{i,j}$.

2. It satisfies the Euler allocation principle, because it is a volatility risk measure.

Remark 56 *The previous analysis has been developed for portfolios of zero-coupon bonds. We consider that it remains valid for bonds with coupons.*

Example 38 *We consider a portfolio of four bonds:*

- *One German bond (value = 12 M€, duration = 8.2 years);*
- *One French bond (value = 15 M€, duration = 7.1 years);*
- *Two Italian bonds (value = 16 and 8 M€, duration = 6.5 and 5.9 years).*

The value of the bonds corresponds to the mark-to-market price at March 1, 2012. At this date, the market-based parameters to characterize the credit risk are those given in Table 4.15.

TABLE 4.15: Market-based parameters (March 1, 2012)

Country	$\mathfrak{s}_i(t)$	$\sigma_i^{\mathfrak{s}}$	$\rho_{i,j}$		
Germany	76 bps	66.0%	1.00		
France	166 bps	70.9%	0.86	1.00	
Italy	356 bps	74.2%	0.73	0.80	1.00

TABLE 4.16: Computing the credit risk measure $\sigma_i^{\mathfrak{c}}$ for one bond

Bond	Country	B_i	D_i	$\sigma_i^{\mathfrak{c}}$
1	Germany	12	8.2	4.11%
2	France	15	7.1	8.36%
3	Italy	16	6.5	17.17%
4	Italy	8	5.9	15.58%

First, we compute the credit risk measure of individual bonds. The results are reported in Table 4.16. For instance, $\sigma_i^{\mathfrak{c}}$ takes the value 4.11% for the German bond[42]. If we consider a portfolio with the first three bonds, we obtain a credit risk measure equal to 9.84% (see Table 4.17). In this portfolio, the German bond represents 27.91% of the weight in the portfolio, but only 9.85% of the credit risk. The case with the four bonds requires special attention, because of the specification of the correlation matrix ρ. It is equal to:

$$\rho = \begin{pmatrix} 1.00 & & & \\ 0.86 & 1.00 & & \\ 0.73 & 0.80 & 1.00 & \\ 0.73 & 0.80 & 1.00 & 1.00 \end{pmatrix}$$

Indeed, the correlation between the two Italian bonds is equal to one, because they are exposed to the same credit spread. We then obtain the results given in Table 4.18.

TABLE 4.17: Credit risk measure of the portfolio with three bonds

Bond	x_i	$\mathcal{MR}_i$	$\mathcal{RC}_i$	$\mathcal{RC}_i^{\star}$
1	27.91	3.48	0.97	9.85
2	34.88	7.65	2.67	27.12
3	37.21	16.67	6.20	63.03
Risk measure			9.84	

[41] We omit dt, which is a constant term.

[42] We have:

$$\sigma_i^{\mathfrak{c}} = 8.2 \times 0.66 \times 0.0076 = 0.04113$$

TABLE 4.18: Credit risk measure of the portfolio with four bonds

Bond	x_i	$\mathcal{MR}_i$	$\mathcal{RC}_i$	$\mathcal{RC}_i^\star$
1	23.53	3.39	0.80	7.45
2	29.41	7.47	2.20	20.55
3	31.37	16.87	5.29	49.52
4	15.69	15.31	2.40	22.47
Risk measure			10.69	

TABLE 4.19: Credit risk measure of the portfolio with the Italian meta-bond

Bond	x_i	$\mathcal{MR}_i$	$\mathcal{RC}_i$	$\mathcal{RC}_i^\star$
1	23.53	3.39	0.80	7.45
2	29.41	7.47	2.20	20.55
$3'$	47.06	16.35	7.70	71.99
Risk measure			10.69	

If there are several bonds of the same country, we can show that the credit risk measure may be computed by grouping the exposures and averaging the duration with respect to the weights. For example, the Italian exposure is 24 M€ and the weighted mean of durations is 6.3 years. Using this meta-bond $3'$, we obtain the results reported in Table 4.19. We verify that they are equivalent to those in Table 4.18.

Let us now apply the previous credit risk measurement to the Citigroup EGBI index. In Table 4.20, we report the weights and risk contributions for the constituents of the EGBI at some dates. In January 2008, Italy represented 22.6% of the eurozone debt, but 42.1% of the risk of the EGBI portfolio. At the same time, the weight of Germany was 24.3% but its risk contribution was only 12.3%. The credit risk measure was very low and was equal to 30 bps. Four years later, we obtain a completely different situation. Indeed, it is equal to 10.7% in January 2012. To better understand these changes, we show the evolution of the credit risk contribution of each country since the beginning of 2008 in Figure 4.15. We first note the impact of the Lehman Brothers default on the risk contribution of safe countries (Austria, France, Germany, etc.). We also observe the increase of Greece's risk contribution since 2008. On April 27, 2010, the country's sovereign debt rating was cut to BB+ by Standard & Poor's. Therefore, Greece lost its investment grade status and exited the EGBI index at its next rebalancing date[43]. After July 2010, there is a significant increase in the credit risk of Portugal, Ireland, Italy and Spain, but it decreases recently for Ireland and Portugal. Since 2011, we also observe the rise of the French risk.

[43]The risk contribution of Greece reached a maximum of 28.5% on April 30, 2010. At the end of June 2010, just before its exit from the index, it remained at the very high level of 26%.

TABLE 4.20: Weights and risk contribution of the EGBI portfolio (in %)

Country	Jan. 08		Jan. 10		Jan. 11		Jan. 12	
	x_i	$\mathcal{RC}_i^\star$	x_i	$\mathcal{RC}_i^\star$	x_i	$\mathcal{RC}_i^\star$	x_i	$\mathcal{RC}_i^\star$
Austria	3.9	1.7	3.8	4.5	4.0	1.9	4.2	2.8
Belgium	6.3	6.7	6.1	4.8	6.3	6.1	6.2	7.5
Finland	1.3	0.4	1.2	0.3	1.3	0.1	1.5	0.3
France	19.9	10.4	20.2	9.6	22.1	11.7	23.5	19.6
Germany	24.3	12.3	21.6	7.2	22.9	5.8	23.4	8.0
Greece	5.2	8.5	5.0	15.6	0.0	0.0	0.0	0.0
Ireland	1.0	1.0	1.9	3.0	2.1	6.2	1.7	2.2
Italy	22.6	42.1	23.1	35.2	23.4	38.3	20.8	40.3
Netherlands	5.5	1.8	5.3	2.1	6.1	1.4	6.5	2.6
Portugal	2.2	2.7	2.4	2.8	2.1	7.4	1.5	2.6
Spain	7.8	12.4	9.5	14.9	9.6	21.1	10.7	14.0
$\mathcal{R}(x)$	0.3		2.8		8.3		10.7	

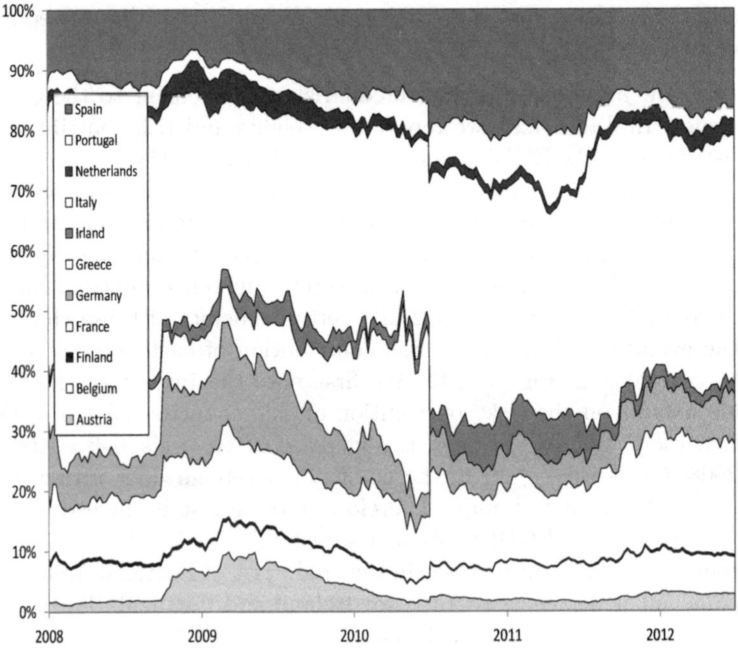

FIGURE 4.15: Dynamics of the risk contributions (EGBI portfolio)

4.3.2.2 Comparing debt-weighted, gdp-weighted and risk-based indexations

As for equities, it is possible to define different alternative-weighted schemes for bonds. The capitalization-weighted indexation corresponds to the debt-weighted method. Let DEBT_i be the mark-to-market debt of the ith country. The weights x_i are then defined as follows:

$$x_i = \frac{\mathrm{DEBT}_i}{\sum_{j=1}^{n} \mathrm{DEBT}_j}$$

For instance, the EGBI index uses this approach with an investment grade constraint. Fundamental indexation is more difficult to define in the case of a bond portfolio. The rationale is to link the weight of one country with its capacity to service the debt. For example, Toloui (2010) suggests using the gross domestic product (or GDP) as a criterion[44]:

> "*An alternative to market-cap weighting is to weight country exposures in global bond indexes by GDP instead of market capitalization. [...] In contrast to market-cap weighting, GDP weighting does not reward countries with high levels of debt issuance: Countries with higher debt-to-GDP levels generally have a lower representation in a GDP-weighted index versus a market-cap-weighted index.*"

In this case, we have:

$$x_i = \frac{\mathrm{GDP}_i}{\sum_{j=1}^{n} \mathrm{GDP}_j}$$

where GDP_i is the gdp of the ith country. More concretely, if we consider the eleven national subindices of the Citigroup EGBI, the gdp-weighted approach consists in fixing at each rebalancing date[45] the weight of the ith subindex in the portfolio as equal to the share of its gdp in the total gdp of the eleven countries. Each subindex is then treated as if it were a bond. This approach has the advantage of keeping the same duration and interest rate risks as in the original index.

Debt-weighted (DEBT-WB) and gdp-weighed (GDP-WB) methods are then special cases of the weight budgeting (WB) approach[46]. Using the EGBI subindices, we have simulated these two schemes and some results are given in Tables 4.21 and 4.22. For the debt-weighted portfolio, we obtain the same results as the EGBI portfolio before the exit of Greece in July 2010. However, keeping Greek bonds in the portfolio increases the sovereign risk measure by

[44] More sophisticated approaches of fundamental indexing are available, but they are more difficult to implement (Arnott *et al.*, 2010; Brodsky *et al.*, 2011).

[45] For all the simulations presented in this section, we rebalance the portfolio on a monthly basis.

[46] This general approach is presented on page 130.

TABLE 4.21: Weights and risk contribution of the DEBT-WB indexation (in %)

Country	Jan. 08		Jan. 10		Jan. 11		Jan. 12	
	x_i	$\mathcal{RC}_i^\star$	x_i	$\mathcal{RC}_i^\star$	x_i	$\mathcal{RC}_i^\star$	x_i	$\mathcal{RC}_i^\star$
Austria	3.9	1.7	3.8	4.5	3.9	1.6	4.2	2.6
Belgium	6.3	6.7	6.1	4.8	6.0	5.0	6.1	7.1
Finland	1.3	0.4	1.2	0.3	1.2	0.1	1.5	0.3
France	19.9	10.4	20.2	9.6	21.2	9.4	23.3	18.4
Germany	24.3	12.3	21.6	7.2	21.9	4.7	23.2	7.5
Greece	5.2	8.5	5.0	15.6	4.3	19.2	1.0	5.3
Ireland	1.0	1.0	1.9	3.0	2.0	5.1	1.7	2.1
Italy	22.6	42.1	23.1	35.2	22.4	30.6	20.6	38.3
Netherlands	5.5	1.8	5.3	2.1	5.9	1.1	6.5	2.5
Portugal	2.2	2.7	2.4	2.8	2.0	6.1	1.5	2.5
Spain	7.8	12.4	9.5	14.9	9.2	17.0	10.6	13.3
$\mathcal{R}(x)$	0.3		2.8		9.7		11.1	

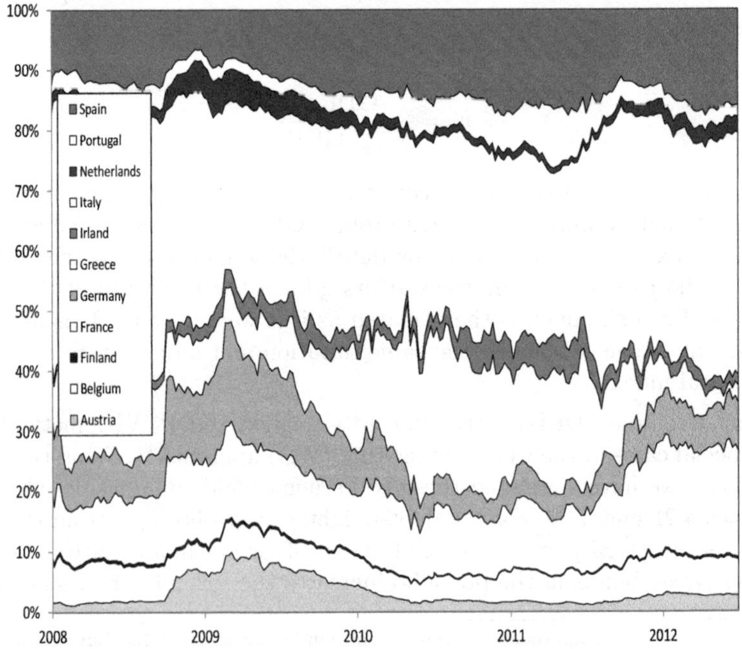

FIGURE 4.16: Dynamics of the risk contributions (DEBT-WB indexation)

TABLE 4.22: Weights and risk contribution of the GDP-WB indexation (in %)

Country	Jan. 08		Jan. 10		Jan. 11		Jan. 12	
	x_i	$\mathcal{RC}_i^\star$	x_i	$\mathcal{RC}_i^\star$	x_i	$\mathcal{RC}_i^\star$	x_i	$\mathcal{RC}_i^\star$
Austria	3.1	1.4	3.1	4.2	3.2	1.5	3.3	1.9
Belgium	3.8	4.3	3.9	3.5	4.0	3.7	4.0	4.3
Finland	2.0	0.7	2.0	0.5	2.1	0.2	2.1	0.4
France	21.3	11.8	21.5	12.1	21.5	11.0	21.7	15.6
Germany	27.4	15.1	27.5	10.8	27.8	6.9	28.1	8.3
Greece	2.6	4.2	2.7	9.1	2.5	12.2	2.3	19.9
Ireland	2.1	2.3	1.8	3.2	1.6	4.7	1.7	2.1
Italy	17.4	32.5	17.2	29.5	17.1	26.7	17.0	29.0
Netherlands	6.5	2.4	6.5	2.9	6.6	1.5	6.5	2.3
Portugal	1.9	2.4	1.9	2.6	1.9	6.5	1.8	3.0
Spain	12.0	22.9	11.9	21.5	11.8	25.2	11.5	13.3
$\mathcal{R}(x)$	0.3		2.5		8.5		11.3	

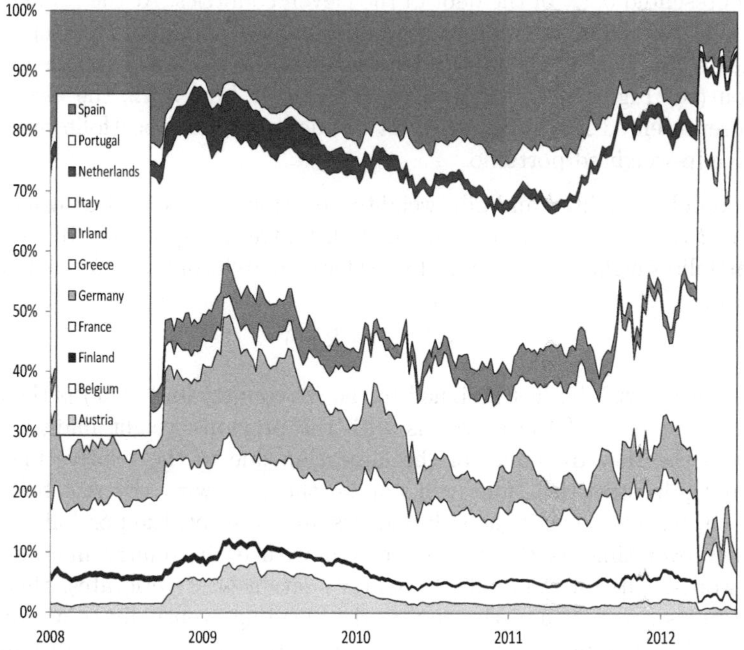

FIGURE 4.17: Dynamics of the risk contributions (GDP-WB indexation)

1% in January 2012. Under gdp-weighted indexation, Greece has a smaller allocation than under debt-weighted indexation. For example, in January 2008, the weight of Greece was 5.2% by market capitalization (Table 4.21) but only 2.6% under gdp-based indexation (Table 4.21). We observe the same phenomenon for Italy and Portugal. However, we note that this type of indexation may also increase the weight of some risky countries. For instance, the weight of Spain is bigger with gdp-weighted indexation than with debt-weighted indexation because the debt-to-gdp ratio is lower than average for this country. We remark that gdp-based indexation produces a more balanced portfolio in terms of risk than debt-based indexation, but that risk is still concentrated in a few countries. Another disappointing result is that the two methods produce similar credit risk measures. For example, it is equal to 11.3% for the gdp-weighted indexation in January 2012 which is close to the 11.1% and 10.7% obtained with the debt-weighted and EGBI portfolios. The changes of risk contributions are reported in Figures 4.16 and Figure 4.17. We note that their dynamics are very similar until September 2011. This is due to the relative inertia of debt and gdp measures. But we observe a divergence between the two indexations since this date. This is because the mark-to-market of the Greek debt has been highly reduced during this period, whereas the Greek gdp has been almost stable. Indeed, the mark-to-market debt was divided by a factor of 5 between January 2012 and June 2012. At the end of June 2012, it only represented 0.2% of the debt of the eleven countries. At the same period, the weight was equal to 2.3% in the gdp-weighted portfolio. By keeping this significant exposure to the Greek debt, most of the risk was explained by this position (see Figure 4.17). Indeed, at the end of June 2012, the Greek allocation represented 2.9% of the risk for the debt-weighted portfolio but 72.6% for the gdp-weighted portfolio.

Both debt-weighted and gdp-weighted portfolios present clear drawbacks. Bruder *et al.* (2011) suggest using the risk budgeting approach. In this case, the portfolio weights are optimized in order to match some given risk budgets. We have:

$$\mathcal{RC}_i = b_i \cdot \mathcal{R}(x)$$

where b_i is the risk budget assigned to the ith country and $\mathcal{R}(x)$ is the credit risk measure. One of the problems with the previous weight budgeting approach is the slow dynamics of the allocation due to the relative inertia of debt and gdp statistics. This will not be the case with the risk budgeting approach, because even if the risk budgets are constant, the portfolio weights still vary over time as the risk associated with each country fluctuates. In other words, when an individual country's situation deteriorates, this is reflected in its spread[47] and thereby its risk, leading eventually to a reduction of its weight. Risk budgeting thus allows investors to control the distribution of risk over time. However, it does not address the problem of how to choose

[47] For instance, the level or the volatility of the spread may go up.

the right risk budgets. We could consider an ERC portfolio, but this indexation ignores liquidity issues. Moreover, having the same risk budget for large and small countries does not make sense. Consequently, Bruder *et al.* (2011) propose to set the risk budget of each country with respect to their economic size. Indeed, if we focus on debt, the risk budget b_i is then proportional to the debt:

$$b_i = \frac{\text{DEBT}_i}{\sum_{j=1}^{n} \text{DEBT}_j}$$

This approach defines the debt risk-based indexation (DEBT-RB). In the same way, the gdp risk-based indexation (GDP-RB) is built using the following risk budgets:

$$b_i = \frac{\text{GDP}_i}{\sum_{j=1}^{n} \text{GDP}_j}$$

The results are reported in Tables 4.23 and 4.24. Contrary to the WB indexation, we first define the risk budgets in order to calibrate the weights. For example, the risk budget of Italy was 22.6% in January 2008 for the DEBT-RB indexation, because Italy represented 22.6% of the debt at this date. Therefore, the risk budgets b_i given in Table 4.23 (resp. Table 4.24) correspond to weights x_i given in Table 4.21 (resp. Table 4.22). We note that we obtain portfolios that are totally different than those obtained with the weight budgeting approach, because of the impact of the sovereign credit risk. Indeed, the DEBT-RB weight of Italy was 10.0% in January 2008, compared to the 22.6% of the EGBI and DEBT-WB portfolios. Another example is the weight of Germany (or other safe countries such as Finland and the Netherlands), which is higher in the risk-budgeting portfolios than in the weight-budgeting portfolios. We also note that the credit risk measure decreases when using the risk parity approach. Moreover, we observe that the portfolio allocation varies a lot over time, depending on the sovereign risk priced by the market (see also Figures 4.18 and 4.19). The DEBT-RB portfolio is further away from the DEBT-WB portfolio in periods of risk aversion, such as November 2011, whereas it is closer when sovereign risk concerns are relatively low, as in July 2009. For example, in January 2008, the Greek debt represented 2.5% of the DEBT-RB allocation in comparison to the residual weight since 2010. The case of Spain is also interesting. Its weight increased at the beginning of 2010, then decreased until mid-2011 and finally recovered at the end of 2011. If we compare DEBT-RB and GDP-RB allocations, we note that they are different from a static point of view, but their dynamics are very similar.

TABLE 4.23: Risk budgets and weights of the DEBT-RB indexation (in %)

Country	Jan. 08 b_i	Jan. 08 x_i	Jan. 10 b_i	Jan. 10 x_i	Jan. 11 b_i	Jan. 11 x_i	Jan. 12 b_i	Jan. 12 x_i
Austria	3.9	6.3	3.8	2.2	3.9	4.4	4.2	3.9
Belgium	6.3	4.2	6.1	5.1	6.0	3.3	6.1	3.6
Finland	1.3	2.6	1.2	3.1	1.2	5.5	1.5	5.3
France	19.9	26.1	20.2	24.5	21.2	19.8	23.3	19.3
Germany	24.3	31.6	21.6	38.5	21.9	43.4	23.2	42.7
Greece	5.2	2.5	5.0	1.1	4.3	0.5	1.0	0.2
Ireland	1.0	0.7	1.9	0.8	2.0	0.4	1.7	0.8
Italy	22.6	10.0	23.1	10.4	22.4	7.3	20.6	7.5
Netherlands	5.5	10.6	5.3	8.8	5.9	12.8	6.5	11.1
Portugal	2.2	1.4	2.4	1.3	2.0	0.3	1.5	0.5
Spain	7.8	3.9	9.5	4.1	9.2	2.4	10.6	5.1
$\mathcal{R}(x)$	0.2		1.8		4.4		7.3	

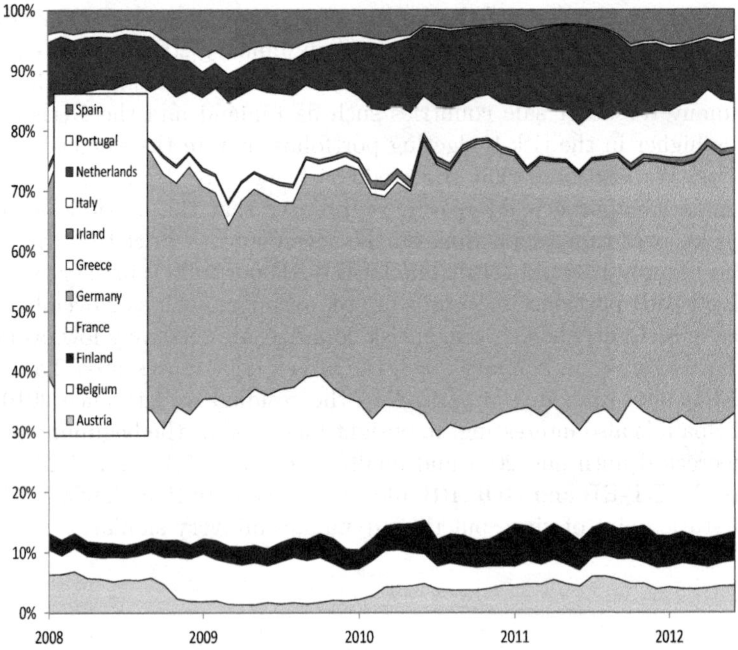

FIGURE 4.18: Evolution of the weights (DEBT-RB indexation)

TABLE 4.24: Risk budgets and weights of the GDP-RB indexation (in %)

Country	Jan. 08		Jan. 10		Jan. 11		Jan. 12	
	b_i	x_i	b_i	x_i	b_i	x_i	b_i	x_i
Austria	3.1	4.8	3.1	1.7	3.2	3.2	3.3	2.9
Belgium	3.8	2.5	3.9	3.0	4.0	1.9	4.0	2.2
Finland	2.0	4.0	2.0	4.6	2.1	8.0	2.1	6.6
France	21.3	26.9	21.5	23.3	21.5	17.8	21.7	16.8
Germany	27.4	33.6	27.5	43.6	27.8	48.0	28.1	47.8
Greece	2.6	1.3	2.7	0.6	2.5	0.3	2.3	0.3
Ireland	2.1	1.4	1.8	0.7	1.6	0.3	1.7	0.8
Italy	17.4	7.7	17.2	7.3	17.1	5.1	17.0	5.9
Netherlands	6.5	11.2	6.5	9.7	6.6	12.6	6.5	10.5
Portugal	1.9	1.2	1.9	1.0	1.9	0.3	1.8	0.6
Spain	12.0	5.4	11.9	4.7	11.8	2.7	11.5	5.5
$\mathcal{R}(x)$	0.2		1.7		3.9		6.8	

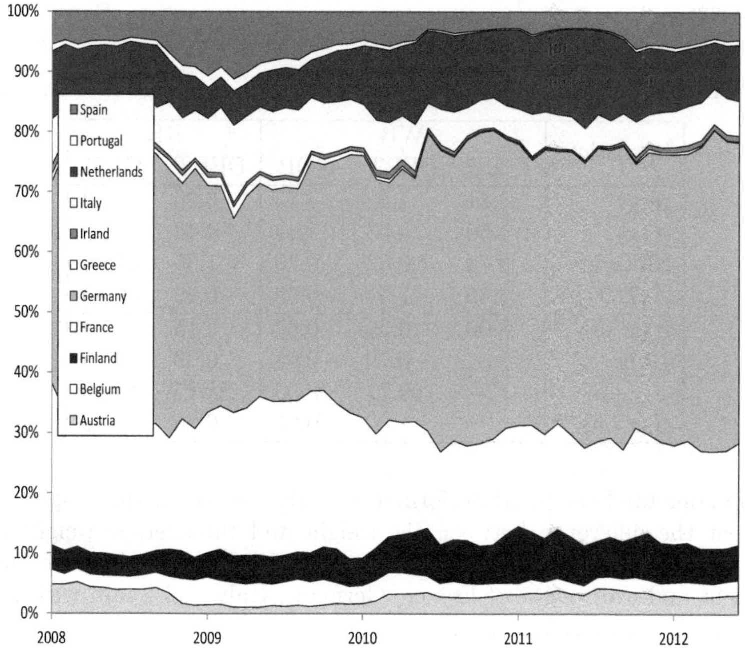

FIGURE 4.19: Evolution of the weights (GDP-RB indexation)

Let us now compare these four weighting methods. In Figure 4.20, we have reported the evolution of the credit risk measure. As expected, we find that risk-based indexations present lower sovereign risk. Indeed, Figure 4.20 is an illustration of the famous inequality[48], which states that the risk measure of the risk budgeting portfolio is always lower than the risk measure of the weight budgeting portfolio if we consider identical budgets b_i:

$$\mathcal{R}(x_{\mathrm{mr}}) \leq \mathcal{R}(x_{\mathrm{rb}}) \leq \mathcal{R}(x_{\mathrm{wb}})$$

More surprising is the fact that using debt or gdp has no real impact on the credit risk measure. We obtain the same conclusion if we consider the risk contribution of GIIPS countries. In Figure 4.21, we observe a substantial difference between the risk budgeting and weight budgeting approaches, but only a small difference between debt and gdp.

This general result is confirmed by the simulated performances. In Figure 4.22, we observe two groups. The first one is composed of EGBI, debt-weighted and gdp-weighted indexations. The second group is composed of risk-based indices. Hence, the main important choice is not between debt and gdp, but between weight and risk budgeting methods. In Table 4.25, we report the main statistics by considering the EGBI index as the benchmark[49]. With RB indexation, we obtain a better performance and also smaller volatility and drawdowns.

TABLE 4.25: Main statistics of bond indexations (Jan. 2008 – Jun. 2012)

Statistics	WB			RB	
	EGBI	DEBT	GDP	DEBT	GDP
$\mu(x)$	4.81	4.52	4.75	6.26	6.43
$\sigma(x)$	4.60	4.63	4.49	4.44	4.49
$\mathrm{SR}(x \mid r)$	0.70	0.63	0.70	1.05	1.07
$\mathcal{MDD}(x)$	−6.80	−7.94	−6.90	−6.29	−6.34
$\sigma(x \mid b)$	0.00	0.35	0.67	2.18	2.48
$\mathrm{IR}(x \mid b)$		−0.79	−0.08	0.63	0.62
$\rho(x \mid b)$		99.71	98.93	88.47	85.10
$\beta(x \mid b)$		1.00	0.96	0.85	0.83

To understand the good performance of the risk budgeting approach, we represent the difference between the weight and the average weight of the entire period for some countries in Figure 4.23. We can then see the underweight and overweight of France, Germany, Italy and Spain with respect

[48]See Equation (2.38) on page 131.

[49]They are the yearly return $\mu(x)$ (in %), the volatility $\sigma(x)$ (in %), the Sharpe ratio $\mathrm{SR}(x \mid r)$, the maximum drawdown $\mathcal{MDD}(x)$ (in %), the volatility of the tracking error $\sigma(x \mid b)$ (in %), the information ratio $\mathrm{IR}(x \mid b)$, the correlation with the benchmark $\rho(x \mid b)$ (in %) and the beta $\beta(x \mid b)$.

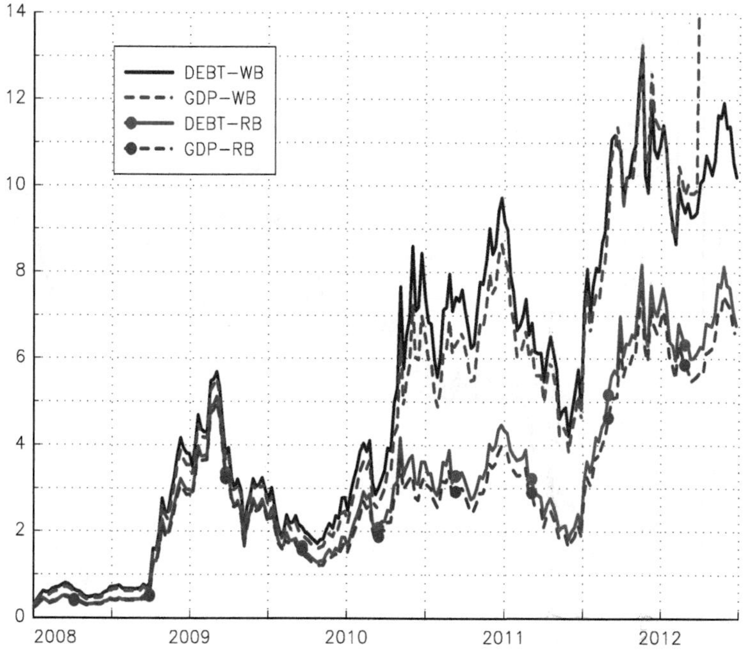

FIGURE 4.20: Dynamics of the credit risk measure (in %)

FIGURE 4.21: Evolution of the GIIPS risk contribution (in %)

FIGURE 4.22: Simulated performance of the bond indexations

to the average allocation. We note that the allocation is more dynamic for the RB approach than for the WB approach. Indeed, the yearly turnover is equal respectively to 11%, 2%, 89% and 85% for the DEBT-WB, GDP-WB, DEBT-RB and GDP-RB methods. In conclusion, the risk budgeting approach benefits from this dynamic process, which deleverages risky exposures when the sovereign risk rises, but takes on more risk when the latter decreases.

In Figure 4.24, we compare the performance of WB and RB indexations with respect to the 205 active funds available in the Morningstar database under the 'Bond Euro Government' category. More precisely, we report the different performance quantiles (10%, 50% and 90%) of these funds and also the performance of the EGBI and DEBT-RB portfolios. We verify the academic rule which states that the average performance of active management is equal to the performance of the benchmark minus the fees[50]. It is also interesting to note that risk-based indexations compare favorably with the top 10% active management funds. This result highlights the pertinence of the risk parity approach, which creates alpha by consistently managing the credit risk in a bond portfolio.

[50]We obtain an implied fees ratio equal to 57 bps per year using the median performance of mutual funds and the EGBI benchmark.

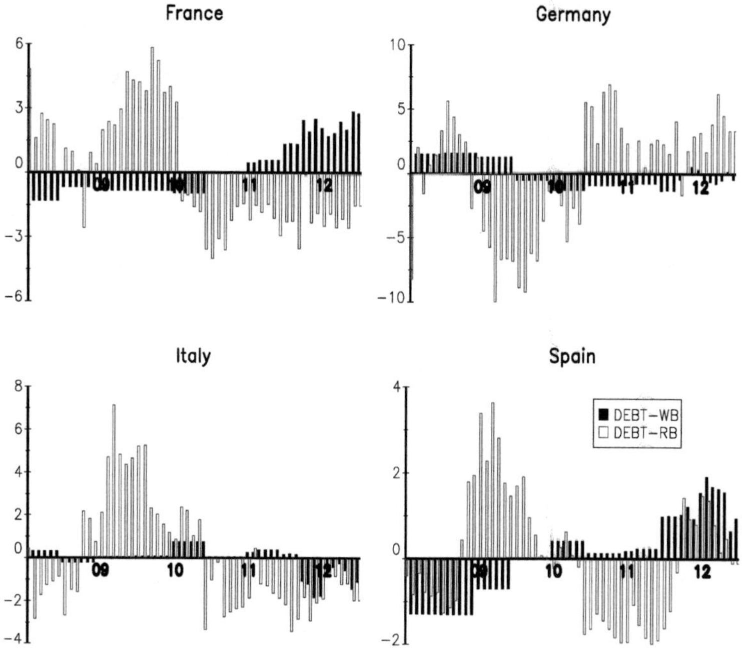

FIGURE 4.23: Comparing the dynamic allocation for four countries

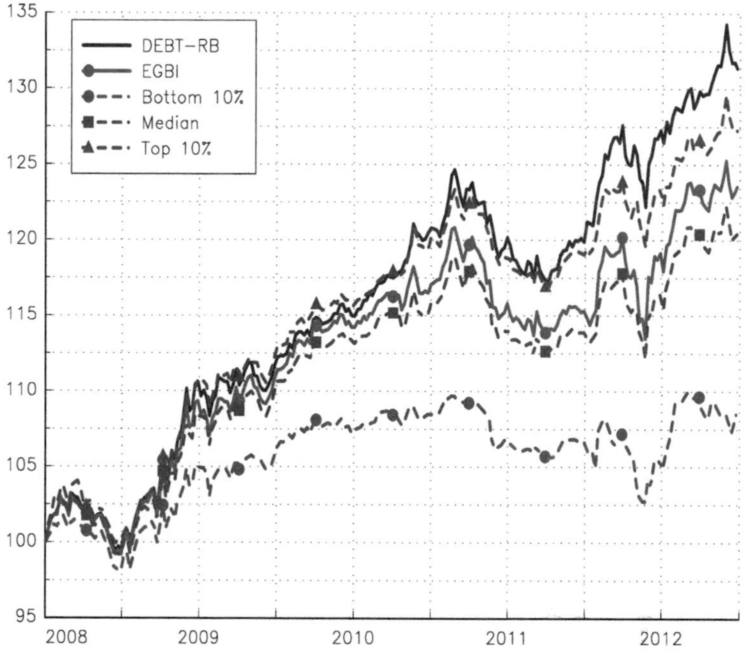

FIGURE 4.24: Comparison with active management

Chapter 5

Risk Parity Applied to Alternative Investments

There are strong links between risk parity and alternative investments. Indeed, hedge funds managers were pioneers in the use of risk parity techniques. For instance, Bridgewater's All Weather fund is generally considered the first risk parity fund. Actually, risk parity allowed some hedge funds, such as AQR Capital Management, to be considered as traditional asset managers. These two examples relate to risk parity strategies as a new form of global macro diversified funds. However, the use of risk parity in alternative investments is not restricted to such a strategy. Thus, CTA and equity market neutral hedge funds have used risk budgeting to size their bets for a very long time[1].

The purpose of this chapter is to show some applications of risk budgeting techniques in the field of alternative investments. It is also an opportunity to consider other risk measures than the volatility, which is not well adapted to these assets.

The first section is dedicated to commodities. We show how to design a long-only commodity portfolio. In particular, we introduce the concept of diversification return, which is a key element to understand the 'risk premium' of commodity futures. This diversification return is generally captured by the equally weighted portfolio, but we show that the ERC portfolio is another candidate to build an allocation that can capture this excess return. Such portfolios can be used as a form of diversified exposure to the commodity markets, but they can also be used as a neutral portfolio of dynamic strategies.

The second section focuses on hedge funds. After showing why risk parity has been extensively used by equity market neutral strategies, we consider the allocation problem between hedge fund strategies. For that, we compare ERC portfolios with RFP portfolios by using non-Gaussian risk measures. The allocation problem is particularly interesting, because the liquidity of these underlyings is much lower than the liquidity observed in equity and bond markets. Liquidity management then implies adding some constraints on the turnover and we will see how to do so in practice.

[1]For instance, the Turtle Trading System, which is certainly one of the most famous trading models and has fascinated many professional and non-professional traders, used risk parity to allocate between the bets (see Chapter 4 of the document "*The Original Turtle Trading Rules*" available on the web site `http://www.dailystocks.com/turtlerules.pdf`).

5.1 Case of commodities

5.1.1 Why investing in commodities is different

Before addressing the issue of risk parity allocation of commodities, we recall some important specific features of this asset class.

5.1.1.1 Commodity futures markets

Contrary to traditional assets, commodities are real assets and not financial assets. Investing in commodities is more difficult, because it requires large infrastructure in terms of storage, transport and market access. However, the development of futures markets has 'financialized' commodities since the 1970s. Commodities, in the form of futures contracts, have become financial assets and are today widely used by both hedge funds and institutional investors[2].

As Routledge *et al.* (2000) explain, commodities present different patterns when compared with other conventional financial assets (stocks, bonds):

- The valuation of futures prices is tricky, meaning that the behavior of the term structure is complex.
- Spot and futures prices are mean-reverting for many commodities.
- Commodity prices are strongly heteroscedastic.

There is no consensus-based model to price commodity futures. First, we must take into account the storage cost s of the commodity. Let S_t be the spot price. The forward price with time-to-delivery T is given by:

$$F_t(T) = S_t e^{(r+s)(T-t)}$$

The physical storage has a cost, but it may constitute some advantages in certain situations (Gibson and Schwartz, 1990). Notably, this is the case when inventories are low and demand is high. The convenience yield c measures this effect, and may also be assimilated to a liquidity premium. We then have:

$$F_t(T) = S_t e^{(r+s-c)(T-t)}$$

This explains why futures prices may be lower than spot prices if $c > r + s$. However, this simple model does not explain the comprehensive dynamics of futures prices (Schwartz, 1997). Futures prices are more sensitive to the intertemporal equilibrium between supply, demand and inventory levels and cannot be reduced to a mathematical formula of arbitrage[3] (Pindyck, 2001).

[2]We generally make the distinction between hedging (producers and consumers) and speculating (traders). A speculator is one who does not produce or use a commodity, but invests in order to make a profit on price changes.

[3]This is particularly true for non-storable commodities such as electricity.

Another specific feature of this asset class is the contango/backwardation aspect of the term structure of futures contacts. In Figure 5.1, we have represented the term structure of crude oil futures for different dates. It may be increasing or decreasing depending on the study period. Contango designs a growing curve whereas backwardation corresponds to a falling curve. These two configurations affect the roll of the futures. Generally, the most liquid contract is the first contract. At the maturity of this contract, we may roll our exposure on the contract with the closest settlement date (or nearby contract). This roll method has an impact on the return of the strategy. Indeed, if the commodity is in backwardation, it produces a gain whereas we observe a loss if it is in contango (see Figure 5.2). A trading strategy based on commodity futures involves a return that depends on the roll method. It is called the roll yield.

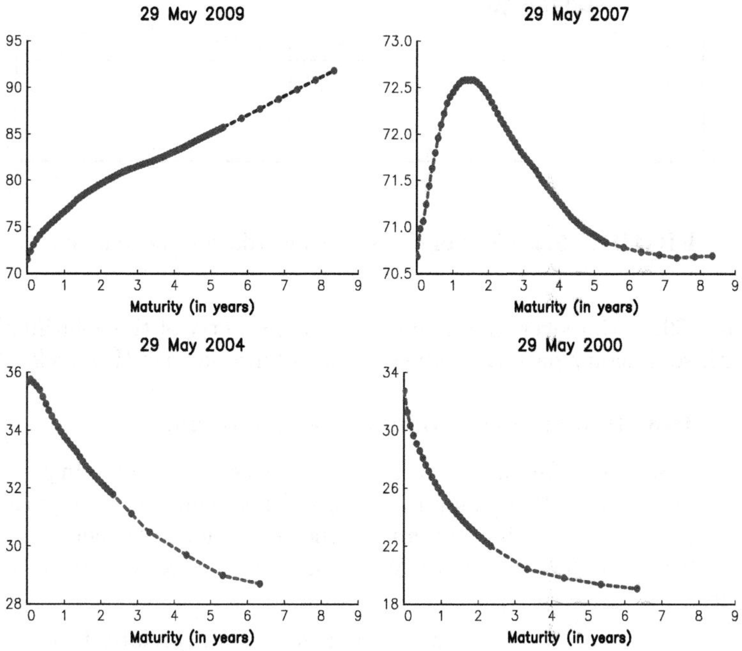

FIGURE 5.1: Term structure of crude oil futures

Econometric analysis of commodity futures shows that futures prices are mean-reverting (Bessembinder *et al.*, 1995). More precisely, it is generally accepted that they exhibit short-term continuation and long-term reversal (Miffre and Rallis, 2007). This short-term property explains the popularity of trend-following strategies in commodity futures markets. Thus, commodity trading advisors (or CTAs) are principally trend followers (Fung and Hsieh, 2001). Another important feature of commodity prices is their statistical properties: they exhibit strong autocorrelation, skewness and kurtosis (Deaton and

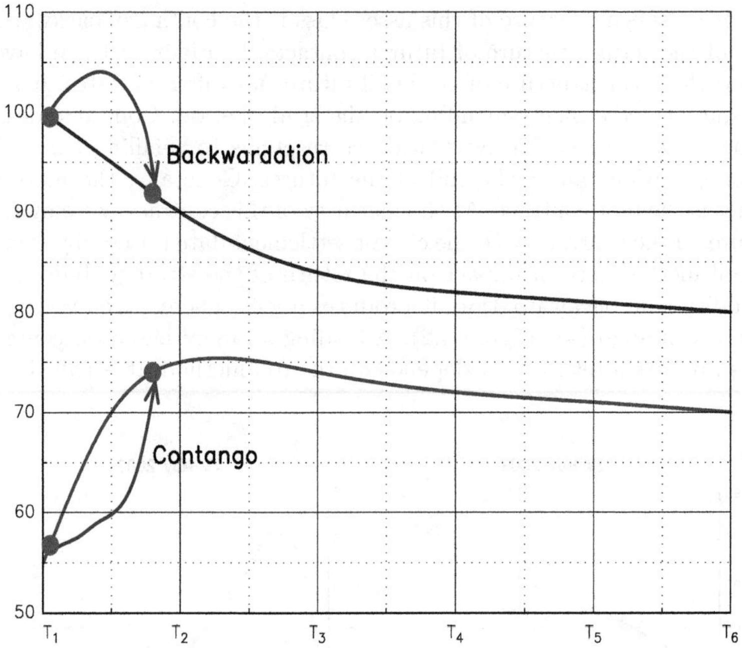

FIGURE 5.2: Contango and backwardation movements

Laroque, 1992). Moreover, the heteroscedastic property of the volatility[4] coupled with seasonality patterns induces some additional risk (Pindyck, 2004).

5.1.1.2 How to define the commodity risk premium

Unlike traditional financial assets, the existence of a commodity risk premium is a question with no simple answers. Thus, contrary to equities and bonds, it cannot be explained by an income (or dividend) generated by the physical storage of the commodity. We must find other reasons if we want to justify it.

The risk premium theory advanced by Keynes (1923) relates futures prices to anticipated future spot prices, arguing that speculators bear risks and should be compensated for their risk-bearing services in the form of a discount, known as normal backwardation. On the other hand, the theory of storage as proposed by Kaldor (1939) postulates that the return from purchasing a commodity at time t and selling it forward for delivery at time T should be equal to the cost of storage minus a convenience yield. The empirical evidences for the existence of time-varying commodity risk premia related to business cycles are strong (Deaton and Laroque, 1992). Typically, periods of strong global demand are associated with low inventories, implying a surge

[4]See e.g. Baillie and Myers (1991) and Duffie *et al.* (2004).

of commodities prices. However, the evidence for a long-run commodity risk premium is patchy. One could therefore argue that over long time horizons commodity returns should be close to a risk-free rate. However, from an economic point of view, the necessity of leaving a share of natural resources for future generations to use, and the growing role of emerging economies could structurally drive prices higher, justifying a risk premium over the long term.

However, recent studies have rekindled the debate, by arguing that roll yield and portfolio diversification justify a commodity risk premium. Indeed, Gorton and Rouwenhorst (2006) construct an equally weighted index of commodity futures for the period 1959-2004 and find an excess of return over T-bills of about 5% per annum. They conclude that this commodity futures risk premium is essentially the same as the historical equity risk premium. This view is supported by different empirical works (see Basu and Miffre, 2009). However, as noted by Erb and Harvey (2006), there is confusion between (individual) commodity risk premium and commodity portfolio return:

> *"A number of studies have argued that commodity futures are an appealing long-only investment class because they have earned a return similar to that of equities. Focusing on the dangers of naive historical extrapolation raises a question, however, about what this historical evidence means. Does the average commodity futures contract have an equity-like return? Our research suggests it does not: The average excess returns of individual commodity futures contracts have been indistinguishable from zero. Might portfolios of commodity futures have equity-like returns? Here, the answer seems to be maybe. A commodity futures portfolio can have equity-like returns if it can achieve a high enough diversification return."*

However, even if a commodity risk premium is difficult to justify based on past performance, this does not mean that commodities are not appropriate in a strategic asset allocation portfolio. Low correlation with bonds and equities and hedging inflation risk are generally the reasons to consider long-only exposure in commodities (Bodie, 1983). Moreover, Hoevenaars *et al.* (2008) show that commodities can serve as diversifying assets in a liability hedging portfolio.

5.1.2 Designing an exposure to the commodity asset class

5.1.2.1 Diversification return

By comparing the compound return (or the geometric return) R_i of asset i with its arithmetic average μ_i, Booth and Fama (1992) obtain this famous formula:

$$R_i \simeq \mu_i - \frac{1}{2}\sigma_i^2$$

The compound return is then approximately equal to the average return minus one-half of the return variance. This formula is valid for one asset but also for a portfolio of n assets rebalanced with constant weights $(x_1, \ldots, x_n)$. Let $\mu(x) = \sum_{i=1}^n x_i \mu_i$ be the weighted average return. Following Willenbrock (2011), we have:

$$\begin{aligned} R(x) + \frac{1}{2}\sigma^2(x) &\simeq \mu(x) \\ &= \sum_{i=1}^n x_i \mu_i \\ &= \sum_{i=1}^n x_i \left(R_i + \frac{1}{2}\sigma_i^2\right) \end{aligned}$$

It follows that:

$$R(x) \simeq \bar{R} + \frac{1}{2}\left(\sum_{i=1}^n x_i \sigma_i^2 - \sigma^2(x)\right)$$

with $\bar{R} = \sum_{i=1}^n x_i R_i$. The difference between the return of the portfolio $R(x)$ and the weighted average return $\bar{R}$ is known as one-half of the portfolio diversification return[5]:

$$\eth(x) = \sum_{i=1}^n x_i \sigma_i^2 - \sigma^2(x)$$

Booth and Fama conclude that diversification increases compound returns by dampening return volatility. Indeed, the diversification return may be viewed as an incremental return provided by a rebalanced portfolio.

Another way to obtain the Booth-Fama formula is to consider that the price process $S_i(t)$ follows a geometric Brownian motion:

$$\mathrm{d}S_i(t) = \mu_i S_i(t)\,\mathrm{d}t + \sigma_i S_i(t)\,\mathrm{d}W_i(t)$$

Using Ito's lemma, we have:

$$\ln S_i(t) - \ln S_i(0) = \left(\mu_i - \frac{1}{2}\sigma_i^2\right)t + \sigma_i W_i(t)$$

The expected return for $t = 1$ is then equal to:

$$\mathbb{E}\left[\ln \frac{S_i(t)}{S_i(0)}\right] = \mu_i - \frac{1}{2}\sigma_i^2$$

[5] This concept is very close to the dispersion measure used in correlation trading:

$$\eth(x) = \sum_{i=1}^n x_i \sigma_i - \sigma(x)$$

We obtain finally:

$$\mathbb{E}[R_i] = \mu_i - \frac{1}{2}\sigma_i^2$$

Let us extend this framework in the multivariate case. We note x_i the weight of the ith asset in the portfolio with $\sum_{i=1}^{n} x_i = 1$. The buy-and-hold strategy consists in buying the assets at time 0 and keeping this portfolio until the maturity T without any rebalancing schemes. In this case, the terminal wealth $V(T)$ is given by:

$$\frac{V(T)}{V(0)} = \sum_{i=1}^{n} x_i \frac{S_i(T)}{S_i(0)}$$

The return $\bar{R}$ of the buy-and-hold strategy between 0 and T is then equal to

$$\begin{aligned} \bar{R} &= \frac{V(T)}{V(0)} - 1 \\ &= \sum_{i=1}^{n} x_i R_i \end{aligned}$$

with R_i the return of asset i. The constant-mix strategy consists in rebalancing portfolio x in order to maintain the exposures proportional to the constant weights $(x_1, \ldots, x_n)$. The dynamics of the wealth $V(t)$ is then:

$$\frac{\mathrm{d}V(t)}{V(t)} = \sum_{i=1}^{n} x_i \frac{\mathrm{d}S_i(t)}{S_i(t)}$$

Ito's lemma gives:

$$\mathrm{d}\ln S_i(t) = \frac{\mathrm{d}S_i(t)}{S_i(t)} - \frac{1}{2}\sigma_i^2\,\mathrm{d}t$$

We deduce then:

$$\mathrm{d}\ln V(t) = \sum_{i=1}^{n} x_i \left(\mathrm{d}\ln S_i(t) + \frac{1}{2}\sigma_i^2\,\mathrm{d}t\right) - \frac{1}{2}\sigma^2(x)\,\mathrm{d}t$$

and:

$$\frac{V(T)}{V(0)} = \exp\left(\sum_{i=1}^{n} x_i \ln\frac{S_i(T)}{S_i(0)} + \frac{1}{2}\left(\sum_{i=1}^{n} x_i\sigma_i^2 - \sigma^2(x)\right)T\right)$$

with $\sigma(x)$ the volatility of the constant-mix portfolio. Let us consider a second-order Taylor expansion. It follows that[6]:

$$\begin{aligned}
\frac{V(T)}{V(0)} &\simeq \left(1+\bar{R}-\frac{1}{2}\sum_{i=1}^{n}x_iR_i^2+\frac{1}{2}\bar{R}^2+\mathcal{O}\left(\bar{R}^3\right)\right)\cdot \\
&\quad \left(1+\frac{T}{2}\left(\sum_{i=1}^{n}x_i\sigma_i^2-\sigma^2(x)\right)+\mathcal{O}\left(\bar{R}^3\right)\right) \\
&\simeq 1+\bar{R}+\frac{1}{2}\left(\bar{R}^2+T\left(\sum_{i=1}^{n}x_i\sigma_i^2-\sigma^2(x)\right)-\sum_{i=1}^{n}x_iR_i^2\right)+\mathcal{O}\left(\bar{R}^3\right)
\end{aligned}$$

Without any loss of generality, we set $T=1$ and we deduce that:

$$R(x)-\bar{R}\simeq\frac{1}{2}\left(\left(\sum_{i=1}^{n}x_i\sigma_i^2-\sigma^2(x)\right)-\left(\sum_{i=1}^{n}x_iR_i^2-\bar{R}^2\right)\right)$$

We recognize the portfolio diversification return for the first term. The second term is the cross-section variance of asset returns:

$$\mathfrak{c}(x)=\sum_{i=1}^{n}x_iR_i^2-\bar{R}^2$$

Even if this formula is only valid when the prices do not deviate largely from their initial values, it completes the result of Booth and Fama. The difference between the return of the constant-mix portfolio $R(x)$ and the return of the buy-and-hold portfolio $\bar{R}$ is approximately equal to one-half of the diversification return minus the cross-section variance of asset returns:

$$R(x)-\bar{R}\simeq\frac{1}{2}\left(\mathfrak{d}(x)-\mathfrak{c}(x)\right)$$

[6] We recall that:

$$R_i=\frac{S_i(T)}{S_i(0)}-1$$

We have:

$$\begin{aligned}
\exp\left(\sum_{i=1}^{n}x_i\ln\frac{S_i(T)}{S_i(0)}\right) &= \exp\left(\sum_{i=1}^{n}x_i\left(R_i-\frac{1}{2}R_i^2+\mathcal{O}\left(\bar{R}^3\right)\right)\right) \\
&\simeq 1+\sum_{i=1}^{n}x_iR_i-\frac{1}{2}\sum_{i=1}^{n}x_iR_i^2+\mathcal{O}\left(\bar{R}^3\right)+ \\
&\quad \frac{1}{2}\left(\sum_{i=1}^{n}x_iR_i-\frac{1}{2}\sum_{i=1}^{n}x_iR_i^2+\mathcal{O}\left(\bar{R}^3\right)\right)^2 \\
&\simeq 1+\bar{R}-\frac{1}{2}\sum_{i=1}^{n}x_iR_i^2+\frac{1}{2}\bar{R}^2+\mathcal{O}\left(\bar{R}^3\right)
\end{aligned}$$

and:

$$\exp\left(\frac{1}{2}\left(\sum_{i=1}^{n}x_i\sigma_i^2-\sigma^2(x)\right)T\right)=1+\frac{T}{2}\left(\sum_{i=1}^{n}x_i\sigma_i^2-\sigma^2(x)\right)+\mathcal{O}\left(\bar{R}^3\right)$$

This second term $\mathfrak{c}(x)$ mitigates then the positive effect of the diversification return. It vanishes when returns between assets are equal. Diversification return is then maximum when the performance contributions of all the assets are similar.

As shown by Erb and Harvey (2006), the good performance of the equally weighted commodity portfolio is explained by this diversification return. However, as noted by Willenbrock (2011), other rebalanced portfolios may generate substantial excess return. In the next section, we question whether the ERC portfolio is a better candidate to build a commodity futures index than the EW portfolio[7].

5.1.2.2 Comparing EW and ERC portfolios

We consider a universe of 12 commodities: crude oil, Brent, natural gas, heating oil, wheat, corn, soybeans, cotton, copper, gold, silver, platinum. All the main sectors are represented (energy, agricultural and metals) in the same proportion. The study period begins in January 1990 and ends in November 2012. For each commodity, we invest in the futures contract with the smallest maturity and we roll it three days before the first notice day (FND). In Table 5.1, we have computed the annualized excess return of each commodity for the entire period. Few of these futures contracts present a positive excess return (natural gas, wheat, corn and cotton). On average, we obtain an excess return equal to −30 bps per year, meaning that the performance of commodity futures is very close to the risk-free rate. This figure confirms the result of Erb and Harvey (2006) and we can conclude that commodity futures have offered no risk premium for over 20 years. We have also reported some quantile statistics about the annualized excess return for a one-year rolling window. We observe that the performance may be very high (more than 200% in one year), but that the loss may also be very high (more than 60% in one year). This dispersion of returns explains the success of the CTA industry and the dominance of momentum strategies in commodity futures markets. It also explains why these financial assets are so volatile (see Table 5.2). For instance, natural gas has a volatility equal to 51% while the volatility of the Brent is 35%. The smallest volatility is obtained for gold with a figure close to 16%. Moreover, volatility can vary from year to year, even for gold futures.

These observations led Maillard *et al.* (2010) to apply the ERC approach to eight light agricultural commodities. They find that the ERC portfolio dominates the EW portfolio both in terms of performance, volatility and Sharpe ratio. They also suggest that the ERC portfolio is a strong candidate to capture the diversification return described in the previous section. Chaves *et al.* (2012) consider a large universe of 28 commodity futures. They find comparable Sharpe ratios between risk parity portfolios and the EW portfolio. In what follows, we consider the previous universe of the 12 commodity futures. The

[7]We recall that, at equilibrium, the performance contributions of all the assets are the same for the ERC portfolio.

TABLE 5.1: Annualized excess return (in %) of commodity futures strategies

	1990 – 2012	One-year rolling window Min	10%	50%	90%	Max
Crude Oil	−4.9	−75.7	−40.4	6.3	89.3	175.1
Brent	−10.2	−69.4	−33.2	13.9	74.5	231.2
Natural Gas	19.2	−83.3	−58.0	−20.0	86.0	360.2
Heating Oil	−8.3	−70.6	−30.7	8.4	76.8	230.3
Wheat	8.3	−64.9	−34.2	−8.1	29.5	146.6
Corn	6.4	−60.5	−30.5	−9.9	35.8	104.9
Soybeans	−3.4	−47.3	−20.9	1.5	40.9	89.8
Cotton	6.9	−63.3	−36.8	−8.9	38.3	185.5
Copper	−6.0	−62.4	−26.0	2.4	58.6	223.1
Gold	−2.9	−25.9	−14.4	1.9	26.3	63.2
Silver	−4.3	−41.3	−19.9	1.5	45.0	158.9
Platinum	−5.0	−55.2	−19.7	5.3	42.7	86.9
Average	−0.3					

TABLE 5.2: Annualized volatility (in %) of commodity futures strategies

	1990 – 2012	One-year rolling window Min	10%	50%	90%	Max
Crude Oil	36.5	15.6	21.4	32.5	48.5	70.3
Brent	34.8	16.2	21.6	30.6	43.9	68.4
Natural Gas	51.0	24.0	36.0	49.2	66.0	73.6
Heating Oil	35.1	19.4	23.1	32.0	43.3	62.9
Wheat	28.2	16.6	18.4	24.4	40.0	51.2
Corn	25.0	12.7	15.7	22.2	33.8	44.8
Soybeans	22.8	11.4	16.7	20.6	31.5	41.4
Cotton	26.0	12.2	18.6	23.4	34.4	42.4
Copper	27.2	13.1	17.6	23.8	35.9	56.5
Gold	16.2	5.9	8.0	14.7	22.4	33.0
Silver	29.0	11.5	16.8	25.7	43.4	54.6
Platinum	21.4	8.5	13.6	19.6	27.3	43.3
Average	29.4					

portfolio is rebalanced in a monthly frequency while we estimate the covariance matrix using daily returns and a one-year rolling window. The results are given in Table 5.3. In Figure 5.3, we also report the risk contributions of the EW portfolio, the weights of the ERC portfolio and the simulated performance of these two strategies. We observe some risk concentrations on the energy sector in the case of the EW portfolio, while the ERC portfolio is overweight on the metals sector. This explains why the EW portfolio has a larger volatility than the ERC portfolio, but also a better performance. Finally, they have similar Sharpe ratios (about 25%) and generate significant excess return contrary to the buy and hold portfolio. We conclude that the ERC portfolio is another good alternative candidate to the EW portfolio in terms of capturing the diversification, although it is not necessarily a better approach.

TABLE 5.3: Main statistics of EW and ERC commodity portfolios

Portfolio	$\hat{\mu}_{1Y}$	$\hat{\sigma}_{1Y}$	SR	$\mathcal{MDD}$	γ_1	γ_2
EW	8.26	16.48	0.28	−57.79	−0.31	5.16
ERC	7.24	14.46	0.25	−56.06	−0.22	4.95

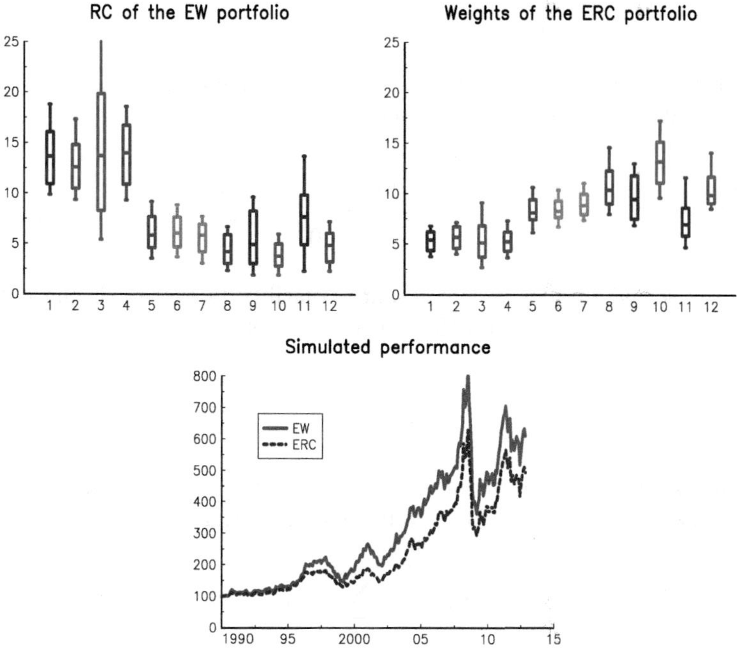

FIGURE 5.3: Simulated performance of EW and ERC commodity portfolios

5.2 Hedge fund strategies

5.2.1 Position sizing

Hedge funds have used risk parity to size the exposures of bets for a long time. This is particularly true for equity market neutral (EMN) strategies. These strategies hold long-short equity positions and size the exposures such that they exhibit low correlations with the equity market. Portfolio construction can be done using the Markowitz framework. The objective is to maximize the expected return $\mu(x) = x^\top \mu$ of the portfolio subject to a constraint of ex-ante volatility $\sigma^\star$ and a constraint of zero-beta:

$$\sum_{i=1}^{n} x_i \beta_i = 0$$

If we note b the benchmark portfolio (or the equity market portfolio), we have:

$$\begin{aligned} \sum_{i=1}^{n} x_i \beta_i &= x^\top \beta \\ &= \frac{x^\top \Sigma b}{b^\top \Sigma b} \end{aligned}$$

The constraint $\sum_{i=1}^{n} x_i \beta_i = 0$ can then be written as a linear restriction $x^\top \Sigma b = 0$. We then obtain a γ-problem:

$$\begin{aligned} x^\star &= \arg\min \frac{1}{2} x^\top \Sigma x - \gamma x^\top \mu \\ \text{u.c.} & \quad \left(b^\top \Sigma\right) x = 0 \end{aligned}$$

where γ is scaled to achieve the ex-ante volatility target $\sigma^\star$. Using a single-factor model, the return of portfolio x at time t is:

$$R_t(x) = \sum_{i=1}^{n} x_i \left(\beta_i R_t(b) + \varepsilon_{i,t}\right)$$

where $R_t(b)$ is the return of the benchmark and $\varepsilon_{i,t}$ is the idiosyncratic risk of the stock i. Because $\sum_{i=1}^{n} x_i \beta_i = 0$, we deduce that:

$$R_t(x) = \sum_{i=1}^{n} x_i \varepsilon_{i,t}$$

The return of the strategy depends then only on specific risks $\varepsilon_{i,t}$ and not on the return of the equity market. This is why the EMN strategy is a pure stock picking strategy. To achieve market neutrality, many fund managers use a pair trading strategy consisting in matching a long position with a short position in two stocks generally of the same sector.

Let us consider the pair trade j. We note β_j^+ and $\varepsilon_{j,t}^+$ (resp. β_j^- and $\varepsilon_{j,t}^-$) the beta and the specific risk of the stock which belongs to the long position (resp. short position). If the two stocks are in the same sector, we may assume that $\beta_j^+ \simeq \beta_j^-$. We obtain then:

$$\begin{aligned} R_t(x) &= \sum_{j=1}^{m} x_j \left(\left(\beta_j^+ - \beta_j^- \right) R_t(b) + \varepsilon_{j,t}^+ - \varepsilon_{j,t}^- \right) \\ &\simeq \sum_{j=1}^{m} x_j \left(\varepsilon_{j,t}^+ - \varepsilon_{j,t}^- \right) \end{aligned}$$

where m is the number of pair trades and x_j is the weight of the pair trade j. A common practice of the market is to calibrate the weights inversely proportional to the volatility of the pair trades:

$$x_j \propto \frac{1}{\sigma_j}$$

This allocation corresponds to the ERC portfolio when we assume a constant correlation between the bets. In the following, we show the rationality of such a choice.

Let $x_+ = \left(x_1^+, \ldots, x_m^+\right)$ and $x_- = \left(x_1^-, \ldots, x_m^-\right)$ be the vectors of weights of the long and short exposures. We partition the covariance matrix of the $2m$ asset returns in the following way:

$$\Sigma = \begin{pmatrix} \Sigma_{++} & \Sigma_{+-} \\ \Sigma_{-+} & \Sigma_{--} \end{pmatrix}$$

The portfolio's volatility is then equal to:

$$\sigma(x) = \sqrt{x_+^\top \Sigma_{++} x_+ + x_-^\top \Sigma_{--} x_- + 2 x_+^\top \Sigma_{+-} x_-}$$

We deduce that the risk contribution of the pair trade $\left(x_j^+, x_j^-\right)$ is:

$$\mathcal{RC}_j = x_j^+ \frac{\partial\, \sigma(x)}{\partial\, x_j^+} + x_j^- \frac{\partial\, \sigma(x)}{\partial\, x_j^-}$$

The first idea how to calibrate the weights x_j^+ and x_j^- is to solve the problem:

$$\left\{ \begin{array}{l} \mathcal{RC}_j = \mathcal{RC}_k \\ x_j^+ > 0 \\ x_j^- < 0 \end{array} \right.$$

However, this problem is under-identified and we have an infinity of solutions.

This is why we must reduce the number of variables. For example, if we impose that $x_j^- = -x_j^+$, we obtain:

$$\begin{cases} \mathcal{RC}_j = \mathcal{RC}_k \\ x_j^+ > 0 \\ x_j^+ + x_j^- = 0 \\ \sigma(x) = \sigma^* \end{cases}$$

We note that we use the volatility target constraint to obtain a unique solution. However, the constraint $\sigma(x) = \sigma^*$ is difficult to manage from a numerical point of view. It is then better to set the weight of the first pair trade to a given constant c:

$$\begin{cases} \mathcal{RC}_j = \mathcal{RC}_k \\ x_j^+ > 0 \\ x_j^+ + x_j^- = 0 \\ x_1^+ = c \end{cases}$$

Let (x_+^c, x_-^c) be the solution of this problem. Because we have $\sigma(\alpha x_+^c, \alpha x_-^c) = \alpha\sigma(x_+^c, x_-^c)$, the risk contribution constraint remains valid if we scale the portfolio. Finding the EMN portfolio such that $\sigma(x) = \sigma^*$ is therefore equivalent to:

$$x_+^\star = \frac{\sigma^*}{\sigma(x^c)} x_+^c \quad \text{and} \quad x_-^\star = \frac{\sigma^*}{\sigma(x^c)} x_-^c$$

If we assume that the assets have the same volatility, the correlation between the pair trades j and k verifies:

$$\rho_{j,k} \propto \rho_{j,k}^{++} + \rho_{j,k}^{--} - \rho_{j,k}^{+-} - \rho_{j,k}^{-+}$$

where $\rho_{j,k}^{++}$ is the correlation between the long asset of the bet j and the long asset of the bet k, $\rho_{j,k}^{+-}$ is the correlation between the long asset of the bet j and the short asset of the bet k, etc. Let us assume a $S+1$ risk factor model with a common risk factor and a specific risk factor for each sector[8]. We note ρ_s the intra-sector asset correlation if the two stocks belong to the same sector s and $\rho^\star$ the inter-sector asset correlation if the two stocks are not members of the same sector. We obtain:

$$\rho_{j,k} \propto \begin{cases} \rho_s + \rho_s - \rho_s - \rho_s & \text{if } (j,k) \in s \\ \rho^* + \rho^* - \rho^* - \rho^* & \text{otherwise} \end{cases}$$

We verify that $\rho_{j,k} = 0$. The market practice then corresponds to the ERC allocation when:

1. The long and short exposures of each pair trade are made in the same sector.

[8]See Remark 13 on page 37 for a presentation of such a model.

2. The volatility of the stocks are the same.

3. The risk factor model corresponds to the sectorial factor model.

From a practical point of view, the first and second assumptions are the most important, because cross-correlations are less sensitive to volatilities than to the sector correlation.

Example 39 *We consider four sectors s_1, s_2, s_3 and s_4. The intra-sector correlations ρ_s are respectively equal to* 80%, 70%, 65% *and* 90% *while the inter-sector correlation ρ^* is equal to* 60%. *The portfolio contains five long-short bets and the pair trades are done in the sectors s_1, s_2, s_3, s_4 and s_4. The asset volatility is equal to* 10%, 40%, 30%, 10% *and* 30% *for the long legs and,* 20%, 20%, 20%, 50% *and* 20% *for the short legs.*

By setting the target volatility of the portfolio to 10%, we obtain the optimized weights reported in Table 5.4. We also indicate the risk contribution of each pair trade. In this example where volatility differences are large, the market practice gives results not too far from the ERC approach.

TABLE 5.4: Calibration of the EMN portfolio

	Market practice			ERC		
j	x_j^+	$\mathcal{RC}_j$	$\mathcal{RC}_j^\star$	x_j^+	$\mathcal{RC}_j$	$\mathcal{RC}_j^\star$
1	42.7%	2.1%	20.8%	38.5%	2.0%	20.0%
2	19.3%	2.5%	24.8%	18.6%	2.0%	20.0%
3	25.1%	3.0%	29.6%	21.5%	2.0%	20.0%
4	13.9%	0.6%	6.5%	18.7%	2.0%	20.0%
5	38.6%	1.8%	18.3%	45.2%	2.0%	20.0%
$\sigma(x)$		10.0%			10.0%	

5.2.2 Portfolio allocation of hedge funds

Hedge fund returns differ substantially from the returns of standard asset classes, making hedge funds of interest to investors seeking to diversify balanced portfolios. Research into hedge fund investing has therefore naturally focused on finding the optimal proportion in which to invest in hedge funds (Cvitanic *et al.*, 2003), identifying hedge fund risk factors (Fung and Hsieh, 1997), and finally constructing optimal hedge fund portfolios (Kat, 2004). During the hedge fund crisis of 2008, funds of hedge funds suffered greatly. In this context, the building of optimal funds of hedge funds is a challenging problem.

Because of the nature of hedge fund investments (dynamic trading strategies, use of derivatives and leverage) and their observable consequences on hedge fund return characteristics (time-varying covariance parameters, high kurtosis of returns distributions), actual research programs are focused on using dynamic specification for the covariance matrix (Giamouridis and Vrontos, 2007), introducing higher moments (Martellini and Ziemann, 2010) or specifying alternative risk measures (Adam *et al.*, 2008). Despite the use of these sophisticated statistical tools, the results are not convincing. On the other side, the use of risk-based approaches such as the MV, ERC or MDP portfolios with the volatility risk measure does not improve the results (Bruder *et al.*, 2011). In this section, we investigate whether or not combining risk-based methods with non-Gaussian risk measures may be a solution.

5.2.2.1 Choosing the risk measure

We consider the Dow Jones Credit Suisse AllHedge index. This index is composed of ten subindices: (1) convertible arbitrage, (2) dedicated short bias, (3) emerging markets, (4) equity market neutral, (5) event driven, (6) fixed income arbitrage, (7) global macro, (8) long-short equity, (9) managed futures and (10) multi-strategy. The study period starts in September 2004 and ends in August 2012.

Following Cont (2001), we present a set of risk measures[9] computed with the monthly returns of the ten subindices in Table 5.5. We observe that the volatility of hedge fund strategies is smaller than the equity volatility. However, some drawdowns may be comparable to those observed in an equity market. We also note that some strategies present high excess kurtosis, larger than 10. If we consider value-at-risk and expected shortfall risk measures at the confidence level 95%, we have a curious phenomenon. The Gaussian (G) value-at-risk is generally higher than the historical (H) value-at-risk, whereas it is the contrary for the expected shortfall. This means that hedge fund strategies exhibit large heavy tails. The volatility is then less adapted to measure the risk of hedge fund strategies than the risk of traditional assets. In this case, we may think that risk parity based on non-Gaussian risk measures can be more appropriate. Based on the previous result, we select the historical expected shortfall and the Cornish-Fisher value-at-risk.

5.2.2.2 Comparing ERC allocations

We build ERC portfolios rebalanced at the end of each month with the three risk measures: the volatility, the historical expected shortfall (ES) and the Cornish-Fisher value-at-risk (CF). Each risk measure $\mathcal{R}(x)$ is estimated using a two-year rolling window and a confidence level equal to 80%. We compare the three ERC strategies with the asset-weighted portfolio[10]. In Figure 5.4, we report the weights while the risk decomposition between the hedge

[9] They are expressed in % except for the skewness γ_1 and the excess kurtosis γ_2.

[10] The data for assets are no longer publicly available from the web site `www.hedgeindex.com`. Therefore, we consider them as constant and use an estimate of the average holdings.

TABLE 5.5: Statistics of monthly returns of hedge fund strategies

Strategy	$\hat{\mu}_{1Y}$	γ_1	γ_2	$\mathcal{MDD}$	VaR (G)	VaR (H)	ES (G)	ES (H)
1	12.2	−3.5	18.0	−47.9	5.6	3.5	7.0	12.5
2	16.3	0.0	0.7	−44.3	8.1	7.6	10.1	10.7
3	14.5	−1.6	5.7	−48.3	6.5	6.2	8.2	11.9
4	10.2	−2.8	16.3	−38.4	5.0	4.4	6.2	10.6
5	8.1	−0.8	1.6	−22.3	3.5	4.3	4.5	6.2
6	10.3	−3.8	24.7	−42.4	5.1	3.9	6.3	11.6
7	9.1	−1.8	8.8	−30.1	4.2	3.6	5.3	7.9
8	9.2	−1.4	3.1	−32.3	4.1	4.5	5.2	7.7
9	10.2	−0.1	−0.8	−9.2	4.3	4.1	5.5	5.1
10	8.0	−2.7	14.4	−35.8	3.5	3.2	4.5	7.9

fund strategies is given in Figure 5.5. We verify that the risk contributions are equal to 10% for each ERC portfolio, even if the risk measure is not the volatility. Figure 5.6 corresponds to the simulated performance and a summary of risk and performance statistics[11] are given in Table 5.6.

TABLE 5.6: Statistics of ERC HF portfolios (Sep. 2006 – Aug. 2012)

	Index	ERC-weighted Vol	ERC-weighted ES	ERC-weighted CF
$\hat{\mu}_{1Y}$ (in %)	0.86	0.23	1.81	1.34
$\hat{\sigma}_{1Y}$ (in %)	7.93	4.85	4.66	5.93
$\mathcal{MDD}$ (in %)	−27.08	−18.22	−16.02	−19.14
γ_1	−2.04	−1.84	−1.37	−1.96
γ_2	6.24	6.88	5.38	8.98
$\bar{\tau}$	0.00	0.89	1.88	2.35
τ^+	0.00	0.41	1.31	0.77
$\mathcal{H}^\star$	0.72	0.29	0.65	0.13
$\mathcal{N}^\star$	1.40	3.05	2.67	5.42
$\mathcal{G}$	0.83	0.65	0.63	0.62
$\mathcal{I}^\star$	1.77	3.98	3.68	3.38

First, we note that ERC strategies reduce the volatility of returns and also

[11] $\hat{\mu}_{1Y}$ is the annualized performance, $\hat{\sigma}_{1Y}$ is the yearly volatility and $\mathcal{MDD}$ is the maximum drawdown observed for the entire period. These statistics are expressed in %. Skewness and excess kurtosis correspond to γ_1 and γ_2. The yearly turnover corresponds to $\bar{\tau}$ while τ^+ is the maximum of the monthly observed turnover. The concentration measures are computed using the risk contributions of the factors: $\mathcal{H}^\star$ is the normalized Herfindahl index, $\mathcal{N}^\star = \mathcal{H}^{-1}$ is the effective number of independent strategies, $\mathcal{G}$ is the Gini index and $\mathcal{I}^\star$ is the diversity measure based on the Shannon entropy.

the drawdown. Indeed, the volatility of the asset-weighted portfolio (or the index) is equal to 7.9% whereas the volatility of ERC strategies lies between 4.7% and 5.9%. The drawdown is reduced by a factor of 30%. However, the excess kurtosis of ERC strategies is not smaller than the excess kurtosis of the index except for the ERC portfolio based on the expected shortfall. Concerning the performance, ERC portfolios with ES and CF risk measures present a better performance than the ERC portfolio with the volatility risk measure or the asset-weighted portfolio. Moreover, the performance of the ERC portfolio with volatility risk measure is disappointing and is smaller than that of the asset-weighted portfolio. It may be tempting to conclude that using a non-Gaussian risk measure improves the ERC strategy when asset returns are leptokurtic or skewed. However, we also observe that using expected shortfall and Cornish-Fisher value-at-risk significantly increases the turnover. This is a serious drawback particularly when we consider hedge fund strategies, as their assets are not liquid.

Remark 57 *The results also depend on the confidence level used when computing the risk measures. We have chosen the level 80% because the two-year rolling window only contains 24 observations. With a higher confidence level, we may face some stability issues when calibrating ERC portfolios.*

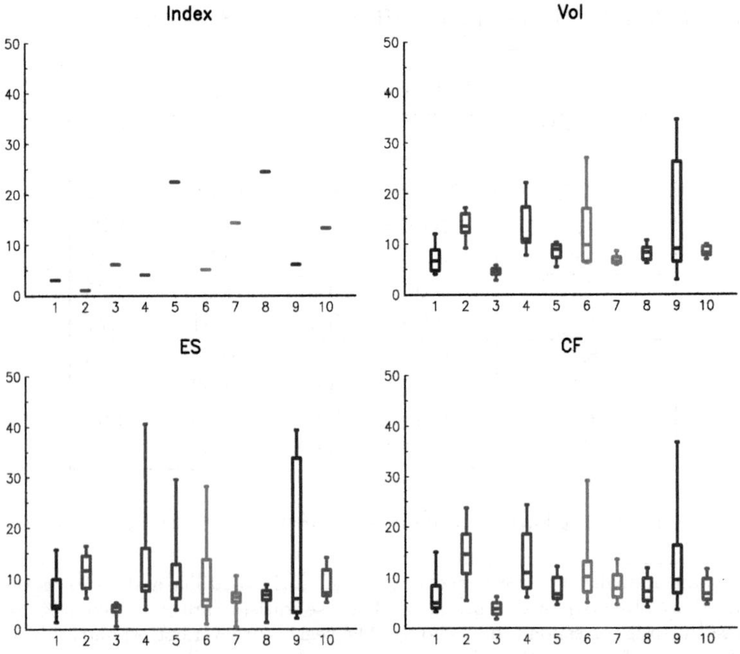

FIGURE 5.4: Weights (in %) of ERC HF portfolios

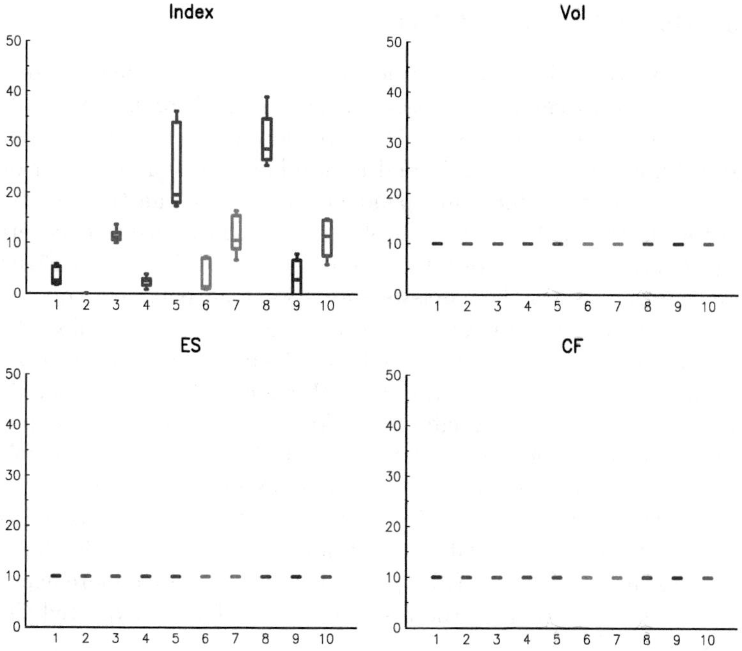

FIGURE 5.5: Risk contributions (in %) of ERC HF portfolios

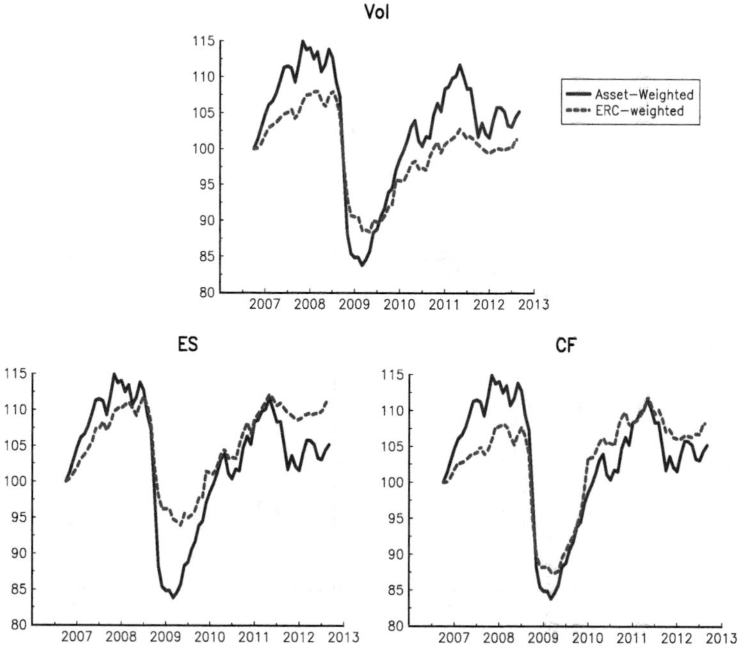

FIGURE 5.6: Simulated performance of ERC HF portfolios

5.2.2.3 Budgeting the risk factors

Let us apply the risk factor approach described on page 135 to these strategies. We build statistical factors based on the principal component analysis of the two-year covariance matrix. PCA is frequently used to classify dynamic strategies (Fung and Hsieh, 1997) and its use holds great interest for our application, because it produces independent factors. We can then easily characterize the degree of diversification of the portfolio by using concentration indices (Meucci, 2009). The results are reported in Table 5.6 and Figure 5.7. We note that the risk of the asset-weighted portfolio is largely concentrated on the first factor. ERC strategies produce more balanced portfolios. We have computed the effective number of freedom $\mathcal{N}^{\star}$ and some concentration measures applied to the risk concentrations with respect to the factors. $\mathcal{N}^{\star}$ takes the value 1.40 for the asset-weighted portfolio meaning that it is exposed to less than two independent risk factors. The ERC portfolio with the volatility risk measure is more diversified because it is exposed to three independent risk factors. However, the most diversified portfolio is the one based on the Cornish-Fisher value-at-risk with more than five independent risk factors. It is also interesting to note that ERC portfolios are on average more exposed to the second risk factor than to the first one, except for the expected shortfall risk measure.

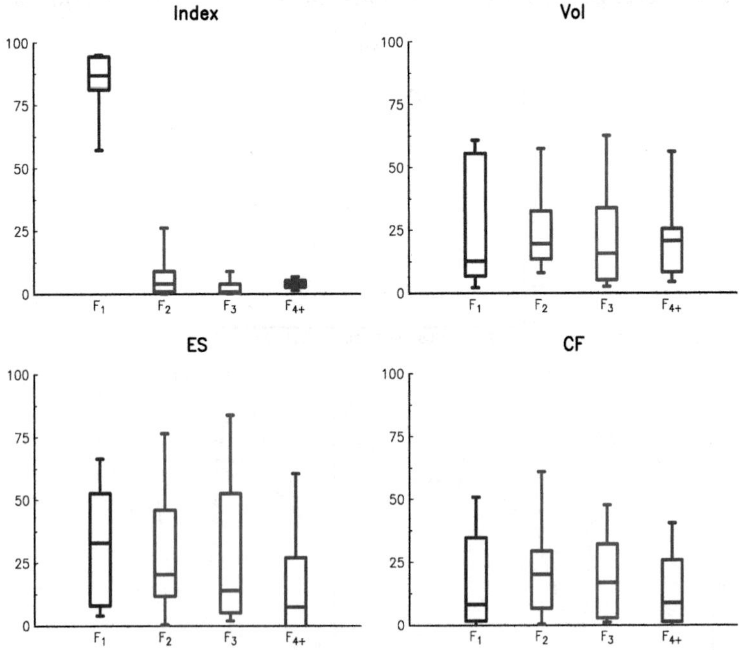

FIGURE 5.7: Risk factor contributions (in %) of ERC HF portfolios

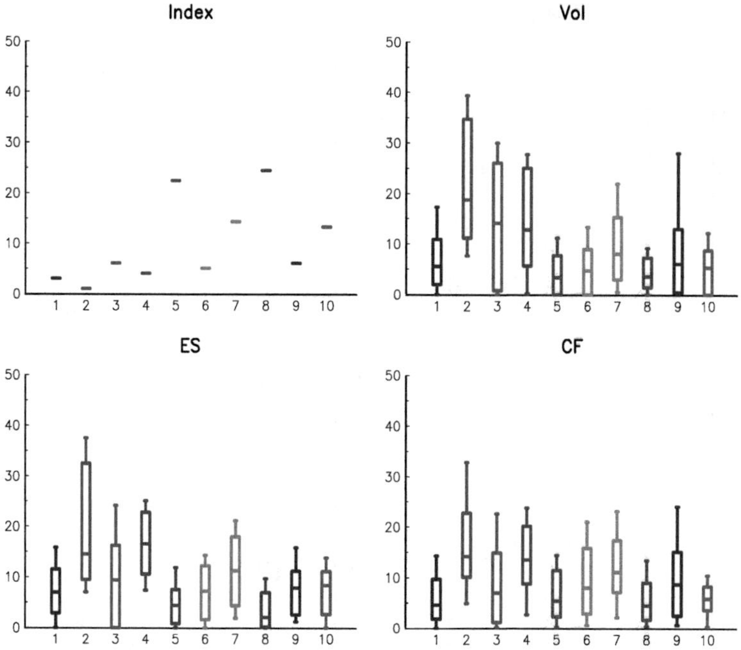

FIGURE 5.8: Weights (in %) of RFP HF portfolios

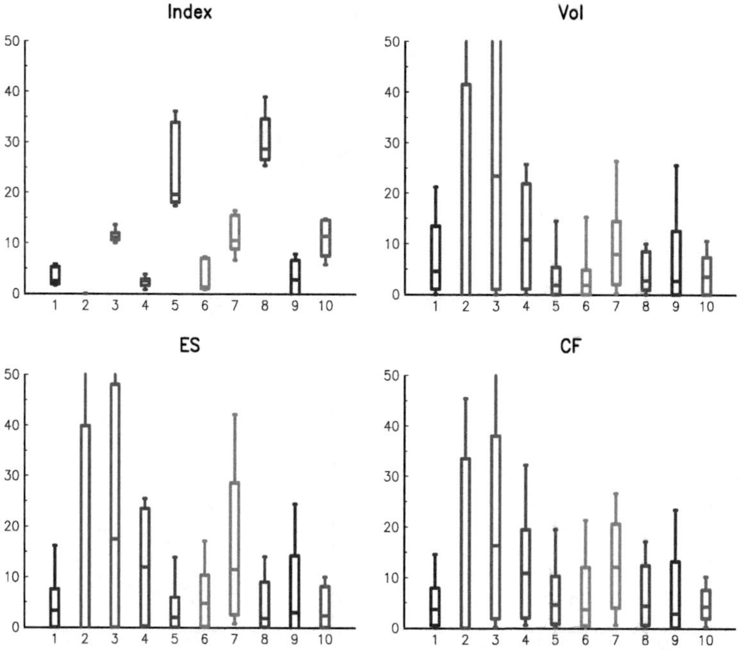

FIGURE 5.9: Risk contributions (in %) of RFP HF portfolios

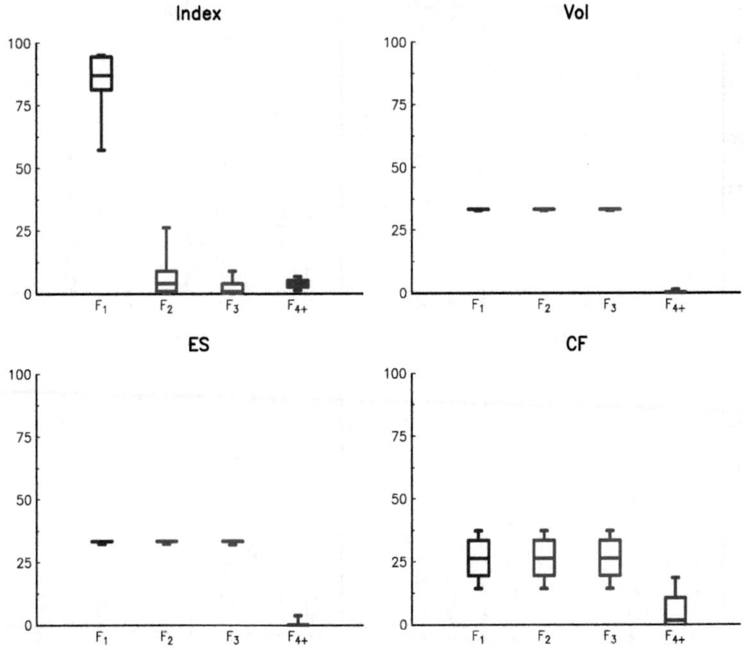

FIGURE 5.10: Risk factor contributions (in %) of RFP HF portfolios

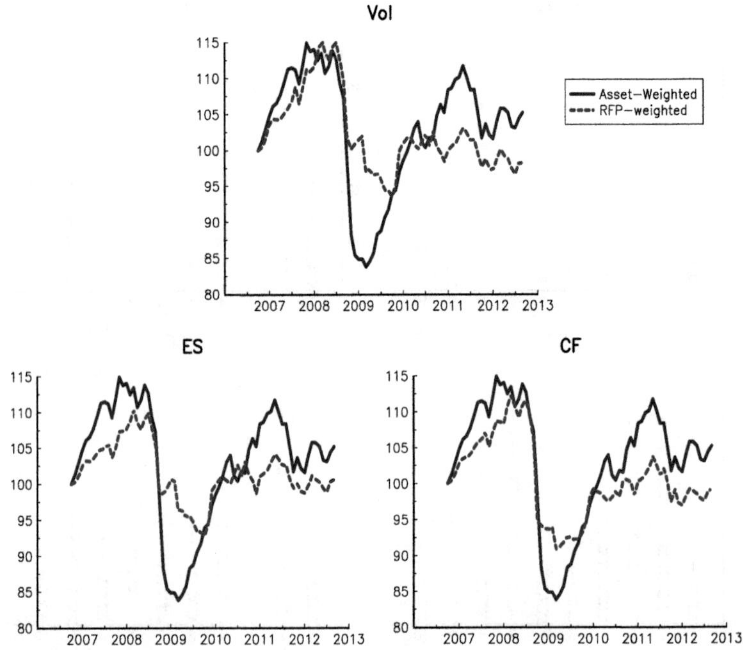

FIGURE 5.11: Simulated performance of RFP HF portfolios

Let us now apply the risk factor parity (RFP) approach. The portfolio is then optimized by minimizing the concentration of the risk contributions with respect to the first three PCA factors. The results are given in Table 5.7 and in Figures 5.8–5.11.

TABLE 5.7: Statistics of RFP HF portfolios (Sep. 2006 – Aug. 2012)

	Asset-weighted	RFP-weighted		
		Vol	ES	CF
$\hat{\mu}_{1Y}$ (in %)	0.86	−0.31	0.10	−0.16
$\hat{\sigma}_{1Y}$ (in %)	7.93	5.98	5.33	6.35
$\mathcal{MDD}$ (in %)	−27.08	−19.23	−15.96	−19.12
γ_1	−2.04	−1.00	−0.70	−2.87
γ_2	6.24	3.87	3.71	14.68
$\bar{\bar{\tau}}$	0.00	3.47	4.84	6.01
τ^+	0.00	1.50	1.63	1.39
$\mathcal{H}^\star$	0.72	0.24	0.22	0.10
$\mathcal{N}^\star$	1.40	3.02	3.01	6.78
$\mathcal{G}$	0.83	0.63	0.52	0.51
$\mathcal{I}^\star$	1.77	3.06	3.54	3.74

If we consider the box plot of risk factor contributions in Figure 5.10, we see that the dispersion between the first three factors is very small in the cases of volatility and expected shortfall risk measures. For the Cornish-Fisher value-at-risk, the dispersion is larger, which indicates that matching risk budgets is more difficult with this risk measure. Another point is the dispersion of the weights. The turnover of these RFP strategies is very high, larger than three. Moreover, the performance lags those of the asset-weighted and ERC portfolios. All these results suggest that the risk factor approach may be suitable for analyzing the risk of a hedge fund portfolio, but is not well designed for allocating between hedge fund strategies.

5.2.2.4 Limiting the turnover

The issue with the previous analysis is that the turnover is too high. Dynamic allocations of hedge fund strategies imply then prohibitive trading costs. We remind that the turnover of the portfolio x with respect to current portfolio x^0 is $\tau(x) = \sum_{i=1}^{n} \left| x_i - x_i^0 \right|$. In order to limit the turnover, we can modify the RB optimization problem (2.22) on page 102 by introducing a non-linear inequality constraint:

$$
\begin{aligned}
x^\star &= \arg\min f(x;b) \\
\text{u.c.} & \left\{ \begin{array}{l} \tau(x) \leq \tau^+ \\ \mathbf{1}^\top x = 1 \\ \mathbf{0} \leq x \leq \mathbf{1} \end{array} \right.
\end{aligned}
$$

where τ^+ is the turnover objective. However, such problem is very difficult to solve numerically.

Another idea to limit the turnover is to find a portfolio between the current allocation x^0 and the optimal RB portfolio $x^\star$. For instance, we can find the constrained RB portfolio such that $\delta \in [0,1]$ and:

$$\left\{ \begin{array}{l} b = \delta b^\star + (1-\delta) b^0 \\ \tau(x) = \tau^+ \end{array} \right.$$

where b^0 and $b^\star$ are the risk budgets of the portfolios x^0 and $x^\star$. Let $x^\star(\delta)$ be the RB portfolio satisfying the risk budgets $b = \delta b^0 + (1-\delta) b^\star$. Finding the constrained RB portfolio is then equivalent to find δ such that $\tau(x^\star(\delta)) = \tau^+$. We generally observe that $\tau(x^\star(\delta))$ is an increasing function of δ. The problem can then be solved numerically using the bisection algorithm.

We can also consider the portfolio $x^\star(\alpha)$ which is a combination of the RB portfolio $x^\star$ and the portfolio x^0:

$$x^\star(\alpha) = \alpha x^\star + (1-\alpha) x^0$$

with $\alpha \in [0,1]$. We have:

$$\begin{aligned} \tau(x^\star(\alpha)) &= \sum_{i=1}^{n} \left| x_i^\star(\alpha) - x_i^0 \right| \\ &= \sum_{i=1}^{n} \left| \alpha x_i^\star + (1-\alpha) x_i^0 - x_i^0 \right| \\ &= \alpha \tau(x^\star) \end{aligned}$$

Limiting the turnover to τ^+ can then be achieved by choosing the following optimal value of α:

$$\alpha = \min\left(\frac{\tau^+}{\tau(x)}, 1 \right)$$

Example 40 *We consider a universe of four assets. We assume that their volatilities are equal to 20%, 15%, 25% and 20%. The correlation of asset returns is given by the following matrix:*

$$\rho = \begin{pmatrix} 1.00 & & & \\ 0.20 & 1.00 & & \\ 0.10 & 0.15 & 1.00 & \\ 0.20 & 0.20 & 0.50 & 1.00 \end{pmatrix}$$

We want to rebalance the current allocation $x^0 = (20\%, 40\%, 5\%, 35\%)$ in order to obtain the ERC portfolio.

We have reported the risk decomposition of the current allocation x^0 and

TABLE 5.8: Risk decomposition of the current allocation x^0

Asset	x_i	$\mathcal{MR}_i$	$\mathcal{RC}_i$	$\mathcal{RC}_i^\star$
1	20.00	10.90	2.18	17.68
2	40.00	10.20	4.08	33.07
3	5.00	12.26	0.61	4.97
4	35.00	15.61	5.46	44.28
Volatility			12.34	

TABLE 5.9: Risk decomposition of the RB portfolio $x^\star$

Asset	x_i	$\mathcal{MR}_i$	$\mathcal{RC}_i$	$\mathcal{RC}_i^\star$
1	25.68	12.08	3.10	25.00
2	33.56	9.24	3.10	25.00
3	18.84	16.46	3.10	25.00
4	21.93	14.14	3.10	25.00
Volatility			12.41	

TABLE 5.10: Risk decomposition of the constrained RB portfolio $x^\star(\delta)$ when $\tau^+ = 5\%$

Asset	x_i	$\mathcal{MR}_i$	$\mathcal{RC}_i$	$\mathcal{RC}_i^\star$
1	20.59	11.06	2.28	18.51
2	39.13	10.11	3.95	32.15
3	6.91	12.92	0.89	7.26
4	33.37	15.51	5.18	42.08
Volatility			12.30	

TABLE 5.11: Risk decomposition of the constrained RB portfolio $x^\star(\alpha)$ when $\tau^+ = 5\%$

Asset	x_i	$\mathcal{MR}_i$	$\mathcal{RC}_i$	$\mathcal{RC}_i^\star$
1	20.73	11.10	2.30	18.71
2	39.17	10.12	3.96	32.24
3	6.77	12.86	0.87	7.08
4	33.33	15.48	5.16	41.96
Volatility			12.29	

the RB portfolio $x^\star$ in Tables 5.8 and 5.9. We then obtain a turnover equal to 39.04%, which is very high. If we limit the turnover to 5%, we obtain the solutions given in Tables 5.10 and 5.11. These solutions correspond to the optimal values $\delta = 11.42\%$ and $\alpha = 12.81\%$. We note that the two portfolios $x^\star(\delta)$ and $x^\star(\alpha)$ are very close. If we consider a higher turnover $\tau^+ = 20\%$, we obtain the same conclusion[12] (see Tables 5.12 and 5.13).

TABLE 5.12: Risk decomposition of the constrained RB portfolio $x^\star(\delta)$ when $\tau^+ = 20\%$

Asset	x_i	$\mathcal{MR}_i$	$\mathcal{RC}_i$	$\mathcal{RC}_i^\star$
1	22.64	11.52	2.61	21.26
2	36.58	9.77	3.57	29.12
3	12.36	14.68	1.81	14.78
4	28.42	15.04	4.28	34.84
Volatility			12.27	

TABLE 5.13: Risk decomposition of the constrained RB portfolio $x^\star(\alpha)$ when $\tau^+ = 20\%$

Asset	x_i	$\mathcal{MR}_i$	$\mathcal{RC}_i$	$\mathcal{RC}_i^\star$
1	22.91	11.61	2.66	21.70
2	36.70	9.80	3.60	29.33
3	12.09	14.55	1.76	14.35
4	28.30	14.99	4.24	34.61
Volatility			12.26	

Remark 58 *Even though limiting the turnover is an issue for illiquid assets like hedge fund strategies, the previous methods may also be used for liquid assets like equities and bonds. These methods are particularly adapted to institutional investors which are very sensitive to this issue.*

[12]The optimal values of δ and α are 48.98% and 51.24%.

Chapter 6

Portfolio Allocation with Multi-Asset Classes

This last chapter is dedicated to the allocation problem when the investment universe is a set of asset classes. In fact, the roots of risk parity come from this asset mix policy problem. What relative proportion of equities and bonds must be held by an institutional investor, for instance a pension fund? If we refer to performances over the past century, a portfolio fully invested in equities offers the best performance (Ibbotson Associates, 2010). However, this high equity risk premium combined with a low risk-free rate and smooth consumption is difficult to reconcile with the typical risk aversion of investors (Mehra and Prescott, 1985). Numerous explanations have been put forward to solve this paradox, known as the equity premium puzzle (see e.g. Odean, 1998; Rabin, 1998; Hirshleifer, 2001; Barberis and Thaler, 2003). One of the main arguments is that time accounting differs from time preference. Many long-term investors evaluate their allocation policy on a yearly basis, which explains why they are more sensitive to losses than gains:

> "*The equity premium puzzle refers to the empirical fact that stocks have outperformed bonds over the last century by a surprisingly large margin. We offer a new explanation based on two behavioral concepts. First, investors are assumed to be 'loss averse', meaning that they are distinctly more sensitive to losses than to gains. Second, even long-term investors are assumed to evaluate their portfolios frequently. We dub this combination 'myopic loss aversion'. Using simulations, we find that the size of the equity premium is consistent with the previously estimated parameters of prospect theory if investors evaluate their portfolios annually*" (Benartzi and Thaler, 1995).

The fact that equities are too risky in short and medium-term horizons then pushes institutional investors to diversify their portfolios, by including sovereign and corporate bonds and alternative investments such as commodities and hedge funds.

This is particularly true for pension funds which face liability constraints. Indeed, pension liabilities modify asset allocation decisions, because the matching of asset and liability durations leads to investment in bonds. In this context, it is not surprising to observe a wide dispersion of the asset allocation between equities and bonds (Antolin, 2008). This dispersion is effective between countries, but also within a country. Even when pension funds have the same constraints and objectives, the stock/bond asset mix policies differ. Nevertheless, some portfolio references have emerged over the past thirty years. Indeed, the 60/40 equity/bond portfolio is an anchor point for many Anglo-Saxon pension funds (Ambachtsheer, 1987). However, with the recent crisis, this constant-mix portfolio has suffered and institutional investors are trying to find a more robust asset mix policy.

The choice of stock/bond allocation is all the more important as many studies have shown that most of the differences in pension fund performances are caused by the asset allocation policy. Brinson *et al.* (1986, 1991) estimate that its contribution is larger than 90%. Ibbotson and Kaplan (2000) find similar levels when they consider the variability in the returns of pension funds over time. In a more recent study, Blake *et al.* (1999) conclude that "*strategic asset allocation accounts for most of the time-series variation in portfolio returns, while market timing and asset selection appear to have been far less important*". Among the different components of a long-term policy, i.e. strategic asset allocation, tactical asset allocation, market timing and asset selection, the reference portfolio is the key choice. In a sense, the risk budgeting approach embodies both of the first two steps. Risk parity portfolio policies may also be an alternative of constant-mix portfolio policies.

Risk parity with multi-asset classes is not limited to the design of diversified funds or strategic asset allocation. In fact, it was first used by hedge funds to build absolute return strategies. The success of Bridgewater's All Weather fund, which is a global macro hedge fund, has led the asset management industry to propose absolute return funds built around the risk parity approach.

In this chapter, we consider these different topics. The first section is dedicated to the construction of diversified funds. The second section deals with long-term investment policy and the third section shows how to use risk parity to develop absolute return strategies.

6.1 Construction of diversified funds

6.1.1 Stock/bond asset mix policy

The diversified funds business has suffered a lot of criticism, both from a theoretical and a practical point of view. In chapter one, we have seen that mean-variance portfolios of risky assets define the set of efficient portfolios. If we introduce a risk-free asset, the efficient frontier becomes a straight line called the capital market line. Along this frontier, optimal portfolios correspond to a combination of the risk-free asset and the tangency portfolio. The separation theorem states that all the investors hold the tangency portfolio.

However, the allocation between the cash and the tangency portfolio differs from one investor to another, because it depends on the investor's risk profile. All these results, obtained in a static framework with two periods, were extended by Merton (1973) to the dynamic case (see Appendix A.3.2 on page 324). In this framework, the part of the tangency portfolio in the investor portfolio depends on the utility function. Generally, we distinguish three profiles depending on the risk tolerance:

- Conservative (low risk tolerance)
- Moderate (medium risk tolerance)
- Aggressive (high risk tolerance)

As the conservative investor has a smaller appetite for risk than the aggressive investor, his portfolio will contain more cash and less risky assets. However, the relative proportions between risky assets are the same for conservative and aggressive investors. In practice, it may be inefficient to pay fees just to leverage or deleverage the tangency portfolio. However, the business of diversified funds is largely based on this framework. Generally, we distinguish three profiles depending on the benchmark of the diversified fund:

- Defensive (20% of equities and 80% of bonds)
- Balanced (50% of equities and 50% of bonds)
- Dynamic (80% of equities and 20% of bonds)

These diversified funds are also called lifestyle funds. The difference between them comes from the relative proportion of stocks and bonds. This means that in practice the composition of the risky portfolio varies according to the investor's risk aversion. This portfolio construction contradicts the separation theorem, which states that the composition of the risky portfolio should be the same for all investors. Indeed, the relationship between investor risk profiles and fund profiles is not simple. The business of diversified funds suggests however that a defensive (resp. balanced or dynamic) fund profile matches the need of a conservative (resp. moderate or aggressive) investor profile. However, we clearly face a gap between the theory and the industry practice.

In Figure 6.1, we have illustrated these different results. The first panel represents the efficient frontier of Markowitz between equities and bonds. By combining the mean-variance optimized portfolio with the highest Sharpe ratio and the risk-free rate, we obtain the capital market line. In the second panel, we have reported the risk/return profile of optimal portfolios corresponding to different values for the risk aversion parameter. The aggressive (A) investor leverages the tangency portfolio in order to take more risk and to expect better performance. For the moderate (M) investor, the optimal portfolio is close to the tangency portfolio. The conservative (C) investor wants

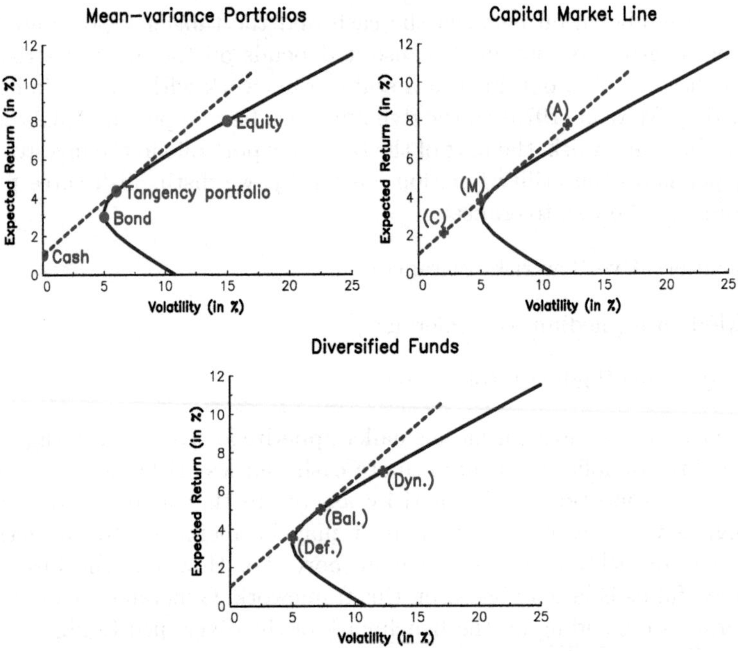

FIGURE 6.1: Asset allocation puzzle of diversification funds

less risk so he allocates his wealth between the tangency portfolio (risky assets) and the risk-free asset. However, this theoretical view of asset allocation is far removed from the lifestyle funds, which are represented in the third panel. Indeed, contrary to the optimal portfolios corresponding to the three investor profiles, the portfolios corresponding to these three lifestyle funds are located on the efficient frontier and not on the capital market line. This paradox, known as the *asset allocation puzzle* (Canner *et al.*, 1997), has resulted in a wealth of literature being written in an attempt to find explanations (see Campbell (2000) for a survey). Even if we could partially solve this problem[1] (Bajeux-Besnainou *et al.*, 2003), the controversy is still relevant today (Campbell and Viceira, 2002).

From the practical point of view, the main criticism concerns the fact that the allocation of diversified funds is not always diversified. Let us illustrate this drawback by simulating the defensive, balanced and dynamic diversified funds with bond and equity asset classes[2] for the period January 2000 – December 2011. In Figure 6.2, we have reported the evolution of the risk contributions of these two asset classes for the diversified funds with a one-year rolling empirical covariance matrix. We note that these risk contributions are time-

[1]See Appendix A.3.3.1 on page 326.

[2]The bond and equity asset classes are represented by the Citigroup WGBI All Maturities index and the MSCI World TR Net index.

varying, especially for defensive and balanced funds. For defensive funds, the bond asset class has a larger risk contribution than the equity asset class. For balanced funds, we obtain the opposite, meaning that even if they are very well balanced in terms of weights, they are certainly not well balanced in terms of risk diversification. For dynamic funds, the risk is almost entirely explained by equities. In a sense, dynamic funds may be viewed as a deleverage of an equity exposure. More important, we note that there is no mapping between fund profiles and volatility regimes (see Figure 6.3). For example, the volatility of a dynamic fund in 2006 is smaller than the volatility of a balanced fund in 2009. We then deduce that the relationship between the investor risk tolerance and the risk profile of diversified funds is blurred.

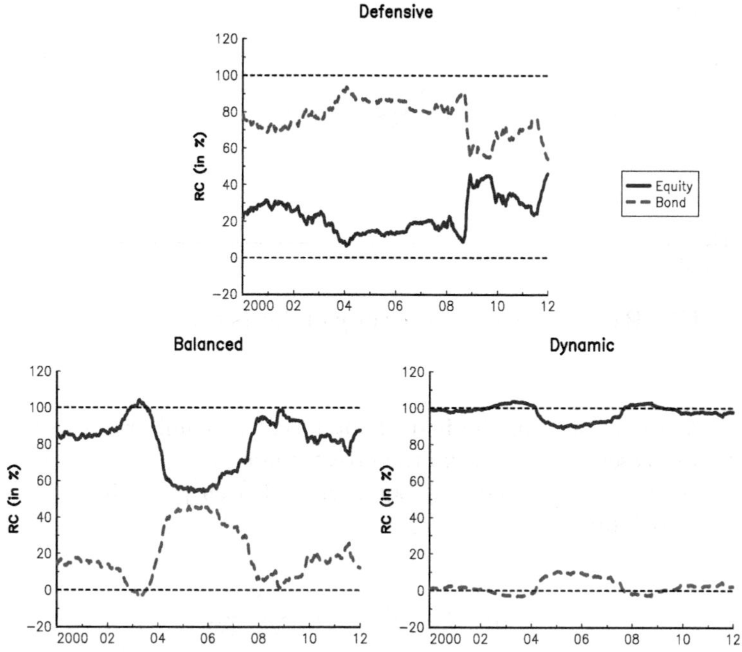

FIGURE 6.2: Equity and bond risk contributions in diversified funds

6.1.2 Growth assets versus hedging assets

6.1.2.1 Are bonds growth assets or hedging assets?

The choice between the three diversified funds is not only a question of risk aversion. In fact, it is related to the financial status of equities and bonds: are they growth assets or hedging assets? To answer this question, we consider the model of Black and Litterman (1992). If the investor holds one of these three funds, he is confident that it is the best investment for him with respect to its portfolio. From his point of view, this investment is then an optimal

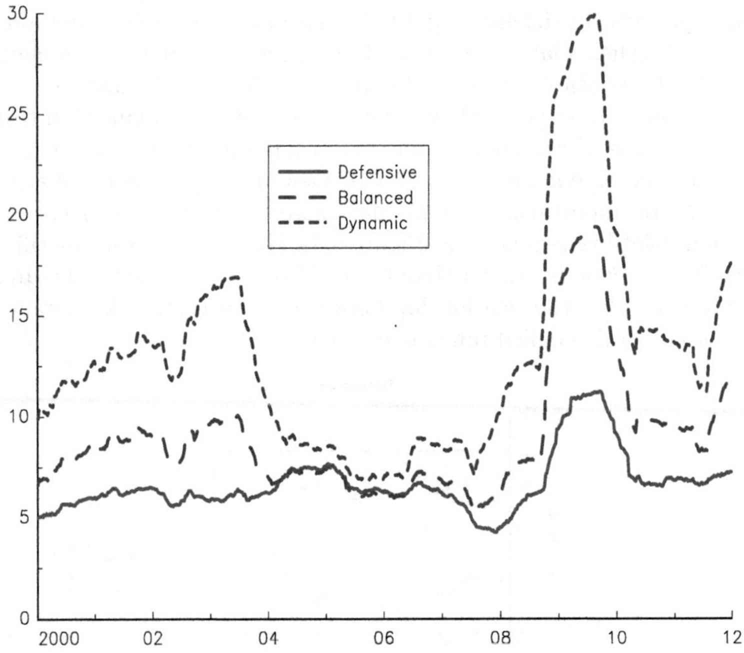

FIGURE 6.3: Realized volatility of diversified funds (in %)

portfolio. We have seen in the first chapter that at equilibrium risk premia are related to asset risks. Let x be the investor portfolio and Σ the covariance matrix of asset returns. Following Equation (1.12) on page 23, the expected (or ex-ante) risk premia are:

$$\tilde{\pi} = \tilde{\mu} - r = \mathrm{SR}\left(x \mid r\right) \frac{\Sigma x}{\sqrt{x^\top \Sigma x}}$$

where $\mathrm{SR}\left(x \mid r\right)$ is the expected Sharpe ratio of the portfolio.

We have calibrated the ex-ante risk premium using a one-year empirical covariance matrix and the assumption that the Sharpe ratio is constant and equal to 25%. The dynamics of the risk premium is given in Figure 6.4 whereas the mean and the standard deviation computed for the entire period are reported in Table 6.1. We note that the equity risk premium increases with the equity weight in the diversified fund. More important, we note that the bond risk premium is very low for the balanced and dynamic funds. Whereas it represents 75% of the equity risk premium in the case of the defensive fund, this ratio becomes 21% and 6% for the two other diversified funds. If we compute the correlation of the equity risk premium variations between the three funds, we obtain:

$$\rho = \begin{pmatrix} 1.00 & & \\ 0.82 & 1.00 & \\ 0.77 & 0.99 & 1.00 \end{pmatrix}$$

For the bond risk premium, we have:

$$\rho = \begin{pmatrix} 1.00 & & \\ 0.73 & 1.00 & \\ 0.38 & 0.88 & 1.00 \end{pmatrix}$$

Whereas the correlation is high for equities, we note that it is lower for bonds, in particular between the risk premia of the defensive and dynamic funds. In this last case, the correlation is equal to 38%.

TABLE 6.1: Mean and standard deviation of the ex-ante risk premium for diversified funds (in %)

Asset	$\hat{\mu}(\tilde{\pi})$			$\hat{\sigma}(\tilde{\pi})$		
	Def.	Bal.	Dyn.	Def.	Bal.	Dyn.
Equity	2.05	3.71	4.02	20.68	28.19	28.26
Bond	1.57	0.77	0.26	4.05	7.37	7.56

All these results show that bonds do not have the same status in the three funds. In the case of the dynamic fund, bonds are not considered as a growth asset, i.e. an investment with the objective to perform and to achieve a return. Bonds are more considered as a hedging asset, i.e. an asset that will protect the portfolio when equities have a negative return. In the case of the balanced fund, bonds become a growth asset and the investor expects a reward from this exposure. These results are confirmed by the distribution of the expected performance contributions in Figure 6.5. In the second chapter[3], we have seen that there is a duality between risk contributions and performance contributions. Indeed, the expected performance attribution is given by the risk contributions. In Figure 6.5, we verify that most of the expected performance comes from the equity exposure in the case of the dynamic fund.

To conclude, an investor considering a defensive fund makes the assumption that bonds may be rewarded and are then a growth asset. If he considers a dynamic fund, bonds are then viewed as a hedging asset and the investor does not expect that they will contribute to the performance of the fund. Contrary to equities, which are growth assets in all the cases, the status of bonds is then singular.

[3]See Section 2.2.3 on page 113.

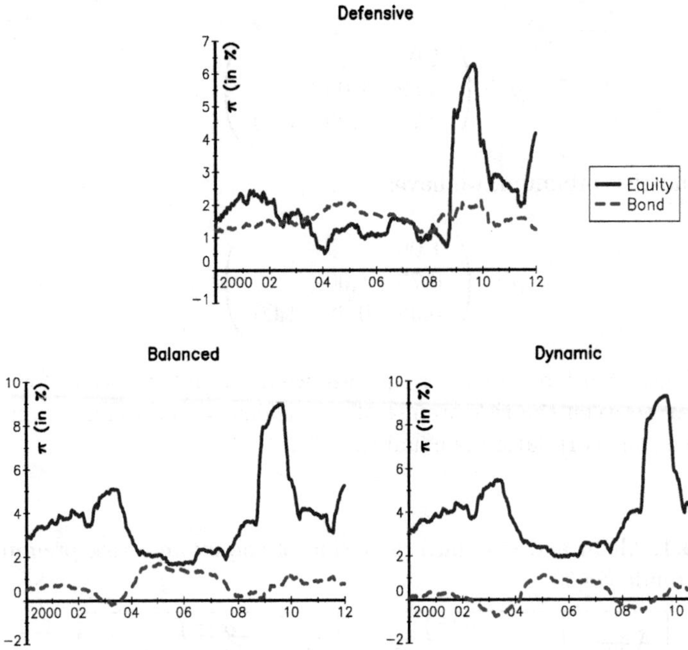

FIGURE 6.4: Equity and bond ex-ante risk premia for diversified funds

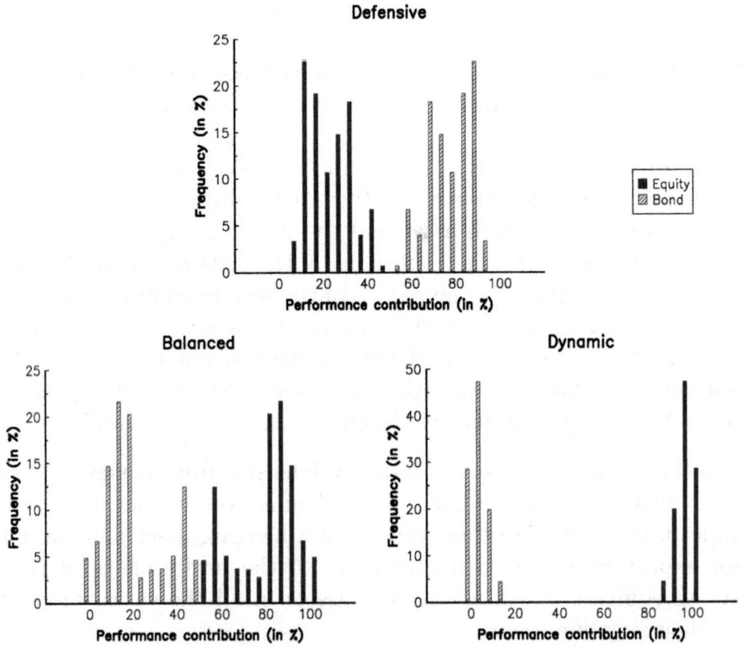

FIGURE 6.5: Histogram of ex-ante performance contributions

6.1.2.2 Analytics of these results

To understand these results, we can rewrite the ex-ante risk premium in the following way:

$$\tilde{\pi} = \mathrm{SR}\left(x \mid r\right) \frac{\partial\, \sigma\left(x\right)}{\partial\, x}$$

where $\sigma\left(x\right)$ is the volatility of portfolio x. In this case, the risk premium of asset i is proportional to its marginal volatility. If one asset has a smaller marginal volatility than another, then it has a smaller expected risk premium. Using the marginal volatility formula, we obtain:

$$\tilde{\pi}_i = \mathrm{SR}\left(x \mid r\right) \frac{\left(x_i \sigma_i^2 + \sigma_i \sum_{j \neq i} x_j \rho_{i,j} \sigma_j\right)}{\sigma\left(x\right)}$$

In the two-asset case, this expression becomes[4]:

$$\tilde{\pi}_i = c\left(x\right) \left(\underbrace{x_i \sigma_i^2}_{\text{variance}} + \underbrace{\rho \sigma_i \sigma_j \left(1 - x_i\right)}_{\text{covariance}} \right)$$

with $c\left(x\right) = \mathrm{SR}\left(x \mid r\right) / \sigma\left(x\right)$ and ρ the cross-correlation between the two asset returns. We note that there are two components in the risk premium. The first one is a variance component and is an increasing function of the volatility and the weight of the asset. The higher its volatility, the higher its risk premium. The second component is a covariance component and depends on the correlation between the asset returns. If the correlation ρ is zero, the covariance component vanishes. The contribution of this component is positive only if the correlation is positive. This means that a low volatility asset could benefit from a high volatility asset in terms of risk premium if the correlation is high[5]. In fact, the risk premium $\tilde{\pi}_i$ is an increasing function of the asset weight[6]. In Figure 6.6, we have reported the evolution of the risk premium of two assets with respect to x_1 when $\sigma_1 = 20\%$, $\sigma_2 = 5\%$ and $\mathrm{SR}\left(x \mid r\right) = 0.25$. If $\rho = 70\%$, the portfolio composition has little impact on the expected risk premia. This is not the case if $\rho = 0$, but risk premia continue to be positive. If $\rho = -40\%$, the risk premium may be negative for the two assets, even for the more volatile asset. This means that an asset is neither a hedging asset

[4]The indices are $i = 1$ and $j = 2$.

[5]This result relates to the model developed by Lucas (1978) in which required risk premia depend on the correlation between asset return and marginal utility of consumption.

[6]Let $x^\star = \sigma_j \left(\sigma_i + \sigma_j\right)^{-1}$. We also have:

$$\lim_{\rho \to 1} \tilde{\pi}_i = \mathrm{SR}\left(x \mid r\right) \cdot \sigma_i$$

and:

$$\lim_{\rho \to -1} \tilde{\pi}_i = \begin{cases} -\mathrm{SR}\left(x \mid r\right) \cdot \sigma_i & \text{if} \quad x_i < x^\star \\ 0 & \text{if} \quad x_i = x^\star \\ \mathrm{SR}\left(x \mid r\right) \cdot \sigma_i & \text{if} \quad x_i > x^\star \end{cases}$$

nor a growth asset in an absolute way. For instance, if $\rho = -40\%$, the first asset is a hedging asset for the portfolio if its weight is small. If its weight is high, the second asset becomes the hedging asset.

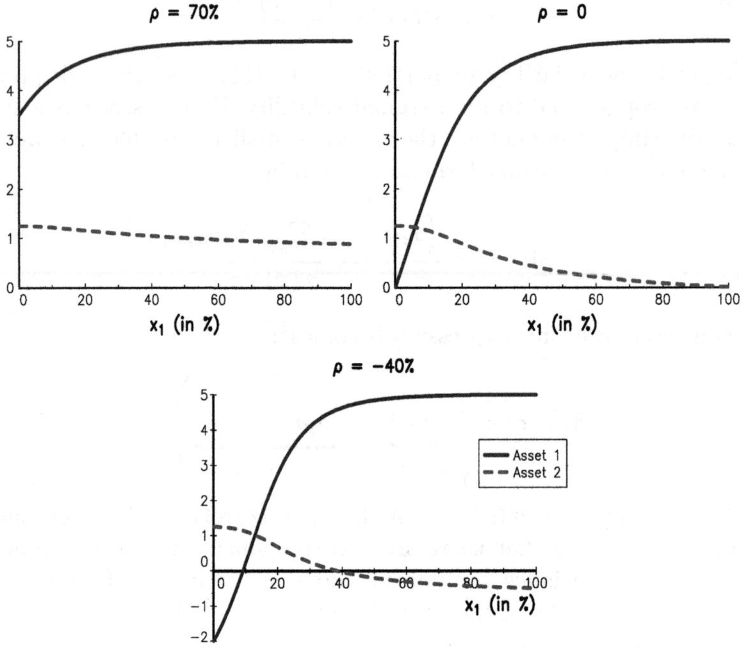

FIGURE 6.6: Influence of the correlation on the expected risk premium

Remark 59 *By analyzing the world market portfolio of stocks and bonds,* Hereil et al. *(2013) find that the ex-ante risk premium of bonds is close to zero. So, the overall financial market estimates that bonds are hedging assets and not growth assets. However, they observe some differences between countries and regions. This result is particularly true for Anglo-Saxon financial markets, but it is less relevant for Japan and Italy. This vision is supported by academic research, which finds a substantial difference (about 5%) between the expected return of equities and bonds (Ibbotson and Chen, 2003). However, this result is generally based on the last two centuries of performance and this view is not yet shared by everyone (Arnott and Bernstein, 2002).*

6.1.3 Risk-balanced allocation

The drawbacks of diversified funds led the investment industry to propose an alternative to these funds. A risk parity fund is an ERC strategy on multi-assets classes:

"*Diversify, but diversify by risk, not by dollars – that is, take a similar amount of risk in equities and in bonds*" (Asness *et al.*, 2012).

Applying this concept to our example is equivalent to building a portfolio where the risk of the bond asset class is equal to the risk of the equity asset class. If we assume that we rebalance the portfolio every month, we obtain the dynamic allocation given in Figure 6.7. We observe that the weights vary significantly over time. Indeed, the weight of bonds ranges between 47.4% and 78.5%, whereas the range is 21.5%–52.6% for equities. On average, the weights are respectively 67.2% and 32.8%. The simulated performance of the risk parity fund is reported in the bottom-right quadrant. It is difficult to compare it to the performance of the diversified funds (bottom-left quadrant), but we note that risk parity and defensive funds are close. We have also reported the simulated performance when the weights are constant and are equal to the average weights of the risk parity fund. Compared to this static fund, the risk parity fund presents an outperformance of 40 bps and a volatility smaller than 35 bps. Moreover, the maximum drawdown is reduced by 2.7%. Sometimes, the risk parity strategy is combined with a leverage effect in order to obtain a more risky profile. For example, if we apply a 10% volatility target, we obtain the performance of the leveraged risk parity fund. It has a volatility similar to the balanced fund, but a better performance (see Table 6.2).

TABLE 6.2: Statistics of diversified and risk parity portfolios

Portfolio	$\hat{\mu}_{1Y}$	$\hat{\sigma}_{1Y}$	SR	$\mathcal{MDD}$	γ_1	γ_2
Defensive	5.41	6.89	0.42	−17.23	0.19	2.67
Balanced	3.68	9.64	0.12	−33.18	−0.13	3.87
Dynamic	1.70	14.48	−0.06	−48.90	−0.18	5.96
Risk parity	5.12	7.29	0.36	−21.22	0.08	2.65
Static	4.71	7.64	0.29	−23.96	0.03	2.59
Leveraged RP	6.67	9.26	0.45	−23.74	0.01	0.78

In practice, risk parity funds use a larger universe than the example presented here. It may include American, European, Japanese and Emerging Market equities, large-cap and small-cap equities, American and European sovereign bonds, inflation-linked bonds, corporate and high yield bonds, etc. Thus, Chaves *et al.* (2011) compare risk parity with other diversified portfolios (60/40 asset mix, equally weighted portfolio and minimum variance portfolio). They find that this investment strategy has some appealing characteristics in terms of Sharpe ratio. However, they also warn investors about these backtests, because they are highly dependent on the study period and the choice of universe:

"*[...] we also find that risk parity is very sensitive to the inclusion decision for assets. The methodology is mute on how many asset*

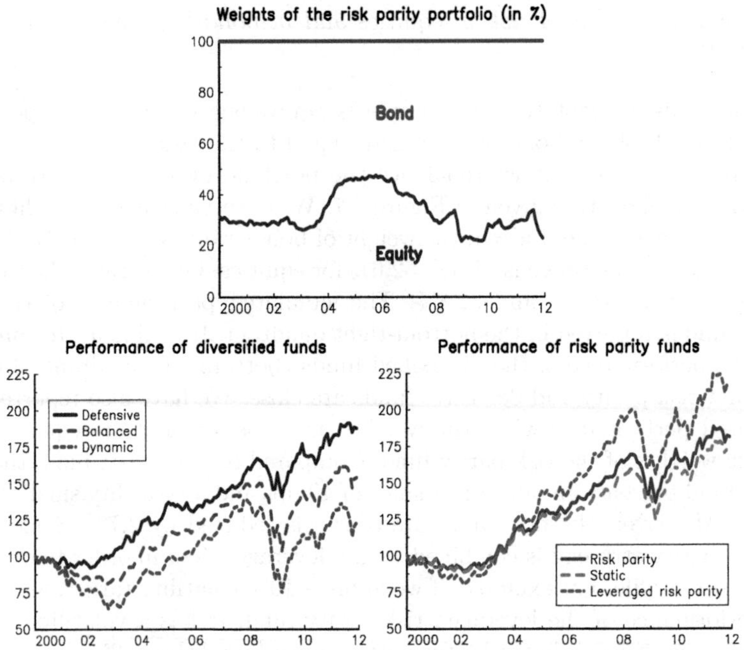

FIGURE 6.7: Backtest of the risk parity strategy

classes and what asset classes to include. This last point is particularly problematic because there is little in the way of theory to guide the asset inclusion decision".

6.1.4 Pros and cons of risk parity funds

Risk parity funds present some advantages, but also some drawbacks (Thiagarajan and Schachter, 2011). In the words of the Financial Times[7], "*risk parity strategy has its critics as well as fans*". The difficulty with this topic is that the debate is generally not scientific and involves people that are in the asset management business.

It is obvious that the risk parity approach is a good way to obtain a well diversified portfolio (Qian, 2011). In fact, it is the essence of risk parity portfolios and it largely explains the success of risk parity funds. It is not a coincidence that they emerge today, in the aftermath of the global financial crisis of 2008. The credit, equity, hedge fund and now sovereign bonds crises have changed the investment behavior of most investors, who have now become more sensitive to risk management.

It is debatable whether or not volatility is a good risk measure (Inker,

[7] *Source*: FTfm, June 10, 2012.

2011), but we have already seen that we can use alternative risk measures. A more important criticism concerns the performance of risk parity funds. Such funds were launched after 2005 and have performed very well. However, they benefit from the strong performance of bonds since 2000. We know that this performance is exceptional and is due to the decrease of interest rates. Today, the situation is not the same. With interest rates being close to zero and the fear of growing inflation, there is no guarantee that the performance of bonds will be the same over the next 10 years as it was over the previous ten years. In this case, risk parity funds may (highly) suffer from rising interest rates.

The second important criticism concerns the use of leverage (Inker, 2011; Sebastian, 2012). Indeed, to achieve a better return, many risk parity funds target a volatility greater than 8% by leveraging the portfolio (Ruban and Melas, 2011). In this case, they meet the huge demand of European investors for diversified growth funds. Paradoxically, the use of leverage is seen as a real advantage by the supporters of risk parity funds. Indeed, Asness *et al.* (2012) estimate, based on the results of Frazzini and Pedersen (2010), that the good performance of risk parity funds can be explained because they overweight less volatile assets and leverage them.

These authors consider an equilibrium model with two periods, n risky assets whose prices and dividends are denoted $P_{i,t}$ and $D_{i,t}$ and m investors who have a given amount of wealth W_j. Let x_j and ϕ_j be the portfolio and the risk aversion of the investor j. At time t, investors maximize their utility function[8]:

$$x_j^\star = \arg\max x_j^\top \mathbb{E}_t \left[P_{t+1} + D_{t+1} - (1+r) P_t\right] - \frac{\phi_j}{2} x_j^\top \Sigma x_j \tag{6.1}$$

with P_{t+1} the vector of future prices, D_{t+1} the vector of future dividends, Σ the covariance matrix of $P_{t+1} + D_{t+1}$ and r the risk-free rate. Frazzini and Pedersen (2010) assume that investors face some borrowing constraints:

$$m_j \left(x_j^\top P_t\right) \leq W_j \tag{6.2}$$

They consider three cases:

1. If $m_j < 1$, the investor must hold some of his wealth in cash.
2. If $m_j = 1$, the investor cannot use leverage because of regulatory constraints or borrowing capacity.
3. If $m_j > 1$, the investor is able to leverage his exposure on risky assets.

The equilibrium between demand and supply implies that:

$$\sum_{j=1}^m x_j = \bar{x} \tag{6.3}$$

[8]See Equation (1.3) on page 6.

where $\bar{x}_i$ is the number of shares outstanding of asset i and $\bar{x} = (\bar{x}_1, \ldots, \bar{x}_n)$. In other words, $\bar{x}$ is also the market capitalization portfolio. The existence of investor constraints will modify the traditional relationship between the risk premium and the beta of asset i:

$$\mathbb{E}_t [R_{i,t+1}] - r = \beta_i (\mathbb{E}_t [R_{t+1} (\bar{x})] - r) \tag{6.4}$$

Using the previous framework, Frazzini and Pedersen (2010) deduce then:

$$\bar{x} = \frac{1}{\phi} \Sigma^{-1} (\mathbb{E}_t [P_{t+1} + D_{t+1}] - (1 + r + \psi) P_t)$$

with $\phi = \left(\sum_{j=1}^m \phi_j^{-1}\right)^{-1}$ and $\psi = \sum_{j=1}^m \phi \phi_j^{-1} \lambda_j m_j$. Let $\beta_i = \beta (e_i \mid \bar{x})$ be the beta of asset i with respect to the market portfolio. Frazzini and Pedersen (2010) show that:

$$\mathbb{E}_t [R_{i,t+1}] - r = \alpha_i + \beta_i (\mathbb{E}_t [R_{t+1} (\bar{x})] - r) \tag{6.5}$$

where $\alpha_i = \psi (1 - \beta_i)$. If we compare this relationship with Equation (6.4), we notice the presence of a new term α_i, which is Jensen's alpha. Finally, Frazzini and Pedersen (2010) conclude that "*the alpha decreases in the beta* β_i" and "*the Sharpe ratio is highest for an efficient portfolio with a beta less than 1 and decreases in* β_i *for higher betas and increases with lower betas*".

Even though this result is not new[9], it does clarify some insights about the relationship between risk premium and beta. To illustrate this model, we consider a financial market with fifteen risky assets. We assume that their volatilities range from 3% to 17% with an increment of 1%. The correlation matrix is constant with $\rho = 50\%$. The Sharpe ratio is the same for all risky assets and is equal to 0.5 while the risk-free return is equal to 2%. We consider five investors and different sets of portfolio constraints m_j:

	Investor j				
Set	1	2	3	4	5
#1	1.0	1.0	1.0	1.0	1.0
#2	0.8	0.8	1.0	1.0	1.0
#3	0.7	0.7	0.7	0.7	1.0
#4	0.5	0.5	0.5	0.5	1.0

For each set, we find the optimized portfolios $x_j^\star$ for the five investors and then deduce the market portfolio $\bar{x}$. We also compute the beta of each asset with respect to the market portfolio and the corresponding alpha. The results are given in Figure 6.8. We verify the negative relationship between β_i and α_i.

Asness *et al.* (2012) use the previous results to formalize a theory of leverage aversion. Because of borrowing constraints, some investors prefer to have a portfolio with a higher expected return than the tangency portfolio[10] (see

[9] Black (1972) has already shown that the slope of the capital market line changes when there are borrowing restrictions.

[10] For instance, some investors may prefer portfolios A, B or C because they cannot leverage the tangency portfolio.

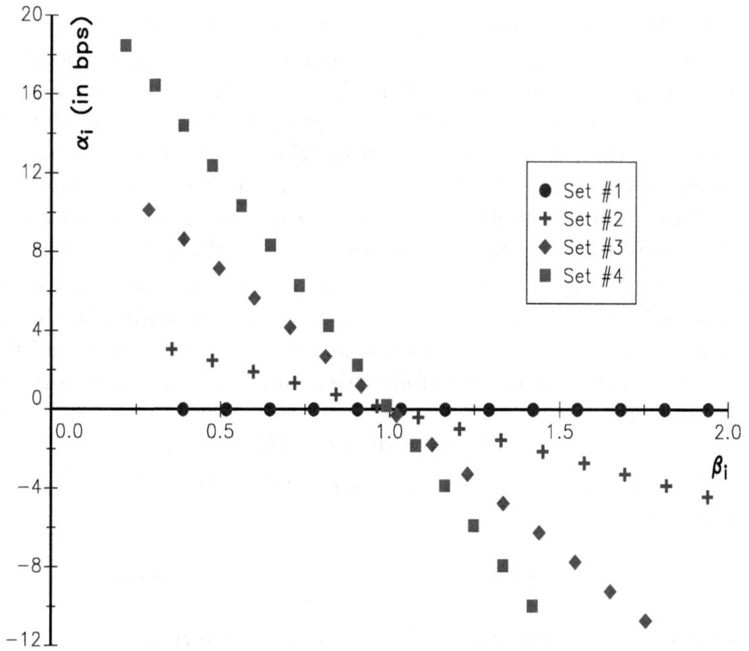

FIGURE 6.8: Relationship between the beta β_i and the alpha α_i in the presence of borrowing constraints

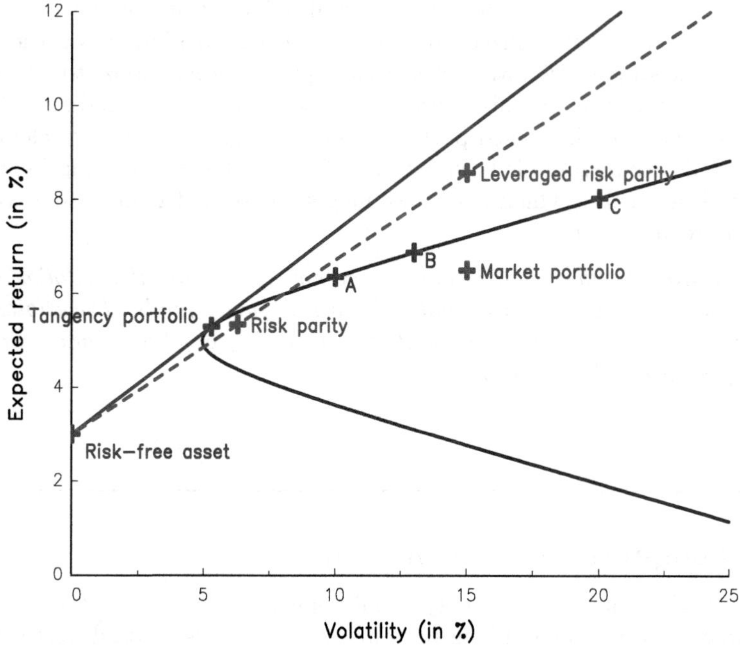

FIGURE 6.9: Impact of leverage aversion on the efficient frontier

Figure 6.9). The market portfolio is then the average of the different suboptimal portfolios, and also does not correspond to the tangency portfolio. In this case, the market portfolio is overweight on riskier assets (equities) and underweight on safer assets (bonds). According to Asness *et al.* (2012), the safer assets are then traded at low prices whereas the expected returns of riskier assets are reduced. In this case, even if a risk parity portfolio is not optimal, it outperforms many portfolios (market portfolio, 60/40 asset mix policy) because it is more invested in safer assets and leverages them.

The problem with this theory is that it ignores the cost of leverage due to gamma trading. However, this issue has been largely studied in the case of portfolio insurance (Perold, 1986; Perold and Sharpe, 1988; Black and Perold, 1992). Let S_t be the value of a portfolio. We assume that S_t follows a geometric Brownian motion:

$$\mathrm{d}S_t = \mu S_t \,\mathrm{d}t + \sigma S_t \,\mathrm{d}W_t$$

We consider a leveraged strategy on this portfolio. Perold (1986) shows that its value is given by:

$$V_t = V_0 \exp\left(\left(r + m(\mu - r) - \frac{1}{2}m^2\sigma^2\right)t + m\sigma W_t\right)$$

with m the leverage coefficient[11]. We can then deduce that:

$$V_t = V_0 \exp\left(\left(r + m(\hat{\mu} - r) - \frac{1}{2}\left(m^2 - m\right)\hat{\sigma}^2\right)t\right)$$

where $\hat{\mu}$ and $\hat{\sigma}$ are the return and the realized volatility of the underlying portfolio. We see that realized volatility has a negative impact on the value of the leveraged portfolio. Moreover, this impact is proportional to $(m^2 - m)$ meaning that it can be substantial if $m \gg 1$. If we add trading costs to these gamma costs, leveraged portfolios present high risks. Nevertheless, risk parity funds use reasonable leverage ($m \leq 3$) and a low rebalancing frequency (weekly or monthly), which is why we may suppose that gamma and trading costs are relatively small.

Remark 60 *The theory of leverage aversion applies to all portfolios which are overweight in safer assets, not only to risk parity funds. The choice of a risk parity allocation rather than a 20/80 (or 30/70) asset mix policy can then not be justified by this theory.*

6.2 Long-term investment policy

Risk parity is an investment style that appeals to sophisticated investors, in particular institutional and long-term investors (as pension funds or sovereign

[11] $m = 1$ corresponds to the unleveraged portfolio.

wealth funds). This may sound curious, because risk parity is completely ignorant of expected returns. However, if predictability of returns is difficult in the short term, there is some evidence of return predictability in the long run (Barberis, 2002). In this context, risk parity may appear to be inappropriate to long-term investors, who have more time than other investors and can ignore the short-term dynamics of the financial markets. Moreover, their investment decisions are generally based on macro-economic scenarios. In this section, we see how risk parity is applied in long-term investment policies, how to reconciliate risk parity with the economic approach of institutional investors and why some investors use it instead of only investing in long-term assets (equities, real estate, infrastructure).

6.2.1 Capturing the risk premia

In order to understand the asset allocation of long-term investors, we consider some case studies. For example, we report the average allocation of European pension funds in Figure 6.10. We note some substantial differences between countries. For example, the weight of bonds is equal to 76% in France whereas it represents only 27.1% in the UK. Another example is the weight of equities, which stands at 12.1% in Spain and 45.8% in the UK. If we calculate the ratio of the weights of cash, bills and bonds in the portfolio, we obtain an average of 55.3% for European countries, with levels ranging from 31% to 78%. However, these figures are misleading, and some large pension funds have a large part of their portfolios in equities. For instance, the 2012 reference portfolio of CPP[12] Investment Board comprises 65% of equities and 35% of bonds. The asset mix policy of CalPERS[13] is close to a 60/40 portfolio (Sharpe, 2010). We obtain similar figures for the strategic asset allocation of the Norwegian Government Pension Fund Global[14] managed by Norges Bank Investment Management (NBIM). The 60/40 asset mix policy has its roots in the 1980s (Ambachtsheer, 1987). As noted by Chaves *et al.* (2011), this portfolio posture is "*a hybrid child of legacy portfolio practice and return targeting. [...] By assuming a 9% equity return and a 6.5% bond return, the 60/40 portfolio conveniently achieves the 8% portfolio return target that is common to most pension funds*".

The objective of pension funds is then to capture the risk premia of traditional asset classes, in particular the equity risk premium. As shown previously, the 60/40 portfolio concentrates its risk in the equity side and uses the bond exposure to hedge part of the equity position[15]. However, the efficiency of this portfolio has been called into question for two reasons. The first concerns the poor performance of this portfolio over the past decade. The second concerns the fact that historical figures are not necessarily reliable indicators of future

[12]Canada Pension Plan.

[13]California Public Employees' Retirement System.

[14]Also known as the Norwegian Petroleum Fund.

[15]See Qian (2011) for a comprehensive analysis of its drawbacks.

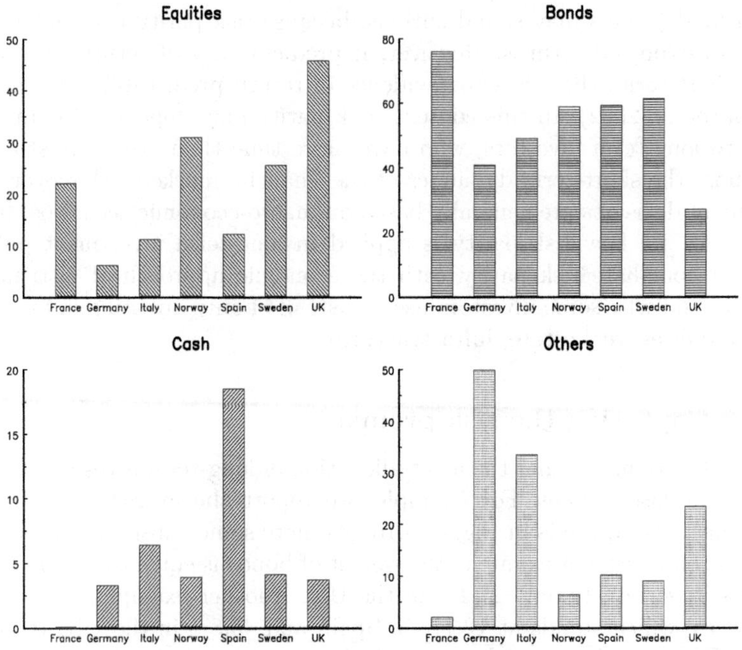

FIGURE 6.10: Average allocation of European pension funds

Source: Investment & Pensions Europe, September 2010.

performance. Some pension funds have therefore decided to abandon this reference portfolio and the 'cult of equity'. Capturing the equity risk premium is no longer the only objective of these pension funds. The risk budgeting approach can help them match their new objectives, namely inflation hedging, capital protection and diversification.

6.2.2 Strategic asset allocation

6.2.2.1 Allocation between asset classes

We have already defined strategic asset allocation (SAA) in the first chapter. In fact, it concerns the choice of equities, bonds, and alternative assets that the investor wishes to hold over the long run. By construction, SAA requires long-term assumptions of asset risk/return characteristics as a key input. This can be achieved by using macroeconomic models and forecasts of structural factors such as population growth, productivity and inflation (Eychenne *et al.*, 2011). Using these inputs, one can obtain a SAA portfolio using a mean-variance optimization procedure. Because of the uncertainty of these inputs and the instability of mean-variance portfolios, some institutional investors prefer to use these figures as a criterion when selecting the asset classes they would like to have in their strategic portfolio and to define the corresponding risk budgets. For instance, such an approach is used by the Danish pension fund ATP. Indeed, it defines its SAA using a risk parity approach. According[16] to Henrik Gade Jepsen, CIO of ATP:

> *"Like many risk practitioners, ATP follows a portfolio construction methodology that focuses on fundamental economic risks, and on the relative volatility contribution from its five risk classes. [...] The strategic risk allocation is 35% equity risk, 25% inflation risk, 20% interest rate risk, 10% credit risk and 10% commodity risk."*

These risk budgets are then transformed into asset class weights. At the end of Q1 2012, the asset allocation of ATP was also 52% in fixed-income, 15% in credit, 15% in equities, 16% in inflation and 3% in commodities[17].

Let us illustrate this process with an example. We consider a universe of nine asset classes: US Bonds 10Y (1), EURO Bonds 10Y (2), Investment Grade Bonds (3), High Yield Bonds (4), US Equities (5), Euro Equities (6), Japan Equities (7), EM Equities (8) and Commodities (9). In Tables 6.3 and 6.4, we indicate the long-run statistics used to compute the strategic asset allocation[18]. Based on these statistics, we assume that the pension fund decides to define, with respect to its constraints, the strategic portfolio according to the risk budgets[19] given in Figure 6.11.

TABLE 6.3: Expected returns and risks for the SAA approach (in %)

	(1)	(2)	(3)	(4)	(5)	(6)	(7)	(8)	(9)
μ_i	4.2	3.8	5.3	10.4	9.2	8.6	5.3	11.0	8.8
σ_i	5.0	5.0	7.0	10.0	15.0	15.0	15.0	18.0	30.0

If we match these risk budgets, we obtain the solution RB given in Table 6.5. Of course, the pension fund may modify this strategic portfolio by using the expected returns. This can be done within Black-Litterman or tracking error frameworks. For example, if we would like to maximize the expected return of the portfolio according to a 1% tracking error with respect to the RB portfolio, we obtain the RB* portfolio given in Table 6.5. We can compare this modified portfolio with the mean-variance optimized (MVO) portfolio which has the same ex-ante volatility. The results are reported in Table 6.5 and in Figure 6.12. First, we note that the two portfolios RB* and MVO are very close in terms of risk/return profile. Second, the RB* portfolio is much more diversified than the MVO portfolio, which concentrates 50% of its risk

[16] *Source*: Investment & Pensions Europe, June 2012, Special Report on Risk Parity.

[17] *Source*: FTfm, June 10, 2012.

[18] These figures are taken from Eychenne *et al.* (2011).

[19] In real life, the calibration of the risk budgets is more complex, because it involves the uncertainty around the return forecasts and the design of long-run economic scenarios.

TABLE 6.4: Correlation matrix of asset returns for the SAA approach (in %)

	(1)	(2)	(3)	(4)	(5)	(6)	(7)	(8)	(9)
(1)	100								
(2)	80	100							
(3)	60	40	100						
(4)	−20	−20	50	100					
(5)	−10	−20	30	60	100				
(6)	−20	−10	20	60	90	100			
(7)	−20	−20	20	50	70	60	100		
(8)	−20	−20	30	60	70	70	70	100	
(9)	0	0	10	20	20	20	30	30	100

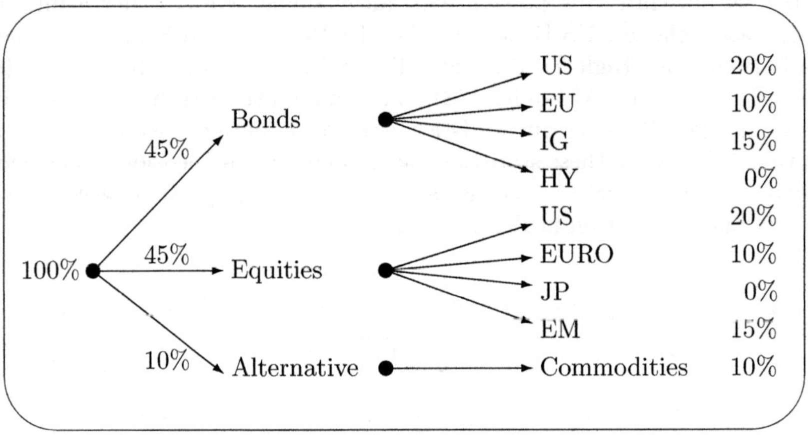

FIGURE 6.11: Risk budgeting policy of the pension fund (SAA approach)

in the EM Equities asset class. As a consequence, the MVO portfolio is too far from the pension fund's objective in terms of risk budgeting to be an acceptable strategic portfolio.

6.2.2.2 Asset classes or risk factor classes

Today, some investors are exploring a new approach combining the risk budgeting approach to define the asset allocation, and the economic approach to define the factors. This approach has already been proposed by Kaya *et al.* (2011) who use two economic factors: growth and inflation. As explained by Eychenne *et al.* (2011), these factors are the two main pillars of strategic asset allocation models. Using their long-run path, we can then define the long-run path for short rates, bonds, equities, high yield, etc. This approach

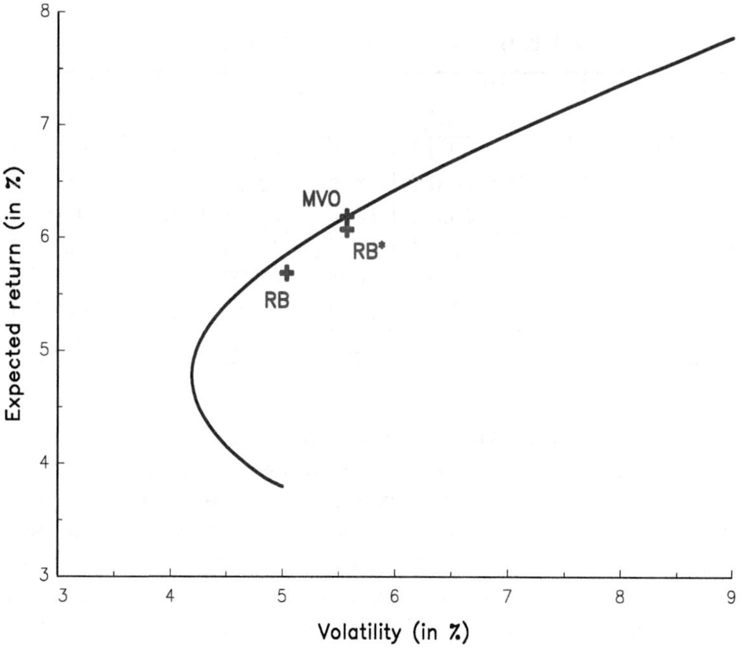

FIGURE 6.12: Strategic asset allocation in Markowitz framework

is appropriate for pension funds with liabilities that are indexed on some economic factors such as inflation.

Following Eychenne *et al.* (2011), we consider seven economic factors grouped into four categories:

1. Activity: gdp and industrial production.

2. Inflation: consumer prices and commodity prices.

3. Interest rate: real interest rate and slope of the yield curve.

4. Currency: real effective exchange rate.

The investment universe is composed of thirteen asset classes classified as follows: equity (US, Euro, UK and Japan), sovereign bonds (US, Euro, UK and Japan), corporate bonds (US, Euro), High yield (US, Euro) and TIPS (US). We assume a linear factor model:

$$R_t = A\mathcal{F}_t + \varepsilon_t \tag{6.6}$$

where R_t is the vector of asset returns and $\mathcal{F}_t$ is the vector of the seven economic factors. We collect quarterly data from Datastream and estimate the model (6.6) using YoY relative variations for the study period Q1 1999 –

TABLE 6.5: Long-term strategic portfolios

Asset class	RB		RB*		MVO	
	x_i	$\mathcal{RC}_i^\star$	x_i	$\mathcal{RC}_i^\star$	x_i	$\mathcal{RC}_i^\star$
(1)	36.8%	20.0%	45.9%	18.1%	66.7%	25.5%
(2)	21.8%	10.0%	8.3%	2.4%	0.0%	0.0%
(3)	14.7%	15.0%	13.5%	11.8%	0.0%	0.0%
(5)	10.2%	20.0%	10.8%	21.4%	7.8%	15.1%
(6)	5.5%	10.0%	6.2%	11.1%	4.4%	7.6%
(8)	7.0%	15.0%	11.0%	24.9%	19.7%	49.2%
(9)	3.9%	10.0%	4.3%	10.3%	1.5%	2.7%

Q2 2012. We can then use the framework given in Chapter two (see Section 2.5 on page 135) to compute the risk decomposition of SAA portfolios with respect to economic factors[20]. For example, we consider the four portfolios given in Table 6.6. The first portfolio is a balanced stock/bond asset mix, the second portfolio represents a defensive allocation with only 20% invested in equities, and the third portfolio represents an aggressive allocation with 80% invested in equities. The fourth portfolio is calibrated such that activity, inflation, interest rates and currency represent respectively 34%, 20%, 40% and 5%. In this scenario, the overall weight on equities sums to 49%, while the weight on bonds sums to 51% with a large position on corporate bonds.

TABLE 6.6: Weights of the SAA portfolios

Asset class	Region	#1	#2	#3	#4
Equity	US	20%	10%	30%	19.0%
	EU	20%	10%	30%	21.7%
	UK	5%		10%	6.2%
	JP	5%		10%	2.3%
Sovereign Bonds	US	10%	20%	10%	5.9%
	EU	5%	15%	10%	
	UK	5%	5%		
	JP	5%	5%		
Corporate Bonds	US	5%	5%		24.1%
	EU	5%	5%		10.7%
High Yield	US	5%	5%		2.6%
	EU	5%	5%		7.5%
TIPS	US	5%	15%		

In Table 6.7, we report the risk contributions of these allocations with respect to our four categories and an additional grouping representing specific

[20]See also Deguest *et al.* (2013) for an application with PCA factors.

risk not explained by the economic factors. We obtain results coherent with financial and economic theories. For example, activity explains a large part of the risk of the aggressive portfolio (#3). The defensive portfolio (#2) concentrates most of the risk on interest rates. Holding a portfolio more exposed to inflation risk implies de-leveraging the exposure on sovereign bonds and TIPS (cf. Portfolio #4).

TABLE 6.7: Risk contributions of SAA portfolios with respect to economic factors

Factor	#1	#2	#3	#4
Activity	36.91%	19.18%	51.20%	34.00%
Inflation	12.26%	4.98%	9.31%	20.00%
Interest rate	42.80%	58.66%	32.92%	40.00%
Currency	7.26%	13.04%	5.10%	5.00%
Residual factors	0.77%	4.14%	1.47%	1.00%

Remark 61 *By considering risk factors, the risk parity approach reintroduces economic scenarios at the heart of the allocation process and reconciles the economic nature of strategic asset allocation with risk budgeting techniques.*

6.2.2.3 Allocation within an asset class

Factor models are generally used to allocate within an asset class. For instance, if we consider the equity asset class, we can compute the sensitivities of the different equity exposures to Fama-French-Carhart factors, but also to country and sector factors. Let us consider the example of an investor who would like to allocate an American equity exposure between capitalization-weighted and risk-based indices. We use the three Fama-French factors (MKT, SMB and HML) and the momentum factor of Carhart. Using the S&P 100 universe, we obtain the results given in Table 6.8. We note that the indices are negatively sensitive to the SMB factors. This is normal, because the S&P 100 (or CW) index is composed of stocks with large capitalization. The beta with respect to the value/growth (or HML) factor is positive for the four risk-based indices. In order to measure the magnitude of these sensitivities, we have reported the (absolute) volatility contributions in Figure 6.13. We see that the MV portfolio is exposed to the SMB and HML factors which represent 15% of its volatility, whereas MDP and MV portfolios are principally exposed to the HML factor (about 6%). Let us now consider long-short portfolios between risk-based indices and the S&P 100 index. In Figure 6.13, we verify that EW and ERC portfolios take small risk with respect to the market factor (MKT) contrary to the MV portfolio.

The framework with economic factors may also be used to help the investors to allocate within this asset class. Using the same data as previously, we obtain the results in Table 6.9. As expected, the capitalization-weighted

TABLE 6.8: Estimate of the loading matrix A (Jan. 1992 – Jun. 2012)

	MKT	SMB	HML	MOM
CW	0.98	−0.26	−0.06	−0.01
EW	0.99	−0.08	0.22	−0.13
MV	0.58	−0.16	0.18	−0.03
MDP	0.80	−0.04	0.21	−0.11
ERC	0.87	−0.11	0.24	−0.09

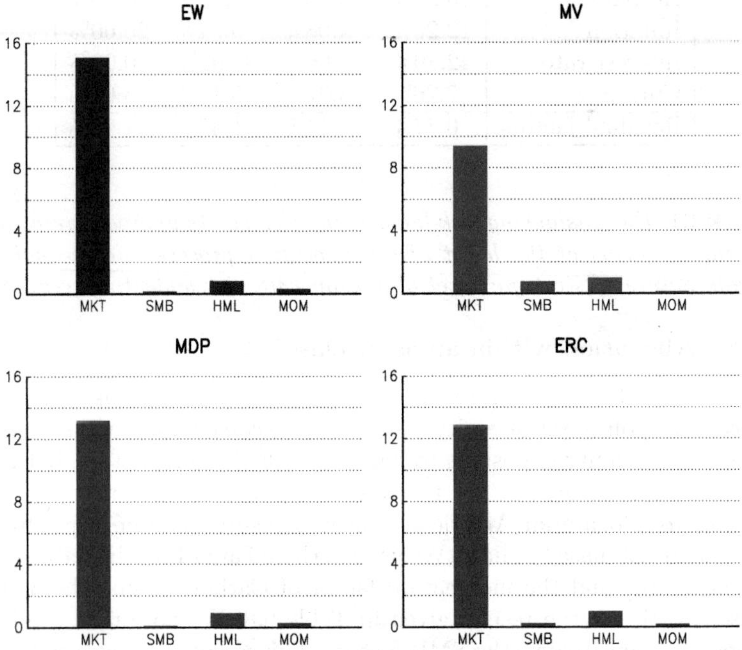

FIGURE 6.13: Volatility decomposition of the risk-based S&P 100 indices

TABLE 6.9: Risk contributions of risk-based S&P 100 indices with respect to economic factors (Q1 1992 – Q2 2012)

Factor	CW	EW	MV	MDP	ERC
Activity	71.7%	70.0%	29.4%	41.8%	62.1%
Inflation	21.8%	16.7%	4.7%	9.4%	9.5%
Interest rate	6.0%	12.7%	64.7%	46.9%	27.5%
Currency	0.6%	0.6%	1.2%	1.9%	0.9%

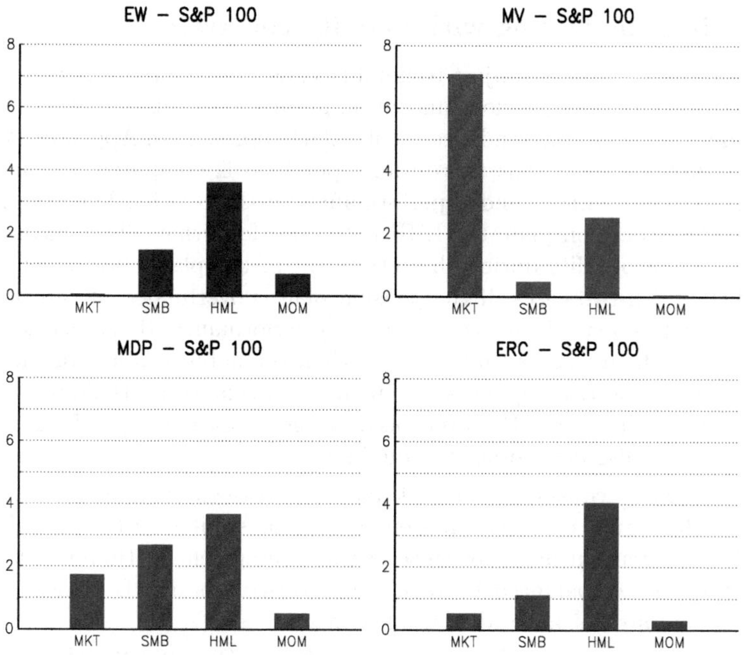

FIGURE 6.14: Volatility decomposition of long-short portfolios

portfolio is more sensitive to the activity than the risk-based indices. The high exposure of the minimum variance (or MV) portfolio to the interest rates raises the question of its behavior: is it an equity-like or a bond-like asset? Certainly a mix of them. In this case, the frontier between asset classes becomes blurred. Therefore, the risk factor parity based on economic factors may help the investor, even though the allocation exercise is difficult.

Remark 62 *As explained previously, strategic asset allocation is performed by considering economic scenarios. As a result, the weights of the SAA portfolio reflect the risks that the investor would like to take in the different economic factors. For instance, if the investor thinks that economic growth will be stronger in region A than in region B, the risk allocated to equities belonging to region A may be larger than the risk allocated to equities belonging to region B. In the end, the decision on the risk to be allocated to the different asset classes is a complex process depending on the forecasts concerning economic growth, inflation, monetary policy, etc. Using an alternative-weighted index in place of the capitalization-weighted index may then induce an allocation that is not coherent with the strategic asset allocation. These considerations explain why allocating between alternative-weighted indices is considered by some sophisticated institutional investors as a SAA choice. Such a decision concerns therefore the top-down approach and not the bottom-up (or asset selection) approach of long-term investment policy.*

6.2.3 Risk budgeting with liability constraints

When the investor is subject to liability constraints, long-term investment policy differs from the previous allocation methods. In this case, the investor must adopt a liability driven investment (LDI) approach. In Appendix A.3.3.3 on page 332, we show that the optimal portfolio is composed of a liability hedging portfolio (LHP) and a performance portfolio (PP). For example, if we consider a pension plan, the LHP is the portfolio that allows payment of future pension benefits during the retirement of the plan's participants. We then distinguish two cases. If the pension plan is overfunded, it can buy the LHP and it has a surplus to generate extra performance. If the pension plan is underfunded, the assets value is lower than the liability value. In this case, it must take some risks by investing in the performance portfolio in order to improve its funding ratio. However, because the performance portfolio is risky, its return may also deteriorate the funding ratio.

Peters (2011) considers that risk parity portfolios offer liability hedging benefits. He argues that they present durations similar to those of defined benefit (DB) pension plans, because of the leverage effect. However, this argument is counter balanced by regulatory constraints of DB plans, because most of them are not allowed to use leverage.

Another route to use risk parity is suggested by Qian (2012). The idea is to consider risk parity portfolios as the performance portfolio. For an overfunded pension plan, investing the surplus in a risk parity portfolio decreases the risk. In this case, the pension fund may expect extra performance, which is more certain than considering the 60/40 asset mix strategy. If the pension plan is underfunded, a risk parity portfolio will be less aggressive than current performance portfolios. This implies a higher recovery time. One may think that the probability of achieve a funding ratio equal to 100% is higher, but it remains an open question.

6.3 Absolute return and active risk parity

We have seen previously that risk parity is an appealing method for building diversified funds. However, in the asset management industry, most risk parity strategies concern absolute return funds, not diversified funds. Based on hedge fund strategies, the objective of an absolute return fund is to offer positive returns every year. If we consider the results in Table 6.2 on page 279, we notice that the risk parity fund faces a drawdown of 21.22%. A pure risk parity strategy can therefore not be the only performance engine when building an absolute return fund.

Active risk parity refers to investment strategies that use risk parity portfolios as the neutral allocation (Bhansali, 2012). In practical terms, this means that if the fund manager has no views, the weights of the active risk parity portfolio are equal to the weights of the risk parity portfolio. On the other hand, if he is positive on equities and negative on bonds, he will increase the weight of equities and decrease the weight of bonds. In such cases, the choice of the risk parity portfolio as the neutral allocation makes a lot of sense, because it is balanced in terms of risk contribution and performance contribution. From an ex-ante point of view, the risk parity portfolio is then a strategy without any active bets. This explains why many absolute return funds use it as the neutral allocation.

Whereas risk parity strategies ignore expected returns, active risk parity strategies reintroduce them in order to generate better performance. However, there are many methods for building an active risk parity strategy. This means that active risk parity and pure risk parity are very different. Indeed, risk parity strategies can be compared with each other, but not with active risk parity. Thus, a fund manager may use discretionary or quantitative bets, he may control or not the deviation of the allocation with a tracking error objective, he may define the deviation with respect to the weights or the risk budgets, short positions may be allowed or not, etc. In this context, the performance of the active risk parity strategy will depend more on the fund manager's ability than on the risk parity portfolio.

To illustrate an active risk parity strategy, we consider our example with the equity and bond asset classes represented by the MSCI World and Citigroup WGBI indices. While the neutral allocation is given by the ERC portfolio, we consider a trend-following investment style. We assume that the expected returns correspond to the one-year trend, which is estimated using a uniform moving average of the past 260 daily returns. We rebalance the portfolio on a weekly frequency. The allocation is done with the Black-Litterman framework by considering a 4% tracking error with respect to the neutral portfolio[21]. We impose that the risky asset weights are positive, but their sum may be less than one, because we target a volatility of 4.5%. The results[22] are given in Table 6.10 and Figure 6.15. We note that active risk parity (ARP) reduces drawdowns with respect to risk parity (RP). Indeed, the maximum drawdown is now 9.35%. We also obtain a better Sharpe ratio (0.67 versus 0.35). We also notice that the turnover τ has highly increased. Compared to the risk parity strategy, the turnover of the active risk parity strategy has been multiplied by a factor larger than sixteen. However, it remains reasonable for an absolute return fund, because we turn the portfolio only five times in the year.

The previous simulation uses only stocks and bonds (S/B). Let us now

[21]Using notations introduced in Section 1.1.5 on page 22, we have $P = I_n$, $Q = \hat{\mu}$, $\Omega = 0.05^2 I_n$ with $\hat{\mu}$ the vector of the one-year trends. We also assume that the Sharpe ratio of the neutral allocation is 0.50 in order to estimate the implied risk premia.

[22]Compared to the results on page 279, we have some small differences for the risk parity strategy, because the rebalancing frequency is not the same. It was monthly previously while it is weekly here.

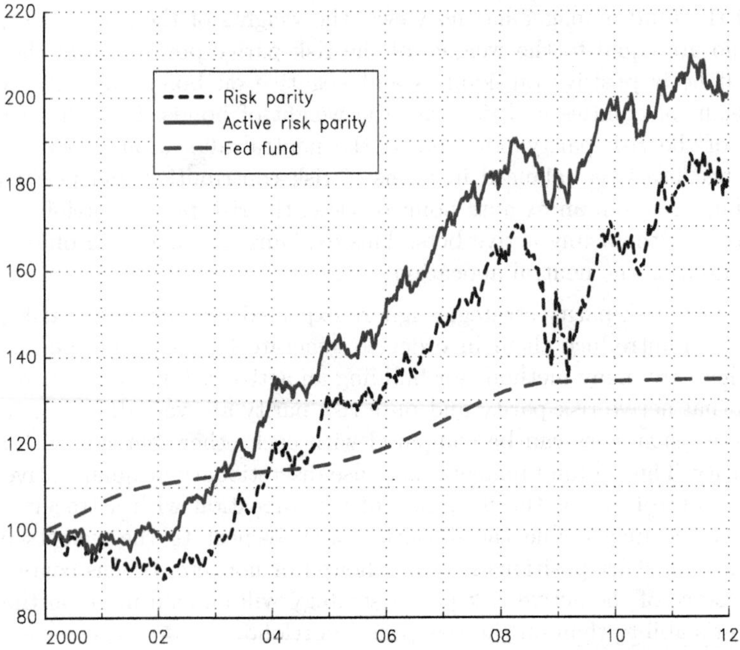

FIGURE 6.15: Simulated performance of the S/B risk parity strategies

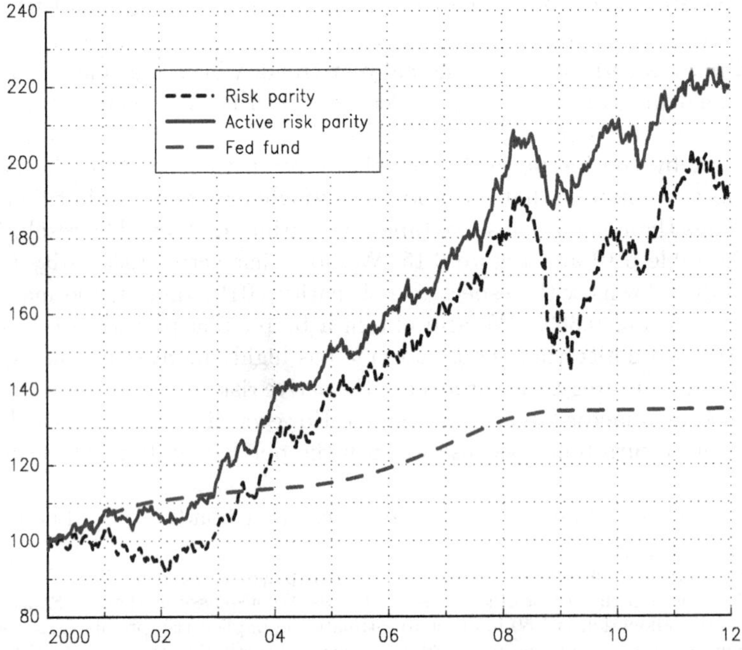

FIGURE 6.16: Simulated performance of the S/B/C risk parity strategies

introduce commodities (S/B/C) by using the DJ UBS Commodity index. In this case, an ERC portfolio may not be pertinent to define the neutral allocation, because commodities do not exhibit risk premium. However, this asset class may be interesting in order to have a hedging exposure on inflation. In this case, a risk budgeting portfolio is more appropriate. If we assume that the risk budgets are respectively 40%, 40% and 20% for bonds, equities and commodities[23], we obtain the results in Table 6.10 and in Figure 6.16. The volatility is close to the previous one, but the return is improved by 1% per year. In the end, the Sharpe ratio is about 0.80, which is a good performance for an absolute return strategy.

TABLE 6.10: Statistics of active risk parity strategies

AC	Strategy	$\hat{\mu}_{1Y}$	$\hat{\sigma}_{1Y}$	SR	$\mathcal{MDD}$	γ_1	γ_2	τ
S/B	RP	5.10	7.30	0.35	−21.39	0.07	2.68	0.30
	ARP	5.99	5.16	0.67	−9.35	0.02	2.11	4.92
S/B/C	RP	5.67	7.36	0.43	−24.55	0.01	3.29	0.39
	ARP	6.82	5.10	0.84	−10.21	0.05	1.93	6.74

Remark 63 *In practice, fund managers use many more asset classes and underlyings to design active risk parity strategies. For instance, they may distinguish between American, European, Japanese and Emerging Market equities. They may consider different bond indexations by countries, by issuers (sovereign, corporate, etc.) or by durations. Commodities may be split between energy, metals and agriculture.*

Remark 64 *In these illustrations, we have chosen to directly modify the weights of the risk parity portfolio to take into account the expected returns. Another route is to modify the risk budgets. The difficulty is then to find a method to link the risk budgets with the expected returns. A more interesting solution is to consider risk measures that depend on expected returns, such as value-at-risk or expected shortfall*[24].

[23] We use the same parameters as previously to perform the allocation with the Black-Litterman model. The only difference comes from the expected returns which are estimated using six-month trends.

[24] See Equations (2.11) and (2.12) on page 80.

Conclusion

Since the seminal work of Harry Markowitz, risk management and asset management have been very closely associated. Asset management was among the first fields of finance to integrate risk concepts in decision-making processes. This integration took place before the emergence of banking risk management with the development of value-at-risk models and financial regulation (Basle II, Solvency II, etc.). Therefore, the Sharpe ratio was already a risk-adjusted performance measure (RAPM) like the Raroc measures which are today widely used by banking and insurance sectors for managing their businesses.

The books of Peter Bernstein (1992, 2007) show admirably how risk management has improved the asset management industry and how investment problems have contributed to the development of risk management. However, it took a long time to spread the concepts of diversification, risk factors and information ratio to institutional investors, although the dot-com bubble and the recent financial crisis have accelerated this learning process.

The development of risk parity approaches marks an important milestone in the deepening of the relationship between risk and asset management. With risk parity, the primary goal is no longer to measure the risk, but to manage it. In this case, risk management and portfolio management tend to converge. The issue of performance management is nevertheless not abandoned. By better managing risk, risk parity approaches hope to manage the portfolio performance better than do traditional approaches. In a sense, investors who adopt risk parity methods acknowledge that performance generation is a difficult task. Financial markets are cruel. How many setbacks and disappointments have there been for each success story in the investment industry?

Chief investment officers of pension funds and institutionals remain investors, but some years ago they started to gradually became more like risk managers for different reasons. The various crises have reduced the flexibility and room for maneuver available to investors. At the same time, their fiduciary obligations have increased. In this context, an investment decision is not a gamble, particularly when subject to substantial liability constraints. So it is no coincidence that risk parity methods are emerging today in the aftermath of a series of black swans (Taleb, 2007): credit crisis, equity crisis, hedge fund crisis, Madoff fraud, sovereign debt crisis, etc. Economic theory teaches us that there is no free lunch and that risk must be rewarded. Modern portfolio theory adds that the risk-return tradeoff must take into account diversification. The risk parity approach is just an application of these two principles. It is not a superior method to generate performance, but provides a set of principles to guide asset allocation, in line with the 'prudent person principle'. People who think that risk parity can be a martingale will be disappointed: it is not a martingale, it is simply risk management.

In this book, we have tried to present a comprehensive analysis of risk budgeting, both from academic and professional viewpoints. The book can help students better understand investment problems. It also shows the importance of academic research for the asset management industry. For professionals, this book is a reminder that investment theory cannot be reduced to an art. It is also a science, despite all its shortcomings. Recent crises highlight the role of risk management and the risk parity approach helps bring about change in the field of asset management. Good knowledge of the foundations of these new techniques is essential to better understand investment principles.

Appendix A

Technical Appendix

In this appendix, we review the different methods to numerically solve some optimization problems. In the first section, we present in particular the quadratic programming problem, which is the core of portfolio optimization as it permits solving Markowitz and Black-Litterman models, as well as constrained least squares problems such as ridge or lasso regression. We also present non-linear optimization problems and focus on the sequential quadratic programming problem. The second section is dedicated to copula functions, which are a powerful tool for risk management and multivariate survival modeling. With copula functions, we can then define non-Gaussian risk measures for risk budgeting problems. Finally, the third section concerns dynamic portfolio optimization. After reviewing the Bellman approach and the seminal work of Merton (1969, 1971), we consider dynamic allocation problems related to long-term investment policy.

A.1 Optimization problems

A.1.1 Quadratic programming problem

A quadratic programming (QP) problem is an optimization problem with a quadratic objective function and linear inequality constraints:

$$\begin{aligned} x^\star &= \arg\min \frac{1}{2} x^\top Q x - x^\top R \\ &\text{u.c. } \quad Sx \leq T \end{aligned} \tag{A.1}$$

where x is a $n \times 1$ vector, Q is a $n \times n$ matrix and R is a $n \times 1$ vector. We note that the system of constraints $Sx \leq T$ allows specifying linear equality constraints[1] $Ax = B$ or weight constraints $x^- \leq x \leq x^+$. Most numerical

[1]This is equivalent to imposing that $Ax \geq B$ and $Ax \leq B$.

packages then consider the following formulation:

$$\begin{aligned} x^\star &= \arg\min \frac{1}{2} x^\top Q x - x^\top R \\ \text{u.c.} & \left\{ \begin{array}{l} Ax = B \\ Cx \leq D \\ x^- \leq x \leq x^+ \end{array} \right. \end{aligned} \tag{A.2}$$

because the problem (A.2) is equivalent to the canonical problem (A.1) with the following system of linear inequalities:

$$\begin{bmatrix} -A \\ A \\ C \\ -I_n \\ I_n \end{bmatrix} x \leq \begin{bmatrix} -B \\ B \\ D \\ -x^- \\ x^+ \end{bmatrix}$$

If the space Ω defined by $Sx \leq T$ is non-empty and if Q is a symmetric positive definite matrix, the solution exists because the function $f(x) = \frac{1}{2} x^\top Q x - x^\top R$ is convex. In the general case where Q is a square matrix, the solution may not exist.

The Lagrange function is also:

$$\mathcal{L}(x; \lambda) = \frac{1}{2} x^\top Q x - x^\top R + \lambda^\top (Sx - T)$$

We deduce that the dual problem is defined by:

$$\begin{aligned} \lambda^\star &= \arg\max \left\{ \inf_x \mathcal{L}(x; \lambda) \right\} \\ \text{u.c.} & \quad \lambda \geq 0 \end{aligned}$$

We note that $\partial_x \mathcal{L}(x; \lambda) = Qx - R + S^\top \lambda$. The solution to the problem $\partial_x \mathcal{L}(x; \lambda) = 0$ is then $x = Q^{-1}\left(R - S^\top \lambda\right)$. We obtain:

$$\begin{aligned} \inf_x \mathcal{L}(x; \lambda) &= \frac{1}{2}\left(R^\top - \lambda^\top S\right) Q^{-1}\left(R - S^\top \lambda\right) - \left(R^\top - \lambda^\top S\right) Q^{-1} R + \\ & \quad \lambda^\top \left(S Q^{-1}\left(R - S^\top \lambda\right) - T\right) \\ &= \frac{1}{2} R^\top Q^{-1} R - \lambda^\top S Q^{-1} R + \frac{1}{2} \lambda^\top S Q^{-1} S^\top \lambda - R^\top Q^{-1} R + \\ & \quad 2 \lambda^\top S Q^{-1} R - \lambda^\top S Q^{-1} S^\top \lambda - \lambda^\top T \\ &= -\frac{1}{2} \lambda^\top S Q^{-1} S^\top \lambda + \lambda^\top \left(S Q^{-1} R - T\right) - \frac{1}{2} R^\top Q^{-1} R \end{aligned}$$

The dual program is another quadratic program:

$$\begin{aligned} \lambda^\star &= \arg\min \frac{1}{2} \lambda^\top \bar{Q} \lambda - \lambda^\top \bar{R} \\ \text{u.c.} & \quad \lambda \geq 0 \end{aligned}$$

with $\bar{Q} = SQ^{-1}S^{\top}$ and $\bar{R} = SQ^{-1}R - T$.

Let us consider the lasso optimization problem:

$$\begin{aligned} x^{\star} &= \arg\min \frac{1}{2} x^{\top} Q x - x^{\top} R + S^{\top} \left| x - y \right| \\ & \text{u.c.} \quad \left\{ \begin{array}{l} Ax = B \\ Cx \geq D \\ x^{-} \leq x \leq x^{+} \end{array} \right. \end{aligned}$$

where y is a given vector. If we use the following decomposition:

$$x_i = y_i + \Delta_i^+ - \Delta_i^-$$

with $\Delta_i^- \geq 0$ and $\Delta_i^+ \geq 0$, we deduce that:

$$|x_i - y_i| = \left| \Delta_i^+ - \Delta_i^- \right| = \Delta_i^+ + \Delta_i^-$$

The objective function becomes:

$$x^{\star} = \arg\min \frac{1}{2} x^{\top} Q x - \left(x^{\top} R - S^{\top} \Delta^{+} - S^{\top} \Delta^{-} \right)$$

Let $\tilde{x} = \left(x_1, \ldots, x_n, \Delta_1^-, \ldots, \Delta_n^-, \Delta_1^+, \ldots, \Delta_n^+ \right)$ be the vector of unknown variables. We obtain an augmented QP problem of dimension $3n$:

$$\begin{aligned} \tilde{x}^{\star} &= \arg\min \frac{1}{2} \tilde{x}^{\top} \tilde{Q} \tilde{x} - \tilde{x}^{\top} \tilde{R} \\ & \text{u.c.} \quad \left\{ \begin{array}{l} \tilde{A}\tilde{x} = \tilde{B} \\ C\tilde{x} \geq D \\ \tilde{x}^{-} \leq \tilde{x} \leq \tilde{x}^{+} \end{array} \right. \end{aligned}$$

with:

$$\begin{aligned} \tilde{Q} &= \begin{pmatrix} Q & \mathbf{0} & \mathbf{0} \\ \mathbf{0} & \mathbf{0} & \mathbf{0} \\ \mathbf{0} & \mathbf{0} & \mathbf{0} \end{pmatrix}, \quad \tilde{R} = \begin{pmatrix} R \\ -S \\ -S \end{pmatrix}, \quad \tilde{A} = \begin{pmatrix} A & \mathbf{0} & \mathbf{0} \\ I_n & I_n & -I_n \end{pmatrix} \\ \tilde{B} &= \begin{pmatrix} B \\ y \end{pmatrix}, \quad \tilde{x}^{-} = \begin{pmatrix} x^{-} \\ \mathbf{0} \\ \mathbf{0} \end{pmatrix} \quad \text{and} \quad \tilde{x}^{+} = \begin{pmatrix} x^{+} \\ x^{+} - x^{-} \\ x^{+} - x^{-} \end{pmatrix} \end{aligned}$$

A.1.2 Non-linear unconstrained optimization

We consider the minimization problem:

$$x^{\star} = \arg\min f(x) \tag{A.3}$$

where $x \in \mathbb{R}^n$. Let $G(x)$ and $H(x)$ be the gradient vector and the Hessian matrix of $f(x)$. The optimum verifies:

$$G(x^{\star}) = \mathbf{0} \tag{A.4}$$

The first-order Taylor expansion of $G(x)$ around the point x_0 is given by:

$$G(x) = G(x_0) + H(x_0)(x - x_0)$$

If x is the solution of Equation (A.4), we obtain $G(x_0) + H(x_0)(x - x_0) = \mathbf{0}$. The Newton-Raphson algorithm uses an iterative process to find this root:

$$x_{k+1} = x_k - H_k^{-1} G_k$$

where k is the iteration index, $G_k = G(x_k)$ and $H_k = H(x_k)$. Starting from an initial point x_0, we find the solution $x^\star$ if the algorithm converges[2]. However, we generally prefer to use the following process:

$$\begin{aligned} x_{k+1} &= x_k - \lambda_k H_k^{-1} G_k \\ &= x_k + \lambda_k d_k \end{aligned}$$

where $\lambda_k > 0$ is a scalar. The difference comes from the introduction of the step length λ_k. Starting from the point x_k, the vector $d_k = -H_k^{-1} G_k$ indicates the direction to reach the maximum. Nonetheless, using a step length equal to 1 is not always optimal. For instance, we could exceed the optimum[3] or the convergence may be very slow. This is why numerical optimization methods use two types of algorithms:

1. An algorithm to approximate the Hessian matrix H_k and to compute the descent d_k.

2. A second algorithm to define the optimal step length λ_k:

$$\lambda_k = \arg \min_{\lambda > 0} f(x_k + \lambda d_k)$$

The Hessian approximation avoids singularity problems which are frequent in the neighborhood of the optimum. Press *et al.* (2007) distinguish two algorithm families to define the descent, namely conjugate gradient and quasi-Newton methods.

- In the case of the conjugate gradient approach, we have:

$$d_{k+1} = -(G_{k+1} - \varrho_{k+1} d_k)$$

For the Polak-Ribiere algorithm, the scalar ϱ is given by:

$$\varrho_{k+1} = \frac{G_{k+1}^\top G_{k+1}}{G_k^\top G_k}$$

whereas for the Fletcher-Reeves algorithm, we have:

$$\varrho_{k+1} = \frac{(G_{k+1} - G_k)^\top G_{k+1}}{G_k^\top G_k}$$

[2]We stop the algorithm when the gradient is close to zero. For example, the stopping rule may be $\max_i |G_{k,i}| \leq \varepsilon$ with ε the allowed tolerance.

[3]This means that f does not necessarily decrease at each iteration.

- For quasi-Newton methods, the direction is defined as follows:

$$d_{k+1} = -\tilde{H}_{k+1} G_{k+1}$$

where $\tilde{H}$ is an approximation of the inverse of the Hessian matrix. Its expression is:

$$\begin{aligned}\tilde{H}_{k+1} &= \tilde{H}_k - \frac{\tilde{H}_k y_k y_k^\top \tilde{H}_k}{y_k^\top \tilde{H}_k y_k} + \frac{s_k s_k^\top}{s_k^\top y_k} + \\ &\quad \beta \left(\tilde{H}_k y_k - \theta_k s_k\right)\left(\tilde{H}_k y_k - \theta_k s_k\right)^\top\end{aligned}$$

with $y_k = G_{k+1} - G_k$, $s_k = x_{k+1} - x_k$ and:

$$\theta_k = \frac{y_k^\top \tilde{H}_k y_k}{s_k^\top y_k}$$

The Davidon, Fletcher and Powell (DFP) algorithm corresponds to $\beta = 0$, whereas the Broyden, Fletcher, Goldfarb and Shanno (BFGS) algorithm is given by:

$$\beta = \frac{1}{y_k^\top \tilde{H}_k y_k}$$

To find the optimal value of λ_k, we employ a simple one-dimension minimization algorithm[4] such as the golden section, Brent's method or the cubic spline approximation (Press *et al.*, 2007).

Remark 65 *Newton's method may also be used to solve non-linear optimization problems with linear constraints:*

$$\begin{aligned} x^\star &= \arg\min f(x) \\ u.c. &\quad Ax = B \end{aligned}$$

Indeed, this constrained problem is equivalent to the following unconstrained problem:

$$y^\star = \arg\min g(y)$$

where $g(y) = f(Cy + D)$, C *is an orthonormal basis for the nullspace of* A, $D = \left(A^\top A\right)^+ A^\top B$ *and* $\left(A^\top A\right)^+$ *is the Moore-Penrose pseudo-inverse of* $A^\top A$. *The solution is then:*

$$x^\star = Cy^\star + D$$

[4]Computing the optimal value of λ_k may be time consuming. In this case, we may also prefer the half method which consists in dividing the test value by one half each time the function fails to decrease – λ then takes the respective values 1, 1/2, 1/4, 1/8, etc. – and to stop when the criteria $f(x_k + \lambda_k d_k) < f(x_k)$ is satisfied.

A.1.3 Sequential quadratic programming algorithm

The sequential quadratic programming (or SQP) algorithm solves this constrained non-linear programming problem:

$$\begin{aligned} x^\star &= \arg\min f(x) \\ & \text{u.c.} \quad \left\{ \begin{array}{l} A(x) = \mathbf{0} \\ B(x) \geq \mathbf{0} \end{array} \right. \end{aligned} \tag{A.5}$$

where $A(x)$ and $B(x)$ are two multi-dimensional non-linear functions. Like Newton's methods, this algorithm is an iterative process:

$$x_{k+1} = x_k + \lambda_k d_k$$

with:

$$\begin{aligned} d_k &= \arg\min \frac{1}{2} d^\top H_k d + d^\top G_k \\ & \text{u.c.} \quad \left\{ \begin{array}{l} \partial_x A(x_k) d + A(x_k) = \mathbf{0} \\ \partial_x B(x_k) d + B(x_k) \geq \mathbf{0} \end{array} \right. \end{aligned}$$

It consists in replacing the non-linear programming problem by a sequence of quadratic programming problems (Boggs and Tolle, 1995). The QP problem corresponds to the second-order Taylor expansion of $f(x)$:

$$f(x_k + \delta) = f(x_k) + \delta^\top G_k + \frac{1}{2} \delta^\top H_k \delta$$

with:

$$\left\{ \begin{array}{l} A(x_k + \delta) = A(x_k) + \partial_x A(x_k) \delta = \mathbf{0} \\ B(x_k + \delta) = B(x_k) + \partial_x B(x_k) \delta \geq \mathbf{0} \end{array} \right.$$

and $\delta = \lambda d$. We can use quasi-Newton methods to approximate the Hessian matrix H_k. However, if we define λ_k as previously:

$$\lambda_k = \min_{\lambda > 0} f(x_k + \lambda d_k)$$

we may face some problems because the constraints $A(x) = \mathbf{0}$ and $B(x) \geq \mathbf{0}$ are not necessarily satisfied. This is why we prefer to specify λ_k as the solution to this one-dimensional minimization problem:

$$\lambda_k = \min_{\lambda > 0} m(x_k + \lambda d_k)$$

where $m(x)$ is the merit function:

$$m(x) = f(x) + p_A \sum_j |A_j(x)| - p_B \sum_j \min(0, B_j(x))$$

We generally choose the penalization weights p_A and p_B as the infinite norm of Lagrange coefficients associated with linear and non-linear constraints (Nocedal and Wright, 2006).

A.1.4 Numerical solutions of the RB problem

We recall that the RB portfolio satisfies a system of non-linear equations: $\mathcal{RC}_i = b_i \mathcal{R}(x)$ for $i = 1, \ldots, n$. We can transform this non-linear system into an optimization problem[5]:

$$\begin{aligned} x^\star &= \arg\min f(x; b) \\ &\text{u.c.} \quad \mathbf{1}^\top x = 1 \quad \text{and} \quad \mathbf{0} \leq x \leq \mathbf{1} \end{aligned} \tag{A.6}$$

For example, we may specify the function $f(x; b)$ as follows:

$$f(x; b) = \sum_{i=1}^{n} \left(\mathcal{RC}_i - b_i \mathcal{R}(x)\right)^2$$

If $x^\star$ is the solution and if $f(x^\star; b) = 0$, this implies that $x^\star$ is the RB portfolio. Our experience shows that a more appropriate function $f(x; b)$ is:

$$f(x; b) = \sum_{i=1}^{n} \sum_{j=1}^{n} \left(\frac{\mathcal{RC}_i}{b_i} - \frac{\mathcal{RC}_j}{b_j} \right)^2$$

The SQP algorithm is appropriate to solve the optimization problem (A.6) with $A(x) = \mathbf{1}^\top x - 1$ and $B(x) = x$. Moreover, using analytical derivatives reduces computational times, in particular when the asset universe is large[6] (more than 100 assets). The convergence can also be improved by deleting the linear constraint $\mathbf{1}^\top x = 1$. Another way of obtaining the RB portfolio is to consider a simplified version of the optimization problem (2.31) presented on page 109:

$$\begin{aligned} x^\star &= \arg\min \mathcal{R}(x) \\ &\text{u.c.} \quad \textstyle\sum_{i=1}^{n} b_i \ln x_i \geq c \end{aligned}$$

with c an arbitrary constant such that $c < \sum_{i=1}^{n} b_i \ln b_i$. In practice, it is faster to solve this problem than the two previous ones[7].

Remark 66 *In the case where the risk measure is the volatility, Chaves et al. (2012) suggest using the Jacobi power method (Golub and Van Loan, 1996). On page 107, we have shown that the RB portfolio satisfies the following system of non-linear equations:*

$$x_i = \frac{b_i / \beta_i}{\sum_{j=1}^{n} b_j / \beta_j}$$

[5]See Equation (2.22) on page 102.

[6]We do not report here the analytical expression of the gradient and Hessian matrices, because doing so would take much space, but they are absolutely necessary in high dimensions. Moreover, we generally observe that the computation time is divided by a factor between n and n^2.

[7]This method is similar to Algorithm 1 presented in Chaves *et al.* (2012).

with β_i the beta of asset i with respect to the RB portfolio. The Jacobi power method consists in iterating the previous formula:

$$x_i^{k+1} = \frac{b_i / \beta_i^k}{\sum_{j=1}^n b_j / \beta_j^k}$$

where k is the iteration index. Here, the β_i^k values are calculated with respect to the portfolio x^k and are used to compute the new weights x^{k+1}. This algorithm converges well with small universes, but it generally fails when the number of assets is large.

Remark 67 *All the previous algorithms require an initial portfolio as input. Choosing a good starting point is then the key point to achieve the convergence when the asset universe is large (more than 100 assets). In the case of the volatility risk measure, closed-form formulas are available when the correlation matrix is $C_n(0)$ or $C_n(1)$. We suggest using one of these two analytical solutions or a mix of them to initialize the algorithm.*

A.2 Copula functions

If we use a correlation matrix to model the dependency of the random vector X, we make the implicit assumption that X is a Gaussian vector. Copula functions are a generalization of the correlation measure when this assumption does not hold.

A.2.1 Definition and main properties

Nelsen (2006) defines a bi-dimensional copula (or a 2-copula) as a function $\mathbf{C}$ which satisfies the following properties:

1. $\operatorname{Dom} \mathbf{C} = [0,1] \times [0,1]$;

2. $\mathbf{C}(0,u) = \mathbf{C}(u,0) = 0$ and $\mathbf{C}(u,1) = \mathbf{C}(1,u) = u$ for all u in $[0,1]$;

3. $\mathbf{C}$ is 2-increasing:

$$\mathbf{C}(v_1,v_2) - \mathbf{C}(v_1,u_2) - \mathbf{C}(u_1,v_2) + \mathbf{C}(u_1,u_2) \geq 0$$

   for all $(u_1,u_2) \in [0,1]^2$, $(v_1,v_2) \in [0,1]^2$ such that $0 \leq u_1 \leq v_1 \leq 1$ and $0 \leq u_2 \leq v_2 \leq 1$.

This definition means that $\mathbf{C}$ is a probability distribution with uniform margins. Let $\mathbf{F}_1$ and $\mathbf{F}_2$ be any two univariate distributions. It is obvious that $\mathbf{F}(x_1,x_2) = \mathbf{C}(\mathbf{F}_1(x_1), \mathbf{F}_2(x_2))$ is a probability distribution with margins $\mathbf{F}_1$

and $\mathbf{F}_2$. We say that it is a distribution with fixed (or given) margins. Conversely, Sklar proved in 1959 that any bivariate distribution $\mathbf{F}$ admits such a representation and that the copula $\mathbf{C}$ is unique provided the margins are continuous. This result is important, because we can associate to each bivariate distribution a copula function.

Example 41 *In Figure A.1, we have built a bivariate probability distribution by considering that the margins are an inverse Gaussian distribution and a beta distribution. The copula function corresponds to the normal copula*[8] *such that its Kendall's tau is equal to 50%.*

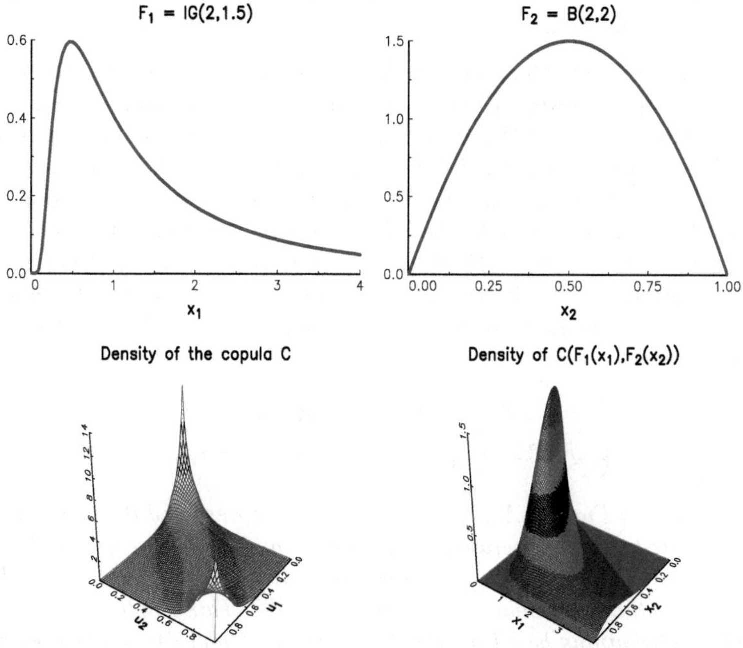

FIGURE A.1: Example of building a bivariate probability distribution with a copula function

Let $\mathbf{C}_1$ and $\mathbf{C}_2$ be two copula functions. We say that $\mathbf{C}_1$ is smaller than $\mathbf{C}_2$ and we note $\mathbf{C}_1 \prec \mathbf{C}_2$ if and only if $\mathbf{C}_1(u_1, u_2) \leq \mathbf{C}_2(u_1, u_2)$ for all $(u_1, u_2) \in [0,1]^2$. This partial order $\prec$ is called the concordance order. We also show that a copula function $\mathbf{C}$ verifies $\mathbf{C}^- \prec \mathbf{C} \prec \mathbf{C}^+$ with $\mathbf{C}^-(u_1, u_2) = \max(u_1 + u_2 - 1, 0)$ and $\mathbf{C}^+(u_1, u_2) = \min(u_1, u_2)$. Let $X = (X_1, X_2)$ be a random vector with distribution $\mathbf{F}$. We define the copula of (X_1, X_2) by the copula of $\mathbf{F}$:

$$\mathbf{F}(x_1, x_2) = \mathbf{C}\langle X_1, X_2\rangle(\mathbf{F}_1(x_1), \mathbf{F}_2(x_2))$$

[8]This copula is defined in the next section.

We can show that:

$$\mathbf{C}\left\langle g_1(X_1), g_2(X_2)\right\rangle = \mathbf{C}\left\langle X_1, X_2\right\rangle$$

if g_1 and g_2 are two non-decreasing functions on $\operatorname{Im} X_1$ and $\operatorname{Im} X_2$. Therefore, the copula function is invariant by strictly increasing transformations of the random variables. We show then the following results: $\mathbf{C}\left\langle X_1, X_2\right\rangle = \mathbf{C}^-$ if there exists a random variable X such that $X_1 = f_1(X)$ and $X_2 = f_2(X)$ with f_1 a non-decreasing function and f_2 a non-increasing function; two random variables X_1 and X_2 are independent if their dependence is the product copula $\mathbf{C}^\perp$ with $\mathbf{C}^\perp(u_1, u_2) = u_1 u_2$; $\mathbf{C}\left\langle X_1, X_2\right\rangle = \mathbf{C}^+$ if there exists a random variable X such that $X_1 = f_1(X)$ and $X_2 = f_2(X)$ with f_1 and f_2 two non-decreasing functions.

These results show that the copula $\mathbf{C}\left\langle X_1, X_2\right\rangle$ is the exhaustive statistic of the dependence between the two random variables X_1 and X_2. The notion of copula generalizes then the notion of correlation in the non-Gaussian world. As the linear correlation, it is interesting to summarize the dependence by a unique value. In this case, we use the concordance measures. Among them, the most famous are Kendall's tau and Spearman's rho. Let us consider a sample $\{(x_1, y_1), \ldots, (x_n, y_n)\}$ of the random vector (X, Y). Kendall's tau is the concordance probability minus the discordance probability of pairs[9]. Spearman's rho is the rank correlation $\operatorname{cor}(\mathbf{F}_X(X), \mathbf{F}_Y(Y))$. Their theoretical expressions are:

$$\begin{cases} \tau = 4 \iint_{[0,1]^2} \mathbf{C}(u_1, u_2)\, \mathrm{d}\mathbf{C}(u_1, u_2) - 1 \\ \varrho = 12 \iint_{[0,1]^2} u_1 u_2 \, \mathrm{d}\mathbf{C}(u_1, u_2) - 3 \end{cases}$$

Example 42 *In Figures A.2 and A.3, we have represented the level curves of bivariate distributions generated by the Frank and Gumbel copulas. For each copula, we consider four sets of margins and the copula function is calibrated in order to obtain a Kendall's tau equal to 50%. This implies that the four bivariate distributions based on the Frank (resp. Gumbel) copula present the same dependency and the smallest exhaustive statistic is given by the first quadrant*[10] *of the graphic. If we summarize this dependency by the Kendall's tau, the distributions generated by the Frank copula are then equivalent to the distributions generated by the Gumbel copula. However, this is not true, which is why such a statistical measure is not exhaustive.*

Remark 68 *The generalization of the previous results to the multivariate case is straightforward. Indeed, a n-copula is a probability distribution function*

[9] We have:

$$\tau = \Pr\{(X_i - X_j)(Y_i - Y_j) > 0\} - \Pr\{(X_i - X_j)(Y_i - Y_j) < 0\}$$

[10] It corresponds to the case when the margins are uniform.

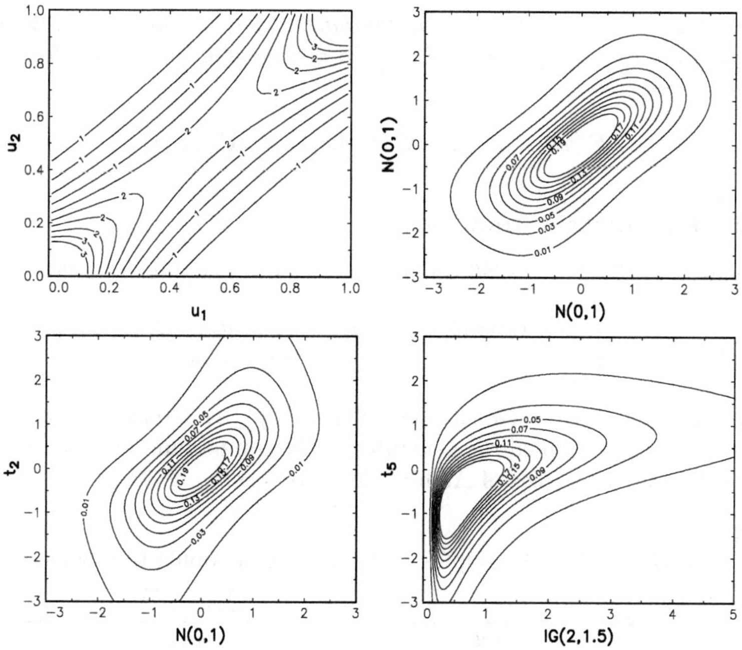

FIGURE A.2: Level curves of bivariate distributions (Frank copula)

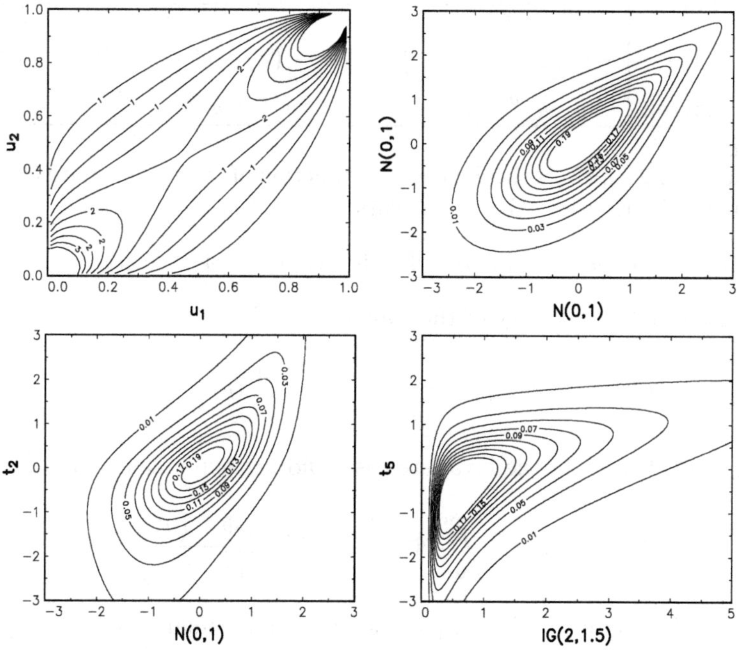

FIGURE A.3: Level curves of bivariate distributions (Gumbel copula)

$\mathbf{C}$ *from* $[0,1]^n$ *to* $[0,1]$ *with uniform margins. It follows that any multivariate probability distribution function* $\mathbf{F}$ *can be decomposed into a copula function and its margins:*

$$\mathbf{F}(x_1,\ldots,x_n) = \mathbf{C}(\mathbf{F}_1(x_1),\ldots,\mathbf{F}_n(x_n))$$

A.2.2 Parametric functions

Genest and MacKay (1986) define Archimedean copulas as follows:

$$\mathbf{C}(u_1,u_2) = \begin{cases} \varphi^{-1}(\varphi(u_1)+\varphi(u_2)) & \text{if } \varphi(u_1)+\varphi(u_2) \leq \varphi(0) \\ 0 & \text{otherwise} \end{cases}$$

with φ a C^2 function which satisfies $\varphi(1) = 0$, $\varphi'(u) < 0$ and $\varphi''(u) > 0$ for all $u \in [0,1]$. $\varphi(u)$ is called the generator of the copula function. In Table A.1, we provide some examples of Archimedean copulas[11].

TABLE A.1: Examples of Archimedean copula functions

Copula	$\varphi(u)$	$\mathbf{C}(u_1,u_2)$
$C^{\perp}$	$-\ln u$	u_1u_2
Gumbel	$(-\ln u)^{\theta}$	$\exp\left(-\left(\tilde{u}_1^{\theta}+\tilde{u}_2^{\theta}\right)^{1/\theta}\right)$
Frank	$-\ln \frac{e^{-\theta u}-1}{e^{-\theta}-1}$	$-\frac{1}{\theta}\ln\left(1+\frac{\left(e^{-\theta u_1}-1\right)\left(e^{-\theta u_2}-1\right)}{e^{-\theta}-1}\right)$
Joe	$-\ln\left(1-(1-u)^{\theta}\right)$	$1-\left(\bar{u}_1^{\theta}+\bar{u}_2^{\theta}-\bar{u}_1^{\theta}\bar{u}_2^{\theta}\right)^{1/\theta}$
Clayton	$u^{-\theta}-1$	$\left(u_1^{-\theta}+u_2^{-\theta}-1\right)^{-1/\theta}$

The normal copula is the dependency function of the multivariate Gaussian probability distribution with a correlation matrix ρ:

$$\mathbf{C}(u_1,\ldots,u_n;\rho) = \Phi_{\rho}\left(\Phi^{-1}(u_1),\ldots,\Phi^{-1}(u_n)\right)$$

We deduce that the density of the copula is:

$$c(u_1,\ldots,u_n,;\rho) = \frac{1}{|\rho|^{\frac{1}{2}}}\exp\left(-\frac{1}{2}\varsigma^{\top}\left(\rho^{-1}-I_n\right)\varsigma\right)$$

with $\varsigma_i = \mathbf{\Phi}^{-1}(u_i)$. In the bivariate case, another expression of $\mathbf{C}$ is:

$$\mathbf{C}(u_1,u_2;\rho) = \int_0^{u_1}\Phi\left(\frac{\Phi^{-1}(u_2)-\rho\Phi^{-1}(u)}{\sqrt{1-\rho^2}}\right)du$$

with ρ the bivariate correlation parameter[12]. In a similar way, the t copula is

[11] We use the notations $\bar{u} = 1-u$ and $\tilde{u} = -\ln u$.

[12] We have $\tau = 2\pi^{-1}\arcsin(\rho)$ and $\varrho = 6\pi^{-1}\arcsin(\rho/2)$.

the dependency function associated with the multivariate Student's t probability distribution:

$$\mathbf{C}(u_1,\ldots,u_n;\rho,\nu)=\mathbf{t}_{\rho,\nu}\left(\mathbf{t}_\nu^{-1}(u_1),\ldots,\mathbf{t}_\nu^{-1}(u_n)\right)$$

We can show that its density is then:

$$c(u_1,\ldots,u_n,;\rho)=|\rho|^{-\frac{1}{2}}\frac{\Gamma\left(\frac{\nu+n}{2}\right)\left[\Gamma\left(\frac{\nu}{2}\right)\right]^n}{\left[\Gamma\left(\frac{\nu+1}{2}\right)\right]^n\Gamma\left(\frac{\nu}{2}\right)}\frac{\left(1+\frac{1}{\nu}\varsigma^\top\rho^{-1}\varsigma\right)^{-\frac{\nu+n}{2}}}{\prod_{i=1}^n\left(1+\frac{\varsigma_i^2}{\nu}\right)^{-\frac{\nu+1}{2}}}$$

with $\varsigma_i=\mathbf{t}_\nu^{-1}(u_i)$.

Example 43 *In Figure A.4, we have simulated 1 024 observations of the normal and Student's t copula using Sobol random numbers.*

We have considered two values for the parameters ρ. When ρ is equal to zero, we note that the t copula produces a dependency between the two random variables, which is different than that observed for the normal copula. Indeed, it places more probability on the corners of the quadrant. When ρ is equal to 90%, we also have the feeling that the dependency is stronger for the t copula than for the normal copula.

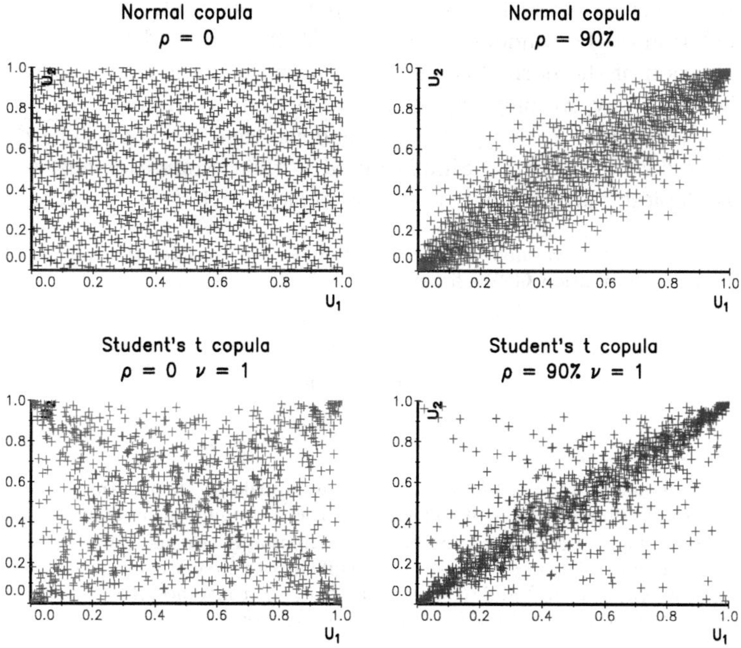

FIGURE A.4: Comparison of normal and t copulas

A.2.3 Simulation of copula models

We consider the simulation problem of the random vector $X = (X_1, \ldots, X_n)$ whose probability distribution is $\mathbf{C}(\mathbf{F}_1(x_1), \ldots, \mathbf{F}_n(x_n))$. This problem is equivalent to simulating the random vector $U = (U_1, \ldots, U_n)$ whose probability distribution is the copula $\mathbf{C}$ and to using the probability integral transform[13]:

$$X = \left(\mathbf{F}_1^{-1}(U_1), \ldots, \mathbf{F}_n^{-1}(U_n)\right)$$

The major difficulty is then to simulate the copula $\mathbf{C}$. We generally distinguish two main methods, which are well adapted to such problems.

A.2.3.1 Distribution approach

Let $\mathbf{F}$ be a multivariate probability distribution function. We have:

$$\mathbf{C}(U_1, \ldots, U_n) = \mathbf{F}\left(\mathbf{F}_1^{-1}(U_1), \ldots, \mathbf{F}_n^{-1}(U_n)\right)$$

To simulate $U = (U_1, \ldots, U_n)$, we also simulate $X = (X_1, \ldots, X_n)$ with distribution $\mathbf{F}$ and apply the transform $U = (\mathbf{F}_1(X_1), \ldots, \mathbf{F}_n(X_n))$. This method is particularly interesting if the distribution $\mathbf{F}$ is easier to simulate than the copula $\mathbf{C}$. For instance, it is the case of the normal copula, because simulating the multivariate normal distribution $\mathcal{N}(\mathbf{0}, \rho)$ is standard. Indeed, we have $\mathcal{N}(\mathbf{0}, \rho) = P\mathcal{N}(\mathbf{0}, I_n)$ with P the Cholesky decomposition matrix of ρ, i.e. the lower triangular matrix such that $PP^\top = \rho$. It follows that, if $x = (x_1, \ldots, x_n)$ is a simulation of the random vector $X \sim \mathcal{N}(\mathbf{0}, \rho)$, $u = (\Phi(x_1), \ldots, \Phi(x_n))$ is a simulation of the normal copula function with the matrix of parameters ρ. In the same way, we may exploit this algorithm to simulate the t copula. If $X \sim \mathcal{N}(\mathbf{0}, \rho)$ and $Y \sim \chi^2_\nu$ are independent, then $Z = X/\sqrt{(Y/\nu)}$ has a Student's t distribution $\mathbf{t}_{\rho,\nu}$. Simulating Z is then straightforward with the previous Cholesky algorithm[14]. It follows that $u = (\mathbf{t}_v(z_1), \ldots, \mathbf{t}_v(z_n))$ is a

[13] Let Y be a random variable with distribution $\mathbf{F}$. We consider the random variable $Z = \mathbf{F}(Y)$ with distribution $\mathbf{G}$. We have:

$$\begin{aligned} \mathbf{G}(z) &= \Pr\{Z \leq z\} \\ &= \Pr\{\mathbf{F}(Y) \leq z\} \\ &= \Pr\left\{Y \leq \mathbf{F}^{-1}(z)\right\} \\ &= \mathbf{F}\left(\mathbf{F}^{-1}(z)\right) \\ &= z \end{aligned}$$

with $\mathbf{G}(0) = 0$ and $\mathbf{G}(1) = 1$. We deduce that the probability distribution of $Z = \mathbf{F}(Y)$ is the uniform distribution $\mathcal{U}_{[0,1]}$. It follows that, if U is a uniform random variable, then $\mathbf{F}^{-1}(U)$ is a random variable with probability distribution $\mathbf{F}$ (Angus, 1994). To simulate a sequence of random variates $\{y_1, \ldots, y_m\}$, it is sufficient to simulate a sequence of uniform random variates $\{u_1, \ldots, u_m\}$ and to set $y_j = \mathbf{F}^{-1}(u_j)$. For instance, if Y is an exponential random variable $\mathcal{E}(\lambda)$ defined by $\mathbf{F}(y) = 1 - \exp(-\lambda y)$, we have:

$$y_j = -\frac{\ln(1 - u_j)}{\lambda}$$

[14] We have $z = (z_1, \ldots, z_n)$ with $z_i = x_i/\sqrt{(y/\nu)}$ and y a simulation of the χ^2_ν distribution.

simulation of the t copula function with ν degrees of freedom and the correlation matrix ρ.

Remark 69 *The distribution approach is appropriate when the copula function comes from a well-known multivariate distribution.*

A.2.3.2 Simulation based on conditional copula functions

Let us first consider the bivariate case. If $U = (U_1, U_2)$ is a random vector with distribution $\mathbf{C}$, we have $\Pr\{U_1 \leq u_1\} = u_1$ and $\mathbf{C}_{2|1}(u_1, u_2) = \partial_{u_1} \mathbf{C}(u_1, u_2) = \Pr\{U_2 \leq u_2 \mid U_1 = u_1\}$. Because $\mathbf{C}(U_1, 1)$ and $\mathbf{C}_{2|1}(u_1, U_2)$ are two independent uniform random variables, we obtain the following algorithm:

1. We draw two independent uniform random variates v_1 and v_2.
2. We set u_1 equal to v_1.
3. We find the root u_2 such that it satisfies the equation $\mathbf{C}_{2|1}(u_1, u_2) = v_2$. In an equivalent manner, u_2 is equal to $h^{-1}(v_2; u_1)$ where $h(u; u_1) = \mathbf{C}_{2|1}(u_1, u)$ is the univariate function which depends on the unknown variable u.

This algorithm is proposed by Genest and MacKay (1986). It is used, for instance, by Genest (1987) to simulate the Frank copula:

$$\mathbf{C}(u_1, u_2; \theta) = -\frac{1}{\theta} \ln \left(1 + \frac{\left(e^{-\theta u_1} - 1\right)\left(e^{-\theta u_2} - 1\right)}{e^{-\theta} - 1} \right)$$

We deduce that:

$$\mathbf{C}_{2|1}(u_1, u_2; \theta) = \frac{\left(e^{-\theta u_2} - 1\right) e^{-\theta u_1}}{\left(e^{-\theta} - 1\right) + \left(e^{-\theta u_1} - 1\right)\left(e^{-\theta u_2} - 1\right)}$$

We finally obtain:

$$\begin{aligned} h^{-1}(u; u_1) &= \left\{u_2 : \mathbf{C}_{2|1}(u_1, u_2; \theta) = u\right\} \\ &= -\frac{1}{\theta} \ln \left(1 + \frac{u\left(e^{-\theta} - 1\right)}{u + (1 - u)\, e^{-\theta u_1}} \right) \end{aligned}$$

More generally, we can use this algorithm to simulate Archimedean copulas. Genest and MacKay (1986) show that it is equivalent to the following steps:

1. We draw two independent uniform random variates v_1 and v_2.
2. If φ is the generator function of the Archimedean copulas, u_1 and u_2 are equal respectively to v_1 and $\varphi^{-1}\left(\varphi\left(\varphi'^{-1}\left(\varphi'\left(\frac{v_1}{v_2}\right)\right)\right) - \varphi(v_1)\right)$.

Remark 70 *In some cases, it is not possible to find an analytical formula for $h^{-1}(v_2; u_1)$ (Joe, 1987, pages 146-147). In such a situation, we must solve the equation $\mathbf{C}_{2|1}(u_1, u_2) = v_2$ with a numerical method.*

The algorithm can be extended to multivariate copulas in a direct way. For instance, we obtain this algorithm in the trivariate case:

1. We draw three independent uniform random variates v_1, v_2 and v_3.

2. u_1 is equal to v_1.

3. u_2 is the root of the non-linear equation:

$$\mathbf{C}_{2|1}(u_1, u_2, 1) = \partial_{u_1} \mathbf{C}(u_1, u_2, 1) = v_2$$

4. u_3 is the root of the non-linear equation:

$$\mathbf{C}_{3|1,2}(u_1, u_2, u_3) = \partial^2_{u_1,u_2} \mathbf{C}(u_1, u_2, u_3) = v_3$$

A.2.4 Copulas and risk management

Copula functions are frequently used in risk management. Indeed, they are particularly well adapted for dependence modeling in the case of credit and operational risks, and they are intensively used to design multivariate stress scenarios (Jouanin *et al.*, 2004).

By decomposing a multivariate distribution into univariate distributions and a non-reducible dependency function, copulas are the appropriate tool to understand the risk aggregation (Embrechts *et al.*, 2002). Coles *et al.* (1999) introduce the quantile-quantile dependence measure defined by:

$$\begin{aligned}
\lambda_U(\alpha) &= \Pr\left\{X_2 > \mathbf{F}_2^{-1}(\alpha) \mid X_1 > \mathbf{F}_1^{-1}(\alpha)\right\} \\
&= \frac{\Pr\left\{X_2 > \mathbf{F}_2^{-1}(\alpha), X_1 > \mathbf{F}_1^{-1}(\alpha)\right\}}{\Pr\left\{X_1 > \mathbf{F}_1^{-1}(\alpha)\right\}} \\
&= \frac{\Pr\left\{\mathbf{F}_2(X_2) > \alpha, \mathbf{F}_1(X_1) > \alpha\right\}}{\Pr\left\{\mathbf{F}_1(X_1) > \alpha\right\}}
\end{aligned}$$

We finally obtain:

$$\lambda_U(\alpha) = \frac{1 - 2\alpha + \mathbf{C}(\alpha, \alpha)}{1 - \alpha}$$

This statistical measure has a natural interpretation in risk management. If we consider two portfolios x_1 and x_2, and if we note L_1 and L_2 the two random losses with distributions $\mathbf{F}_1$ and $\mathbf{F}_2$, we have:

$$\begin{aligned}
\lambda_U(\alpha) &= \Pr\left\{L_2 > \mathbf{F}_2^{-1}(\alpha) \mid L_1 > \mathbf{F}_1^{-1}(\alpha)\right\} \\
&= \Pr\left\{L_2 > \mathrm{VaR}_\alpha(x_2) \mid L_1 > \mathrm{VaR}_\alpha(x_1)\right\}
\end{aligned}$$

$\lambda_U(\alpha)$ is then the probability that the loss of one portfolio exceeds its value-at-risk knowing that the loss of the other portfolio has already exceeded its value-at-risk. It is remarkable that $\lambda_U(\alpha)$ depends only on the copula function between L_1 and L_2.

Let us consider a bivariate copula $\mathbf{C}$ such that the limit:

$$\lim_{\alpha \to 1^-} \frac{\bar{\mathbf{C}}(\alpha, \alpha)}{1-\alpha} = \lim_{\alpha \to 1^-} \frac{1 - 2\alpha + \mathbf{C}(\alpha, \alpha)}{1-\alpha} = \lambda_U$$

exists. In this case, we say that $\mathbf{C}$ has an upper tail dependence if $\lambda_U \in (0, 1]$ and it has no upper tail dependence if $\lambda_U = 0$ (Joe, 1997). We have $\lambda_U = 0$ for the copulas $\mathbf{C}^{\perp}$ and $\mathbf{C}^{-}$, and $\lambda_U = 1$ for the copula $\mathbf{C}^{+}$. In the case of the Gumbel copula, we obtain $\lambda_U = 2 - 2^{1/\theta}$, whereas $\lambda_U = 0$ in the case of the Clayton copula. An interesting result concerns the difference between the normal and t copulas in terms of risk aggregation. The normal copula does not correlate extreme quantiles ($\lambda_U = 0$ if $\rho \neq 1$) whereas the correlation is always positive for the t copula (Demarta and McNeil, 2005):

$$\lambda_U = 2 - 2\mathbf{t}_{\nu+1}\left(\left(\frac{(\nu+1)(1-\rho)}{(1+\rho)}\right)^{1/2}\right) > 0 \quad \text{if} \quad \rho > -1$$

In Figure A.5, we have reported the quantile-quantile dependence measure $\lambda_U(\alpha)$ for the normal copula. We note that when α is close to one, $\lambda_U(\alpha)$ takes very small values even if the correlation is high. We do not observe such a phenomenon with the t_1 copula (see Figure A.6). We conclude that if these two copulas are equivalent from the point of view of correlation, they do not aggregate risks in the same way. In particular, with high confidence levels, the normal copula does not correlate value-at-risk of different portfolios. Everything happens as if they are independent, implying that the global risk measure may be largely underestimated.

Remark 71 *We can also define a lower quantile-quantile dependence measure as follows:*

$$\begin{aligned} \lambda_L(\alpha) &= \Pr\left\{X_2 < \mathbf{F}_2^{-1}(\alpha) \mid X_1 < \mathbf{F}_1^{-1}(\alpha)\right\} \\ &= \frac{\mathbf{C}(\alpha, \alpha)}{\alpha} \end{aligned}$$

The limit is called the lower tail dependence:

$$\lambda_L = \lim_{\alpha \to 0^+} \frac{\mathbf{C}(\alpha, \alpha)}{\alpha}$$

This measure plays an important role in multivariate survival modeling and credit risk. For radially symmetric copulas, we have $\lambda_L = \lambda_U$. This is the case of $\mathbf{C}^{\perp}$, $\mathbf{C}^{-}$, $\mathbf{C}^{+}$, the normal copula and the t copula. Contrary to the previous results, we can also show that the Gumbel copula does not exhibit a lower tail dependence, whereas the Clayton copula has a lower tail dependence ($\lambda_L = 2^{-1/\theta}$).

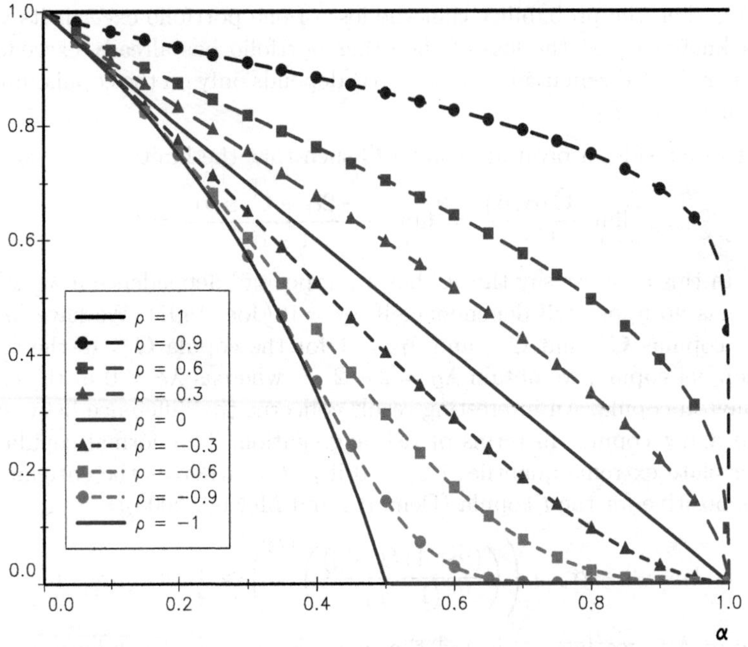

FIGURE A.5: Quantile-quantile dependence measure for the normal copula

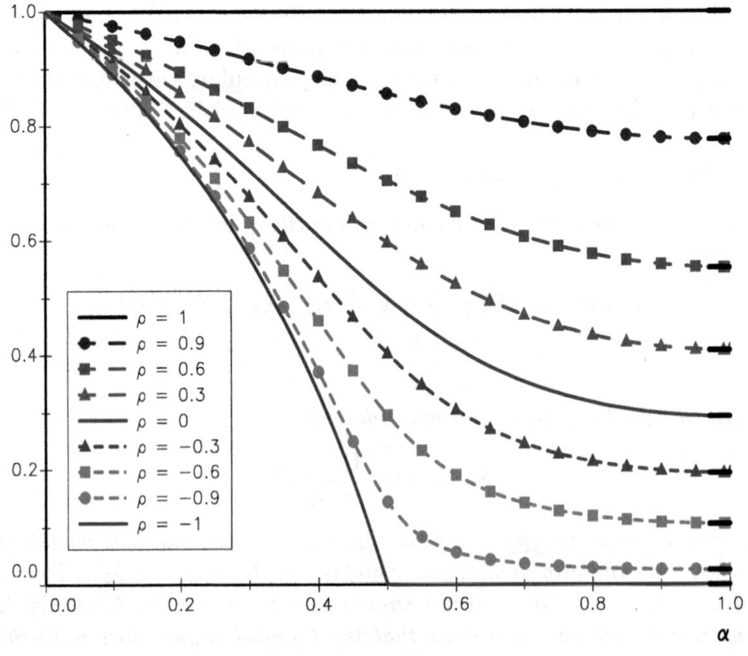

FIGURE A.6: Quantile-quantile dependence measure for the t_1 copula

Remark 72 *The concept of tail dependence is related to the multivariate extreme value theory. Let us consider the order statistic:*

$$X_{m:m} = \max(X_1, \ldots, X_m)$$

with X_i iid random variables. The univariate extreme value theory permits to characterize the asymptotic distribution of $X_{m:m}$ when m tends to infinity. We can show that the limit distribution is a non-degenerate distribution which can only be a Gumbel, a Frechet or a Weibull distribution. The multivariate extreme value theory studies the asymptotic behavior of the random vector $\left(X_{m:m}^{(1)}, \ldots, X_{m:m}^{(n)}\right)$ where $X_{m:m}^{(i)} = \max\left(X_1^{(i)}, \ldots, X_m^{(i)}\right)$ and $X_i^{(j)}$ are iid random variables for each index i. By definition, the multivariate extreme value distribution of $\left(X_{m:m}^{(1)}, \ldots, X_{m:m}^{(n)}\right)$ verifies a copula representation, where the margins are those given by the univariate extreme value theory. Tawn (1990) shows also that the copula of these extreme values must satisfy the following property:

$$\mathbf{C}\left(u_1^t, \ldots, u_n^t\right) = \mathbf{C}^t(u_1, \ldots, u_n)$$

for all $t > 0$. In this case, such a copula is called an extreme value copula. By construction, this extreme value copula is only related to the copula between the non-ordering random variables $\left(X_i^{(1)}, \ldots, X_i^{(n)}\right)$. In particular, we can show that these two copulas have the same upper tail dependence. Moreover, if $\lambda_U = 0$ for the copula, it necessarily implies that this extreme value copula is the independent copula $\mathbf{C}^{\perp}$ (Joe, 1997). For instance, if we consider the normal copula, we have $\lambda_U = 0$, meaning that extreme values are independent.

A.2.5 Multivariate survival modeling

Another important field using copulas is the modeling of multivariate survival functions. To correlate default times, the simplest method is to build a multivariate survival function based on univariate survival functions and a copula function. This construction method is particularly useful in credit risk modeling and was first introduced by Li (2000) in the case of managing a credit portfolio.

Let $\mathbf{S}(t_1, \ldots, t_n) = \Pr\{\tau_1 > t_1, \ldots, \tau_n > t_n\}$ be a multivariate survival function[15]. We note $\mathbf{S}_i$ the ith margin of $\mathbf{S}$:

$$\mathbf{S}_i(t) = \Pr\{\tau_1 > 0, \ldots, \tau_{i-1} > 0, \tau_i > t, \tau_{i+1} > 0, \ldots, \tau_n > 0\}$$

We can show that $\mathbf{S}$ admits a copula representation:

$$\mathbf{S}(t_1, \ldots, t_n) = \breve{\mathbf{C}}(\mathbf{S}_1(t_1), \ldots, \mathbf{S}_n(t_n))$$

where $\breve{\mathbf{C}}$ is a copula function, which is unique if the margins are continuous.

[15] We note $\tau_1, \ldots, \tau_n$ the default times.

This copula is called a survival copula function. It can be deduced from the distribution copula $\mathbf{C}$. In the case $n = 2$, we have:

$$\begin{aligned}
\mathbf{S}(t_1, t_2) &= \Pr\{\tau_1 > t_1, \tau_2 > t_2\} \\
&= 1 - \mathbf{F}_1(t_1) - \mathbf{F}_2(t_2) + \mathbf{F}(t_1, t_2) \\
&= \mathbf{S}_1(t_1) + \mathbf{S}_2(t_2) - 1 + \mathbf{C}(1 - \mathbf{S}_1(t_1), 1 - \mathbf{S}_2(t_2)) \\
&= \breve{\mathbf{C}}(\mathbf{S}_1(t_1), \mathbf{S}_2(t_2))
\end{aligned}$$

with:

$$\breve{\mathbf{C}}(u_1, u_2) = u_1 + u_2 - 1 + \mathbf{C}(1 - u_1, 1 - u_2)$$

In the general case, we have:

$$\breve{\mathbf{C}}(u_1, \ldots, u_n) = \bar{\mathbf{C}}(1 - u_1, \ldots, 1 - u_n)$$

with:

$$\bar{\mathbf{C}}(u_1, \ldots, u_n) = \sum_{i=0}^{n} \left[(-1)^i \sum_{\mathbf{v}(u_1, \ldots, u_n) \in \mathcal{Z}(n-i, n, 1)} \mathbf{C}(v_1, \ldots, v_n) \right]$$

where $\mathcal{Z}(m, n, \epsilon)$ is the set $\{\mathbf{v} \in [0,1]^n \mid v_i \in \{u_i, \epsilon\}, \sum_{i=1}^{n} \mathbb{1}_{\epsilon}\{v_i\} = m\}$. Therefore, we obtain for $n = 3$:

$$\begin{aligned}
\breve{\mathbf{C}}(u_1, u_2, u_3) = \; & u_1 + u_2 + u_3 - 2 + \mathbf{C}(1 - u_1, 1 - u_2) + \\
& \mathbf{C}(1 - u_1, 1 - u_3) + \mathbf{C}(1 - u_2, 1 - u_3) \\
& -\mathbf{C}(1 - u_1, 1 - u_2, 1 - u_3)
\end{aligned}$$

Here are the main properties obtained by Georges *et al.* (2001) about $\breve{\mathbf{C}}$:

- A copula is said to be radially symmetric (or RS) if and only if $\breve{\mathbf{C}} = \mathbf{C}$. For instance, $\mathbf{C}^-$, $\mathbf{C}^\perp$ and $\mathbf{C}^+$ are RS copula functions. We can also show that the Frank, normal and t copula functions are radially symmetric, but not the Gumbel and Clayton copulas.
- In the bivariate case, if $\mathbf{C}_1 \succ \mathbf{C}_2$ then $\breve{\mathbf{C}}_1 \succ \breve{\mathbf{C}}_2$. This result does not hold if $n > 2$.
- We have $\lambda_U \left\langle \breve{\mathbf{C}} \right\rangle = \lambda_L \langle \mathbf{C} \rangle$ and $\lambda_L \left\langle \breve{\mathbf{C}} \right\rangle = \lambda_U \langle \mathbf{C} \rangle$. This implies that the lower and upper tail dependence measures are equal if the copula is radially symmetric.

An important class of survival copulas is the frailty models popularized by Oakes (1989). The idea is to introduce the dependence between the survival times $\tau_1, \ldots, \tau_n$ by using a random variable W. Given the frailty variable W with distribution $\mathbf{G}$, the survival times are assumed to be independent:

$$\Pr\{\tau_1 > t_1, \ldots, \tau_n > t_n \mid W = w\} = \prod_{i=1}^{n} \Pr\{\tau_i > t_i \mid W = w\}$$

We then have:

$$\begin{aligned}\mathbf{S}(t_1,\ldots,t_n \mid w) &= \prod_{i=1}^{n}\mathbf{S}_i(t_i \mid w)\\ &= \prod_{i=1}^{n}\chi_i^w(t_i)\end{aligned}$$

where $\chi_i(t)$ is the survival function using Cox's proportional hazard model:

$$\chi_i(t)=\exp(-\boldsymbol{\Lambda}_i(t))=\exp\left(-\int_0^t \lambda_i(s)\,\mathrm{d}s\right)$$

The unconditional survival function is also:

$$\mathbf{S}(t_1,\ldots,t_n)=\mathbb{E}\left[\mathbf{S}(t_1,\ldots,t_n \mid w)\right]$$

We deduce that:

$$\mathbf{S}(t_1,\ldots,t_n)=\int\prod_{i=1}^{n}\left[\chi_i(t_i)\right]^w\,\mathrm{d}\mathbf{G}(w)$$

Marshall and Olkin (1988) propose a general representation of these frailty models. Let $\mathbf{F}_1,\ldots,\mathbf{F}_n$ and $\mathbf{G}$ be n univariate probability distributions and a probability distribution of dimension n. We note by ψ and ψ_i the Laplace transform of $\mathbf{G}$ and the margins $\mathbf{G}_i$. Let $\mathbf{C}$ be a copula function. If $\mathbf{H}_i(x)=\exp\left(-\psi_i^{-1}(\mathbf{F}_i(x))\right)$, then the function defined as follows:

$$\mathbf{F}(x_1,\ldots,x_n)=\int\cdots\int\mathbf{C}\left(\left[\mathbf{H}_1(x_1)\right]^{w_1},\ldots,\left[\mathbf{H}_n(x_n)\right]^{w_n}\right)\mathrm{d}\mathbf{G}(w_1,\ldots,w_n)$$

is a multivariate distribution with margins $\mathbf{F}_1,\ldots,\mathbf{F}_n$. Let us consider the special case where the margins $\mathbf{G}_i$ are the same, $\mathbf{G}$ is the upper Frechet distribution and $\mathbf{C}$ is the product copula $\mathbf{C}^{\perp}$. We obtain:

$$\begin{aligned}\mathbf{F}(x_1,\ldots,x_n) &= \int\prod_{i=1}^{n}\left[\mathbf{H}_i(x_i)\right]^{w_1}\,\mathrm{d}\mathbf{G}_1(w_1)\\ &= \int\exp\left(-w_1\sum_{i=1}^{n}\psi_1^{-1}(\mathbf{F}_i(x_i))\right)\mathrm{d}\mathbf{G}_1(w_1)\\ &= \psi_1\left(\psi_1^{-1}(\mathbf{F}_1(x_1))+\ldots+\psi_1^{-1}(\mathbf{F}_n(x_n))\right)\end{aligned}$$

The corresponding copula function is then:

$$\mathbf{C}(u_1,\ldots,u_n)=\psi_1\left(\psi_1^{-1}(u_1)+\ldots+\psi_1^{-1}(u_n)\right)$$

We deduce that a frailty survival copula is an Archimedean copula where the generator function φ is the inverse Laplace transform of a random variable W. For example, the Clayton copula is a frailty copula with a Laplace transform applied to a Gamma random variable[16]. The Gumbel copula is another frailty copula[17].

A.3 Dynamic portfolio optimization

A.3.1 Stochastic optimal control

In this section, we present the main results of stochastic optimal control[18]. In particular, we consider the Bellman approach and the martingale approach which are now standard tools to solve many dynamic financial models.

A.3.1.1 Bellman approach

We note $x(t)$ and $v(t)$ the state variable and the control variable. The goal of dynamic programming is to choose the control variable $v(t)$ in order to maximize the expected utility:

$$\begin{aligned} \max \quad & \mathbb{E}_0\left[\int_0^T \mathcal{U}(t, x(t), v(t))\, \mathrm{d}t + g(x(T))\right] \qquad \text{(A.7)} \\ \text{u.c.} \quad & \begin{cases} \mathrm{d}x(t) = f(t, x(t), v(t))\, \mathrm{d}t + \sigma(t, x(t), v(t))\, \mathrm{d}W(t) \\ x(0) = x_0 \\ v(t, x) \in \mathcal{V} \end{cases} \end{aligned}$$

where $W(t)$ is a Brownian motion, $\mathcal{U}$ is the utility function, g is the bequest function and $\mathcal{V}$ is the set of constraints on the control variable. The state variable $x(t)$ is a diffusion process and we know its initial value. Let $\mathcal{J}(t, x)$ be the function such that:

$$\mathcal{J}(t, x) = \max_v \mathbb{E}_t\left[\int_t^{t+h} \mathcal{U}(s, x(s), v(s))\, \mathrm{d}s + \mathcal{J}(t+h, x(t+h))\right]$$

When $h \to 0$ and using Ito's lemma, we obtain the Hamilton-Jacobi-Bellman (HJB) equation:

$$\frac{\partial \mathcal{J}}{\partial t}(t, x) + \max_v \left[\mathcal{U}(t, x, v) + f(t, x, v)\frac{\partial \mathcal{J}}{\partial x}(t, x) + \frac{1}{2}\sigma^2(t, x, v)\frac{\partial^2 \mathcal{J}}{\partial x^2}(t, x)\right] = 0$$

[16]We have $\psi(x) = (1 + x)^{-1/\theta}$.

[17]The Laplace transform corresponds to a positive stable random variable: $\psi(x) = \exp\left(-x^{1/\theta}\right)$.

[18]Readers may refer to Prigent (2007) or Pham (2009) for a comprehensive study of such problems.

with the terminal condition:

$$\mathcal{J}(T,x) = g(x)$$

We define the Hamiltonian function:

$$H(t,x,v,p,q) = \mathcal{U}(t,x,v) + pf(t,x,v) + \frac{1}{2}q\sigma^2(t,x,v)$$

To solve Problem (A.7), we proceed in two steps:

1. We investigate a solution $\mathcal{J}(t,x)$ of the HJB equation. Thus, we compute for each date t and each state x the optimal control value $v^\star(t,x) = v \in \mathcal{V}$, which maximizes the Hamiltonian function. The HJB equation becomes:

$$\begin{aligned} \frac{1}{2}\sigma^2(t,x,v^\star(t,x))\frac{\partial^2 \mathcal{J}}{\partial x^2}(t,x) + f(t,x,v^\star(t,x))\frac{\partial \mathcal{J}}{\partial x}(t,x) + & \\ \frac{\partial \mathcal{J}}{\partial t}(t,x) + \mathcal{U}(t,x,v^\star(t,x)) &= 0 \end{aligned}$$

2. We solve the stochastic differential equation:

$$\mathrm{d}x(t) = f(t,x(t),v^\star(t,x))\,\mathrm{d}t + \sigma(t,x(t),v^\star(t,x))\,\mathrm{d}W(t)$$

and we deduce the optimal control:

$$v^\star(t) = v^\star(t,x(t))$$

Remark 73 *The previous framework is valid in the multivariate case where $x(t) \in \mathbb{R}^n$, $v(t) \in \mathbb{R}^m$, $W(t) \in \mathbb{R}^k$ with $\mathbb{E}\left[W(t)W(t)^\top\right] = \rho t$, ρ is a $k \times k$ correlation matrix, $f(t,x(t),v(t))$ is a $n \times 1$ vector and $\sigma(t,x(t),v(t))$ is a $n \times k$ matrix. The Hamiltonian function becomes:*

$$\begin{aligned} H(t,x,v,p,q) &= \mathcal{U}(t,x,v) + \langle p, f(t,x,v)\rangle + \\ & \frac{1}{2}\operatorname{tr}\left(\sigma(t,x(t),v(t))^\top q\sigma(t,x(t),v(t))\,\rho\right) \end{aligned}$$

where q is a $n \times n$ matrix with:

$$q_{i,j} = \frac{\partial^2 \mathcal{J}}{\partial x_i \partial x_j}(t,x)$$

A.3.1.2 Martingale approach

In the martingale approach[19], the stochastic control problem is:

$$\max \mathbb{E}_{\mathbb{P}}\left[\mathcal{U}(x(T))\right]$$

[19]The martingale approach has its roots in the article by Brennan and Solanki (1981) and was formulated by Pliska (1986) and Karatzas *et al.* (1987).

with the budget constraint:

$$x_0 = \mathbb{E}_{\mathbb{Q}}\left[e^{-\int_0^T r_t \,\mathrm{d}t} x(T)\right]$$

This constraint means that the investor buys a contingent claim $x(T)$ whose price $\mathbb{E}_{\mathbb{Q}}\left[\exp\left(-\int_0^T r_t \,\mathrm{d}t\right) x(T)\right]$ is equal to its initial wealth x_0. If we suppose that interest rates are not stochastic, we use the risk-neutral probability $\mathbb{Q}$ to evaluate the contingent claim whereas the investor uses the historical probability $\mathbb{P}$ to maximize his terminal wealth. In this case, the optimal solution $x^\star(T)$ satisfies the following relationship:

$$\mathcal{U}'(x^\star(T)) = \lambda \frac{\mathrm{d}\mathbb{Q}}{\mathrm{d}\mathbb{P}} \tag{A.8}$$

where λ is the Lagrange coefficient associated with the budget constraint[20]. In the event of stochastic interest rates, we need to consider a change of numéraire (Geman *et al.*, 1995) and we use the forward probability measure $\mathbb{Q}^T$ instead of the risk-neutral probability measure $\mathbb{Q}$.

A.3.2 Portfolio optimization in continuous-time

The extension of the Markowitz model in a dynamic case was formulated by Merton (1969). We consider an investor who allocates his wealth $x(t)$ between a risky asset $S(t)$ and a risk-free asset $B(t)$. Their dynamics are given by the following equations:

$$\left\{\begin{array}{l} \mathrm{d}S(t) = \mu S(t)\,\mathrm{d}t + \sigma S(t)\,\mathrm{d}W(t) \\ \mathrm{d}B(t) = rB(t)\,\mathrm{d}t \end{array}\right.$$

We deduce that:

$$\begin{aligned} \mathrm{d}x(t) &= \alpha(t)x(t)\frac{\mathrm{d}S(t)}{S(t)} + (1-\alpha(t))x(t)\frac{\mathrm{d}B(t)}{B(t)} \\ &= (\alpha(t)(\mu - r) + r)x(t)\,\mathrm{d}t + \alpha(t)\sigma x(t)\,\mathrm{d}W(t) \end{aligned}$$

[20] If we note $Z(t)$ the stochastic process defined by:

$$Z(t) = \frac{\mathrm{d}\mathbb{Q}(t)}{\mathrm{d}\mathbb{P}(t)}$$

the optimization problem becomes:

$$\max \mathbb{E}_{\mathbb{P}}[\mathcal{U}(x(T))] \quad \text{u.c.} \quad \mathbb{E}_{\mathbb{P}}[Z(T)x(T)] = \frac{x_0}{B(0,T)}$$

with $B(0,T)$ the zero-coupon price of maturity T. We may reformulate the optimization problem as follows:

$$\max \inf_{\lambda} \left\{\mathbb{E}_{\mathbb{P}}[\mathcal{U}(x(T))] - \lambda\left(\mathbb{E}_{\mathbb{P}}[Z(T)x(T)] - \frac{x_0}{B(0,T)}\right)\right\}$$

The solution then satisfies:

$$x^\star(T) = (\mathcal{U}')^{-1}(\lambda Z(T))$$

The objective of the investor is to maximize the expected utility of his terminal wealth:

$$\max \mathbb{E}_0 \left[\mathcal{U} \left(x \left(T \right) \right) \right]$$

The HJB equation is then:

$$\frac{\partial \mathcal{J}}{\partial t} \left(t, x \right) + \max_{\alpha \in \mathbb{R}} \left[\left(\alpha \left(\mu - r \right) + r \right) x \frac{\partial \mathcal{J}}{\partial x} \left(t, x \right) + \frac{1}{2} \alpha^2 \sigma^2 x^2 \frac{\partial^2 \mathcal{J}}{\partial x^2} \left(t, x \right) \right] = 0$$

The maximization of the Hamiltonian function gives[21]:

$$\alpha^\star \left(t \right) = - \frac{\left(\mu - r \right)}{\sigma^2} \frac{\partial_x \mathcal{J} \left(t, x \right)}{x \partial_x^2 \mathcal{J} \left(t, x \right)}$$

Another expression of the solution $\alpha^\star \left(t \right)$ is:

$$\alpha^\star \left(t \right) = \frac{\text{SR}}{\mathfrak{R} \left(t, x \right) \cdot \sigma}$$

where SR is the Sharpe ratio of the risky asset and $\mathfrak{R} \left(t, x \right)$ is the relative risk aversion of the investor:

$$\mathfrak{R} \left(t, x \right) = - \frac{x \partial_x^2 \mathcal{J} \left(t, x \right)}{\partial_x \mathcal{J} \left(t, x \right)}$$

In this model, the risky allocation $\alpha^\star$ is equal to the Sharpe ratio divided by the product of the volatility and the relative risk aversion. We deduce that the optimal solution verifies the following PDE:

$$\frac{\partial \mathcal{J}}{\partial t} \left(t, x \right) + r x \frac{\partial \mathcal{J}}{\partial x} \left(t, x \right) - \frac{1}{2} \frac{\left(\mu - r \right)^2}{\sigma^2} \frac{\left(\partial_x \mathcal{J} \left(t, x \right) \right)^2}{\partial_x^2 \mathcal{J} \left(t, x \right)} = 0$$

with the terminal condition $\mathcal{J} \left(T, x \right) = \mathcal{U} \left(x \right)$. In the general case, we solve this problem using the numerical algorithm of finite differences. In the case of the CRRA utility function $\mathcal{U} \left(x \right) = \left(1 - \gamma \right)^{-1} x^{1 - \gamma}$ with $\gamma > 0$, we postulate solution to the form $\mathcal{J} \left(t, x \right) = h \left(t \right) g \left(x \right)$ (Demange and Rochet, 2005). The HJB equation becomes:

$$\left\{ \begin{array}{l} h' \left(t \right) + c h \left(t \right) = 0 \\ h \left(T \right) = 1 \end{array} \right.$$

with:

$$c = r \left(1 - \gamma \right) + \frac{1}{2} \frac{\left(\mu - r \right)^2}{\sigma^2} \frac{\left(1 - \gamma \right)}{\gamma}$$

[21] The first-order condition is:

$$\left(\mu - r \right) x \frac{\partial \mathcal{J}}{\partial x} \left(t, x \right) + \alpha \sigma^2 x^2 \frac{\partial^2 \mathcal{J}}{\partial x^2} \left(t, x \right) = 0$$

The solution is $\mathcal{J}(t,x) = e^{c(T-t)}(1-\gamma)^{-1} x^{1-\gamma}$ because $h(t) = e^{c(T-t)}$. We finally obtain:

$$\alpha^\star(t) = \frac{1}{\gamma}\frac{(\mu - r)}{\sigma^2}$$

The risky allocation is constant over time. It corresponds to the famous constant-mix portfolio. We can generalize this result with several assets (Merton, 1971). We show that the optimal dynamic portfolio is the tangency portfolio and that the CAPM theory remains valid (see Prigent (2007) for a comprehensive presentation of these results).

A.3.3 Some extensions of the Merton model

The Merton model was extended to different problems concerning long-term investment solutions, in particular for pension plans. We generally distinguish two main categories. In a defined contribution (DC) plan, the level of the pension is linked to the performance of the fund and the risk is then supported by the employees. In a defined benefit (DB) plan, the pension level is guaranteed at inception by the sponsor of the pension plan. The objective of the pension fund is then to generate sufficient performance to match these liability constraints. In this context, investment solutions differ from one category to another. Many DC participants choose to invest in diversified funds, typically lifestyle and lifecycle funds whereas the trustee of the DB plan considers the liability driven investment (LDI) approach to manage the fund performance.

A.3.3.1 Lifestyle funds

In Chapter 6, we have seen that lifestyle funds are diversified funds based on the constant-mix strategy. We generally distinguish three profiles depending on the stock/bond asset mix policy[22]:

Profile	Stock	Bond
Defensive	20%	80%
Balanced	50%	50%
Dynamic	80%	20%

The choice of the profile depends on the investor risk aversion. However, in the Merton model, this parameter changes the allocation between the risk-free asset and the risky portfolio, but does not modify the composition of the risk portfolio. To solve this asset allocation puzzle[23], we need to adapt the framework of the Merton model.

Bajeux-Besnainou *et al.* (2003) consider a model with three funds: a money

[22] In real life, these allocations of course differ from one asset manager to another and may include other asset classes (commodities, real estate, etc.) for a small proportion.

[23] See page 272 for more details.

market fund $C(t)$, a bond fund $B(t)$ and an equity fund $S(t)$. The price dynamics are given by:

$$\begin{cases} \mathrm{d}C(t) = r(t) C(t)\ \mathrm{d}t \\ \mathrm{d}B(t) = \mu_B(t) B(t)\ \mathrm{d}t + \sigma_B B(t)\ \mathrm{d}W(t) \\ \mathrm{d}S(t) = \mu_S(t) S(t)\ \mathrm{d}t + \sigma_S S(t)\ \mathrm{d}W_S(t) \end{cases}$$

where the instantaneous interest rate $r(t)$ is assumed to follow the Vasicek model:

$$\mathrm{d}r(t) = a(b - r(t))\ \mathrm{d}t - \sigma\,\mathrm{d}W(t)$$

The expected returns are $\mu_B(t) = r(t) + \pi_B$ and $\mu_S(t) = r(t) + \pi_S$ with π_B and π_S the bond and equity risk premia. If we assume that the bond fund has a constant duration D, the volatility σ_B is equal to:

$$\sigma_B = \sigma \frac{1 - e^{-aD}}{a}$$

Let $\rho_{S,B}$ be the stock/bond correlation. We have:

$$\sigma_S\,\mathrm{d}W_S(t) = \underbrace{\sqrt{1 - \rho_{S,B}^2}\sigma_S}_{\sigma_1}\,\mathrm{d}W'(t) + \underbrace{\rho_{S,B}\sigma_S}_{\sigma_2}\,\mathrm{d}W(t)$$

The parameters of this model are then the interest rate parameters a, b and σ, the bond fund coefficients π_B and D and the equity fund coefficients π_S, σ_S and $\rho_{S,B}$. Let $x(t)$ be the wealth at time t. We have:

$$\frac{\mathrm{d}x(t)}{x(t)} = \alpha_C(t) \frac{\mathrm{d}C(t)}{C(t)} + \alpha_B(t) \frac{\mathrm{d}B(t)}{B(t)} + \alpha_S(t) \frac{\mathrm{d}S(t)}{S(t)}$$

The investor maximizes the utility of his terminal wealth:

$$\max \mathbb{E}_{\mathbb{P}}\left[\mathcal{U}(x(T))\right]$$

with the budget constraint:

$$x_0 = \mathbb{E}_{\mathbb{Q}}\left[e^{-\int_0^T r_t\,\mathrm{d}t} x(T)\right]$$

We have seen that the optimal solution $x^\star(T)$ satisfies the relationship:

$$\frac{\mathrm{d}\mathbb{Q}}{\mathrm{d}\mathbb{P}} = \frac{\mathcal{U}'(x^\star(T))}{\lambda}$$

with λ the Lagrange coefficient associated with the budget constraint. In the case of the CRRA utility function $\mathcal{U}(x) = x^{1-\gamma}/(1-\gamma)$, Bajeux-Besnainou *et al.* (2003) find that the optimal allocations are:

$$\begin{cases} \alpha_C^\star(t) = 1 - \alpha_B^\star(t) - \alpha_S^\star(t) \\ \alpha_B^\star(t) = \left(1 - \gamma^{-1}\right) \sigma_B^{-1} \sigma(t, T) + \gamma^{-1} h_B \\ \alpha_S^\star(t) = \gamma^{-1} h_S \end{cases}$$

with $h_B = (\sigma_1 \sigma_B)^{-1}(\lambda' \sigma_1 - \lambda \sigma_2)$, $h_S = \sigma_1^{-1}\lambda$, $\sigma(t,T) = \sigma a^{-1}\left(1 - e^{-a(T-t)}\right)$, $\lambda' = \sigma_B^{-1}\pi_B$ and $\lambda = \sigma_1^{-1}\left(\pi_S - \pi_B \sigma_B^{-1}\sigma_2\right)$.

We consider the following parameters: $a = 10\%$, $b = 3\%$, $\sigma = 4\%$, $\pi_B = 1\%$, $D = 10$ years, $\pi_S = 5\%$, $\sigma_S = 20\%$ and $\rho_{S,B} = 20\%$. In order to calibrate the risk aversion parameter γ, we minimize the L^2 norm between the target allocation of the lifestyle fund and the model allocation $(\alpha_C^\star, \alpha_B^\star, \alpha_S^\star)$. The results are reported in Table A.2. In the case of the safety profile[24], we obtain a quasi-infinite risk aversion, which is coherent with the financial theory (Wachter, 2003). The defensive profile corresponds to $\hat{\gamma} = 7.20$ and the equity allocation is 17.50%. For the dynamic profile, the parameter $\hat{\gamma}$ is lower and is equal to 1.55. However, we note that $\hat{\gamma}$ is sensitive to the parameter values. Indeed, if $\rho_{S,B} = -20\%$, we obtain the results given in Table A.3.

TABLE A.2: Calibration of the lifestyle fund profiles ($T = 10$ years, $\rho_{S,B} = 20\%$)

Profile	$\alpha_C^\star$	$\alpha_B^\star$	$\alpha_S^\star$	$\hat{\gamma}$
Safety	−0.03	99.88	0.15	867
Defensive	−3.02	85.52	17.50	7.20
Balanced	−8.55	59.07	49.48	2.55
Dynamic	−14.07	32.61	81.46	1.55

TABLE A.3: Calibration of the lifestyle fund profiles ($T = 10$ years, $\rho_{S,B} = -20\%$)

Profile	$\alpha_C^\star$	$\alpha_B^\star$	$\alpha_S^\star$	$\hat{\gamma}$
Safety	−0.08	99.93	0.15	867
Defensive	−10.59	90.62	19.98	6.72
Balanced	−26.16	76.82	49.34	2.72
Dynamic	−41.72	63.03	78.70	1.71

We consider the previous parameter values and three risk profiles: $\gamma = 1.5$, $\gamma = 3$ and $\gamma = 7$. In Figure A.7, we measure the sensitivity of the equity allocation $\alpha_S^\star$ with respect to the parameters. We note that the most important parameters are the risk premium and the volatility. By definition, the risk premium is a long-term parameter, which remains relatively constant over time. By contrast, volatility is a short-term parameter which significantly changes from one period to another. To be coherent with the investor risk aversion, we need to have a dynamic allocation, which is not the case of the constant-mix strategy. This justifies the existence of flexible funds, which change the allocation according to market conditions.

[24] This profile consists in investing all the wealth in money market and bond funds.

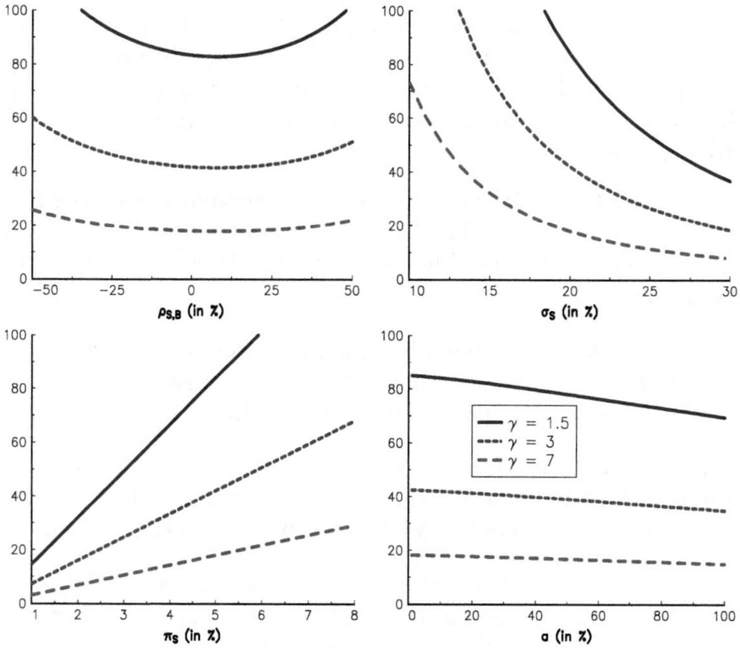

FIGURE A.7: Sensitivity of the equity allocation $\alpha_S^\star$ (in %) in lifestyle funds

A.3.3.2 Lifecycle funds

In the Merton model, the choice of the optimal risky portfolio does not depend on the time horizon. This means that the relative proportion of risky assets (stocks, bonds, etc.) remains the same for investors with different levels of risk aversion. The only element that changes is the absolute proportion of risk-free assets. However, this theoretical perspective differs greatly from the 'popular advice' on portfolio allocation, which is summarized by Vanguard founder John C. Bogle in the old rule of thumb: *you should hold your age in bonds and the rest in equities*. There are many popular justifications given for this, but most of them do not make economic sense. For instance, it is generally accepted that "*stocks are less risky over a young person's long investment horizon*" (Jagannathan and Kocherlakota, 1996). However, if we assume that the asset returns are stationary, investing 80% in stocks during the first 20 years and reducing this weight to 30% in the following 20 years is equivalent to investing 30% in the first 20 years and 80% in the following 20 years.

Nevertheless, these allocation rules may be perfectly justified when taking into account the individual's life cycle (Bodie *et al.*, 1992; Viceira, 2001; Cocco *et al.*, 2005; Munk and Sørensen, 2010), and more specifically his savings component. From a professional point of view, these allocation methods are intensively used in target-date (or lifecycle) funds, which are designed to maximize the pension benefits of individuals during retirement. These funds have been gaining in popularity since the early 21st century. One of the reasons for this is the major shift from defined benefit (DB) toward defined contribution (DC) pension plans and the transfer of investment risk from the corporate sector to households.

Optimal allocation of lifecycle funds may be obtained using the intertemporal model of Merton (1971) in which one introduces a stochastic permanent contribution. This is the main difference between the design of a mutual fund and a target-date fund:

- In a mutual fund, the individual makes an initial investment and seeks to maximize the net asset value for a given horizon. Generally, the investment horizon for such funds is three or five years.
- In a target-date fund, the individual makes an initial investment and continues to contribute throughout his working life. The investor's objective is then to maximize his pension benefits during retirement.

Bruder *et al.* (2012) consider a target-date fund with maturity T owned by only one investor. Its value at time t is denoted by $x(t)$. It can be invested in a risky portfolio $S(t)$ with a proportion $\alpha(t)$ and in a zero-coupon bond $B(t,T)$ with a proportion $1-\alpha(t)$. Since the investor has some income or private wealth and would like to use the target-date fund for retirement pension benefits, he regularly contributes to the fund. We note his contribution as $\pi(t)$. The dynamics of the target-date fund is then:

$$\frac{\mathrm{d}x(t)}{x(t)} = \alpha(t)\frac{\mathrm{d}S(t)}{S(t)} + (1-\alpha(t))\frac{\mathrm{d}B(t,T)}{B(t,T)} + \frac{\pi(t)\,\mathrm{d}t}{x(t)}$$

We assume that $\pi(t) = p(t)Q(t)$ where $p(t)$ is the average contribution behavior for the representative agent and $Q(t)$ is a random factor related to contribution uncertainty. The state variable $Q(t)$ is important, because the investor does not know exactly what his contribution will be in the future. We then have[25]:

$$\left\{\begin{array}{l} \mathrm{d}S(t) = \mu(t)S(t)\,\mathrm{d}t + \sigma(t)S(t)\,\mathrm{d}W^S(t) \\ \mathrm{d}Q(t) = \theta(t)Q(t)\,\mathrm{d}t + \zeta(t)Q(t)\,\mathrm{d}W^Q(t) \\ \mathrm{d}B(t,T) = r(t)B(t,T)\,\mathrm{d}t + \Gamma(t,T)B(t,T)\,\mathrm{d}W^B(t) \end{array}\right.$$

The objective of the investor is to maximize his utility at time T:

$$\alpha^\star(t) = \arg\max \mathbb{E}_t\left[\mathcal{U}(x(T))\right]$$

[25] Let $W(t) = \left(W^S(t), W^Q(t), W^B(t)\right)$ be the vector of Brownian motions. We assume that:

$$\mathbb{E}\left[W(t)W(t)^\top\right] = \begin{pmatrix} 1 & \rho_{S,Q} & \rho_{S,B} \\ \rho_{S,Q} & 1 & \rho_{Q,B} \\ \rho_{S,B} & \rho_{Q,B} & 1 \end{pmatrix}\mathrm{d}t$$

Using the zero-coupon bond $B(t,T)$ of maturity T as the numéraire (Geman *et al.*, 1995), Bruder *et al.* (2012) show that the optimal exposure is given by this relationship[26]:

$$\alpha^*(t) = -\frac{\mu(t,T)\,\partial_x \mathcal{J}(t,x,q) + \rho^\star \sigma(t,T)\,\zeta(t,T)\,q\partial^2_{x,q}\mathcal{J}(t,x,q)}{x\sigma^2(t,T)\,\partial^2_x \mathcal{J}(t,x,q)} \tag{A.9}$$

with $\mathcal{J}(t,x,q) = \sup_\alpha \mathbb{E}_t\left[\mathcal{U}(x(T)) \mid x(t,T) = x, Q(t,T) = q\right]$.

If we suppose the investor does not contribute to the fund ($\pi_t = 0$) and if we suppose the parameters are constant, we retrieve the Merton solution:

$$\alpha^*(t) = -\frac{(\mu - r)}{\sigma^2}\frac{\partial_x \mathcal{J}(t,x)}{x\partial^2_x \mathcal{J}(t,x)}$$

If there is no uncertainty on the future contribution of the investor ($Q_t = 1$) and if we assume that the interest rate is nul, the solution in the case of the CRRA utility function is:

$$\alpha^*(t) = \frac{(\mu - r)}{\gamma\sigma^2} + \frac{\mu\int_t^T \pi(s)\,\mathrm{d}s}{\gamma\sigma^2 x(t)}$$

Therefore, the optimal exposure depends on the future contributions to be made by the investor. This result was already found by Merton (1971) when he introduced non-capital gain income:

> "*[...] one finds that, in computing the optimal decision rules, the individual capitalizes the lifetime flow of wage income at the market (risk-free) rate of interest and then treats the capitalized value as an addition to the current stock of wealth.*"

In the general case, Solution (A.9) is computed numerically using finite differences. However, this solution is highly dynamic because it is a function of the time t, and of the state variables $x(t)$ and $Q(t)$:

$$\alpha^*(t) = \alpha(t, x(t), Q(t))$$

In practice, target-date funds do not use $\alpha^*(t)$ to define the allocation, but the glide path $g^*(t)$, which is the expected dynamic allocation:

$$g^*(t) = \mathbb{E}_0\left[\alpha^*(t)\right]$$

[26]The parameters are:

$$\begin{aligned}
\mu(t,T) &= \mu(t) - r(t) + \Gamma^2(t,T) - \rho_{S,B}\sigma(t)\Gamma(t,T)\\
\sigma(t,T) &= \sqrt{\sigma^2(t) + \Gamma^2(t,T) - 2\rho_{S,B}\sigma(t)\Gamma(t,T)}\\
\theta(t,T) &= \theta(t) - r(t) + \Gamma^2(t,T) - \rho_{Q,B}\zeta(t)\Gamma(t,T)\\
\zeta(t,T) &= \sqrt{\zeta^2(t) + \Gamma^2(t,T) - 2\rho_{Q,B}\zeta(t)\Gamma(t,T)}
\end{aligned}$$

while the forward correlation is given by:

$$\rho^\star = \frac{\rho_{S,Q}\sigma(t)\zeta(t) - \rho_{S,B}\sigma(t)\Gamma(t,T) - \rho_{Q,B}\zeta(t)\Gamma(t,T) + \Gamma^2(t,T)}{\sigma(t,T)\zeta(t,T)}$$

In Figure A.8, we reproduce some results of Bruder *et al.* (2012) and show the impact of the parameters on the glide path. We verify that the proportion of equities decreases over time. Young people then have a higher proportion invested in equities than old people from the current wealth viewpoint, but they have a similar proportion if we include future savings in the reference wealth.

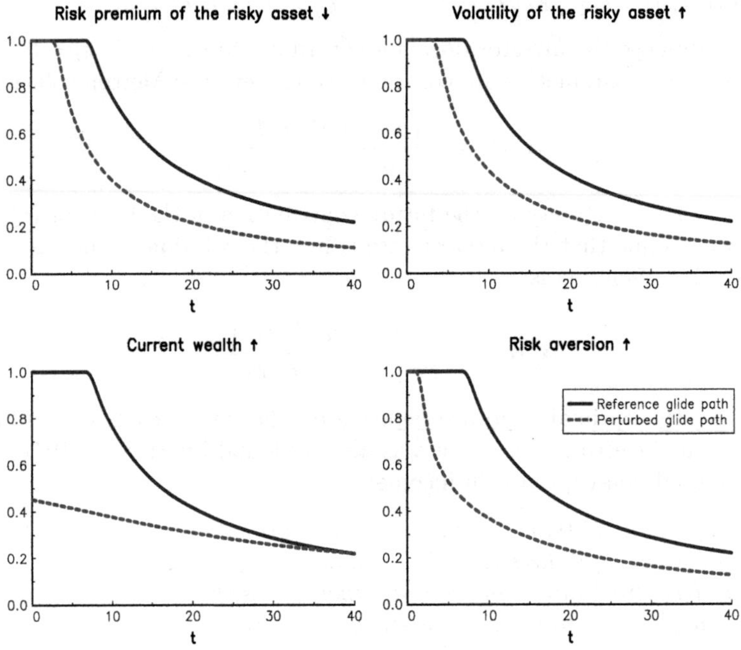

FIGURE A.8: Influence of the parameters on the glide path of target-date funds

A.3.3.3 Liability driven investment

LDI is an allocation method that takes into account liability constraints. It is particularly adapted to manage defined benefit (DB) pension plans. By computing the actualized value of the liabilities and the different sensitivities to the market parameters (interest rate, inflation, etc.), we are able to define the liability hedging portfolio (LHP) of the pension fund. We then consider two cases:

1. If the asset value is larger than the LHP value, the pension fund must buy the LHP. Indeed, the liability structure is perfectly hedged with this portfolio composed of zero-coupons, inflation-linked bonds, etc. The pension fund may also invest the difference between the asset value and the LHP value into risky assets to generate surplus performance.

2. If the asset value is lower than the LHP value, the pension fund must also invest in risky assets in order to generate additional performance and bridge the funding gap. The LDI approach allows then to define the optimal strategy, i.e. the allocation between the LHP and the risky portfolio.

Let $A(t)$ and $L(t)$ be the asset value of the fund and the present value of liabilities. We define the funding ratio as:

$$f(t) = \frac{A(t)}{L(t)}$$

The objective of the pension fund is to maximize the utility function at a given maturity:

$$\max \mathbb{E}_{\mathbb{P}}\left[\mathcal{U}\left(f(T)\right)\right]$$

under the budget constraint:

$$A_0 = \mathbb{E}_{\mathbb{Q}}\left[e^{-\int_0^T r_t \,\mathrm{d}t} A(T)\right]$$

As seen previously, the budget constraint means that the pension fund buys a contingent claim $A(T)$ whose price $\mathbb{E}_{\mathbb{Q}}\left[\exp\left(-\int_0^T r_t \,\mathrm{d}t\right) A(T)\right]$ is equal to its initial wealth, i.e. the asset value A_0. Following the martingale approach, we deduce that the optimal solution $A^\star(T)$ satisfies the following relationship:

$$\frac{\mathrm{d}\mathbb{Q}}{\mathrm{d}\mathbb{P}} = \frac{\mathcal{U}'\left(f^\star(T)\right)}{\lambda}$$

where λ is the Lagrange coefficient associated with the budget constraint. Bruder *et al.* (2010) show that solving this problem is equivalent to considering the Merton model when the risk-free asset is the LHP portfolio and the risky asset is the tangency portfolio, also called the performance portfolio (PP). To find the optimal allocation, we must specify the utility function. As noted by Martellini and Milhau (2012), it must satisfy a number of constraints:

- The pension fund cannot accept having a funding ratio $f(t)$ smaller than a given value f^-. This constraint implies that the solvency of the pension fund is never guaranteed below this ratio. f^- is then the minimum acceptable performance.

- The pension fund has no particular interest in taking more risk if $f(t)$ is larger than a given value f^+. f^+ is then the ultimate performance goal and is generally equal to 100% when the pension plan is underfunded. If $f(t) = f^+$, the asset value is sufficient to buy the LHP, implying that the liability structure is then hedged.

The utility function then has the following form:

$$\mathcal{U}(f) = \begin{cases} -\infty & \text{if } f < f^- \\ U(f) & \text{if } f \in [f^-, f^+] \\ U(f^+) & \text{if } f > f^+ \end{cases}$$

In Figure A.9, we illustrate the utility function $\mathcal{U}(f)$ when $f^- = 70\%$, $f^+ = 100\%$ and U is a CRRA function with parameter γ. The optimal solution is then equal to:

$$A^\star(t) = (1 - \alpha^\star(t))\, x_{\text{LHP}}(t) + \alpha^\star(t)\, x_{\text{PP}}(t)$$

where $x_{\text{LHP}}(t)$ and $x_{\text{PP}}(t)$ are the values of the LHP and PP portfolios and $\alpha^\star(t)$ is the proportion of the asset value invested in the performance portfolio. The case of the CRRA utility function is considered by Martellini and Milhau (2012), who find analytical solutions. In Figure A.10, we report the optimal allocation $\alpha^\star(t)$ for a pension fund with a funding ratio equal to 85%. Its objective is to reach a funding ratio f^+ equal to 100% with the constraint that it is always larger than 70%. We assume that the risk premium and the volatility of the performance portfolio are 7.50% and 15%. The allocation $\alpha^\star(t)$ decreases when the funding ratio is close to the minimum acceptable performance. In this case, the pension fund reduces its risk, because of the solvency constraint. We also note that the optimal allocation depends on the maturity T. If the pension fund wants to achieve its objective in one year, it must take more risk than if its investment horizon is five years.

Remark 74 *When the pension plan is overfunded, i.e. when it has more assets than liabilities, the minimum acceptable performance f^- is larger than 100%. By using the LDI approach, the pension fund is always guaranteed to remain overfunded:*

$$1 \leq f^- \leq f(t) \leq f^+$$

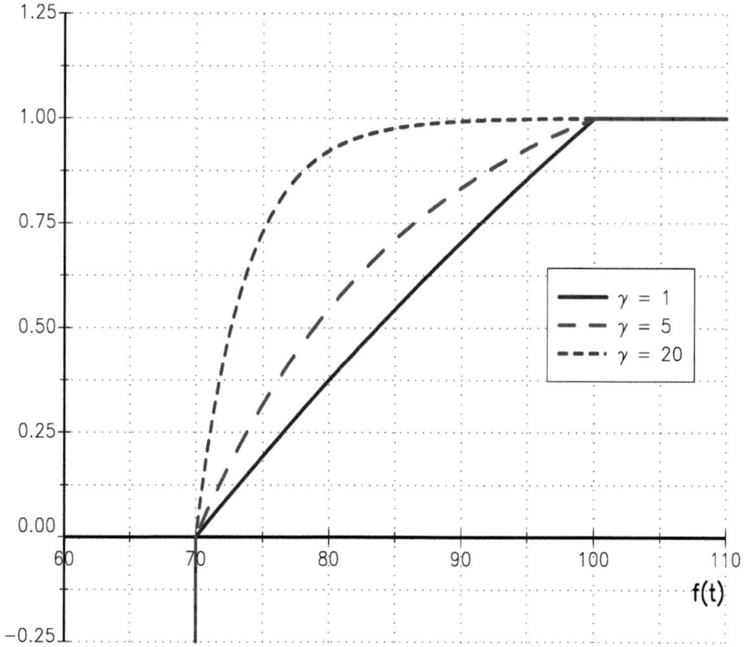

FIGURE A.9: Example of the LDI utility function

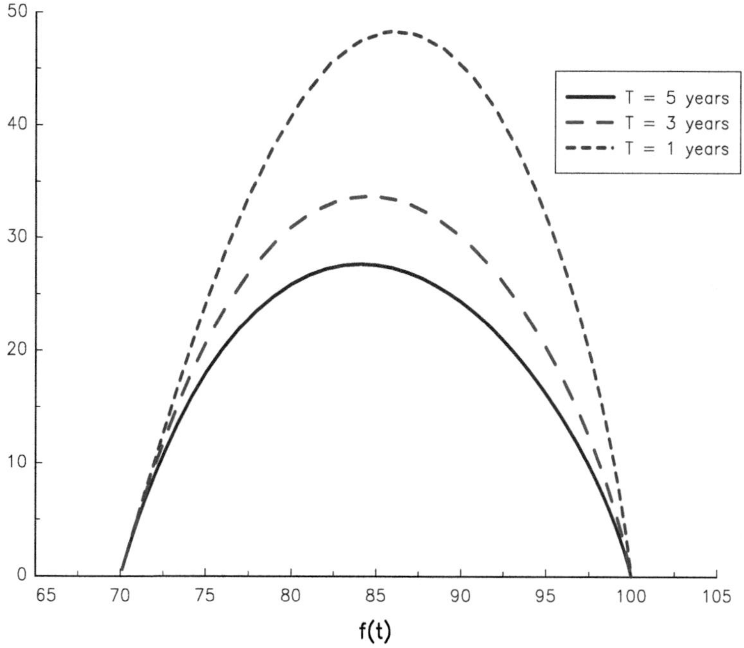

FIGURE A.10: Optimal exposure $\alpha^{\star}(t)$ (in %) in the LDI portfolio

Appendix B

Tutorial Exercises

The solutions of the tutorial exercises are available at the following web page:

`http://www.thierry-roncalli.com/riskparitybook.html`

B.1 Exercises related to modern portfolio theory

B.1.1 Markowitz optimized portfolios

We consider three assets whose volatilities are 15%, 15% and 5% and expected returns are 10%, 10% and 5%. The correlation matrix is defined as follows:

$$\rho = \begin{pmatrix} 100\% & & \\ 50\% & 100\% & \\ 20\% & 20\% & 100\% \end{pmatrix}$$

1. Find the minimum variance portfolio.
2. Find the optimal portfolio which has an ex-ante volatility equal to 5%.
3. Find the optimal portfolio which has an ex-ante volatility equal to 10%.
4. Comment on these results.
5. We impose a minimum exposure of 8% for the first asset ($x_1 \geq 8\%$).
   (a) Calculate the solution to Questions 1, 2 and 3 by taking into account this new constraint.
   (b) Define the dual quadratic problem associated to each of the three problems.
   (c) For each problem, compute the Lagrange coefficient corresponding with the minimum exposure constraint.
   (d) Comment on these results.
6. Why is there no solution to Question 2 when the minimum exposure for the first asset is set to 20%? Calculate analytically the maximum value of the lower bound x_1^- ($x_1 \geq x_1^-$) such that Question 2 has a solution.

B.1.2 Variations on the efficient frontier

We consider an investment universe of four assets. We assume that their expected returns are equal to 5%, 6%, 8% and 6%, and their volatilities are equal to 15%, 20%, 25% and 30%. The correlation matrix is:

$$\rho = \begin{pmatrix} 100\% & & & \\ 10\% & 100\% & & \\ 40\% & 70\% & 100\% & \\ 50\% & 40\% & 80\% & 100\% \end{pmatrix}$$

We note x_i the weight of the ith asset in the portfolio. We only impose that the sum of the weights is equal to 100%.

1. Represent the efficient frontier[1].

2. Calculate the minimum variance portfolio. What are its expected return and its volatility?

3. Calculate the optimal portfolio which has an ex-ante volatility $\sigma^\star$ equal to 10%. Same question if $\sigma^\star = 15\%$ and $\sigma^\star = 20\%$.

4. We note $x^{(1)}$ the minimum variance portfolio and $x^{(2)}$ the optimal portfolio with $\sigma^\star = 20\%$. We consider the set of portfolios $x^{(\alpha)}$ defined by the relationship:
$$x^{(\alpha)} = (1 - \alpha)\, x^{(1)} + \alpha x^{(2)}$$
In the previous efficient frontier, place the portfolios $x^{(\alpha)}$ when α is equal to -0.5, -0.25, 0, 0.1, 0.2, 0.5, 0.7 and 1. What do you observe? Comment on this result.

5. Repeat Questions 3 and 4 by considering the constraint $0 \leq x_i \leq 1$. Explain why we do not retrieve the same observation.

6. We now include in the investment universe a fifth asset corresponding to the risk-free asset. Its return is equal to 3%.

   (a) Define the expected return vector and the covariance matrix of asset returns.

   (b) Deduce the efficient frontier by solving directly the quadratic problem.

   (c) What is the shape of the efficient frontier? Comment on this result.

   (d) Choose two arbitrary portfolios $x^{(1)}$ and $x^{(2)}$ of this efficient frontier. Deduce the Sharpe ratio of the tangency portfolio.

   (e) Calculate then the composition of the tangency portfolio from $x^{(1)}$ and $x^{(2)}$.

---

[1]Consider the following values of γ: -1, -0.5, -0.25, 0, 0.25, 0.5, 1 and 2.

7. We consider the general framework with n risky assets whose vector of expected returns is μ and the covariance matrix of asset returns is Σ while the return of the risk-free asset is r. We note $\tilde{x}$ the portfolio invested in the $n+1$ assets. We have:

$$\tilde{x} = \begin{pmatrix} x \\ x_r \end{pmatrix}$$

with x the weight vector of risky assets and x_r the weight of the risk-free asset. We impose the following constraint:

$$\sum_{i=1}^{n} \tilde{x}_i = \sum_{i=1}^{n} x_i = 1$$

   (a) Define $\tilde{\mu}$ and $\tilde{\Sigma}$ the vector of expected returns and the covariance matrix of asset returns associated with the $n+1$ assets.

   (b) By using the Markowitz ϕ-problem, retrieve the *Separation Theorem* of Tobin.

B.1.3 Sharpe ratio

1. We consider two risky assets with returns R_1 and R_2. We assume that:

$$R = \begin{pmatrix} R_1 \\ R_2 \end{pmatrix} \sim \mathcal{N}\left(\begin{pmatrix} \mu_1 \\ \mu_2 \end{pmatrix}, \begin{pmatrix} \sigma_1^2 & \rho\sigma_1\sigma_2 \\ \rho\sigma_1\sigma_2 & \sigma_2^2 \end{pmatrix} \right)$$

   (a) Let r be the return of the risk-free asset. Define the Sharpe ratio SR_i of each asset i.

   (b) Let $x = (x_1, x_2)$ be a portfolio composed of the two risky assets. Give the expression of the Sharpe ratio $\mathrm{SR}(x \mid r)$.

   (c) We assume that $x_1 + x_2 = 1$ and the second asset corresponds to the risk-free asset. Show that we have:

$$\mathrm{SR}(x \mid r) = \begin{cases} -\mathrm{SR}_1 & \text{if } x_1 < 0 \\ +\mathrm{SR}_1 & \text{if } x_1 > 0 \end{cases}$$

2. We consider an equally weighted portfolio with n assets[2]. Let $R = (R_1, \ldots, R_n)$ be the vector of asset returns. We assume that $R \sim \mathcal{N}(\mu, \Sigma)$ with $\mu = (\mu_1, \ldots, \mu_n)$, $\Sigma = (\Sigma_{i,j})$ and[3] $\Sigma_{i,j} = \rho_{i,j}\sigma_i\sigma_j$. We study the case when the asset returns are not correlated ($\rho_{i,j} = 0$ if $i \neq j$).

   (a) Give the expression of the Sharpe ratio of the portfolio $x = (x_1, \ldots, x_n)$.

[2] We have $x_i = n^{-1}$.

[3] We have of course $\rho_{i,i} = 1$.

(b) Show also that the Sharpe ratio of the portfolio is a linear combination of the Sharpe ratios of the assets:

$$\mathrm{SR}(x \mid r) = \sum_{i=1}^{n} w_i \, \mathrm{SR}_i$$

(c) Verify that the weights are in the range $[0, 1]$:

$$0 < w_i < 1$$

(d) We consider a portfolio with five assets. In the following table, we report two sets of parameters #1 and #2 for modeling these assets:

	i	1	2	3	4	5
#1	σ_i	20%	20%	30%	10%	30%
	SR_i	0.40	0.35	0.30	0.70	0.40
#2	σ_i	15%	15%	20%	10%	50%
	SR_i	0.10	0.15	0.05	0.05	0.90

For each set of parameters, calculate the weights w_i and the corresponding Sharpe ratio of the portfolio. Why are these results surprising? How do you explain this fact?

3. We consider the framework of Question 2, but we now assume that the correlation is uniform ($\rho_{i,j} = \rho$ if $i \neq j$) and the volatilities are the same ($\sigma_{i,j} = \sigma$).

(a) Give the expression of the Sharpe ratio $\mathrm{SR}(x \mid r)$ of portfolio x.

(b) Show then that the Sharpe ratio $\mathrm{SR}(x \mid r)$ of portfolio x is proportional to the average Sharpe ratio of the assets:

$$\mathrm{SR}(x \mid r) = w \cdot \left(\frac{1}{n} \sum_{i=1}^{n} \mathrm{SR}_i \right)$$

(c) We set $\rho = 50\%$. How many assets do we need to obtain a Sharpe ratio larger than 25% compared to the average Sharpe ratio of the assets?

(d) Same question if $\rho = 80\%$.

(e) Comment on these results.

4. We consider a fund of hedge funds manager, whose objective is to achieve a performance of Libor + 400 bps with a volatility of 4%. We assume that the management and performance fees of the FoF are 1% per year and 10% above Libor.

(a) What is the gross performance objective of the FoF? What is the objective of the hedge funds portfolio in terms of Sharpe ratio?

(b) We use the framework of Question 3. Compute the coefficient w for different values of n and ρ.

(c) Do you think that the performance objective can be reached if we assume that the average Sharpe ratio of single hedge funds is 0.5 and if the correlation between these hedge funds is larger than 20%?

B.1.4 Beta coefficient

1. We consider an investment universe of n assets with:

$$R = \begin{pmatrix} R_1 \\ \vdots \\ R_n \end{pmatrix} \sim \mathcal{N}(\mu, \Sigma)$$

The weights of the market portfolio (or the benchmark) are $b = (b_1, \ldots, b_n)$.

(a) Define the beta β_i of asset i with respect to the market portfolio.

(b) Let X_1, X_2 and X_3 be three random variables. Show that:

$$\operatorname{cov}(c_1 X_1 + c_2 X_2, X_3) = c_1 \operatorname{cov}(X_1, X_3) + c_2 \operatorname{cov}(X_2, X_3)$$

(c) We consider the asset portfolio $x = (x_1, \ldots, x_n)$ such that $\sum_{i=1}^n x_i = 1$. What is the relationship between the beta $\beta(x \mid b)$ of the portfolio and the betas β_i of the assets?

(d) Calculate the beta of the portfolios $x^{(1)}$ and $x^{(2)}$ with the following data:

i	1	2	3	4	5
β_i	0.7	0.9	1.1	1.3	1.5
$x_i^{(1)}$	0.5	0.5			
$x_i^{(2)}$	0.25	0.25	0.5	0.5	−0.5

2. We assume that the market portfolio is the equally weighted portfolio[4].

(a) Show that $\sum_{i=1}^n \beta_i = n$.

(b) We consider the case $n = 3$. Show that $\beta_1 \geq \beta_2 \geq \beta_3$ implies $\sigma_1 \geq \sigma_2 \geq \sigma_3$ if $\rho_{i,j} = 0$.

(c) What is the result if the correlation is uniform $\rho_{i,j} = \rho$?

(d) Find a general example such that $\beta_1 > \beta_2 > \beta_3$ and $\sigma_1 < \sigma_2 < \sigma_3$.

[4]We have $b_i = n^{-1}$.

(e) Do we have $\sum_{i=1}^{n} \beta_i < n$ or $\sum_{i=1}^{n} \beta_i > n$ if the market portfolio is not equally weighted?

3. We search a market portfolio $b \in \mathbb{R}^n$ such that the betas are the same for all the assets: $\beta_i = \beta_j = \beta$.

   (a) Show that there is an obvious solution which satisfies $\beta = 1$.

   (b) Show that this solution is unique and corresponds to the minimum variance portfolio.

4. We assume that $b \in [0,1]^n$.

   (a) Show that if one asset has a beta greater than one, there exists another asset which has a beta smaller than one.

   (b) We consider the case $n = 3$. Find a covariance matrix Σ and a market portfolio b such that one asset has a negative beta.

5. We report the return $R_{i,t}$ and $R_t(b)$ of asset i and market portfolio b at different dates:

t	1	2	3	4	5	6
$R_{i,t}$	−22	−11	−10	−8	13	11
$R_t(b)$	−26	−9	−10	−10	16	14
t	7	8	9	10	11	12
$R_{i,t}$	21	13	−30	−6	−5	−5
$R_t(b)$	14	15	−22	−7	−11	2
t	13	14	15	16	17	18
$R_{i,t}$	19	−17	2	−24	25	−7
$R_t(b)$	15	−15	−1	−23	15	−6

   (a) Estimate the beta of the asset.

   (b) What is the proportion of the asset volatility explained by the market?

B.1.5 Tangency portfolio

We consider an investment universe with four assets. We assume that their expected returns are 15%, 10%, 8% and 6%, and that their volatilities are 15%, 10%, 7% and 5%. The correlation matrix is given by:

$$\rho = \begin{pmatrix} 100\% & & & \\ 50\% & 100\% & & \\ 20\% & 20\% & 100\% & \\ 0\% & 0\% & 0\% & 100\% \end{pmatrix}$$

We restrict the analysis to long-only portfolios x meaning that $\sum_{i=1}^{4} x_i = 1$ and $x_i \geq 0$.

1. In what follows, we characterize the tangency portfolio.

   (a) Calculate this portfolio when the return r of the risk-free asset is equal to 2%.

   (b) Same question if $r = 3\%$.

   (c) Same question if $r = 4\%$.

   (d) How do you explain the weight differences of Solutions (a), (b) and (c)?

2. We consider that the benchmark b is equal to $(60\%, 30\%, 10\%, 0\%)$.

   (a) Find the portfolios which have the smallest and highest tracking error with respect to the benchmark.

   (b) Calculate the portfolio which maximizes the information ratio.

   (c) Repeat Questions (a) and (b) by considering the supplementary constraint $x_i \in [10\%, 50\%]$.

   (d) Does the dominance order based on the information ratio imply the dominance order based on the Sharpe ratio? Justify your answer.

B.1.6 Information ratio

1. We consider a universe of n assets. The vector of expected returns and the covariance matrix are noted μ and Σ. Let b (resp. x) be the weights of the benchmark (resp. the portfolio).

   (a) Compute the volatility of the tracking error $\sigma(x \mid b)$.

   (b) Compute the correlation between the portfolio and the benchmark $\rho(x, b)$.

   (c) What is the relationship between $\rho(x, b)$ and $\sigma(x \mid b)$?

   (d) Show that the volatility of the tracking error is bounded:

   $$|\sigma(x) - \sigma(b)| \leq \sigma(x \mid b) \leq \sigma(x) + \sigma(b)$$

   (e) Explain why $\sigma(x \mid b)$ may be high even if $\rho(x, b)$ is close to one.

2. We define the preference ordering as follows:

   $$x \succ y \Leftrightarrow \mathrm{IR}(x \mid b) \geq \mathrm{IR}(y \mid b)$$

   (a) We suppose that $\sigma(x \mid b) = \sigma(y \mid b)$. What is the rationale of the preference ordering?

   (b) We suppose that $\sigma(x \mid b) \neq \sigma(y \mid b)$. What is the rationale of the preference ordering?

(c) We have $\mu(x \mid b) = 5\%$, $\sigma(x \mid b) = 5\%$, $\mu(y \mid b) = 2\%$ and $\sigma(y \mid b) = 3\%$. Build a portfolio z having the same tracking error volatility as portfolio y, but which dominates it.

3. We now suppose that it is not possible to perfectly replicate the benchmark b and we note x_0 the best tracker, i.e. the tracker having the smallest tracking error volatility with respect to b.

   (a) We consider a linear combination of the tracker x_0 and the portfolio x:

   $$z = (1 - \alpha) x_0 + \alpha x$$

   Calculate the information ratio $\text{IR}(z \mid b)$ as a function of $\mu(x_0 \mid b)$, $\mu(x \mid b)$, $\sigma(x_0 \mid b)$, $\sigma(x \mid b)$ and $\sigma(x \mid x_0)$.

   (b) We consider the application of Question 2(c). We have $\mu(x_0 \mid b) = -2\%$, $\sigma(x_0 \mid b) = 1\%$ and $\sigma(x \mid x_0) = 3\%$. Build a portfolio z having the same tracking error volatility as portfolio y. Calculate $\text{IR}(z \mid b)$.

   (c) Comment on this result.

B.1.7 Building a tilted portfolio

We consider an investment universe of four assets. We assume that their expected returns are equal to 10%, -10%, 0 and 5%, and that their volatilities are equal to 15%, 20%, 25% and 30%. The correlation matrix is given by:

$$\rho = \begin{pmatrix} 100\% & & & \\ 50\% & 100\% & & \\ 40\% & 30\% & 100\% & \\ 10\% & 10\% & 10\% & 100\% \end{pmatrix}$$

1. Calculate the ERC portfolio.

2. We restrict the analysis to long-only portfolios x ($\sum_{i=1}^{4} x_i = 1$ and $x_i \geq 0$).

   (a) Find the optimized tilted portfolio by considering a constraint of 1% tracking error volatility with respect to the ERC portfolio.

   (b) Same question if the constraint of tracking error volatility is 5% and 10%.

   (c) Find the optimized portfolio in a heuristic manner when the tracking error volatility is 35%. Justify your answer.

3. Repeat Questions 2(a), 2(b) and 2(c) assuming that the weights may be negative. Comment on these results.

B.1.8 Implied risk premium

We consider an investment universe of n risky assets. We note μ and Σ the vector of expected returns and the covariance matrix of asset returns. The risk-free rate is r.

1. We consider the quadratic utility function:

$$\mathcal{U}(x) = x^\top (\mu - r) - \frac{\phi}{2} x^\top \Sigma x$$

with x the vector of portfolio weights.

(a) Show that the optimal portfolio is a linear function of the risk premium $\pi = \mu - r$ at the equilibrium.

(b) Let x_0 be the investor's portfolio. Calculate the implied risk premium. Comment on this result.

(c) The investor assumes an ex-ante Sharpe ratio $\mathrm{SR}(x_0 \mid r)$ for his portfolio. Show that the risk aversion parameter ϕ satisfies the following relationship:

$$\phi = \frac{\mathrm{SR}(x_0 \mid r)}{\sqrt{x_0^\top \Sigma x_0}}$$

(d) Deduce then that the implied risk premium of asset i is a linear function of its marginal volatility.

(e) What is the economic interpretation of the previous relationship?

(f) Find a new expression of the Sharpe ratio in terms of marginal volatilities.

2. We assume that the investor's portfolio is the market portfolio.

(a) Deduce from the previous analysis the CAPM relationship.

(b) Find a new interpretation of the beta coefficient in terms of risk premium.

3. We assume that the correlations are positive: $\rho_{i,j} \geq 0$.

(a) Show that the implied risk premium is bounded:

$$0 \leq \tilde{\pi}_i \leq \mathrm{SR}(x \mid r) \cdot \sigma_i$$

with $\mathrm{SR}(x \mid r)$ the Sharpe ratio of the portfolio.

(b) In which case is $\tilde{\pi}_i$ equal to the upper bound?

(c) Is it possible that $\tilde{\pi}_i$ reaches the lower bound?

(d) What becomes the previous analysis if the correlations may be negative?

4. We consider a numerical application with the following correlation matrix:

$$\rho = \begin{pmatrix} 1.00 & & \\ 0.50 & 1.00 & \\ 0.25 & 0.00 & 1.00 \end{pmatrix}$$

The volatilities are 25%, 20% and 15%.

   (a) The investor's portfolio is $x = (25\%, 25\%, 50\%)$. Find the implied risk premium of each asset when we have $\mathrm{SR}(x \mid r) = 0.50$. Calculate the beta coefficient with the traditional formula if we assume that the market portfolio is x. Verify that the beta coefficient is equal to the ratio between the risk premium of the asset and the risk premium of the portfolio.

   (b) Same question when $x = (5\%, 5\%, 90\%)$.

   (c) Same question when $x = (100\%, 0\%, 0\%)$.

   (d) Comment on these results.

B.1.9 Black-Litterman model

We consider a universe of three assets. Their volatilities are 20%, 20% and 15%. The correlation matrix of asset returns is:

$$\rho = \begin{pmatrix} 1.00 & & \\ 0.50 & 1.00 & \\ 0.20 & 0.60 & 1.00 \end{pmatrix}$$

We would like to implement a trend-following strategy. For that, we estimate the trend of each asset and the volatility of the trend. We obtain the following results:

Asset	1	2	3
$\hat{\mu}$	10%	−5%	15%
$\sigma(\hat{\mu})$	4%	2%	10%

We assume that the neutral portfolio is the equally weighted portfolio.

1. Find the optimal portfolio if the constraint of the tracking error volatility is set to 1%, 2%, 3%, 4% and 5%.

2. In order to tilt the neutral portfolio, we now consider the Black-Litterman model. The risk-free rate is set to 0.

   (a) Find the implied risk premium of the assets if we target a Sharpe ratio equal to 0.50. What is the value of ϕ?

   (b) How does one incorporate a trend-following strategy in the Black-Litterman model? Give the P, Q and Ω matrices.

(c) Calculate the conditional expectation $\bar{\mu} = \mathbb{E}[\mu \mid P\mu = Q + \varepsilon]$ if we assume that $\Gamma = \tau\Sigma$ and $\tau = 0.01$.

(d) Find the Black-Litterman optimized portfolio.

3. We would like to compute the Black-Litterman optimized portfolio, corresponding to a 3% tracking error volatility.

   (a) What is the Black-Litterman portfolio when $\tau = 0$ and $\tau = +\infty$?

   (b) Using the previous results, apply the bisection algorithm and find the Black-Litterman optimized portfolio, which corresponds to a 3% tracking error volatility.

   (c) Compare the relationship between $\sigma(x \mid b)$ and $\mu(x \mid b)$ of the Black-Litterman model with the one of the tracking error model. Comment on these results.

B.1.10 Portfolio optimization with transaction costs

We consider a universe of six assets. Their expected returns are 5%, 6%, 7%, 8%, 0% and 12% whereas their volatilities are 5%, 5%, 8%, 8%, 9% and 18%. We assume a constant correlation matrix with a uniform correlation equal to 25%. At the date $t = 0$, the investment portfolio $x^{(0)}$ corresponds to the equally weighted portfolio.

1. (a) Find the optimal long-only portfolios when the target volatility is respectively 4.0%, 4.5%, 5.0%, 5.5% and 6.0%. Calculate the turnover of these portfolios with respect to $x^{(0)}$.

   (b) Draw the relationship between the tracking error volatility and the turnover between the optimized portfolio and $x^{(0)}$. Comment on this result.

   (c) Compare the unconstrained efficient frontier with the constrained efficient frontier when we impose that the turnover $\tau\left(x \mid x^{(0)}\right)$ is smaller than 10% (resp. 20% and 40%).

   (d) Compute the optimized portfolio when we target a volatility equal to 5% and when we impose that the maximum turnover τ^{+} is equal to 10% (resp. 20%, 40% and 80%).

   (e) We now suppose that $x^{(0)} = \mathbf{e}_5$. At each rebalancing date, the goal of the portfolio manager is to optimize the return and to reduce the volatility by 50 bps. Moreover, the maximum turnover is limited to 30% at each rebalancing of the portfolio. How many rebalancing passes do we need to obtain a portfolio located in the unconstrained efficient frontier?

2. (a) Let c_i^{-} and c_i^{+} be the bid and ask trading costs. Find the Markowitz γ-program which takes into account these transaction costs. Write the corresponding QP problem.

(b) Calculate the optimal portfolio corresponding to an ex-ante volatility of 5% if we take into account the transaction costs, which are $c_i^- = 2\%$ and $c_i^+ = 1\%$. Compare this solution to the optimal portfolio obtained without transaction costs.

(c) Why is it more complicated to consider transaction costs for long-short portfolios? Extend the previous analysis in this case.

B.1.11 Impact of constraints on the CAPM theory

1. Let $\pi(x \mid x^\star) = \beta(x \mid x^\star)(\mu(x^\star) - r)$ be the excess return of portfolio x explained by its beta with respect to $x^\star$. $\pi(x \mid x^\star)$ is also called the beta return of the portfolio x. We note $x^\star$ the unconstrained tangency portfolio, i.e. the true market portfolio.

(a) Give the CAPM relationship between the risk premium of asset i and the excess return of the market portfolio $x^\star$.

(b) We suppose that the investor uses a market portfolio x, which does not correspond to the tangency portfolio $x^\star$. Decompose the risk premium of asset i as a sum of two components:

$$\mu_i - r = \pi(\mathbf{e}_i \mid x) + \delta_i(x^\star, x)$$

where $\pi(\mathbf{e}_i \mid x)$ is the beta return explained by the wrong market portfolio x. How do you interpret the deviation $\delta_i(x^\star, x)$?

(c) In which case does the beta return overestimate the risk premium of asset i?

(d) Show that the first-order condition of the tangency portfolio optimization program is:

$$\frac{\partial_x \mu(x^\star) - r\mathbf{1}}{\mu(x^\star) - r} = \frac{\partial_x \sigma(x^\star)}{\sigma(x^\star)}$$

Verify that the tangency portfolio weights are:

$$x^\star = \frac{\Sigma^{-1}(\mu - r\mathbf{1})}{\mathbf{1}^\top \Sigma^{-1}(\mu - r\mathbf{1})}$$

(e) We consider a utility maximization program with the following utility function:

$$\mathcal{U}(x) = x^\top(\mu - r\mathbf{1}) - \frac{\phi}{2} x^\top \Sigma x$$

Calculate the optimal utility. Deduce that there is a value ϕ such that the solution is the tangency portfolio.

2. We consider a universe of four assets. We assume that their expected returns are equal to 7%, 7%, 4% and 6%, and their volatilities are equal to 20%, 15%, 4% and 9%. The correlation matrix of asset returns is:

$$\rho = \begin{pmatrix} 100\% & & & \\ 80\% & 100\% & & \\ 30\% & 20\% & 100\% & \\ 20\% & 0\% & 0\% & 100\% \end{pmatrix}$$

The return r of the risk-free asset is equal to 2%.

(a) Calculate the unconstrained tangency portfolio $x^\star$. Verify that the asset risk premium is explained by the beta return.

(b) We impose that $x_i \geq 0$. Calculate the constrained portfolio x. Decompose the risk premium $\mu_i - r$ into the beta return $\pi(\mathbf{e}_i \mid x)$ and the deviation $\delta_i(x^\star, x)$. Comment on this result.

(c) Same question if we impose that $x_i \geq 10\%$.

(d) Find a portfolio such that the beta return of the fourth asset overestimates the risk premium.

B.1.12 Generalization of the Jagannathan-Ma shrinkage approach

1. We consider a universe of five assets. Their volatilities are equal to 15%, 20%, 25%, 25% and 15% whereas the correlation matrix of asset returns is:

$$\rho = \begin{pmatrix} 100\% & & & & \\ 50\% & 100\% & & & \\ 30\% & 20\% & 100\% & & \\ 50\% & 40\% & 80\% & 100\% & \\ 40\% & 50\% & 70\% & 50\% & 100\% \end{pmatrix}$$

(a) We consider the Markowitz problem:

$$\begin{aligned} \min \quad & \frac{1}{2} x^\top \Sigma x \\ \text{u.c.} \quad & \begin{cases} \mathbf{1}^\top x = 1 \\ \mu^\top x \geq \mu^\star \\ x \in \mathcal{C} \end{cases} \end{aligned}$$

where $\mathcal{C}$ is the set of weights constraints. We define two optimized portfolios:

- The first one is the unconstrained portfolio $x^\star$ or $x^\star(\mu, \Sigma)$ with $\mathcal{C} = \mathbb{R}^n$;

- The second one is the constrained portfolio $\tilde{x}$ or $\tilde{x}(\mu, \Sigma)$ when the weight of asset i is between a lower bound x_i^- and an upper bound x_i^+:

$$\mathcal{C} = \left\{x \in \mathbb{R}^n : x_i^- \leq x_i \leq x_i^+\right\}$$

Recall the main result of Jagannathan and Ma (2003).

(b) Find the unconstrained minimum variance portfolio.

(c) Find the minimum variance portfolio if we impose that $0 \leq x_i \leq 40\%$. Deduce the implied shrinkage covariance matrix.

(d) Find the minimum variance portfolio if we impose that $3\% \leq x_i \leq 40\%$. Calculate an approximated value of the portfolio volatility and compare it with the analytical value.

2. We now consider:

$$\mathcal{C} = \{x \in \mathbb{R}^n : Cx \geq D\}$$

(a) Write the first-order condition of the optimization problem and show that the constrained solution is the solution to the unconstrained problem $x^\star\left(\mu, \tilde{\Sigma}\right)$ if we define $\tilde{\Sigma}$ as follows:

$$\tilde{\Sigma} = \Sigma - \left(C^\top \lambda \mathbf{1}^\top + \mathbf{1}\lambda^\top C\right)$$

where λ is the vector of Lagrange coefficients associated to the constraint $Cx \geq D$.

(b) Show that $\tilde{\Sigma}$ is a symmetric matrix. In which cases is $\tilde{\Sigma}$ a positive definite matrix? Comment on these results.

(c) Demonstrate that the result obtained when we impose lower and upper bounds is a special case of the present framework.

(d) Find the minimum variance portfolio when we impose the following constraints:

$$\begin{cases} x_1 + x_2 \leq 40\% \\ x_4 \geq 10\% \end{cases}$$

Deduce the implied shrinkage covariance matrix.

3. We now consider:

$$\mathcal{C} = \{x \in \mathbb{R}^n : Ax = B\}$$

(a) Show that equality constraints may be handled using the previous framework.

(b) Find the minimum variance portfolio when we impose the following constraints:

$$\begin{cases} x_1 + x_2 \leq 50\% \\ x_4 = x_5 \end{cases}$$

Deduce the implied shrinkage covariance matrix.

B.2 Exercises related to the risk budgeting approach

B.2.1 Risk measures

1. We denote $\mathbf{F}$ the cumulative probability distribution of the loss L.

   (a) Give the mathematical definition of the value-at-risk and expected shortfall risk measures.

   (b) Show that:

$$\mathrm{ES}(\alpha) = \frac{1}{1-\alpha} \int_{\alpha}^{1} \mathbf{F}^{-1}(t) \, \mathrm{d}t$$

   (c) We assume that L follows a Pareto distribution $\mathcal{P}(\theta; x_{-})$ defined by:

$$\Pr\{L \leq x\} = 1 - \left(\frac{x}{x_{-}}\right)^{-\theta}$$

   where $x \geq x_{-}$ and $\theta > 1$. Calculate the moments of order one and two. Interpret the parameters x_{-} and θ. Calculate $\mathrm{ES}(\alpha)$ and show that:

$$\mathrm{ES}(\alpha) > \mathrm{VaR}(\alpha)$$

   (d) Calculate the expected shortfall when L is a Gaussian random variable $\mathcal{N}(\mu, \sigma)$. Show that:

$$\Phi(x) = -\frac{\phi(x)}{x^1} + \frac{\phi(x)}{x^3} + \ldots$$

   Deduce that:

$$\mathrm{ES}(\alpha) \rightarrow \mathrm{VaR}(\alpha) \text{ when } \alpha \rightarrow 1$$

   (e) Comment on these results in a risk management perspective.

2. Let $\mathcal{R}(L)$ be a risk measure of the loss L.

   (a) Is $\mathcal{R}(L) = \mathbb{E}[L]$ a coherent risk measure?

   (b) Same question if $\mathcal{R}(L) = \mathbb{E}[L] + \sigma(L)$.

3. We assume that the probability distribution $\mathbf{F}$ of the loss L is defined by:

$$\Pr\{L = \ell_i\} = \begin{cases} 20\% & \text{if } \ell_i = 0 \\ 10\% & \text{if } \ell_i \in \{1, 2, 3, 4, 5, 6, 7, 8\} \end{cases}$$

   (a) Calculate $\mathrm{ES}(\alpha)$ for $\alpha = 50\%$, $\alpha = 75\%$ and $\alpha = 90\%$.

   (b) Let us consider two losses L_1 and L_2 with the same distribution $\mathbf{F}$. Build a joint distribution of (L_1, L_2) which does not satisfy the subadditivity property when $\mathcal{R}(L) = \mathbf{F}^{-1}(\alpha)$.

B.2.2 Weight concentration of a portfolio

1. We consider the Lorenz curve defined by:

$$\begin{array}{rcl} [0,1] & \longrightarrow & [0,1] \\ x & \longmapsto & \mathbb{L}(x) \end{array}$$

We assume that $\mathbb{L}$ is an increasing function and $\mathbb{L}(x) > x$.

(a) Represent graphically the function $\mathbb{L}$ and define the Gini coefficient $\mathcal{G}$ associated with $\mathbb{L}$.

(b) We set $\mathbb{L}_\alpha(x) = x^\alpha$ with $\alpha \geq 0$. Is the function $\mathbb{L}_\alpha$ a Lorenz curve? Calculate the Gini coefficient $\mathcal{G}(\alpha)$ with respect to α. Deduce $\mathcal{G}(0)$, $\mathcal{G}\left(\frac{1}{2}\right)$ and $\mathcal{G}(1)$.

2. Let w be a portfolio of n assets. We suppose that the weights are sorted in a descending order: $w_1 \geq w_2 \geq \ldots \geq w_n$.

(a) We define $\mathbb{L}_w(x)$ as follows:

$$\mathbb{L}_w(x) = \sum_{j=1}^{i} w_j \quad \text{if} \quad \frac{i}{n} \leq x < \frac{i+1}{n}$$

with $\mathbb{L}_w(0) = 0$. Is the function $\mathbb{L}_w$ a Lorenz curve ? Calculate the Gini coefficient with respect to the weights w_i. In which cases does $\mathcal{G}$ take the values 0 and 1?

(b) The definition of the Herfindahl index is:

$$\mathcal{H} = \sum_{i=1}^{n} w_i^2$$

In which cases does $\mathcal{H}$ take the value 1? Show that $\mathcal{H}$ reaches its maximum when $w_i = n^{-1}$. What is the interpretation of this result?

(c) We set $\mathcal{N} = \mathcal{H}^{-1}$. What does the statistic $\mathcal{N}$ mean?

3. We consider an investment universe of five assets. We assume that their asset returns are not correlated. The volatilities are given in the table below:

σ_i	2%	5%	10%	20%	30%
$w_i^{(1)}$		10%	20%	30%	40%
$w_i^{(2)}$	40%	20%		30%	10%
$w_i^{(3)}$	20%	15%	25%	35%	5%

(a) Find the minimum variance portfolio $w^{(4)}$.

(b) Calculate the Gini and Herfindahl indices and the statistic $\mathcal{N}$ for the four portfolios $w^{(1)}$, $w^{(2)}$, $w^{(3)}$ and $w^{(4)}$.

(c) Comment on these results. What differences do you make between portfolio concentration and portfolio diversification?

B.2.3 ERC portfolio

1. We note Σ the covariance matrix of asset returns.

   (a) What is the risk contribution $\mathcal{RC}_i$ of asset i with respect to portfolio x?

   (b) Define the ERC portfolio.

   (c) Calculate the variance of the risk contributions. Define an optimization program to compute the ERC portfolio. Find an equivalent maximization program based on the L^2 norm.

   (d) Let $\beta_i(x)$ be the beta of asset i with respect to portfolio x. Show that we have the following relationship in the ERC portfolio:

   $$x_i \beta_i(x) = x_j \beta_j(x)$$

   Propose a numerical algorithm to find the ERC portfolio.

   (e) We suppose that the volatilities are 15%, 20% and 25% and that the correlation matrix is:

   $$\rho = \begin{pmatrix} 100\% & & \\ 50\% & 100\% & \\ 40\% & 30\% & 100\% \end{pmatrix}$$

   Compute the ERC portfolio using the beta algorithm.

2. We now suppose that the return of asset i satisfies the CAPM model:

   $$R_i = \beta_i R_m + \varepsilon_i$$

   where R_m is the return of the market portfolio and ε_i is the idiosyncratic risk. We note $\varepsilon = (\varepsilon_1, \ldots, \varepsilon_n)$. We assume that $R_m \perp \varepsilon$, $\operatorname{var}(R_m) = \sigma_m^2$ and $\operatorname{cov}(\varepsilon) = D = \operatorname{diag}(\tilde{\sigma}_1^2, \ldots, \tilde{\sigma}_n^2)$.

   (a) Calculate the risk contribution $\mathcal{RC}_i$.

   (b) We assume that $\beta_i = \beta_j$. Show that the ERC weight x_i is a decreasing function of the idiosyncratic volatility $\tilde{\sigma}_i$.

   (c) We assume that $\tilde{\sigma}_i = \tilde{\sigma}_j$. Show that the ERC weight x_i is a decreasing function of the sensitivity β_i to the common factor.

   (d) We consider the numerical application: $\beta_1 = 1$, $\beta_2 = 0.9$, $\beta_3 = 0.8$, $\beta_4 = 0.7$, $\tilde{\sigma}_1 = 5\%$, $\tilde{\sigma}_2 = 5\%$, $\tilde{\sigma}_3 = 10\%$, $\tilde{\sigma}_4 = 10\%$, and $\sigma_m = 20\%$. Find the ERC portfolio.

B.2.4 Computing the Cornish-Fisher value-at-risk

1. Let $X \sim \mathcal{N}(0,1)$. Show that the even moments of X are given by the following relationship:

$$\mathbb{E}\left[X^{2n}\right] = (2n-1)\,\mathbb{E}\left[X^{2n-2}\right]$$

with $n \in \mathbb{N}$. Calculate the odd moments of X.

2. We consider a long position on a call option. The actual price S_t of the underlying asset is equal to 100 dollars, whereas the delta and the gamma of the option are respectively equal to 50% and 2%. We assume that the annual return of the asset is a Gaussian distribution with an annual volatility equal to 32.25%.

   (a) Calculate the Gaussian daily value-at-risk using the delta approximation with a 99% confidence level.

   (b) Calculate the Gaussian daily value-at-risk by considering the delta-gamma approximation.

   (c) Deduce the Cornish-Fisher daily value-at-risk.

3. Let $X \sim \mathcal{N}(\mu, I)$ and $Y = X^\top A X$ with A a symmetric square matrix.

   (a) We recall that:

$$\begin{aligned}
\mathbb{E}[Y] &= \mu^\top A \mu + \operatorname{tr}(A) \\
\mathbb{E}\left[Y^2\right] &= \mathbb{E}^2[Y] + 4\mu^\top A^2 \mu + 2\operatorname{tr}\left(A^2\right)
\end{aligned}$$

   Deduce the moments of $Y = X^\top A X$ when $X \sim \mathcal{N}(\mu, \Sigma)$.

   (b) We suppose that $\mu = \mathbf{0}$. We recall that:

$$\begin{aligned}
\mathbb{E}\left[Y^3\right] &= (\operatorname{tr}(A))^3 + 6\operatorname{tr}(A)\operatorname{tr}\left(A^2\right) + 8\operatorname{tr}\left(A^3\right) \\
\mathbb{E}\left[Y^4\right] &= (\operatorname{tr}(A))^4 + 32\operatorname{tr}(A)\operatorname{tr}\left(A^3\right) + 12\left(\operatorname{tr}\left(A^2\right)\right)^2 + \\
&\quad 12(\operatorname{tr}(A))^2\operatorname{tr}\left(A^2\right) + 48\operatorname{tr}\left(A^4\right)
\end{aligned}$$

   Compute the moments, the skewness and the excess kurtosis of $Y = X^\top A X$ when $X \sim \mathcal{N}(\mathbf{0}, \Sigma)$.

4. We consider a portfolio $x = (x_1, \ldots, x_n)$ of options. We assume that the vector of daily asset returns is distributed according to the Gaussian distribution $\mathcal{N}(\mathbf{0}, \Sigma)$. We note Δ and Γ the vector of deltas and the matrix of gammas.

   (a) Calculate the Gaussian daily value-at-risk using the delta approximation. Define the analytical expression of the risk contributions.

(b) Calculate the Gaussian daily value-at-risk by considering the delta-gamma approximation.

(c) Calculate the Cornish-Fisher daily value-at-risk when assuming that the portfolio is delta neutral.

(d) Calculate the Cornish-Fisher daily value-at-risk in the general case by only considering the skewness.

5. We consider a portfolio composed of 50 options in a first asset, 20 options in a second asset and 20 options in a third asset. We assume that the gamma matrix is:

$$\Gamma = \begin{pmatrix} 4.0\% & & \\ 1.0\% & 1.0\% & \\ 0.0\% & -0.5\% & 1.0\% \end{pmatrix}$$

The actual price of the assets is normalized and is equal to 100. The daily volatility levels of the assets are respectively 1%, 1.5% and 2% whereas the correlation matrix of asset returns is:

$$\rho = \begin{pmatrix} 100\% & & \\ 50\% & 100\% & \\ 25\% & 15\% & 100\% \end{pmatrix}$$

(a) Compare the different methods to compute the daily value-at-risk with a 99% confidence level if the portfolio is delta neutral.

(b) Same question if we now consider that the deltas are equal to 50%, 40% and 60%. Compute the risk decomposition in the case of the delta and delta-gamma approximations. What do you notice? Calculate the ERC portfolio considering the delta approximation.

B.2.5 Risk budgeting when risk budgets are not strictly positive

We consider an investment universe of six assets. Their volatilities are equal to 20%, 20%, 15%, 15%, 10% and 10%. The objective is to compute the weights of the risk budgeting portfolio with the following risk budgets: 40%, 30%, 30%, 0%, 0% and 0%.

1. We assume that the correlation matrix is:

$$\rho = \begin{pmatrix} 1.00 & & & & & \\ 0.70 & 1.00 & & & & \\ 0.80 & 0.50 & 1.00 & & & \\ -0.30 & -0.20 & -0.30 & 1.00 & & \\ -0.20 & -0.10 & -0.10 & 0.20 & 1.00 & \\ -0.40 & -0.10 & -0.10 & 0.20 & 0.00 & 1.00 \end{pmatrix}$$

(a) Find a solution such that all the weights are strictly positive. Compute the risk decomposition of the portfolio. Deduce the number of solutions that satisfy the risk budgeting problem.

(b) Find a solution in the form $(x_1, x_2, x_3, 0, 0, 0)$ with $x_1 > 0$, $x_2 > 0$ and $x_3 > 0$. Compute the risk decomposition of the portfolio. Deduce the number of solutions that satisfy the risk budgeting problem.

(c) Calculate all the other solutions.

2. We assume that the correlation matrix is:

$$\rho = \begin{pmatrix} 1.00 & & & & & \\ 0.70 & 1.00 & & & & \\ 0.80 & 0.50 & 1.00 & & & \\ 0.30 & 0.20 & 0.30 & 1.00 & & \\ -0.20 & -0.10 & -0.10 & -0.20 & 1.00 & \\ 0.40 & 0.10 & 0.10 & 0.20 & 0.00 & 1.00 \end{pmatrix}$$

(a) Find a solution in the form $(x_1, x_2, x_3, 0, 0, 0)$ with $x_1 > 0$, $x_2 > 0$ and $x_3 > 0$. Compute the risk decomposition of the portfolio. Deduce the number of solutions that satisfy the risk budgeting problem.

(b) Calculate all the other solutions.

3. We assume that the correlation matrix is:

$$\rho = \begin{pmatrix} 1.00 & & & & & \\ 0.70 & 1.00 & & & & \\ 0.80 & 0.50 & 1.00 & & & \\ 0.30 & 0.20 & 0.30 & 1.00 & & \\ 0.20 & 0.10 & 0.10 & 0.20 & 1.00 & \\ 0.40 & 0.10 & 0.10 & 0.20 & 0.00 & 1.00 \end{pmatrix}$$

(a) How many solutions are there to the risk budgeting problem?

(b) Calculate them.

4. We consider Question 1 by assuming now that the volatilities of the last three assets are equal to 2%. Compute all the solutions. Comment on these results.

B.2.6 Risk parity and factor models

We assume that the asset returns are driven by the linear factor model:

$$R_t = A\mathcal{F}_t + \varepsilon_t$$

with $\mathcal{F}_t \perp \varepsilon_t$, $\operatorname{cov}(\mathcal{F}_t) = \Omega$ and $\operatorname{cov}(\varepsilon_t) = D = \operatorname{diag}(\tilde{\sigma}_1^2, \ldots, \tilde{\sigma}_n^2)$. For the numerical application, we consider a universe of six assets with:

$$A = \begin{pmatrix} 0.9 & -0.2 & 0.5 \\ 1.1 & -0.5 & -0.3 \\ 1.2 & -0.3 & 0.2 \\ 0.8 & 0.1 & -0.7 \\ 0.5 & 0.5 & 0.5 \\ 1.4 & 0.4 & -0.2 \end{pmatrix}$$

We suppose that the three common factors are uncorrelated with $\sigma(\mathcal{F}_{1,t}) = 20\%$, $\sigma(\mathcal{F}_{2,t}) = 5\%$ and $\sigma(\mathcal{F}_{3,t}) = 8\%$. The idiosyncratic asset volatilities are respectively equal to 4%, 5%, 8%, 5%, 10% and 7%.

1. We consider the portfolio $x = (20\%, 10\%, 15\%, 5\%, 30\%, 20\%)$.

   (a) Decompose the variance σ_i^2 of the asset return with respect to common and specific factors. Calculate the correlation between asset returns.

   (b) Calculate the Moore-Penrose inverse A^+. Find an analytical expression of A^+. Calculate also the matrices B^+, $\tilde{B}$ and $\tilde{B}^+$ defined on page 141. Find an analytical expression of B^+.

   (c) Calculate the weights y and $\tilde{y}$ of common and specific factors. Decompose asset exposures into an exposure to common factors and a specific exposure.

   (d) Perform the risk allocation of portfolio x with respect to the risk factors.

   (e) Find a new parameterization of the linear factor model by eliminating the specific factors ε_t. Calculate then the risk allocation of portfolio x with respect to this new parameterization. Comment on these results.

2. We note $b = (b_1, b_2, b_3)$ the vector of risk factor budgets[5].

   (a) The risk factor budgets are 80%, 10% and 10%. Is it possible to find a long-only portfolio that matches these risk budgets? How does one build a portfolio which has a small sensitivity to the first risk factor?

   (b) Find the RB portfolio that satisfies the budgets 10%, 40% and 40% in terms of risk factors.

   (c) Find the RB portfolio that satisfies the budgets 5%, 90% and 5% in terms of risk factors.

   (d) Find the RB portfolio that satisfies the budgets 5%, 5% and 90% in terms of risk factors.

   (e) Comment on these results.

[5] b_j is then the risk budget assigned to the risk factor $\mathcal{F}_j$.

B.2.7 Risk allocation with the expected shortfall risk measure

1. We consider a portfolio composed of n assets. We assume that asset returns $R = (R_1, \ldots, R_n)$ are normally distributed with $R \sim \mathcal{N}(\mu, \Sigma)$. We note $L(x)$ the loss of the portfolio.

   (a) Find the distribution of $L(x)$.

   (b) Define the expected shortfall $\mathrm{ES}_\alpha(L)$. Calculate its expression in the present case.

   (c) Calculate the risk contribution $\mathcal{RC}_i$ of asset i. Deduce that the expected shortfall verifies the Euler allocation principle.

2. We consider three assets. Their expected returns are 5%, 8% and 3% whereas their volatilities are equal to 12%, 15% and 5%. The correlation matrix is defined as follows:

$$\rho = \begin{pmatrix} 100\% & & \\ 25\% & 100\% & \\ 0\% & -20\% & 100\% \end{pmatrix}$$

   In what follows, we consider that the risk measure corresponds to the expected shortfall with a confidence level of 99%.

   (a) Compute the risk allocation of the long-short portfolio $x = (30\%, 30\%, 40\%)$.

   (b) Find the ERC portfolio.

   (c) Find the long-only risk budgeting portfolio that satisfies the risk budgets $b = (70\%, 20\%, 10\%)$.

   (d) Compute the risk allocation of the long-short portfolio $x = (80\%, 50\%, -30\%)$.

   (e) Find a long-short ERC portfolio. Show that this solution is not unique.

   (f) Find three long-short portfolios that satisfy the risk budgets $b = (70\%, 20\%, 10\%)$.

   (g) Comment on these results.

3. We now consider the general case when asset returns are not necessarily normally distributed.

   (a) Define the risk contribution $\mathcal{RC}_i$ of asset i.

   (b) Find the joint distribution $(R, R(x))$ in the Gaussian case presented in Question 1. Calculate then the risk contributions with the formula given in Question 3(a).

(c) We consider a sample of T observations. Explain how to estimate empirically the risk contributions.

(d) We consider a portfolio with three assets. We suppose that the standardized return R_i of asset i follows a Student's t distribution:

$$\frac{R_i - \mu_i}{\sigma_i} \sim t_{\nu_i}$$

For the numerical application, we assume that the parameters μ_i are equal to 5%, 8% and 3% whereas the parameters σ_i are equal to 12%, 15% and 5%. The number of degrees of freedom ν_i is equal to 4 for the first two assets and to 2 for the third asset. We consider that the dependence between asset returns is given by a normal copula with the following matrix of parameters:

$$\rho = \begin{pmatrix} 100\% & & \\ 25\% & 100\% & \\ 0\% & -20\% & 100\% \end{pmatrix}$$

Using Monte Carlo simulations, compute the expected shortfall with a confidence level of 99% and the corresponding risk allocation when the portfolio is $x = (30\%, 30\%, 40\%)$. Comment on these results.

B.2.8 ERC optimization problem

1. We consider four assets. Their volatilities are equal to 10%, 15%, 20% and 25% whereas the correlation matrix of asset returns is:

$$\rho = \begin{pmatrix} 100\% & & & \\ 60\% & 100\% & & \\ 40\% & 40\% & 100\% & \\ 30\% & 30\% & 20\% & 100\% \end{pmatrix}$$

(a) Find the long-only minimum variance, ERC and equally weighted portfolios.

(b) We consider the following portfolio optimization problem:

$$\begin{aligned} x^\star(c) &= \arg\min \sqrt{x^\top \Sigma x} \qquad \text{(B.1)} \\ &\text{u.c.} \quad \begin{cases} \sum_{i=1}^n \ln x_i \geq c \\ \mathbf{1}^\top x = 1 \\ x \geq \mathbf{0} \end{cases} \end{aligned}$$

with Σ the covariance matrix of asset returns. We note λ_c and λ_0 the Lagrange coefficients associated with the constraints $\sum_{i=1}^n \ln x_i \geq c$ and $\mathbf{1}^\top x = 1$. Write the Lagrange function of the optimization problem. Deduce then an equivalent optimization problem that is easier to solve than Problem (B.1).

(c) Represent the relationship between λ_c and $\sigma(x^\star(c))$, c and $\sigma(x^\star(c))$ and $\mathcal{I}^\star(x^\star(c))$ and $\sigma(x^\star(c))$ where $\mathcal{I}^\star(x)$ is the diversity index of the weights.

(d) Represent the relationship between λ_c and $\mathcal{I}^\star(\mathcal{RC})$, c and $\mathcal{I}^\star(\mathcal{RC})$ and $\mathcal{I}^\star(x^\star(c))$ and $\mathcal{I}^\star(\mathcal{RC})$ where $\mathcal{I}^\star(\mathcal{RC})$ is the diversity index of the risk contributions.

(e) Draw the relationship between $\sigma(x^\star(c))$ and $\mathcal{I}^\star(\mathcal{RC})$. Identify the ERC portfolio.

2. We now consider a slight modification of the previous optimization problem:

$$\begin{aligned} x^\star(c) &= \arg\min \sqrt{x^\top \Sigma x} \\ &\text{u.c.} \quad \begin{cases} \sum_{i=1}^n \ln x_i \geq c \\ x \geq \mathbf{0} \end{cases} \end{aligned} \tag{B.2}$$

(a) Why does the optimization problem (B.1) not define the ERC portfolio?

(b) Find the optimized portfolio of the optimization problem (B.2) when c is equal to -10. Calculate the corresponding risk allocation.

(c) Same question if $c = 0$.

(d) Demonstrate then that the solution to the second optimization problem is:

$$x^\star(c) = \exp\left(\frac{c - c_{\text{erc}}}{n}\right) x_{\text{erc}}$$

where $c_{\text{erc}} = \sum_{i=1}^n \ln x_{\text{erc},i}$. Comment on this result.

(e) Show that there exists a scalar c such that the Lagrange coefficient λ_0 of the optimization problem (B.1) is equal to zero. Deduce then that the volatility of the ERC portfolio is between the volatility of the long-only minimum variance portfolio and the volatility of the equally weighted portfolio:

$$\sigma(x_{\text{mv}}) \leq \sigma(x_{\text{erc}}) \leq \sigma(x_{\text{ew}})$$

B.2.9 Risk parity portfolios with skewness and kurtosis

1. We consider a universe of three assets[6]. The moment matrices calculated with weekly returns are the following:

$$M_1 = \begin{pmatrix} 22.5124 \\ 9.9389 \\ 8.7444 \end{pmatrix} \times 10^{-4}$$

[6] In fact, these three assets correspond to the MSCI World TR Net index, the Citigroup WGBI All Maturities index and the DJ UBS Commodity index. We have calibrated the statistics on the period July 2009 – June 2012.

$$M_2 = \begin{pmatrix} 7.0343 & 0.2931 & 4.5068 \\ 0.2931 & 0.7641 & 0.3564 \\ 4.5068 & 0.3564 & 6.2416 \end{pmatrix} \times 10^{-4}$$

$$M_3 = \left(\begin{matrix} -6.5540 & 0.8548 & -4.3059 & 0.8548 & -0.2178 \\ 0.8548 & -0.2178 & 0.3454 & -0.2178 & -0.0179 \\ -4.3059 & 0.3454 & -3.6789 & 0.3454 & -0.1675 \end{matrix} \right. \cdots$$

$$\left. \begin{matrix} 0.3454 & -4.3059 & 0.3454 & -3.6789 \\ -0.1675 & 0.3454 & -0.1675 & 0.2505 \\ 0.2505 & -3.6789 & 0.2505 & -3.8670 \end{matrix} \right) \times 10^{-6}$$

$$M_4 = \left(\begin{matrix} 2.099 & 0.022 & 1.202 & 0.022 & 0.049 & 0.023 & 1.202 \\ 0.022 & 0.049 & 0.023 & 0.049 & 0.006 & 0.032 & 0.023 \\ 1.202 & 0.023 & 0.959 & 0.023 & 0.032 & 0.040 & 0.959 \end{matrix} \right. \cdots$$

$$\begin{matrix} 0.023 & 0.959 & 0.022 & 0.049 & 0.023 & 0.049 & 0.006 \\ 0.032 & 0.040 & 0.049 & 0.006 & 0.032 & 0.006 & 0.017 \\ 0.040 & 0.985 & 0.023 & 0.032 & 0.040 & 0.032 & 0.007 \end{matrix} \cdots$$

$$\begin{matrix} 0.032 & 0.023 & 0.032 & 0.040 & 1.202 & 0.023 & 0.959 \\ 0.007 & 0.032 & 0.007 & 0.050 & 0.023 & 0.032 & 0.040 \\ 0.050 & 0.040 & 0.050 & 0.092 & 0.959 & 0.040 & 0.985 \end{matrix} \cdots$$

$$\left. \begin{matrix} 0.023 & 0.032 & 0.040 & 0.959 & 0.040 & 0.985 \\ 0.032 & 0.007 & 0.050 & 0.040 & 0.050 & 0.092 \\ 0.040 & 0.050 & 0.092 & 0.985 & 0.092 & 1.531 \end{matrix} \right) \times 10^{-6}$$

We wish to calculate the risk contributions of the equally weighted portfolio by considering a value-at-risk measure and a 95% confidence level.

(a) Compute the mean, the volatility, the skewness and the excess kurtosis of asset returns.

(b) Compute the first four moments of the portfolio PnL Π.

(c) Deduce the mean, the volatility, the skewness and the excess kurtosis of the portfolio loss L.

(d) Compute the risk allocation with respect to the Gaussian value-at-risk.

(e) What is the result if we use the Cornish-Fisher value-at-risk?

2. (a) Compute the ERC portfolio by considering the Gaussian value-at-risk.

(b) Same question using the Cornish-Fisher value-at-risk. Comment on these results.

(c) What do you notice if the confidence level is set to 99%?

B.3 Exercises related to risk parity applications

B.3.1 Computation of heuristic portfolios

We consider a universe of five assets. Their expected returns are 6%, 10%, 6%, 8% and 12% whereas their volatilities are equal to 10%, 20%, 15%, 25% and 30%. The correlation matrix of asset returns is defined as follows:

$$\rho = \begin{pmatrix} 100\% & & & & \\ 60\% & 100\% & & & \\ 40\% & 50\% & 100\% & & \\ 30\% & 30\% & 20\% & 100\% & \\ 20\% & 10\% & 10\% & -50\% & 100\% \end{pmatrix}$$

We assume that the risk-free rate is equal to 2%.

1. We consider unconstrained portfolios. For each portfolio, compute the risk decomposition.
   (a) Find the tangency portfolio.
   (b) Determine the equally weighted portfolio.
   (c) Compute the minimum variance portfolio.
   (d) Calculate the most diversified portfolio.
   (e) Find the ERC portfolio.
   (f) Compare the expected return $\mu(x)$, the volatility $\sigma(x)$ and the Sharpe ratio $\mathrm{SR}(x \mid r)$ of the different portfolios. Calculate then the tracking error volatility $\sigma(x \mid b)$, the beta $\beta(x \mid b)$ and the correlation $\rho(x \mid b)$ if we assume that the benchmark b is the tangency portfolio.
2. Same questions if we impose the long-only portfolio constraint.

B.3.2 Equally weighted portfolio

We note Σ the covariance matrix of n asset returns. In what follows, we consider the equally weighted portfolio based on the universe of these n assets.

1. Let $\Sigma_{i,j} = \rho_{i,j}\sigma_i\sigma_j$ be the elements of the covariance matrix Σ.
   (a) Compute the volatility $\sigma(x)$ of the EW portfolio.
   (b) Let $\sigma_0(x)$ and $\sigma_1(x)$ be the volatility of the EW portfolio when the asset returns are respectively independent and perfectly correlated. Calculate $\sigma_0(x)$ and $\sigma_1(x)$.

(c) We assume that the volatilities are the same. Find the expression of the portfolio volatility with respect to the mean correlation $\bar{\rho}$. What is the value of $\sigma(x)$ when $\bar{\rho}$ is equal to zero? What is the value of $\sigma(x)$ when n tends to $+\infty$?

(d) We assume that the correlations are uniform ($\rho_{i,j} = \rho$). Find the expression of the portfolio volatility as a function of $\sigma_0(x)$ and $\sigma_1(x)$. Comment on this result.

2. (a) Compute the normalized risk contributions $\mathcal{RC}_i^\star$ of the EW portfolio.

   (b) Deduce the risk contributions $\mathcal{RC}_i^\star$ when the asset returns are respectively independent and perfectly correlated[7].

   (c) Show that the risk contribution $\mathcal{RC}_i$ is proportional to the ratio between the mean correlation of asset i and the mean correlation of the asset universe when the volatilities are the same.

   (d) We assume that the correlations are uniform ($\rho_{i,j} = \rho$). Show that the risk contribution $\mathcal{RC}_i$ is a weighted average of $\mathcal{RC}_{0,i}^\star$ and $\mathcal{RC}_{1,i}^\star$.

3. We suppose that the return of asset i satisfies the CAPM:

$$R_i = \beta_i R_m + \varepsilon_i$$

where R_m is the return of the market portfolio and ε_i is the specific risk. We note $\beta = (\beta_1, \ldots, \beta_n)$ and $\varepsilon = (\varepsilon_1, \ldots, \varepsilon_n)$. We assume that $R_m \perp \varepsilon$, $\operatorname{var}(R_m) = \sigma_m^2$ and $\operatorname{cov}(\varepsilon) = D = \operatorname{diag}(\tilde{\sigma}_1^2, \ldots, \tilde{\sigma}_n^2)$.

   (a) Calculate the volatility of the EW portfolio.

   (b) Calculate the risk contribution $\mathcal{RC}_i$.

   (c) Show that $\mathcal{RC}_i$ is approximately proportional to β_i if the number of assets is large. Illustrate this property using a numerical example.

B.3.3 Minimum variance portfolio

We consider unconstrained minimum variance portfolios, meaning that weights can be positive or negative.

1. We suppose that the correlation is uniform $\rho_{i,j} = \rho$.

   (a) Find the expression of the minimum variance portfolio.

   (b) What is the solution when $\rho = 1$? What is the condition to have positive weights for all the assets?

   (c) What is the solution when $\rho = 0$?

[7] We note them $\mathcal{RC}_{0,i}^\star$ and $\mathcal{RC}_{1,i}^\star$.

(d) What is the lower bound ρ^- of the uniform correlation such that the covariance matrix is positive definite? Calculate the analytical solution of the MV portfolio. Comment on this result.

(e) Deduce that there exists a correlation $\rho^\star > 0$ such that the weights are all positive if $\rho \leq \rho^\star$. Calculate the analytical expression $\rho^\star$.

(f) We consider three sets of parameters:

n	σ_1	σ_2	σ_3	σ_4	σ_5	σ_6
4	10%	15%	20%	25%		
5	19%	20%	21%	22%	20%	
6	2%	20%	40%	60%	80%	100%

For each set of parameters, draw the relationship between $\inf x_i^\star$ and ρ. Postulate two rules about the behavior of $\rho^\star$.

(g) For each set of parameters, determine the MV portfolio when the uniform correlation is equal to ρ^-, 0, $\rho^\star$ and 1.

2. We suppose that the return of asset i satisfies the CAPM:

$$R_i = \beta_i R_m + \varepsilon_i$$

where R_m is the return of the market portfolio and ε_i is the specific risk. We note $\beta = (\beta_1, \ldots, \beta_n)$ and $\varepsilon = (\varepsilon_1, \ldots, \varepsilon_n)$. We assume that $R_m \perp \varepsilon$, $\operatorname{var}(R_m) = \sigma_m^2$ and $\operatorname{cov}(\varepsilon) = D = \operatorname{diag}(\tilde{\sigma}_1^2, \ldots, \tilde{\sigma}_n^2)$.

(a) Give the expression of the asset return covariance matrix Σ. Using the Sherman-Morrison-Woodbury formula, calculate Σ^{-1}.

(b) Recall the analytical expression of the minimum variance portfolio. Calculate it in the case of the CAPM.

(c) Show that the solution may be written as:

$$x_i^\star = \frac{\sigma^2(x^\star)}{\tilde{\sigma}_i^2}\left(1 - \frac{\beta_i}{\beta^\star}\right)$$

with $\beta^\star$ a scalar to determine.

(d) In which case is the optimal weight $x_i^\star$ positive? Deduce that if the betas are the same, the weights are positive for all the assets. What can we say about the case of equal volatilities?

(e) We consider the following parameter values:

i	1	2	3	4	5	6
β_i	0.70	0.80	0.90	1.00	1.20	1.20
$\tilde{\sigma}_i$	0.05	0.10	0.10	0.05	0.05	0.20

Calculate the minimum variance portfolio when σ_m takes respectively the following values: 5%, 10%, 15%, 20% and 25%.

B.3.4 Most diversified portfolio

We consider a universe of n assets. We note $\sigma = (\sigma_1, \ldots, \sigma_n)$ the vector of volatilities and Σ the covariance matrix.

1. In what follows, we consider non-constrained optimized portfolios.

   (a) Define the diversification ratio $\mathcal{DR}(x)$ by considering a general risk measure $\mathcal{R}(x)$. How can one interpret this measure from a risk allocation perspective?

   (b) We assume that the weights of the portfolio are positive. Show that $\mathcal{DR}(x) \geq 1$ for all risk measures satisfying the Euler allocation principle. Find an upper bound of $\mathcal{DR}(x)$.

   (c) We now consider the volatility risk measure. Calculate the upper bound of $\mathcal{DR}(x)$.

   (d) What is the most diversified portfolio (or MDP)? In which case does it correspond to the tangency portfolio? Deduce the analytical expression of the MDP and calculate its volatility.

   (e) Demonstrate then that the weights of the MDP are in some sense proportional to $\Sigma^{-1}\sigma$.

2. We suppose that the return of asset i satisfies the CAPM:

$$R_i = \beta_i R_m + \varepsilon_i$$

   where R_m is the return of the market portfolio and ε_i is the specific risk. We note $\beta = (\beta_1, \ldots, \beta_n)$ and $\varepsilon = (\varepsilon_1, \ldots, \varepsilon_n)$. We assume that $R_m \perp \varepsilon$, $\operatorname{var}(R_m) = \sigma_m^2$ and $\operatorname{cov}(\varepsilon) = D = \operatorname{diag}\left(\tilde{\sigma}_1^2, \ldots, \tilde{\sigma}_n^2\right)$.

   (a) Compute the correlation $\rho_{i,m}$ between the asset return and the market return. Deduce the relationship between the specific risk $\tilde{\sigma}_i$ and the total risk σ_i of asset i.

   (b) Show that the solution of the MDP may be written as:

$$x_i^\star = \mathcal{DR}(x^\star) \frac{\sigma_i \sigma(x^\star)}{\tilde{\sigma}_i^2} \left(1 - \frac{\rho_{i,m}}{\rho^\star}\right) \tag{B.3}$$

   with $\rho^\star$ a scalar to be determined.

   (c) In which case is the optimal weight $x_i^\star$ positive?

   (d) Are the weights of the MDP a decreasing or an increasing function of the specific risk $\tilde{\sigma}_i$?

3. In this question, we illustrate that the MDP may be very different than the minimum variance portfolio.

(a) In which case does the MDP coincide with the minimum variance portfolio?

(b) We consider the following parameter values:

i	1	2	3	4
β_i	0.80	0.90	1.10	1.20
$\tilde{\sigma}_i$	0.02	0.05	0.15	0.15

with $\sigma_m = 20\%$. Calculate the unconstrained MDP with Formula (B.3). Compare it with the unconstrained MV portfolio. What is the result if we consider a long-only portfolio?

(c) We assume that the volatility of the assets is 10%, 10%, 50% and 50% whereas the correlation matrix of asset returns is:

$$\rho = \begin{pmatrix} 1.00 & & & \\ 0.90 & 1.00 & & \\ 0.80 & 0.80 & 1.00 & \\ 0.00 & 0.00 & -0.25 & 1.00 \end{pmatrix}$$

Calculate the (unconstrained and long-only) MDP and MV portfolios.

(d) Comment on these results.

B.3.5 Risk allocation with yield curve factors

1. Let Σ be the covariance matrix of n asset returns.

(a) Let λ_i be the eigenvalue associated with the ith eigenvector v_i. Write the eigendecomposition associated with a general square matrix A. Calculate $\operatorname{tr}(A)$ and $\det(A)$ with respect to the eigenvalues.

(b) We assume that the eigenvalues of Σ are all distinct. Show that the eigenvectors are orthogonal. Deduce the eigendecomposition of Σ.

(c) For a correlation matrix C, Morrison (1967) found that the lower bound of the first eigenvalue is:

$$\lambda_1 \geq 1 + (n-1)\,\bar{\rho}$$

where $\bar{\rho}$ is the mean correlation. Show that the lower bound is reached when the correlation matrix is uniform: $C = C_n(\rho)$. Deduce also a lower bound of the percentage of variance explained by the first eigenvector. Calculate the lower bound when $\bar{\rho}$ is equal to 10%, 20%, 50%, 70% and 90% and when n is equal to 2, 3, 5 and 10 assets.

(d) We consider four assets. Their volatilities are 18%, 20%, 22% and 25% whereas the correlation matrix is:

$$C = \begin{pmatrix} 100\% & & & \\ 50\% & 100\% & & \\ 20\% & 30\% & 100\% & \\ 10\% & 40\% & 30\% & 100\% \end{pmatrix}$$

Find the eigendecomposition of the covariance matrix. What is the interpretation of the first and last eigenvectors in terms of optimized portfolios?

(e) Find the eigendecomposition of the correlation matrix. What is the proportion of variance explained by the first eigenvector? Compare it with the lower bound of Morrison.

(f) Using the eigendecomposition of the covariance matrix, propose a four-factor model to explain the asset returns. Same question if we impose only two risk factors. What is the interpretation of these factors?

2. We consider the US yield curve at the date of June 30, 2012. The zero-coupon rates $R_t(T)$ are equal to 0.493%, 0.626%, 0.960%, 1.349% and 1.794% for the maturities 1Y, 3Y, 5Y, 7Y and 10Y. We have also computed the covariance of two-week variations $\Delta_h R_t(T)$ of zero-coupon rates[8] for the period January 2003 – June 2012. The volatilities of $\Delta_h R_t(T)$ expressed in bps are respectively equal to 14.400, 18.815, 20.543, 20.809 and 20.792 whereas the correlation matrix between the variations $\Delta_h R_t(T)$ is:

$$C = \begin{pmatrix} 100.000\% & & & & \\ 80.316\% & 100.000\% & & & \\ 67.518\% & 96.176\% & 100.000\% & & \\ 59.340\% & 91.023\% & 98.467\% & 100.000\% & \\ 51.114\% & 84.085\% & 94.442\% & 98.557\% & 100.000\% \end{pmatrix}$$

In what follows, the risk measure corresponds to the Gaussian value-at-risk for a 99% confidence level and a holding period of two weeks.

(a) Calculate the risk allocation of portfolio x composed by one zero-coupon for each maturity.

(b) Perform the principal component analysis of the covariance matrix. What is the interpretation of the first three factors? Why don't we retrieve exactly the results reported in Figure 4.2 on page 197?

(c) Calculate then the risk allocation of the portfolio x with respect to the PCA factors.

[8] We have $\Delta_h R_t(T) = R_t(T) - R_{t-h}(T)$ where h is equal to ten trading days.

(d) Find the ERC portfolio[9]. Compute the risk allocation with respect to the PCA factors.

(e) We now suppose that the composition of the portfolio is proportional to the eigenvector v_i of the covariance matrix for $i = 1, \ldots, 5$. Compute the risk allocation with respect to the PCA factors. Comment on these results.

B.3.6 Credit risk analysis of sovereign bond portfolios

We consider portfolios of French, German, Italian and Spanish bonds. We use the framework presented in Chapter 4 (Section 4.3.2 on page 220) to measure their credit risk. At the date of July 1, 2011, the levels of CDS $\mathfrak{s}_i$, their volatilities $\sigma_i^{\mathfrak{s}}$ and the correlations $\rho_{i,j}$ between the CDS are:

	$\mathfrak{s}_i$ (in bps)	$\sigma_i^{\mathfrak{s}}$ (in %)	$\rho_{i,j}$ (in %) FR	DE	IT	ES
France	79	57	100			
Germany	42	49	65	100		
Italy	179	61	67	70	100	
Spain	267	59	64	67	83	100

1. We consider two portfolios, each composed of four bonds with the following characteristics:

	#1 x_i	#1 D_i	#2 x_i	#2 D_i
FR	10	6.2	8	7.1
DE	12	7.5	8	6.9
IT	8	6.4	12	6.5
ES	7	5.8	14	8.3

x_i corresponds to the notional of the bond i (expressed in billions of dollars) whereas D_i is the duration measured in years.

(a) Compute the credit risk measure $\mathcal{R}\left(x^{(1)}\right)$ of Portfolio #1 and the risk contributions.

(b) Compute the credit risk measure $\mathcal{R}\left(x^{(2)}\right)$ of Portfolio #2 and the risk contributions.

(c) Let $x^{(1+2)}$ be the merge of Portfolios #1 and #2. What becomes the credit correlation matrix of the merged portfolios? Deduce the credit risk measure $\mathcal{R}\left(x^{(1+2)}\right)$ of the merged portfolio. Compare it to the sum $\mathcal{R}\left(x^{(1)}\right) + \mathcal{R}\left(x^{(2)}\right)$. Comment on this result.

[9]The portfolio is normalized such that it contains one zero-coupon bond of 10-year maturity.

(d) Let $x^{(3)}$ be a portfolio which presents similar characteristics as $x^{(1+2)}$. For this purpose, we replace all the bonds of the same country by a meta-bond. We then have:

$$x_i^{(3)} = x_i^{(1)} + x_i^{(2)}$$

and:

$$D_i^{(3)} = \frac{x_i^{(1)} D_i^{(1)} + x_i^{(2)} D_i^{(2)}}{x_i^{(1)} + x_i^{(2)}}$$

How do you interpret the characteristics of the meta-bonds? Compute the credit risk measure $\mathcal{R}\left(x^{(3)}\right)$ and the risk contributions. Comment on these results. What hypothesis could you postulate?

(e) Comment on the numerical results you obtained for the three portfolios.

2. Let x be a portfolio of $n+1$ assets. We note Σ the associated covariance matrix[10]. We suppose that the last two assets are perfectly correlated: $\rho_{n,n+1} = 1$. We would like to build a portfolio y with n assets such that:

$$\left\{\begin{array}{l} \mathcal{RC}_1(y) = \mathcal{RC}_1(x) \\ \vdots \\ \mathcal{RC}_{n-1}(y) = \mathcal{RC}_{n-1}(x) \\ \mathcal{RC}_n(y) = \mathcal{RC}_n(x) + \mathcal{RC}_{n+1}(x) \end{array}\right. \tag{B.4}$$

We note Σ' the covariance matrix[11] of the n assets belonging to portfolio y.

(a) Compute the risk contribution $\mathcal{RC}_i(x)$ with respect to portfolio x for the $n+1$ assets.

(b) Compute the risk contribution $\mathcal{RC}_i(y)$ with respect to portfolio y for the n assets.

(c) What conditions must satisfy y_i, σ'_i and $\rho'_{i,j}$ to verify the system (B.4)?

(d) Show that there are an infinite number of solutions. Find two particular solutions. Which solution is the best from a financial point of view?

(e) We consider five assets. We suppose that their volatilities are 15%, 20%, 25%, 30% and 20%. The correlation matrix is:

$$\rho = \begin{pmatrix} 1.00 & & & & \\ 0.50 & 1.00 & & & \\ 0.10 & 0.80 & 1.00 & & \\ 0.30 & 0.20 & 0.40 & 1.00 & \\ 0.30 & 0.20 & 0.40 & 1.00 & 1.00 \end{pmatrix}$$

[10] We have $\Sigma_{i,j} = \rho_{i,j}\sigma_i\sigma_j$.
[11] We have $\Sigma'_{i,j} = \rho'_{i,j}\sigma'_i\sigma'_j$.

This means that the fourth and fifth assets are perfectly correlated. Compute the risk decomposition when portfolio x is $(20\%, 30\%, 10\%, 10\%, 30\%)$. Calculate the risk decomposition of two equivalent portfolios y with four assets that satisfy the system (B.4). Find an equivalent portfolio such that $y_4 = 80\%$.

(f) What can we say about the postulated hypothesis of Question 1(d)?

3. We consider the portfolios defined in Question 1. We rebalance each portfolio $x^{(j)}$ in order to obtain a RB portfolio $y^{(j)}$ that matches the following risk budgets $(20\%, 20\%, 30\%, 30\%)$. We suppose that the notional of the RB portfolio does not change, meaning that:

$$\sum y_i^{(j)} = \sum x_i^{(j)}$$

   (a) Find the RB portfolio $y^{(1)}$ corresponding to the portfolio $x^{(1)}$.
   (b) Find the RB portfolio $y^{(2)}$ corresponding to the portfolio $x^{(2)}$.
   (c) Calculate the risk decomposition of the merged portfolio $y^{(1+2)} = y^{(1)} + y^{(2)}$. Comment on this result.
   (d) We would like to build a RB portfolio $y^{(4)}$ such that the risk budgets for each country are 20% (France), 20% (Germany), 30% (Italy) and 30% (Spain). Show that there are many solutions. Using meta-bonds, propose a solution that makes sense from a financial point of view. Calculate the RB portfolio $y^{(4)}$. Conclude.

B.3.7 Risk contributions of long-short portfolios

1. We consider a universe of two assets. We note σ_1 and σ_2 their volatilities whereas ρ corresponds to the correlation.

   (a) Write the risk contributions of the portfolio (x_1, x_2).
   (b) We suppose that the exposure x_1 on the first asset is fixed. In which cases is the risk contribution of the second asset negative? Comment on these results.
   (c) We consider the following numerical values: $\sigma_1 = 18\%$, $\sigma_2 = 10\%$ and $x_1 = 1$. Draw the relationship between x_2 and $\mathcal{RC}_2$ when ρ takes respectively the values 90%, 50%, 0% and -90%.

2. We consider a universe of six assets. The volatilities are 20%, 25%, 30%, 35%, 10% and 10% whereas the correlation matrix is:

$$\rho = \begin{pmatrix} 100\% & & & & & \\ 60\% & 100\% & & & & \\ 50\% & 40\% & 100\% & & & \\ 30\% & 20\% & 30\% & 100\% & & \\ 60\% & 50\% & 20\% & 20\% & 100\% & \\ 50\% & 40\% & 30\% & 30\% & 40\% & 100\% \end{pmatrix}$$

(a) We consider the portfolio $x = (1, -1, 1, -1, 1, -1)$. Calculate the risk decomposition.

(b) We observe that the previous portfolio is composed of three long-short exposures. Find the covariance matrix of the long-short assets. Deduce the risk decomposition.

(c) Calculate the ERC portfolio when we consider the long-short assets. Compare the solution with the one obtained when assuming no correlation between the long-short assets.

(d) Deduce then the risk allocation with respect to the six assets. Comment on this result.

(e) We would like to build a long-short portfolio that satisfies the following characteristics: $x_1 x_2 < 0$, $x_3 x_4 < 0$, $x_5 x_6 < 0$, $\mathcal{RC}_1 = \mathcal{RC}_2$, $\mathcal{RC}_3 = \mathcal{RC}_4$, $\mathcal{RC}_5 = \mathcal{RC}_6$, $\mathcal{RC}_1 + \mathcal{RC}_2 = \mathcal{RC}_3 + \mathcal{RC}_4$ and $\mathcal{RC}_1 + \mathcal{RC}_2 = \mathcal{RC}_5 + \mathcal{RC}_6$. Do you think that such a solution is possible? Justify your answer.

B.3.8 Risk parity funds

1. We consider a universe of three asset classes[12] which are stocks (S), bonds (B) and commodities (C). We have computed the one-year historical covariance matrix of asset returns for different dates and we obtain the following results[13]:

	31/12/1999			31/12/2002		
σ_i	12.40	5.61	12.72	20.69	7.36	13.59
$\rho_{i,j}$	100.00			100.00		
	−5.89	100.00		−36.98	100.00	
	−4.09	−7.13	100.00	22.74	−13.12	100.00
	30/12/2005			31/12/2007		
σ_i	7.97	7.01	16.93	12.94	5.50	14.54
$\rho_{i,j}$	100.00			100.00	−25.76	
	29.25	100.00		−25.76	100.00	
	15.75	15.05	100.00	31.91	6.87	100.00
	31/12/2008			31/12/2010		
σ_i	33.03	9.73	29.00	16.73	6.88	16.93
$\rho_{i,j}$	100.00			100.00		
	−16.26	100.00		15.31	100.00	
	47.31	9.13	100.00	64.13	15.46	100.00

[12] In fact, we use the MSCI World index, the Citigroup WGBI index and the DJ UBS Commodity index to represent these asset classes.
[13] All the numbers are expressed in %.

(a) Compute the weights and the volatility of the risk parity[14] (RP) portfolios for the different dates.
(b) Same question by considering the ERC portfolio.
(c) What do you notice about the volatility of RP and ERC portfolios? Explain these results.
(d) Find the analytical expression of the volatility $\sigma(x)$, the marginal risk $\mathcal{MR}_i$, the risk contribution $\mathcal{RC}_i$ and the normalized risk contribution $\mathcal{RC}_i^\star$ in the case of RP portfolios.
(e) Compute the normalized risk contributions of the previous RP portfolios. Comment on these results.

2. We consider four parameter sets of risk budgets:

Set	b_1	b_2	b_3
#1	45%	45%	10%
#2	70%	10%	20%
#3	20%	70%	10%
#4	25%	25%	50%

(a) Compute the RB portfolios for the different dates.
(b) Compute the implied risk premium $\tilde{\pi}_i$ of the assets for these portfolios if we assume a Sharpe ratio equal to 0.40.
(c) Comment on these results.

B.3.9 The Frazzini-Pedersen model

1. Frazzini and Pedersen (2010) consider an equilibrium model with two periods, n risky assets whose prices and dividends are denoted $P_{i,t}$ and $D_{i,t}$ and m investors who have a given amount of wealth W_j. Let x_j and ϕ_j be the portfolio and the risk aversion of the investor j. At time t, investors maximize their utility function:

$$x_j^\star = \arg\max x_j^\top \mathbb{E}_t \left[P_{t+1} + D_{t+1} - (1+r) P_t\right] - \frac{\phi_j}{2} x_j^\top \Sigma x_j$$

with P_{t+1} the vector of future prices, D_{t+1} the vector of future dividends, Σ the covariance matrix of $P_{t+1} + D_{t+1}$ and r the risk-free rate. Frazzini and Pedersen (2010) assume that investors face some borrowing constraints:

$$m_j \left(x_j^\top P_t\right) \leq W_j$$

They consider three cases. If $m_j < 1$, the investor must hold some of his wealth in cash. If $m_j = 1$, the investor cannot use leverage because of regulatory constraints or borrowing capacity. If $m_j > 1$, the investor is able to leverage his exposure on risky assets.

[14]Here, risk parity refers to the ERC portfolio when we do not take into account the correlations.

(a) Write the first-order condition and deduce the analytical expression of x_j.

(b) Let $\bar{x}$ be the market capitalization portfolio. The equilibrium between demand and supply implies that:

$$\sum_{j=1}^{m} x_j = \bar{x}$$

where $\bar{x}_i$ is the number of shares outstanding of asset i and $\bar{x} = (\bar{x}_1, \ldots, \bar{x}_n)$. Show that we have at the equilibrium:

$$\bar{x} = \frac{1}{\phi}\Sigma^{-1}\left(\mathbb{E}_t\left[P_{t+1} + D_{t+1}\right] - (1 + r + \psi) P_t\right)$$

where ϕ and ψ are two scalars to determine.

(c) Deduce the asset price $P_{i,t}$ at the equilibrium.

(d) Let $\beta_i = \beta(e_i \mid \bar{x})$ be the beta of asset i with respect to the market portfolio. Show that:

$$\mathbb{E}_t\left[R_{i,t+1}\right] - r = \alpha_i + \beta_i\left(\mathbb{E}_t\left[R_{t+1}(\bar{x})\right] - r\right)$$

where $\alpha_i = \psi(1 - \beta_i)$.

(e) Comment on this result.

2. We consider four assets. Their expected returns are equal to 5%, 6%, 8% and 6% while their volatilities are equal to 15%, 20%, 25% and 20%. The correlation matrix of asset returns is given by the following matrix:

$$C = \begin{pmatrix} 1.00 & & & \\ 0.10 & 1.00 & & \\ 0.20 & 0.60 & 1.00 & \\ 0.40 & 0.50 & 0.50 & 1.00 \end{pmatrix}$$

The risk-free rate is set to 2%.

(a) Compute the optimal value of the risk aversion ϕ. Deduce the tangency portfolio $x^\star$, the beta coefficients β_i and the implied risk premia.

(b) We consider two investors with $m_1 = 100\%$ and $m_2 = 50\%$. Their risk aversion is equal to the value obtained in the previous question. Compute the optimal portfolios x_1 and x_2 of these two investors if we assume that they have the same wealth. Deduce the market portfolio $\bar{x}$. Compute then α_i and β_i for the different assets.

(c) Comment on these results.

B.3.10 Dynamic risk budgeting portfolios

1. We consider a universe of n assets. We note r, μ and Σ the risk-free rate, the vector of expected returns and the covariance matrix of asset returns. Moreover, we assume that the risk premium $\pi_i = \mu_i - r$ of asset i is positive.

   (a) Retrieve the formula of the tangency portfolio. What is the solution when the correlations are equal to zero?

   (b) Define the risk contributions of portfolio x.

   (c) We note $(b_1, \ldots, b_n)$ the risk budgets. Find the RB portfolio when the correlations are equal to zero. What is the condition on the risk budgets to obtain the tangency portfolio?

2. We assume now that the risk budgets vary over time. We specify the risk budget of asset i at time t as follows:

$$b_i(t) = b_i(\infty) \frac{\pi_i^\alpha(t)\, \sigma_i^\gamma(\infty)}{\pi_i^\beta(\infty)\, \sigma_i^\delta(t)}$$

   where α, β, γ and δ are four scalar parameters and:

$$b_i(\infty) = \frac{\pi_i^2(\infty)}{\sigma_i^2(\infty)}$$

   $\pi_i(t)$ (resp. $\sigma_i(t)$) corresponds to the risk premium (resp. the volatility) at time t whereas $\pi_i(\infty)$ (resp. $\sigma_i(\infty)$) is the risk premium (resp. the volatility) in the long run.

   (a) Calculate $b_i(t)$ when $\alpha = \beta = \gamma = \delta = 0$. Same question if $\alpha = \beta = \gamma = \delta = 2$.

   (b) In which case does the RB portfolio correspond to the long-run tangency portfolio?

   (c) We consider a universe of four uncorrelated assets with the following characteristics:

i	$\pi_i(\infty)$	$\sigma_i(\infty)$	$\pi_i(t)$	$\sigma_i(t)$
1	2%	10%	1%	12%
2	3%	15%	2%	12%
3	4%	20%	5%	21%
4	5%	25%	6%	24%

   Calculate the long-run tangency portfolio and the long-run risk budgets $b_i(\infty)$. We assume that $\alpha = \beta = \gamma = \delta = \theta$ with $\theta \in [0, 2]$. Draw the relationships between the parameter θ, the risk budgets $b_i(t)$ and the RB weights $x_i(t)$. Comment on these results.

3. We consider risk parity strategies with equity and bond asset classes represented by the MSCI World and Citigroup WGBI indices. The portfolio is rebalanced on a weekly basis using a one-year rolling covariance matrix. The expected returns correspond to the one-year trend, which is estimated using a uniform moving average of the past 260 daily returns. To simulate the dynamic risk parity strategy, we consider the following definition of the risk budgets:

$$b_i(t) \propto b_i(\infty)(1 + c_i(t))^{\alpha}$$

where $c_i(t) > -1$.

   (a) Backtest[15] the pure risk parity strategy based on the ERC portfolio and compare it with the dynamic risk parity strategy when α is set to 1 and:

$$c_i(t) = \min\left(\max\left(-\frac{1}{2}, \frac{\pi_i(t)}{\sigma_i(t)}\right), \frac{3}{2}\right)$$

   (b) Compare the weights of the dynamic risk parity with those given by the tangency portfolio. Comment on these results.

   (c) We consider the dynamic risk parity strategy when α is set to 2 and:

$$c_i(t) = \min(\max(-30\%, \pi_i(t)), 30\%)$$

   (d) Comment on these results.

[15]The study period begins in January 2000 and ends in December 2011.

Bibliography

[1] Abramowitz M. and Stegun I.A. (Eds) (1970), *Handbook of Mathematical Functions*, Ninth edition, Dover.

[2] Acerbi C. (2002), Spectral Measures of Risk: A Coherent Representation of Subjective Risk Aversion, *Journal of Banking & Finance*, 26(7), pp. 1505-1518.

[3] Acerbi C. and Tasche D. (2002), On the Coherence of Expected Shortfall, *Journal of Banking & Finance*, 26(7), pp. 1487-1503.

[4] Adam A., Houkari M. and Laurent J-P. (2008), Spectral Risk Measures and Portfolio Selection, *Journal of Banking & Finance*, 32(9), pp. 1870-1882.

[5] Ahn D-H., Dittmar R.F. and Gallant A.R. (2002), Quadratic Term Structure Models: Theory and Evidence, *Review of Finance Studies*, 15(1), pp. 243-288.

[6] Ambachtsheer K.P. (1987), Pension Fund Asset Allocation: In Defense of a 60/40 Equity/Debt Asset Mix, *Financial Analysts Journal*, 43(5), pp. 14-24.

[7] Amenc N., Goltz F., Martellini L. and Retkowsky P. (2010), Efficient Indexation: An Alternative to Cap-Weighted Indices, *Journal of Investment Management*, 9(4), pp 1-23.

[8] Ang A., Hodrick R.J., Xing Y. and Zhang Y. (2006), The Cross-Section of Volatility and Expected Returns, *Journal of Finance*, 61(1), pp. 259-299.

[9] Angus J.E. (1994), The Probability Integral Transform and Related Results, *SIAM Review*, 36(4), pp. 652-654.

[10] Antolin P. (2008), Pension Fund Performance, OECD Working Papers on Insurance and Private Pensions, 20, *SSRN*, `www.ssrn.com/abstract=1368816`.

[11] Arnott R.D. (2010), Debt Be Not Proud, *Journal of Indexes*, 13(6), pp. 10-17.

[12] Arnott R.D. and Bernstein P.L. (2002), What Risk Premium Is "Normal"?, *Financial Analysts Journal*, 58(2), pp. 64-85.

[13] Arnott R.D. and Hsu J.C. (2008), Noise, CAPM and the Size and Value effects, *Journal of Investment Management*, 6(1), pp. 68-78.

[14] Arnott R.D., Hsu J.C., Li F. and Shepherd S.D. (2010), Valuation-Indifferent Weighting for Bonds, *Journal of Portfolio Management*, 36(3), pp. 117-130.

[15] Arnott R.D., Hsu J.C. and Moore P. (2005), Fundamental Indexation, *Financial Analysts Journal*, 61(2), pp. 83-99.

[16] Arnott R.D., Kalesnik V., Moghtader P. and Scholl C. (2010), Beyond Cap Weight: The Empirical Evidence for a Diversified Beta, *Journal of Indexes*, 13(1), pp. 16-29.

[17] Artzner A., Delbaen F., Eber J.-M. and Heath D. (1999), Coherent Measures of Risk, *Mathematical Finance*, 9(3), pp. 203-228.

[18] Asness C.S., Frazzini A. and Pedersen L.H. (2012), Leverage Aversion and Risk Parity, *Financial Analysts Journal*, 68(1), pp. 47-59.

[19] Azzalini A. (1985), A Class of Distributions Which Includes the Normal Ones, *Scandinavian Journal of Statistics*, 12(2), pp. 171-178.

[20] Azzalini A. (1986), Further Results on A Class of Distributions Which Includes the Normal Ones, *Statistica*, 46, pp. 199-208.

[21] Azzalini A. and Dalla Valle A. (1996), The Multivariate Skew-Normal Distribution, *Biometrika*, 83(4), pp. 715-726.

[22] Baillie R.T. and Myers R.J. (1991), Bivariate Garch Estimation of the Optimal Commodity Futures Hedge, *Journal of Applied Econometrics*, 6(2), pp. 109-124.

[23] Bajeux-Besnainou I., Jordan J.V. and Portait R. (2003), Dynamic Asset Allocation for Stocks, Bonds, and Cash, *Journal of Business*, 76(2), pp. 263-287.

[24] Bansal R., Dittmar R. and Kiku D. (2009), Cointegration and Consumption Risks in Asset Returns, *Review of Financial Studies*, 22(3), pp. 1343-1375.

[25] Bansal R. and Yaron A. (2004), Risks for the Long Run: A Potential Resolution of Asset Pricing Puzzles, *Journal of Finance*, 59(4), pp. 1481-1509.

[26] Barberis N. (2000), Investing for the Long Run When Returns Are Predictable, *Journal of Finance*, 55(1), pp. 225-264.

[27] Barberis N. and Thaler R.H. (2003), A Survey of Behavioral Finance, in G.M. Constantinides, M. Harris and R.M. Stulz (Eds), *Handbook of the Economics of Finance*, 1(B), Elsevier, pp. 1053-1128.

[28] Barras L., Scaillet O. and Wermers R. (2010), False Discoveries in Mutual Fund Performance: Measuring Luck in Estimated Alphas, *Journal of Finance*, 65(1), pp. 179-216.

[29] Basu D. and Miffre J. (2009), Capturing the Risk Premium of Commodity Futures: The Role of Hedging Pressure, *SSRN*, www.ssrn.com/abstract=1340873.

[30] Bauwens L., Laurent S. and Rombouts J.V.K. (2006), Multivariate GARCH Models: A Survey, *Journal of Applied Econometrics*, 21(1), pp. 79-109.

[31] Benartzi S. and Thaler R.H. (1995), Myopic Loss Aversion and the Equity Premium Puzzle, *Quarterly Journal of Economics*, 110(1), pp. 73-92.

[32] Benartzi S. and Thaler R.H. (2001), Naive Diversification Strategies in Defined Contribution Saving Plans, *American Economic Review*, 91(1), pp. 79-98.

[33] Bera A.K. and Park S.Y. (2008), Optimal Portfolio Diversification using the Maximum Entropy Principle, *Econometric Reviews*, 27(4-6), pp. 484-512.

[34] Berkelaar A.B., Kobor A. and Tsumagari M. (2006), The Sense and Nonsense of Risk Budgeting, *Financial Analysts Journal*, 62(5), pp. 63-75.

[35] Bernstein P.L. (1992), *Capital Ideas: The Improbable Origins of Modern Wall Street*, Free Press.

[36] Bernstein P.L. (2007), *Capital Ideas Evolving*, Wiley.

[37] Bessembinder H., Coughenour J.F., Seguin P.J. and Smoller M.M. (1995), Mean Reversion in Equilibrium Asset Prices: Evidence from the Futures Term Structure, *Journal of Finance*, 50(1), pp. 361-375.

[38] Bhansali V. (2012), Active Risk Parity, *Journal of Investing*, 21(3), pp. 88-92.

[39] Bhansali V., Davis J., Rennison G., Hsu J.C. and Li F. (2012), The Risk in Risk Parity: A Factor Based Analysis of Asset Based Risk Parity, *Journal of Investing*, 21(3), pp. 102-110.

[40] Black F. (1972), Capital Market Equilibrium with Restricted Borrowing, *Journal of Business*, 45(3), pp. 444-455.

[41] BLACK F. (1993), Beta and Return, *Journal of Portfolio Management*, 20(1), pp. 8-18.

[42] BLACK F., JENSEN M.C. and SCHOLES M.S. (1972), The Capital Asset Pricing Model: Some Empirical Tests, in M.C. Jensen (Ed.), *Studies in the Theory of Capital Markets*, Praeger Publishers Inc., pp. 79-121.

[43] BLACK F. and LITTERMAN R.B. (1991), Asset Allocation: Combining Investor Views with Market Equilibrium, *Journal of Fixed Income*, 1(2), pp. 7-18.

[44] BLACK F. and LITTERMAN R.B. (1992), Global Portfolio Optimization, *Financial Analysts Journal*, 48(5), pp. 28-43.

[45] BLACK F. and PEROLD A.F. (1992), Theory of Constant Proportion Portfolio Insurance, *Journal of Economic Dynamic and Control*, 16(3-4), pp. 402-426.

[46] BLAKE C.R., ELTON E.J. and GRUBER M.J. (1993), The Performance of Bond Mutual Funds, *Journal of Business*, 66(3), pp. 371-403.

[47] BLAKE D., LEHMANN B.N. and TIMMERMANN A. (1999), Asset Allocation Dynamics and Pension Fund Performance, *Journal of Business*, 72(4), pp. 429-461.

[48] BLITZ D.C. and SWINKELS L. (2008), Fundamental Indexation: An Active Value Strategy in Disguise, *Journal of Asset Management*, 9(5), pp. 264-269.

[49] BLITZ D.C. and VAN VLIET P. (2007), The Volatility Effect, *Journal of Portfolio Management*, 34(1), pp. 102-113.

[50] BODIE Z. (1983), Commodity Futures as a Hedge Against Inflation, *Journal of Portfolio Management*, 9(3), pp. 12-17.

[51] BODIE Z., MERTON R.C. and SAMUELSON W.F. (1992), Labor Supply Flexibility and Portfolio Choice in a Life Cycle Model, *Journal of Economic Dynamics and Control*, 16(3-4), pp. 427-449.

[52] BOGGS P.T. and TOLLE J.W. (1995), Sequential Quadratic Programming, *Acta Numerica*, 4, pp. 1-51.

[53] BOLLERSLEV T. (1986), Generalized Autoregressive Conditional Heteroscedasticity, *Journal of Econometrics*, 31(3), pp. 307-327.

[54] BOOTH D.G. and FAMA E.F. (1992), Diversification Returns and Asset Contributions, *Financial Analyst Journal*, 48(3), pp. 26-32.

[55] BOUDT K., PETERSON B.G. and CROUX C. (2008), Estimation and Decomposition of Downside Risk for Portfolios with Non-Normal Returns, *Journal of Risk*, 11(2), pp. 79-103.

[56] Breeden D.T. (1979), An Intertemporal Asset Pricing Model with Stochastic Consumption and Investment Opportunities, *Journal of Financial Economics*, 7(3), pp. 265-296.

[57] Brennan M.J. and Schwartz E.S. (1982), An Equilibrium Model of Bond Pricing and a Test of Market Efficiency, *Journal of Financial and Quantitative Analysis*, 17(3), pp. 301-329.

[58] Brennan M.J., Schwartz E.S. and Lagnado R. (1997), Strategic Asset Allocation, *Journal of Economic Dynamics and Control*, 21(8-9), pp. 1377-1403.

[59] Brennan M.J. and Solanki R. (1981), Optimal Portfolio Insurance, *Journal of Financial and Quantitative Analysis*, 16(3), pp. 279-300.

[60] Brinson G.P., Hood L.R. and Beebower G.L. (1986), Determinants of Portfolio Performance, *Financial Analysts Journal*, 42(4), pp. 39-44.

[61] Brinson G.P., Singer B.D. and Beebower G.L. (1991), Determinants of Portfolio Performance II: An Update, *Financial Analysts Journal*, 47(3), pp. 40-48.

[62] Britten-Jones M. (1999), The Sampling Error in Estimates of Mean-Variance Efficient Portfolio Weights, *Journal of Finance*, 54(2), pp. 655-671.

[63] Broadie M. (1993), Computing Efficient Frontiers using Estimated Parameters, *Annals of Operations Research*, 45(1), pp. 21-58.

[64] Brodsky B., Flannery G. and Mesrour S. (2011), Introducing the BlackRock Sovereign Risk Index, *BlackRock Investment Institute, White Paper*.

[65] Bruder B., Culerier L. and Roncalli T. (2012), How to Design Target-Date Funds?, *Lyxor White Paper Series*, 9, www.lyxor.com.

[66] Bruder B., Dao T-L., Richard J-C. and Roncalli T. (2011), Trend Filtering Methods for Momentum Strategies, *Lyxor White Paper Series*, 8, www.lyxor.com.

[67] Bruder B., Darolles S., Koudiraty A. and Roncalli T. (2011), Portfolio Allocation of Hedge Funds, *Lyxor White Paper Series*, 5, www.lyxor.com.

[68] Bruder B., Hereil P. and Roncalli T. (2011), Managing Sovereign Credit Risk, *Journal of Indexes Europe*, 1(4), pp. 20-27.

[69] Bruder B., Hereil P. and Roncalli T. (2011), Managing Sovereign Credit Risk in Bond Portfolios, *SSRN*, www.ssrn.com/abstract=1957050.

[70] Bruder B., Jamet G. and Lasserre G. (2010), Beyond Liability-Driven Investment: New Perspectives on Defined Benefit Pension Fund Management, *Lyxor White Paper Series*, 2, www.lyxor.com.

[71] Bruder B., Richard J-C. and Roncalli T. (2013), Regularization of Portfolio Allocation, *Lyxor Research Paper*, www.lyxor.com.

[72] Bruder B. and Roncalli T. (2012), Managing Risk Exposures using the Risk Budgeting Approach, *SSRN*, www.ssrn.com/abstract=2009778.

[73] Campbell J.Y. (2000), Asset Pricing at the Millennium, *Journal of Finance*, 55(4), pp. 1515-1567.

[74] Campbell J.Y., Chan Y.L. and Viceira L.M. (2003), A Multivariate Model of Strategic Asset Allocation, *Journal of Financial Economics*, 67(1), pp. 41-80.

[75] Campbell J.Y. and Shiller R.J. (1988), The Dividend-Price Ratio and Expectations of Future Dividends and Discount Factors, *Review of Financial Studies*, 1(3), pp. 195-228.

[76] Campbell J.Y. and Shiller R.J. (1988), Stock Prices, Earnings, and Expected Dividends, *Journal of Finance*, 43(3), pp. 661-676.

[77] Campbell J.Y. and Viceira L.M. (2002), *Strategic Asset Allocation*, Oxford University Press.

[78] Candelon B., Hurlin C. and Tokpavi S. (2012), Sampling Error and Double Shrinkage Estimation of Minimum Variance Portfolios, *Journal of Empirical Finance*, 19(4), pp. 511-527.

[79] Canner N., Mankiw N.G. and Weil D.N. (1997), An Asset Allocation Puzzle, *American Economic Review*, 87(1), pp. 181-191.

[80] Carhart M.M. (1997), On Persistence in Mutual Fund Performance, *Journal of Finance*, 52(1), pp. 57-82.

[81] Carroll R.B., Perry T., Yang H. and Ho A. (2001), A New Approach to Component VaR, *Journal of Risk*, 3(3), pp. 57-67.

[82] Carvalho R. Leote de, Lu X. and Moulin P. (2012), Demystifying Equity Risk-Based Strategies: A Simple Alpha plus Beta Description, *Journal of Portfolio Management*, 38(3), pp. 56-70.

[83] Chan K.C., Karolyi G.A., Longstaff F.A. and Sanders A.B. (1992), An Empirical Comparison of Alternative Models of the Short-Term Interest Rate, *Journal of Finance*, 47(3), pp. 1209-1227.

[84] Chaves D.B., Hsu J.C., Li F. and Shakernia O. (2011), Risk Parity Portfolio vs. Other Asset Allocation Heuristic Portfolios, *Journal of Investing*, 20(1), pp. 108-118.

[85] Chaves D.B., Hsu J.C., Li F. and Shakernia O. (2012), Efficient Algorithms for Computing Risk Parity Portfolio Weights, *Journal of Investing*, 21(3), pp. 150-163.

[86] Chen N-F. (1991), Financial Investment Opportunities and the Macroeconomy, *Journal of Finance*, 46(2), pp. 529-554.

[87] Cheung W. (2010), The Black-Litterman Model Explained, *Journal of Asset Management*, 11(4), pp. 229-243.

[88] Chou R.Y. (1988), Volatility Persistence and Stock Valuations: Some Empirical Evidence using GARCH, *Journal of Applied Econometrics*, 3(4), pp. 279-294.

[89] Choueifaty Y. and Coignard Y. (2008), Toward Maximum Diversification, *Journal of Portfolio Management*, 35(1), pp. 40-51.

[90] Choueifaty Y., Froidure T. and Reynier J. (2011), Properties of the Most Diversified Portflio, *SSRN*, www.ssrn.com/abstract=1895459.

[91] Chow T., Hsu J.C., Kalesnik V. and Little B. (2011), A Survey of Alternative Equity Index Strategies, *Financial Analysts Journal*, 67(5), pp. 37-57.

[92] Clarke R.G., de Silva H. and Thorley S. (2006), Minimum-Variance Portfolios in the U.S. Equity Market, *Journal of Portfolio Management*, 33(1), pp. 10-24.

[93] Clarke R.G., de Silva H. and Thorley S. (2010), Minimum Variance Portfolio Composition, *SSRN*, www.ssrn.com/abstract=1549949.

[94] Clarke R.G., de Silva H. and Thorley S. (2012), Risk Parity, Maximum Diversification, and Minimum Variance: An Analytic Perspective, *SSRN*, www.ssrn.com/abstract=1977577.

[95] Cocco J.F., Gomes F.J. and Maenhout P.J. (2005), Consumption and Portfolio Choice over the Life Cycle, *Review of Financial Studies*, 18(2), pp. 491-533.

[96] Cochrane J.H. (2001), *Asset Pricing*, Princeton University Press.

[97] Coles S., Heffernan J. and Tawn J. (1999), Dependence Measures for Extreme Value Analyses, *Extremes*, 2(4), pp. 339-365.

[98] Cont R. (2001), Empirical Properties of Asset Returns: Stylized Facts and Statistical Issues, *Quantitative Finance*, 1(2), pp. 223-236.

[99] Cox J.C., Ingersoll J.E. and Ross S.A. (1985), A Theory of The Term Structure of Interest Rates, *Econometrica*, 53(2), pp. 385-407.

[100] CVITANIĆ J., LAZRAK A., MARTELLINI L. and ZAPATERO F. (2003), Optimal Allocation to Hedge Funds: An Empirical Analysis, *Quantitative Finance*, 3(1), pp. 28-39.

[101] DALIO R. (2004), Engineering Targeted Returns and Risks, *Bridgewater Associates Working Paper*.

[102] DAROLLES S., EYCHENNE K. and MARTINETTI S. (2010), Time-varying Risk Premiums and Business Cycle: A Survey, *Lyxor White Paper Series*, 4, www.lyxor.com.

[103] DAROLLES S., GOURIEROUX C. and JAY E. (2012), Robust Portfolio Allocation with Systematic Risk Contribution Restrictions, *SSRN*, www.ssrn.com/abstract=2192399.

[104] DEATON A. and LAROQUE G. (1992), On the Behaviour of Commodity Prices, *Review of Economic Studies*, 59(1), pp. 1-23.

[105] DEGUEST R., MARTELLINI L. and MEUCCI A. (2013), Risk Parity and Beyond: From Asset Allocation to Risk Allocation Decisions, *EDHEC Working Paper*.

[106] DEMANGE G. and ROCHET J-C. (2005), *Méthodes Mathématiques de la Finance*, Third edition, Economica.

[107] DEMARTA S. and MCNEIL A.J. (2005), The t Copula and Related Copulas, *International Statistical Review*, 73(1), pp. 111-129.

[108] DEMEY P., MAILLARD S. and RONCALLI T. (2010), Risk-Based Indexation, *Lyxor White Paper Series*, 1, www.lyxor.com.

[109] DEMIGUEL V., GARLAPPI L., NOGALES F.J. and UPPAL R. (2009), A Generalized Approach to Portfolio Optimization: Improving Performance by Constraining Portfolio Norms, *Management Science*, 55(5), pp. 798-812.

[110] DEMIGUEL V., GARLAPPI L. and UPPAL R. (2009), Optimal Versus Naive Diversification: How Inefficient is the 1/N Portfolio Strategy?, *Review of Financial Studies*, 22(5), pp. 1915-1953.

[111] DENAULT M. (2001), Coherent Allocation of Risk Capital, *Journal of Risk*, 4(1), pp. 1-34.

[112] DETZLER M.L. (1999), The Performance of Global Bond Mutual Funds, *Journal of Banking & Finance*, 23(8), pp. 1195-1217.

[113] DI CESARE A. (2006), Do Market-based Indicators Anticipate Rating Agencies? Evidence for International Banks, *Economic Notes*, 35(1), pp. 121-150.

[114] DICKEY D.A. and FULLER W.A. (1981), Likelihood Ratio Statistics for Autoregressive Time Series with a Unit Root, *Econometrica*, 49(4), pp. 1057-1072.

[115] DIMSON E. and MUSSAVIAN M. (1999), Three Centuries of Asset Pricing, *Journal of Banking & Finance*, 23(12), pp. 1745-1769.

[116] DUFFIE D., GRAY S. and HOANG P. (2004), Volatility in Energy Prices, in V. Kaminski (Ed.), *Managing Energy Risk Price: The New Challenges and Solutions*, Third edition, Risk Books.

[117] DUFFIE D. and KAN R. (1996), A Yield-Factor Model of Interest Rates, *Mathematical Finance*, 6(4), pp. 379-406.

[118] ELTON E.J., GRUBER M.J., DAS S. and HLAVKA M. (1993), Efficiency with Costly Information: A Reinterpretation of Evidence from Managed Portfolios, *Review of Financial Studies*, 6(1), pp. 1-22.

[119] EMBRECHTS P., MCNEIL A.J. and STRAUMANN D. (2002), Correlation and Dependency in Risk Management: Properties and Pitfalls, in M.A.H. Dempster (Ed.), *Risk Management: Value at Risk and Beyond*, Cambridge University Press.

[120] EMMER S. and TASCHE D. (2005), Calculating Credit Risk Capital Charges with the One-Factor Model, *Journal of Risk*, 7(2), pp. 85-101.

[121] ENGLE R.F. (1982), Autoregressive Conditional Heteroscedasticity with Estimates of the Variance of United Kingdom Inflation, *Econometrica*, 50(4), pp. 987-1007.

[122] ENGLE R.F. and GRANGER C.W.J. (1987), Co-Integration and Error Correction: Representation, Estimation, and Testing, *Econometrica*, 55(2), pp. 251-276.

[123] EPPERLEIN E. and SMILLIE A. (2006), Cracking VaR with Kernels, *Risk*, 19(8), pp. 70-74.

[124] ERB C.B. and HARVEY C.R. (2006), The Strategic and Tactical Value of Commodity Futures, *Financial Analysts Journal*, 62(2), pp. 69-97.

[125] ESTRADA J. (2008), Fundamental Indexation and International Diversification, *Journal of Portfolio Management*, 34(3), pp. 93-109.

[126] EYCHENNE K., MARTINETTI S. and RONCALLI T. (2011), Strategic Asset Allocation, *Lyxor White Paper Series*, 6, www.lyxor.com.

[127] FAMA E.F. (1965), The Behavior of Stock-Market Prices, *Journal of Business*, 38(1), pp. 34-105.

[128] FAMA E.F. (1970), Efficient Capital Markets: A Review of Theory and Empirical Work, *Journal of Finance*, 25(2), pp. 383-417.

[129] FAMA E.F. and FRENCH K.R. (1992), The Cross-Section of Expected Stock Returns, *Journal of Finance*, 47(2), pp. 427-465.

[130] FAMA E.F. and FRENCH K.R. (1993), Common Risk Factors in the Returns on Stocks and Bonds, *Journal of Financial Economics*, 33(1), pp. 3-56.

[131] FAMA E.F. and FRENCH K.R. (1996), The CAPM is Wanted, Dead or Alive, *Journal of Finance*, 51(5), pp. 1947-1958.

[132] FAMA E.F. and FRENCH K.R. (2004), The Capital Asset Pricing Model: Theory and Evidence, *Journal of Economic Perspectives*, 18(3), pp. 25-46.

[133] FÖLLMER H. and SCHIED A. (2002), Convex Measures of Risk and Trading Constraints, *Finance and Stochastics*, 6(4), pp. 429-447.

[134] FRAZZINI A. and PEDERSEN L.H. (2010), Betting Against Beta, *NBER Working Paper*, 16601.

[135] FUNG W. and HSIEH D.A. (1997), Empirical Characteristics of Dynamic Trading Strategies: The Case of Hedge Funds, *Review of Financial Studies*, 10(2), pp. 275-302.

[136] FUNG W. and HSIEH D.A. (2001), The Risk in Hedge Fund Strategies: Theory and Evidence from Trend Followers, *Review of Financial Studies*, 14(2), pp. 313-341.

[137] GARMAN M. (1996), Improving on VaR, *Risk*, 9(5), pp. 61-63.

[138] GARMAN M. (1997), Taking VaR to Pieces, *Risk*, 10(10), pp. 70-71.

[139] GENEST C. (1987), Frank's Family of Bivariate Distributions, *Biometrika*, 74(3), pp. 549-555.

[140] GENEST C. and MACKAY J. (1986), The Joy of Copulas: Bivariate Distributions with Uniform Marginals, *American Statistician*, 40(4), pp. 280-283.

[141] GENEST C. and MACKAY R.J. (1986), Copules archimédiennes et familles de lois bidimensionnelles dont les marges sont données, *Canadian Journal of Statistics*, 14(2), pp. 145-159.

[142] GENTON M.G., HE L. and LIU X. (2001), Moments of Skew-Normal Random Vectors and Their Quadratic Forms, *Statistics & Probability Letters*, 51(4), pp. 319-325.

[143] GEORGES P., LAMY A-G., NICOLAS E., QUIBEL G. and RONCALLI T. (2001), Multivariate Survival Modelling: A Unified Approach with Copulas, *SSRN*, `www.ssrn.com/abstract=1032559`.

[144] GEMAN H., EL KAROUI N. and ROCHET J-C. (1995), Changes of Numéraire, Changes of Probability Measure and Option Pricing, *Journal of Applied Probability*, 32(2), pp. 443-458.

[145] GIAMOURIDIS D. and VRONTOS I.D. (2007), Hedge Fund Portfolio Construction: A Comparison of Static and Dynamic Approaches, *Journal of Banking & Finance*, 31(1), pp. 199-217.

[146] GIBSON R. and SCHWARTZ E.S. (1990), Stochastic Convenience Yield and the Pricing of Oil Contingent Claims, *Journal of Finance*, 45(3), pp. 959-976.

[147] GLASSERMAN P. (2005), Measuring Marginal Risk Contributions in Credit Portfolios, *Journal of Computational Finance*, 9(2), pp. 1-41.

[148] GOETZMANN W.N. and IBBOTSON R.G. (1994), Do Winners Repeat?, *Journal of Portfolio Management*, 20(2), pp. 9-18.

[149] GOLTZ F. and CAMPANI C.H. (2011), A Review of Corporate Bond Indices: Construction Principles, Return Heterogeneity, and Fluctuations in Risk Exposures, *EDHEC Risk Working Paper*.

[150] GOLUB G.H. and VAN LOAN C.F. (1996), *Matrix Computations*, Third edition, John Hopkins University Press.

[151] GORDY M.B. (2003), A Risk-Factor Model Foundation for Ratings-Based Bank Capital Rules, *Journal of Financial Intermediation*, 12(3), pp. 199-232.

[152] GORTON G. and ROUWENHORST K.G. (2006), Facts and Fantasies about Commodity Futures, *Financial Analysts Journal*, 62(2), pp. 47-68.

[153] GOURIEROUX C., LAURENT J-P. and SCAILLET O. (2000), Sensitivity analysis of Values at Risk, *Journal of Empirical Finance*, 7(3-4), pp. 225-245.

[154] GREEN R.C. (1986), Positively Weighted Portfolios on the Minimum-Variance Frontier, *Journal of Finance*, 41(5), pp. 1051-1068.

[155] GREEN R.C. and HOLLIFIELD B. (1992), When Will Mean-Variance Efficient Portfolios Be Well Diversified?, *Journal of Finance*, 47(5), pp. 1785-1809.

[156] GRINBLATT M. and TITMAN S. (1989), Portfolio Performance Evaluation: Old Issues and New Insights, *Review of Financial Studies*, 2(3), pp. 393-421.

[157] GRINBLATT M. and TITMAN S. (1992), The Persistence of Mutual Fund Performance, *Journal of Finance*, 47(5), pp. 1977-1984.

[158] GRINOLD R.C. and KAHN R.N. (2000), *Active Portfolio Management: A Quantitative Approach for Providing Superior Returns and Controlling Risk*, Second edition, McGraw-Hill.

[159] HAGAN P.S., KUMAR D., LESNIEWSKI A.S. and WOODWARD D.E. (2002), Managing Smile Risk, *Wilmott Magazine*, September, 1, pp. 84-108.

[160] HALLERBACH W.G. (2003), Decomposing Portfolio Value-at-Risk: A General Analysis, *Journal of Risk*, 5(2), pp. 1-18.

[161] HASTIE T., TIBSHIRANI R. and FRIEDMAN J. (2009), *The Elements of Statistical Learning*, Second edition, Springer.

[162] HAUGEN R.A. and BAKER N.L. (1991), The Efficient Market Inefficiency of Capitalization-weighted Stock Portfolios, *Journal of Portfolio Management*, 17(3), pp. 35-40.

[163] HAYASHI T. and YOSHIDA N. (2005), On Covariance Estimation of Non-synchronously Observed Diffusion Processes, *Bernoulli*, 11(2), pp. 359-379.

[164] HE G. and LITTERMAN R. (1999), The Intuition Behind Black-Litterman Model Portfolios, Goldman Sachs Asset Management, *SSRN*, `www.ssrn.com/abstract=334304`.

[165] HEATH D., JARROW R. and MORTON A. (1992), Bond Pricing and the Term Structure of Interest Rates: A New Methodology for Contingent Claims Valuation, *Econometrica*, 60(1), pp. 77-105.

[166] HEMMINKI J. and PUTTONEN V. (2008), Fundamental Indexation in Europe, *Journal of Asset Management*, 8(6), pp. 401-405.

[167] HENDRICKS D., PATEL J. and ZECKHAUSER R. (1993), Hot Hands in Mutual Funds: Short-Run Persistence of Relative Performance, 1974-1988, *Journal of Finance*, 48(1), pp. 93-130.

[168] HEREIL P. and RONCALLI T. (2011), Measuring the Risk Concentration of Investment Portfolios, *Bloomberg Brief – Risk*, July, pp. 8-9.

[169] HEREIL P., LAPLANTE J. and RONCALLI T. (2013), Multi-Asset Indices, *Journal of Indexes Europe*, 3(1), pp. 8-16.

[170] HIRSHLEIFER D. (2001), Investor Psychology and Asset Pricing, *Journal of Finance*, 56(4), pp. 1533-1597.

[171] HOERL A.E. and KENNARD R.W. (1970), Ridge Regression: Biased Estimation for Nonorthogonal Problems, *Technometrics*, 12(1), pp. 55-67.

[172] HOEVENAARS R.P.M.M., MOLENAAR R.D.J., SCHOTMAN P.C. and STEENKAMP T.B.M. (2008), Strategic Asset Allocation with Liabilities: Beyond Stocks and Bonds, *Journal of Economic Dynamics and Control*, 32(9), pp. 2939-2970.

[173] HOTCHKISS E.S. and RONEN T. (2002), The Informational Efficiency of the Corporate Bond Market: An Intraday Analysis, *Review of Financial Studies*, 15(5), pp. 1325-1354.

[174] HSIEH H.H., HODNETT K. and van RENSBURG P. (2012), Fundamental Indexation For Global Equities: Does Firm Size Matter?, *Journal of Applied Business Research*, 28(1), pp. 105-114.

[175] HSU J.C. (2006), Cap-Weighted Portfolios are Sub-Optimal Portfolios, *Journal of Investment Management*, 4(3), pp. 44-53.

[176] HUIJ J. and DERWALL J. (2008), Hot Hands in Bond Funds, *Journal of Banking & Finance*, 32(4), pp. 559-572.

[177] HULL J. and WHITE A. (1993), One-Factor Interest-Rate Models and the Valuation of Interest-Rate, *Journal of Financial and Quantitative Analysis*, 28(2), pp. 235-254.

[178] IBBOTSON R.G. and CHEN P. (2003), Long-Run Stock Returns: Participating in the Real Economy, *Financial Analysts Journal*, 59(1), pp. 88-98.

[179] IBBOTSON R.G. and KAPLAN P.D. (2000), Does Asset Allocation Policy Explain 40, 90, or 100 Percent of Performance?, *Financial Analysts Journal*, 56(1), pp. 26-33.

[180] IDZOREK T. (2007), A Step-by-step Guide to the Black-Litterman Model, in S. Satchell (Ed.), *Forecasting Expected Returns in the Financial Markets*, Academic Press.

[181] ILMANEN A. (2003), Stock-Bond Correlations, *Journal of Fixed Income*, 13(2), pp. 55-66.

[182] ILMANEN A. (2011), *Expected Returns: An Investor's Guide to Harvesting Market Rewards*, Wiley.

[183] INKER B. (2011), The Dangers of Risk Parity, *Journal of Investing*, 20(1), pp. 90-98.

[184] JAGANNATHAN R. and MA T. (2003), Risk Reduction in Large Portfolios: Why Imposing the Wrong Constraints Helps, *Journal of Finance*, 58(4), pp. 1651-1684.

[185] JAGANNATHAN R. and KOCHERLAKOTA N.R. (1996), Why Should Older People Invest Less in Stocks Than Younger People?, *Quarterly Review, Federal Reserve Bank of Minneapolis*, 20(3), pp. 11-23.

[186] JEGADEESH N. and TITMAN S. (1993), Returns to Buying Winners and Selling Losers: Implications for Stock Market Efficiency, *Journal of finance*, 48(1), pp. 65-91.

[187] JENSEN M.C. (1968), The Performance of Mutual Funds in the Period 1945-1964, *Journal of Finance*, 23(2), pp. 389-416.

[188] JOE H. (1997), *Multivariate Models and Dependence Concepts*, Monographs on Statistics and Applied Probability, 73, Chapmann & Hall.

[189] JONDEAU E. and ROCKINGER M. (2006), The Copula-GARCH Model of Conditional Dependencies: An International Stock Market Application, *Journal of International Money and Finance*, 25(5), pp. 827-853.

[190] JONDEAU E. and ROCKINGER M. (2006), Optimal Portfolio Allocation Under Higher Moments, *European Financial Management*, 12(1), pp. 29-55.

[191] JORION P. (1988), Bayes-Stein Estimation for Portfolio Analysis, *Journal of Financial and Quantitative Analysis*, 21(3), pp. 279-292.

[192] JORION P. (1992), Portfolio Optimization in Practice, *Financial Analysts Journal*, 48(1), pp. 68-74.

[193] JORION P. (2003), Portfolio Optimization with Tracking-Error Constraints, *Financial Analysts Journal*, 59(5), pp. 70-82.

[194] JOUANIN J-F., RIBOULET G. and RONCALLI T. (2004), Financial Applications of Copula Functions, in G. Szegö (Ed.), *Risk Measures for the 21st Century*, Wiley.

[195] JURCZENKO E., MICHEL T. and TEÏLETCHE J. (2013), Generalized Risk-Based Investing, *SSRN*, www.ssrn.com/abstract=2205979.

[196] KALDOR N. (1939), Speculation and Economic Stability, *Review of Economic Studies*, 7(1), pp. 1-27.

[197] KALKBRENER M. (2005), An Axiomatic Approach to Capital Allocation, *Mathematical Finance*, 15(3), pp. 425-437.

[198] KAPLAN P.D. (2008), Why Fundamental Indexation Might—or Might Not—Work, *Financial Analysts Journal*, 64(1), pp. 32-39.

[199] KARATZAS I., LEHOCZKY J.P and SHREVE S.E. (1987), Optimal Portfolio and Consumption Decisions for a Small Investor on a Finite Horizon, *SIAM Journal of Control and Optimization*, 25(6), pp. 1557-1586.

[200] KAT H.M. (2004), In Search of the Optimal Fund of Hedge Funds, *Journal of Wealth Management*, 6(4), pp. 43-51.

[201] KATZ S. (1974), The Price and Adjustment Process of Bonds to Rating Reclassifications: A Test of Bond Market Efficiency, *Journal of Finance*, 29(2), pp. 551-559.

[202] KAYA H., LEE W. and WAN Y. (2011), Risk Budgeting With Asset Class and Risk Class, *Neuberger Berman Working Paper.*

[203] KEYNES J.M. (1923), Some Aspects of Commodity Markets, *Manchester Guardian Commercial, European Reconstruction Series*, 13, pp. 784-786.

[204] KOTHARI S.P., SHANKEN J. and SLOAN R.G. (1995), Another Look at the Cross-Section of Expected Stock Returns, *Journal of Finance*, 50(1), pp. 185-224.

[205] LALOUX L., CIZEAU P., BOUCHAUD J-P. and POTTERS M. (1999), Noise Dressing of Financial Correlation Matrices, *Physical Review Letters*, 83(7), pp. 1467-1470.

[206] LEDOIT O. and WOLF M. (2003), Improved Estimation of the Covariance Matrix of Stock Returns With an Application to Portfolio Selection, *Journal of Empirical Finance*, 10(5), pp. 603-621.

[207] LEDOIT O. and WOLF M. (2004), Honey, I Shrunk the Sample Covariance Matrix, *Journal of Portfolio Management*, 30(4), pp. 110-119.

[208] LEE W. (2011), Risk Based Asset Allocation: A New Answer To An Old Question?, *Journal of Portfolio Management*, 37(4), pp. 11-28.

[209] LEHMANN B.N. (1990), Fads, Martingales and Market Efficiency, *Quarterly Journal of Economics*, 105(1), pp. 1-28.

[210] LEHMANN B.N. and MODEST D.M. (1987), Mutual Fund Performance Evaluation: A Comparison of Benchmarks and Benchmark Comparisons, *Journal of Finance*, 42(2), pp. 233-265.

[211] LETTAU M. and LUDVIGSON S. (2001), Consumption, Aggregate Wealth, and Expected Stock Returns, *Journal of Finance*, 56(3), pp. 815-849.

[212] LI D. (2000), On Default Correlation: A Copula Function Approach, *Journal of Fixed Income*, 9(4), pp. 43-54.

[213] LITTERMAN, R.B. (1996), Hot Spots and Hedges, *Goldman Sachs Risk Management Series.*

[214] LITTERMAN R.B. (2003), *Modern Investment Management: An Equilibrium Approach*, Wiley.

[215] LITTERMAN R.B. and SCHEINKMAN J.A. (1991), Common Factors Affecting Bond Returns, *Journal of Fixed Income*, 1(1), pp. 54-61.

[216] Lindberg C. (2009), Portfolio Optimization When Expected Stock Returns are Determined by Exposure to Risk, *Bernoulli*, 15(2), pp. 464-474.

[217] Lintner J. (1965), The Valuation of Risk Assets and the Selection of Risky Investments in Stock Portfolios and Capital Budgets, *Review of Economics and Statistics*, 47(1), pp. 13-37.

[218] Lo A.W. and MacKinlay A.C. (1990), When are Contrarian Profits due to Stock Market Overreaction?, *Review of Financial Studies*, 3(2), pp. 175-205.

[219] Lo A.W. and Patel P.N. (2008), 130/30: The New Long-Only, *Journal of Portfolio Management*, 34(2), pp. 12-38.

[220] Lohre H., Neugebauer U. and Zimmer C. (2012), Diversified Risk Parity Strategies for Equity Portfolio Selection, *Journal of Investing*, 21(3), pp. 111-128.

[221] Longstaff F.A., Pan J., Pedersen L.H. and Singleton K.J. (2011), How Sovereign is Sovereign Credit Risk?, *American Economic Journal: Macroeconomics*, 3(2), pp. 75-103.

[222] Lucas R.E. (1978), Asset Prices in an Exchange Economy, *Econometrica*, 46(6), pp. 1429-1445.

[223] Maillard S., Roncalli T. and Teïletche J. (2010), The Properties of Equally Weighted Risk Contribution Portfolios, *Journal of Portfolio Management*, 36(4), pp. 60-70.

[224] Malkiel B.G. (2003), Passive Investment Strategies and Efficient Markets, *European Financial Management*, 9(1), pp. 1-10.

[225] Marshall A.W. and Olkin I. (1988), Families of Multivariate Distributions, *Journal of the American Statistical Association*, 83(403), pp. 834-841.

[226] Markowitz H. (1952), Portfolio Selection, *Journal of Finance*, 7(1), pp. 77-91.

[227] Markowitz H. (1956), The Optimization of a Quadratic Function Subject to Linear Constraints, *Naval Research Logistics Quarterly*, 3(1-2), pp. 111-133.

[228] Martellini L. (2008), Toward the Design of Better Equity Benchmarks, *Journal of Portfolio Management*, 34(4), pp. 34-41.

[229] Martellini L. and Milhau V. (2012), Dynamic Allocation Decisions in the Presence of Funding Ratio Constraints, *Journal of Pension Economics and Finance*, 11(4), pp. 549-580.

[230] MARTELLINI L., PRIAULET P. and PRIAULET S. (2003), *Fixed-Income Securities: Valuation, Risk Management and Portfolio Strategies*, Wiley.

[231] MARTELLINI L. and ZIEMANN V. (2010), Improved Estimates of Higher-Order Comoments and Implications for Portfolio Selection, *Review of Financial Studies*, 23(4), pp. 1467-1502.

[232] MEHRA R. and PRESCOTT E.C. (1985), The Equity Premium: A Puzzle, *Journal of Monetary Economics*, 15(2), pp. 145-161.

[233] MERTON R.C. (1969), Lifetime Portfolio Selection under Uncertainty: The Continuous-Time Case, *Review of Economics and Statistics*, 51(3), pp. 247-257.

[234] MERTON R.C. (1971), Optimum Consumption and Portfolio Rules in a Continuous-Time Model, *Journal of Economic Theory*, 3(4), pp. 373-413.

[235] MERTON R.C. (1972), An Analytic Derivation of the Efficient Portfolio Frontier, *Journal of Financial and Quantitative Analysis*, 7(4), pp. 1851-1872.

[236] MERTON R.C. (1973), An Intertemporal Capital Asset Pricing Model, *Econometrica*, 41(5), pp. 867-887.

[237] MERTON R.C. (1974), On the Pricing of Corporate Debt: The Risk Structure of Interest Rates, *Journal of Finance*, 29(2), pp. 449-470.

[238] MERTON R.C. (1980), On Estimating the Expected Return on the Market: An Exploratory Investigation, *Journal of Financial Economics*, 8(4), pp. 323-361.

[239] MEUCCI A. (2005), *Risk and Asset Allocation*, Springer.

[240] MEUCCI A. (2006), Beyond Black-Litterman: Views on Non-Normal Markets, *Risk*, 19(2), pp. 87-92.

[241] MEUCCI A. (2007), Risk Contributions from Generic User-defined Factors, *Risk*, 20(6), pp. 84-88.

[242] MEUCCI A. (2009), Managing Diversification, *Risk*, 22(5), pp. 74-79.

[243] MICHAUD R.O. (1989), The Markowitz Optimization Enigma: Is 'Optimized' Optimal?, *Financial Analysts Journal*, 45(1), pp. 31-42.

[244] MICHAUD R.O. (1998), *Efficient Asset Management: A Practical Guide to Stock Portfolio Management and Asset Allocation*, Financial Management Association, Survey and Synthesis Series, HBS Press.

[245] MIFFRE J. and RALLIS G. (2007), Momentum Strategies in Commodity Futures Markets, *Journal of Banking & Finance*, 31(6), pp. 1863-1886.

[246] MUNK C. and SØRENSEN C. (2010), Dynamic Asset Allocation with Stochastic Income and Interest Rates, *Journal of Financial Economics*, 96(3), pp. 433-462.

[247] NBIM (2012), Alternatives to a Market-value-weighted Index, *NBIM Discussion Note*, 7-2012, www.nbim.no.

[248] NELSEN R.B. (2006), *An Introduction to Copulas*, Second edition, Springer.

[249] NELSON C.R. and SIEGEL A.F. (1987), Parsimonious Modeling of Yield Curves, *Journal of Business*, 60(4), pp. 473-489.

[250] NOCEDAL J. and WRIGHT S.J. (2006), *Numerical Optimization*, Second edition, Springer.

[251] OAKES D. (1989), Bivariate Survival Models Induced by Frailties, *Journal of the American Statistical Association*, 84(406), pp. 487-493.

[252] ODEAN T. (1998), Are Investors Reluctant to Realize Their Losses?, *Journal of Finance*, 53(5), pp. 1775-1798.

[253] PEROLD A.F. (1986), Constant Proportion Portfolio Insurance, *Harvard Business School*, Manuscript.

[254] PEROLD A.F. (2007), Fundamentally Flawed Indexing, *Financial Analysts Journal*, 63(6), pp. 31-37.

[255] PEROLD A.F. and SHARPE W.F. (1988), Dynamic Strategies for Asset Allocation, *Financial Analysts Journal*, 44(1), pp. 16-27.

[256] PETERS E.E. (2011), Balancing Asset Growth and Liability Hedging through Risk Parity, *Journal of Investing*, 20(1), pp. 128-136.

[257] PHAM H. (2009), *Continuous-time Stochastic Control and Optimization with Financial Applications*, Stochastic Modelling and Applied Probability, 61, Springer.

[258] PINDYCK R.S. (2001), The Dynamics of Commodity Spot and Futures Markets: A Primer, *The Energy Journal*, 22(3), pp. 1-29.

[259] PINDYCK R.S. (2004), Volatility and Commodity Price Dynamics, *Journal of Futures Markets*, 24(11), pp. 1029-1047.

[260] PLISKA S.R. (1986), A Stochastic Calculus Model of Continuous Trading: Optimal Portfolios, *Mathematics of Operations Research*, 11(2), pp. 371-382.

[261] POTERBA J.M. and SUMMERS L.H. (1988), Mean Reversion in Stock Prices: Evidence and Implications, *Journal of Financial Economics*, 22(1), pp. 27-59.

[262] PRESS W.H., TEUKOLSKY S.A., VETTERLING W.T. and FLANNERY B.P. (2007), *Numerical Recipes: The Art of Scientific Computing*, Third edition, Cambridge University Press.

[263] PRIGENT J-L. (2007), *Portfolio Optimization and Performance Analysis*, Chapman & Hall.

[264] PYKHTIN M. (2004), Multi-Factor Adjustment, *Risk*, 17(3), pp. 85-90.

[265] QIAN E. (2005), Risk Parity Portfolios, *PanAgora Research Paper.*

[266] QIAN E. (2006), On the Financial Interpretation of Risk Contribution: Risk Budgets Do Add Up, *Journal of Investment Management*, 4(4), pp. 1-11.

[267] QIAN E. (2011), Risk Parity and Diversification, *Journal of Investing*, 20(1), pp. 119-127.

[268] QIAN E. (2012), Pension Liabilities and Risk Parity, *Journal of Investing*, 21(3), pp. 93-101.

[269] RABIN M. (1998), Psychology and Economics, *Journal of Economic Literature*, 36(1), pp. 11-46.

[270] RAHL L. (Ed.) (2000), *Risk Budgeting: A New Approach to Investing*, Risk Books.

[271] REILLY F.K., KAO G.W. and WRIGHT D.J. (1992), Alternative Bond Market Indexes, *Financial Analysts Journal*, 48(3), pp. 44-58.

[272] REINHART C.M. and ROGOFF K.S. (2009), *This Time Is Different: Eight Centuries of Financial Folly*, Princeton University Press.

[273] ROLL R. (1977), A Critique of the Asset Pricing Theory's Tests Part I: On Past and Potential Testability of the Theory, *Journal of Financial Economics*, 4(2), pp. 129-176.

[274] ROLL R. (1992), A Mean/Variance Analysis of Tracking Error, *Journal of Portfolio Management*, 18(4), pp. 13-22.

[275] RONCALLI T. (2009), *La Gestion des Risques Financiers*, Second edition, Economica.

[276] RONCALLI T. (2010), *La Gestion d'Actifs Quantitative*, Economica.

[277] RONCALLI T. and WEISANG G. (2012), Risk Parity Portfolios with Risk Factors, *SSRN*, `www.ssrn.com/abstract=2155159`.

[278] ROUTLEDGE B.R., SEPPI D.J. and SPATT C.S. (2000), Equilibrium Forward Curves for Commodities, *Journal of Finance*, 55(3), pp. 1297-1338.

[279] RUBAN O. and MELAS M. (2011), Constructing Risk Parity Portfolios: Rebalance, Leverage, or Both?, *Journal of Investing*, 20(1), pp. 99-107.

[280] RYAN R.J. and FABOZZI F.J. (2002), Rethinking Pension Liabilities and Asset Allocation, *Journal of Portfolio Management*, 28(4), pp. 7-15.

[281] SATCHELL S. and SCOWCROFT A. (2000), A Demystification of the Black-Litterman Model: Managing Quantitative and Traditional Portfolio Construction, *Journal of Asset Management*, 1(2), pp. 138-150.

[282] SCAILLET O. (2004), Nonparametric Estimation and Sensitivity Analysis of Expected Shortfall, *Mathematical Finance*, 14(1), pp. 115-129.

[283] SCHERER B. (2002), Portfolio Resampling: Review and Critique, *Financial Analysts Journal*, 58(6), pp. 98-109.

[284] SCHERER B. (2007), *Portfolio Construction & Risk Budgeting*, Third edition, Risk Books.

[285] SCHERER B. (2011), A Note on the Returns from Minimum Variance Investing, *Journal of Empirical Finance*, 18(4), pp. 652-660.

[286] SCHWARTZ E.S. (1997), The Stochastic Behavior of Commodity Prices: Implications for Valuation and Hedging, *Journal of Finance*, 52(3), pp. 923-973.

[287] SEBASTIAN M. (2012), Risk Parity and the Limits of Leverage, *Journal of Investing*, 21(3), pp. 79-87.

[288] SHALIT H. and YITZHAKI S. (1984), Mean-Gini, Portfolio Theory, and the Pricing of Risky Assets, *Journal of Finance*, 39(5), pp. 1449-1468.

[289] SHARPE W.F. (1964), Capital Asset Prices: A Theory of Market Equilibrium under Conditions of Risk, *Journal of Finance*, 19(3), pp. 425-442.

[290] SHARPE W.F. (1966), Mutual Fund Performance, *Journal of Business*, 39(1), pp. 119-138.

[291] SHARPE W.F. (1992), Asset Allocation: Management Style and Performance Measurement, *Journal of Portfolio Management*, 18(2), pp. 7-19.

[292] SHARPE W.F. (2010), Adaptive Asset Allocation Policies, *Financial Analysts Journal*, 66(3), pp. 45-59.

[293] SIRACUSANO L. (2007), A Fundamental Challenge, *Journal of Indexes*, 10(5), pp. 24-35.

[294] SKLAR A. (1959), Fonctions de répartition à n dimensions et leurs marges, *Publications de l'Institut de Statistique de l'Université de Paris*, 8(1), pp. 229-231.

[295] STEVENS G.V.G. (1998), On the Inverse of the Covariance Matrix in Portfolio analysis, *Journal of Finance*, 53(5), pp. 1821-1827.

[296] STOCK J.H. and WATSON M.W. (1989), New Indexes of Coincident and Leading Economic Indicators, *NBER Macroeconomics Annual*, 4, pp. 351-394.

[297] SUTCLIFFE C. (2005), The Cult of the Equity for Pension Funds: Should it Get the Boot?, *Journal of Pension Economics and Finance*, 4(1), pp. 57-85.

[298] TALEB N.N. (2007), *The Black Swan: The Impact of the Highly Improbable*, Random House.

[299] TASCHE D. (2002), Expected Shortfall and Beyond, *Journal of Banking & Finance*, 26(7), pp. 1519-1533.

[300] TASCHE D. (2008), Capital Allocation to Business Units and Sub-Portfolios: The Euler Principle, in A. Resti (Ed.), *Pillar II in the New Basel Accord: The Challenge of Economic Capital*, Risk Books, pp. 423-453.

[301] TAWN J.A. (1988), Bivariate Extreme Value Theory: Models and Estimation, *Biometrika*, 75(3), pp. 397-415.

[302] TAWN J.A. (1990), Modelling Multivariate Extreme Value Distributions, *Biometrika*, 77(2), pp. 245-253.

[303] THIAGARAJAN S.R. and SCHACHTER B. (2011), Risk Parity: Rewards, Risks, and Research Opportunities, *Journal of Investing*, 20(1), pp. 78-89.

[304] TIBSHIRANI R. (1996), Regression Shrinkage and Selection via the Lasso, *Journal of the Royal Statistical Society B*, 58(1), pp. 267-288.

[305] TOBIN J. (1958), Liquidity Preference as Behavior Towards Risk, *Review of Economic Studies*, 25(2), pp. 65-86.

[306] TOLOUI R. (2010), Time To Rethink Bond Indexes?, *Journal of Indexes*, 13(5), pp. 26-29.

[307] TREYNOR J.L. (1965), How to Rate Management of Investment Funds, *Harvard Business Review*, 43, pp. 63-75.

[308] TU J. and ZHOU G. (2011), Markowitz Meets Talmud: A Combination of Sophisticated and Naive Diversification Strategies?, *Journal of Financial Economics*, 99(1), pp. 204-215.

[309] TÜTÜNCÜ R.H. and KOENIG M. (2004), Robust Asset Allocation, *Annals of Operations Research*, 132(1-4), pp. 157-187.

[310] VARIAN H.R. (1992), *Microeconomic Analysis*, Third edition, W.W. Norton & Company.

[311] VASICEK O. (1977), An Equilibrium Characterization of the Term Structure, *Journal of Financial Economics*, 5(2), pp. 177-188.

[312] VASICEK O. (2002), Loan Portfolio Value, *Risk*, 15(12), pp. 160-162.

[313] VICEIRA L.M. (2001), Optimal Portfolio Choice for Long-Horizon Investors with Nontradable Labor Income, *Journal of Finance*, 56(2), pp. 433-470.

[314] WACHTER J.A. (2003), Risk Aversion and Allocation to Long-term Bonds, *Journal of Economic Theory*, 112(2), pp. 325-333.

[315] WALKSHÄUSL C. and LOBE S. (2010), Fundamental Indexing Around the World, *Review of Financial Economics*, 19(3), pp. 117-127.

[316] WILDE T. (2001), IRB Approach Explained, *Risk*, 14(5), pp. 87-90.

[317] WILLENBROCK S. (2011), Diversification Return, Portfolio Rebalancing, and the Commodity Return Puzzle, *Financial Analysts Journal*, 67(4), pp. 42-49.

[318] WINDCLIFF H. and BOYLE P.P. (2004), The 1/n Pension Investment Puzzle, *North American Actuarial Journal*, 8(3), pp. 32-45.

[319] WOOD P. and EVANS R. (2003), Fundamental Profit Based Equity Indexation, *Journal of Indexes*, 5(2), pp. 1-7.

[320] YITZHAKI S. (1982), Stochastic Dominance, Mean Variance, and Gini's Mean Difference, *American Economic Review*, 72(1), pp. 178-185.

[321] ZANGARI P. (1996), A VaR Methodology for Portfolios That Include Options, *RiskMetrics Monitor*, First quarter, pp. 4-12.

Subject Index

W

Y

Z

Author Index

I

J

K

L

Q

R

S

T

U

V

W

X

Y

Z